ENCYCLOPEDIA OF

ENVIRONMENT AND SOCIETY

ENCYCLOPEDIA OF

ENVIRONMENT AND SOCIETY

PAUL ROBBINS
GENERAL EDITOR

Volume Five

A SAGE Reference Publication

For information:

SAGE Publications, Inc.
2455 Teller Road
Thousand Oaks, California 91320
E-mail: order@sagepub.com

SAGE Publications Ltd.
1 Oliver's Yard
55 City Road
London EC1Y 1SP
United Kingdom

SAGE Publications India Pvt. Ltd.
B 1/I 1 Mohan Cooperative Industrial Area
Mathura Road, New Delhi 110 044
India

SAGE Publications Asia-Pacific Pte. Ltd.
33 Pekin Street #02-01
Far East Square
Singapore 048763

Library of Congress Cataloging-in-Publication Data

Encyclopedia of environment and society / Paul Robbins, general editor.
p. cm. — (A Sage reference publication)
Includes index.
ISBN 978-1-4129-2761-1 (cloth)
1. Social ecology—Encyclopedias. 2. Human ecology—Social aspects—Encyclopedias. 3. Social change—Environmental aspects—Encyclopedias. I. Robbins, Paul, 1967–

HM856.E53 2007
304.203—dc22 2007021378

This book is printed on acid-free paper.
07 08 09 10 11 10 9 8 7 6 5 4 3 2 1

Photo credits are on page I–93 Volume 5.

GOLSON BOOKS, LTD.

President and Editor	J. Geoffrey Golson
Creative Director	Mary Jo Scibetta
Managing Editor	Susan Moskowitz
Copyeditors	Martha Whitt
	Mary Le Rouge
	Janelle Schiecke
	Jennifer Bussey
Layout Editors	Stephanie Larson
	Oona Patrick
Proofreaders	Jacqueline F. Brownstein
	Julie Grady
Indexer	J S Editorial LLC

SAGE REFERENCE

Vice President and Publisher	Rolf Janke
Project Editor	Tracy Alpern
Cover Production	Michelle Kenny
Marketing Manager	Carmel Withers
Editorial Assistant	Michele Thompson

ENVIRONMENT AND SOCIETY

Contents

Encyclopedia of Environment and Society

About the General Editor

PAUL ROBBINS was raised in Denver, Colorado, but has lived in India, New England, the U.S. Midwest, and the deserts of the U.S. Southwest. He received his Ph.D. in Geography from Clark University in 1996 and is currently Professor in the Department of Geography and Regional Development at the University of Arizona. His research centers on the relationships between individuals (homeowners, hunters, professional foresters), environmental actors (lawns, elk, mesquite trees), and the institutions that connect them. Working with interdisciplinary teams in the fields of biology, economics, climatology, and entomology, his projects have examined chemical use in the suburban United States, elk management in Montana, forest product collection in New England, wolf conservation in India, and mosquito borne illness and management of insect hazards in the US Southwest. His expertise includes the fields of conservation policy, grasslands ecology, and institutional ethnography. He is author of *Political Ecology: A Critical Introduction* (2004) and *Lawn People: How Grass Weeds and Chemicals Make Us Who We Are* (2007) and has served as an editor for the journal *Geoforum*.

Advisory Board

Maps

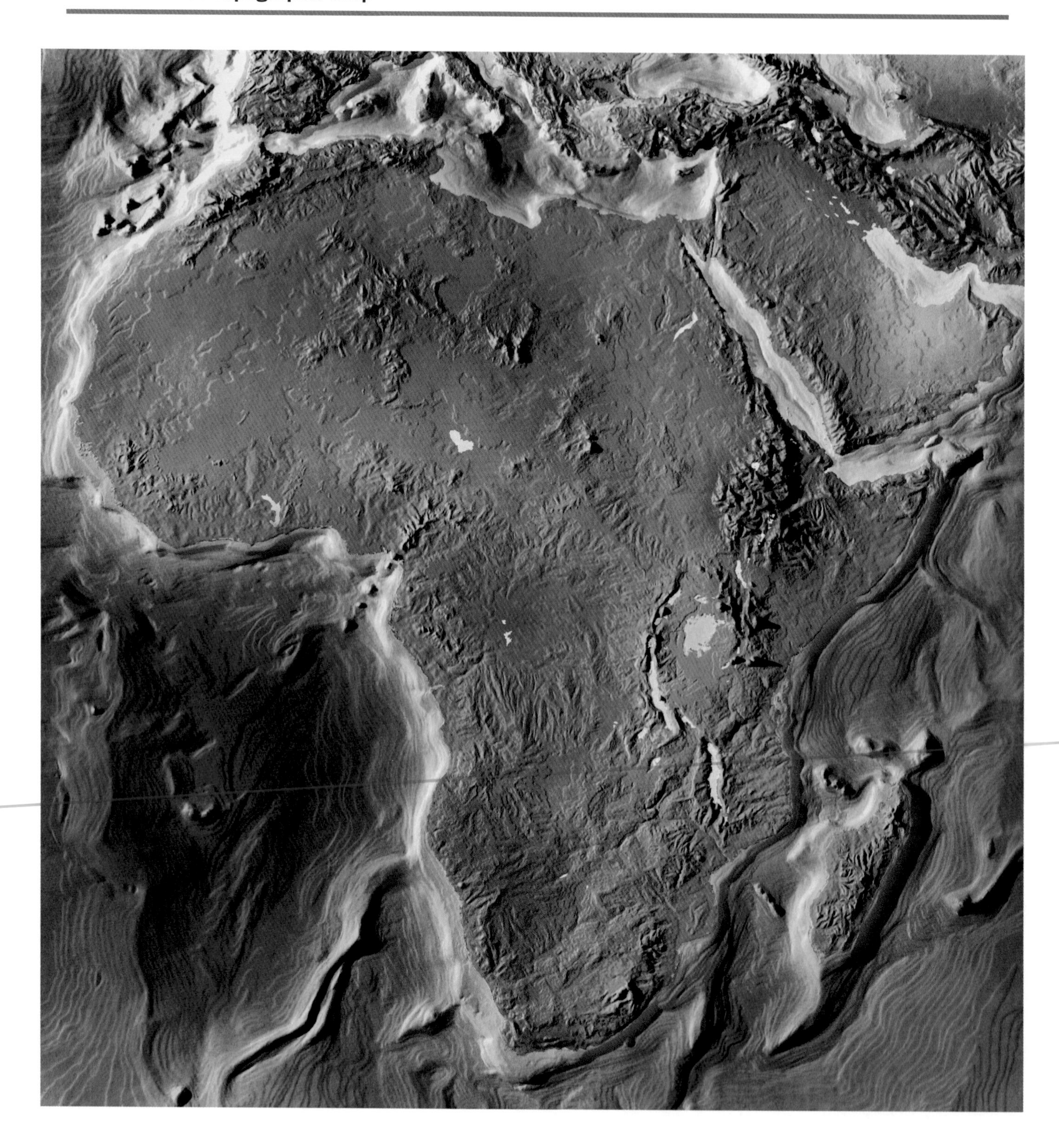

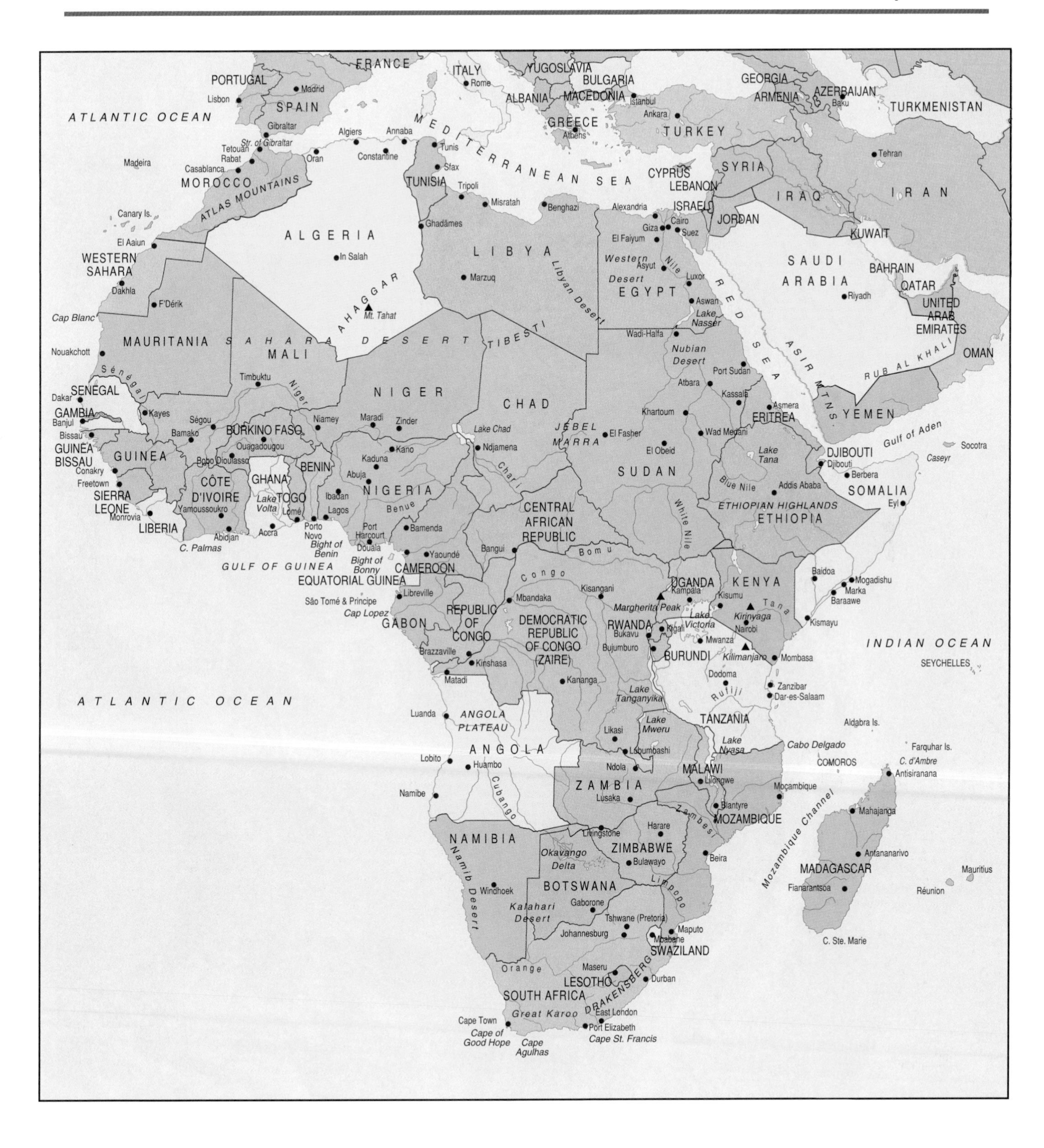
FRANCE
PORTUGAL
Lisbon
Madrid
SPAIN
ATLANTIC OCEAN
Gibraltar
Str. of Gibraltar
Tetouan
Rabat
Casablanca
Madeira
MOROCCO
ATLAS MOUNTAINS
Algiers
Annaba
Oran
Constantine
Tunis
Sfax
TUNISIA
Tripoli
Misratah
Benghazi
ITALY
Rome
YUGOSLAVIA
BULGARIA
ALBANIA
MACEDONIA
Istanbul
GREECE
Athens
Ankara
TURKEY
GEORGIA
ARMENIA
AZERBAIJAN
Baku
TURKMENISTAN
Tehran
MEDITERRANEAN SEA
CYPRUS
LEBANON
SYRIA
IRAQ
IRAN
ISRAEL
JORDAN
KUWAIT
Alexandria
Cairo
Giza
Suez
El Faiyum
Canary Is.
El Aaiun
WESTERN SAHARA
Dakhla
F'Dérik
Cap Blanc
ALGERIA
In Salah
AHAGGAR
Mt. Tahat
Ghadâmes
LIBYA
Marzuq
Libyan Desert
Western Desert
Asyut
Nile
Luxor
EGYPT
Aswan
Lake Nasser
RED SEA
SAUDI ARABIA
Riyadh
BAHRAIN
QATAR
UNITED ARAB EMIRATES
OMAN
RUB AL KHALI
ASIR MTNS.
YEMEN
MAURITANIA
Nouakchott
SAHARA DESERT
MALI
TIBESTI
Wadi-Halfa
Nubian Desert
Port Sudan
Atbara
Kassala
Asmera
ERITREA
Sénégal
Dakar
SENEGAL
Timbuktu
Niger
NIGER
CHAD
GAMBIA
Banjul
Kayes
Ségou
Bamako
Niamey
Maradi
Zinder
Bissau
GUINEA BISSAU
BURKINO FASO
Ouagadougou
Lake Chad
Ndjamena
JEBEL MARRA
El Fasher
Khartoum
Wad Medani
El Obeid
SUDAN
Lake Tana
DJIBOUTI
Djibouti
Gulf of Aden
Socotra
Caseyr
Berbera
GUINEA
Conakry
Freetown
SIERRA LEONE
Monrovia
LIBERIA
Bobo Dioulasso
CÔTE D'IVOIRE
Yamoussoukro
Abidjan
C. Palmas
GHANA
Lake Volta
TOGO
Lomé
Accra
BENIN
Porto Novo
Kano
Kaduna
Abuja
NIGERIA
Ibadan
Lagos
Benue
Port Harcourt
Chari
CENTRAL AFRICAN REPUBLIC
Blue Nile
Addis Ababa
ETHIOPIAN HIGHLANDS
ETHIOPIA
SOMALIA
Eyl
White Nile
Bight of Benin
Douala
Bight of Bonny
Bamenda
Yaoundé
CAMEROON
Bangui
Bomu
GULF OF GUINEA
EQUATORIAL GUINEA
São Tomé & Principe
Libreville
Cap Lopez
GABON
Congo
Mbandaka
Kisangani
UGANDA
Kampala
KENYA
Baidoa
Mogadishu
Marka
Baraawe
Kisumu
Tana
Margherita Peak
Kirinyaga
Nairobi
Kismayu
REPUBLIC OF CONGO
DEMOCRATIC REPUBLIC OF CONGO (ZAIRE)
RWANDA
Kigali
Bukavu
Lake Victoria
Mwanza
Bujumbura
BURUNDI
Kilimanjaro
Mombasa
INDIAN OCEAN
SEYCHELLES
Brazzaville
Kinshasa
Matadi
Kananga
Dodoma
Lake Tanganyika
Rufiji
Zanzibar
Dar-es-Salaam
ATLANTIC OCEAN
Luanda
ANGOLA PLATEAU
TANZANIA
Aldabra Is.
Likasi
Lake Mweru
Lake Nyasa
Cabo Delgado
Farquhar Is.
ANGOLA
Lubumbashi
COMOROS
C. d'Ambre
Lobito
Huambo
Ndola
MALAWI
Lilongwe
Antsiranana
Cubango
ZAMBIA
Lusaka
Moçambique
Namibe
Blantyre
Zambezi
MOZAMBIQUE
Mozambique Channel
Mahajanga
NAMIBIA
Livingstone
Harare
Okavango Delta
ZIMBABWE
Bulawayo
Beira
Antananarivo
MADAGASCAR
Mauritius
Namib Desert
Windhoek
BOTSWANA
Limpopo
Fianarantsoa
Réunion
Kalahari Desert
Gaborone
Tshwane (Pretoria)
Johannesburg
Maputo
Mbabane
SWAZILAND
C. Ste. Marie
Orange
Maseru
LESOTHO
DRAKENSBERG
Durban
SOUTH AFRICA
Great Karoo
East London
Cape Town
Port Elizabeth
Cape of Good Hope
Cape Agulhas
Cape St. Francis

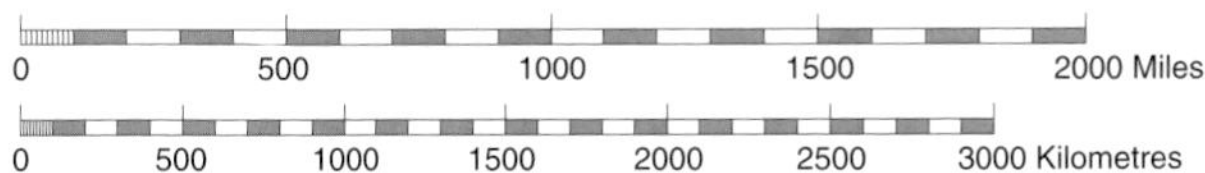
0 500 1000 1500 2000 Miles
0 500 1000 1500 2000 2500 3000 Kilometres

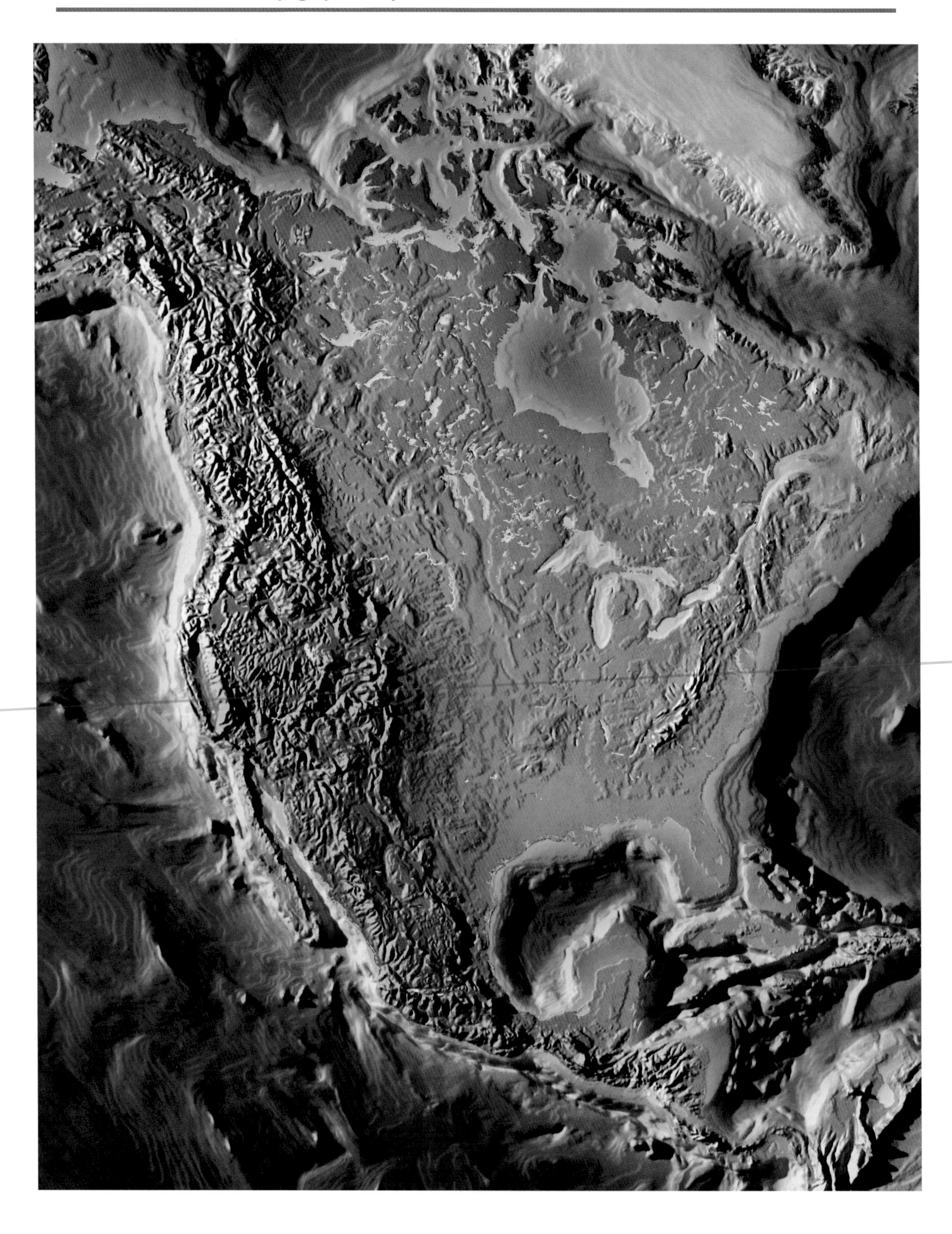

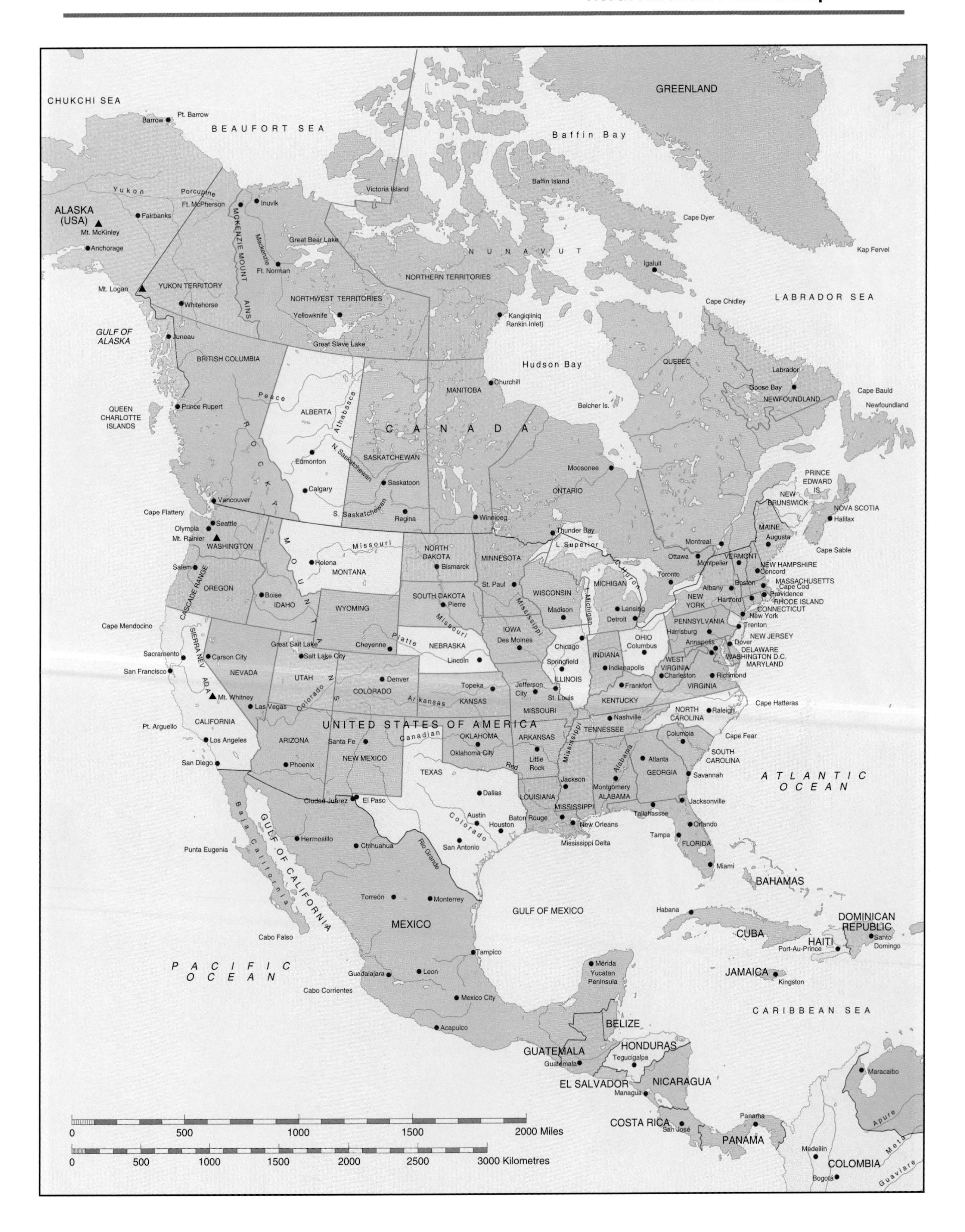
CHUKCHI SEA
Barrow
Pt. Barrow
BEAUFORT SEA
GREENLAND
Baffin Bay
Baffin Island
Victoria Island
Yukon
Porcupine
Ft. McPherson
Inuvik
ALASKA (USA)
Fairbanks
Mt. McKinley
Anchorage
MCKENZIE MOUNTAINS
Mackenzie
Great Bear Lake
Ft. Norman
NUNAVUT
Cape Dyer
Igaluit
Kap Fervel
NORTHERN TERRITORIES
Mt. Logan
YUKON TERRITORY
Whitehorse
NORTHWEST TERRITORIES
Yellowknife
Kangiqliniq Rankin Inlet)
Cape Chidley
LABRADOR SEA
GULF OF ALASKA
Juneau
Great Slave Lake
BRITISH COLUMBIA
Hudson Bay
QUEBEC
Labrador
Churchill
MANITOBA
Goose Bay
NEWFOUNDLAND
Cape Bauld
Newfoundland
Peace
QUEEN CHARLOTTE ISLANDS
Prince Rupert
ALBERTA
Athabasca
Belcher Is.
CANADA
ROCKY MOUNTAINS
N. Saskatchewan
Edmonton
SASKATCHEWAN
Moosonee
Saskatoon
Calgary
ONTARIO
PRINCE EDWARD IS.
NEW BRUNSWICK
NOVA SCOTIA
Vancouver
S. Saskatchewan
Regina
Halifax
Cape Flattery
Winnipeg
Seattle
Thunder Bay
Olympia
Mt. Rainier
WASHINGTON
Missouri
NORTH DAKOTA
L. Superior
MAINE
Augusta
Montreal
Cape Sable
Salem
Helena
Bismarck
MINNESOTA
Ottawa
Montpelier
VERMONT
NEW HAMPSHIRE
Concord
MONTANA
L. Huron
Toronto
CASCADE RANGE
OREGON
St. Paul
MICHIGAN
Albany
Boston
MASSACHUSETTS
Cape Cod
Providence
RHODE ISLAND
Boise
IDAHO
SOUTH DAKOTA
Pierre
WISCONSIN
NEW YORK
Hartford
WYOMING
Mississippi
Madison
L. Michigan
Lansing
CONNECTICUT
New York
Cape Mendocino
Detroit
PENNSYLVANIA
Trenton
IOWA
Des Moines
OHIO
Columbus
Harrisburg
NEW JERSEY
Great Salt Lake
Cheyenne
Platte
NEBRASKA
Chicago
Annapolis
Dover
DELAWARE
WASHINGTON D.C.
MARYLAND
Sacramento
SIERRA NEVADA
Carson City
Salt Lake City
Lincoln
Springfield
INDIANA
WEST VIRGINIA
San Francisco
NEVADA
UTAH
Denver
ILLINOIS
Indianapolis
Charleston
Richmond
COLORADO
Topeka
Jefferson City
St. Louis
Frankfort
VIRGINIA
Mt. Whitney
Las Vegas
Arkansas
KANSAS
KENTUCKY
Cape Hatteras
MISSOURI
NORTH CAROLINA
Raleigh
Pt. Arguello
CALIFORNIA
UNITED STATES OF AMERICA
Nashville
TENNESSEE
Los Angeles
ARIZONA
Santa Fe
Canadian
OKLAHOMA
ARKANSAS
Columbia
Cape Fear
Oklahoma City
SOUTH CAROLINA
San Diego
Phoenix
NEW MEXICO
Little Rock
Atlanta
Red
TEXAS
GEORGIA
Savannah
ATLANTIC OCEAN
Jackson
Alabama
Montgomery
ALABAMA
Dallas
LOUISIANA
Ciudad Juárez
El Paso
MISSISSIPPI
Jacksonville
Baja California
GULF OF CALIFORNIA
Austin
Baton Rouge
Tallahassee
Houston
New Orleans
Orlando
Hermosillo
Chihuahua
Rio Grande
San Antonio
Mississippi Delta
Tampa
FLORIDA
Punta Eugenia
Miami
BAHAMAS
Torreón
Monterrey
GULF OF MEXICO
Habana
DOMINICAN REPUBLIC
MEXICO
CUBA
HAITI
Santo Domingo
Cabo Falso
Port-Au-Prince
Tampico
Mérida
Yucatan Peninsula
PACIFIC OCEAN
Guadalajara
Leon
JAMAICA
Kingston
Cabo Corrientes
Mexico City
CARIBBEAN SEA
Acapulco
BELIZE
HONDURAS
GUATEMALA
Guatemala
Tegucigalpa
Maracaibo
EL SALVADOR
NICARAGUA
Managua
Panama
COSTA RICA
San José
PANAMA
Apure
Meta
Medellín
COLOMBIA
Bogotá
Guaviare
0
500
1000
1500
2000 Miles
0
500
1000
1500
2000
2500
3000 Kilometres

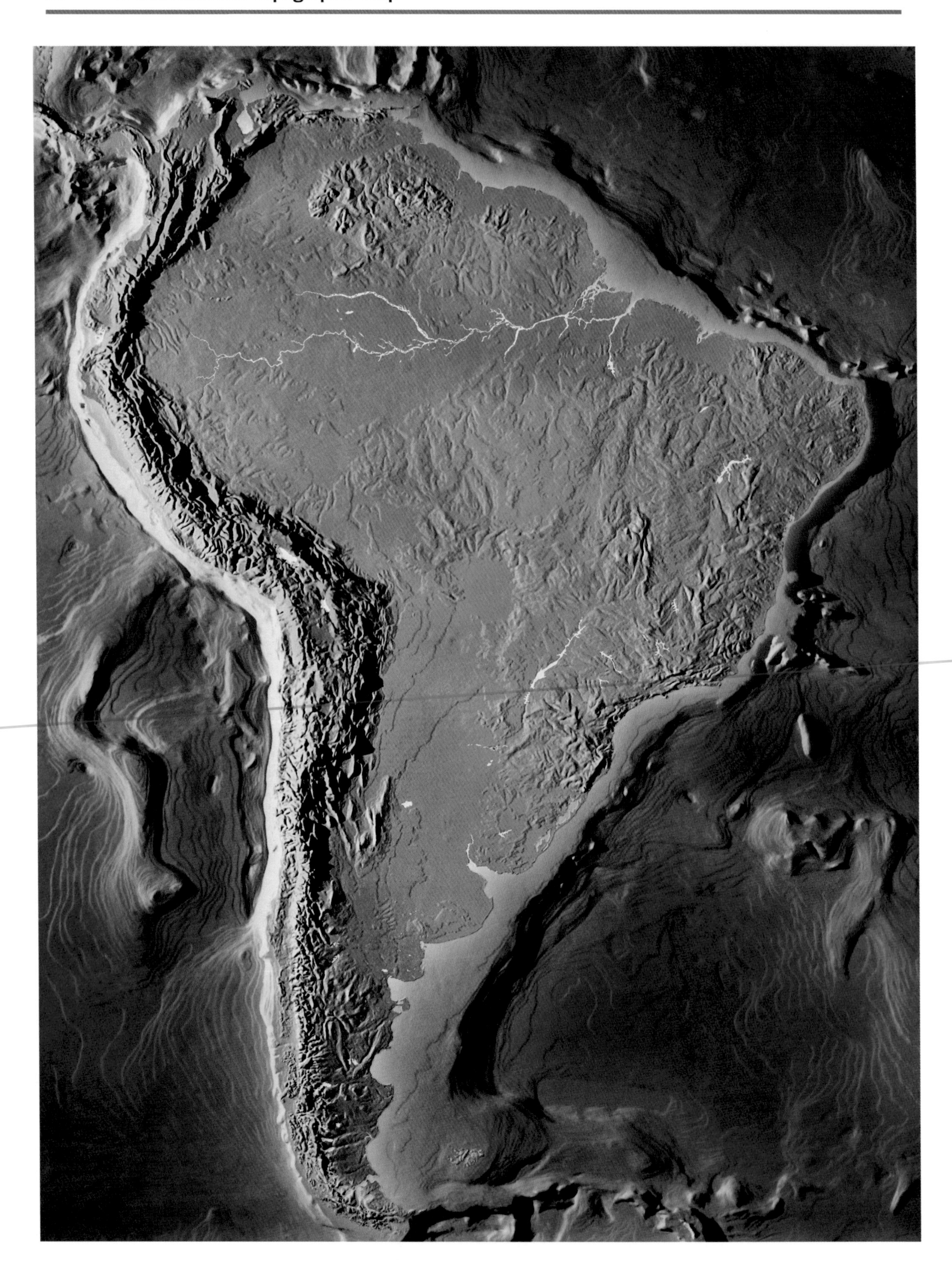

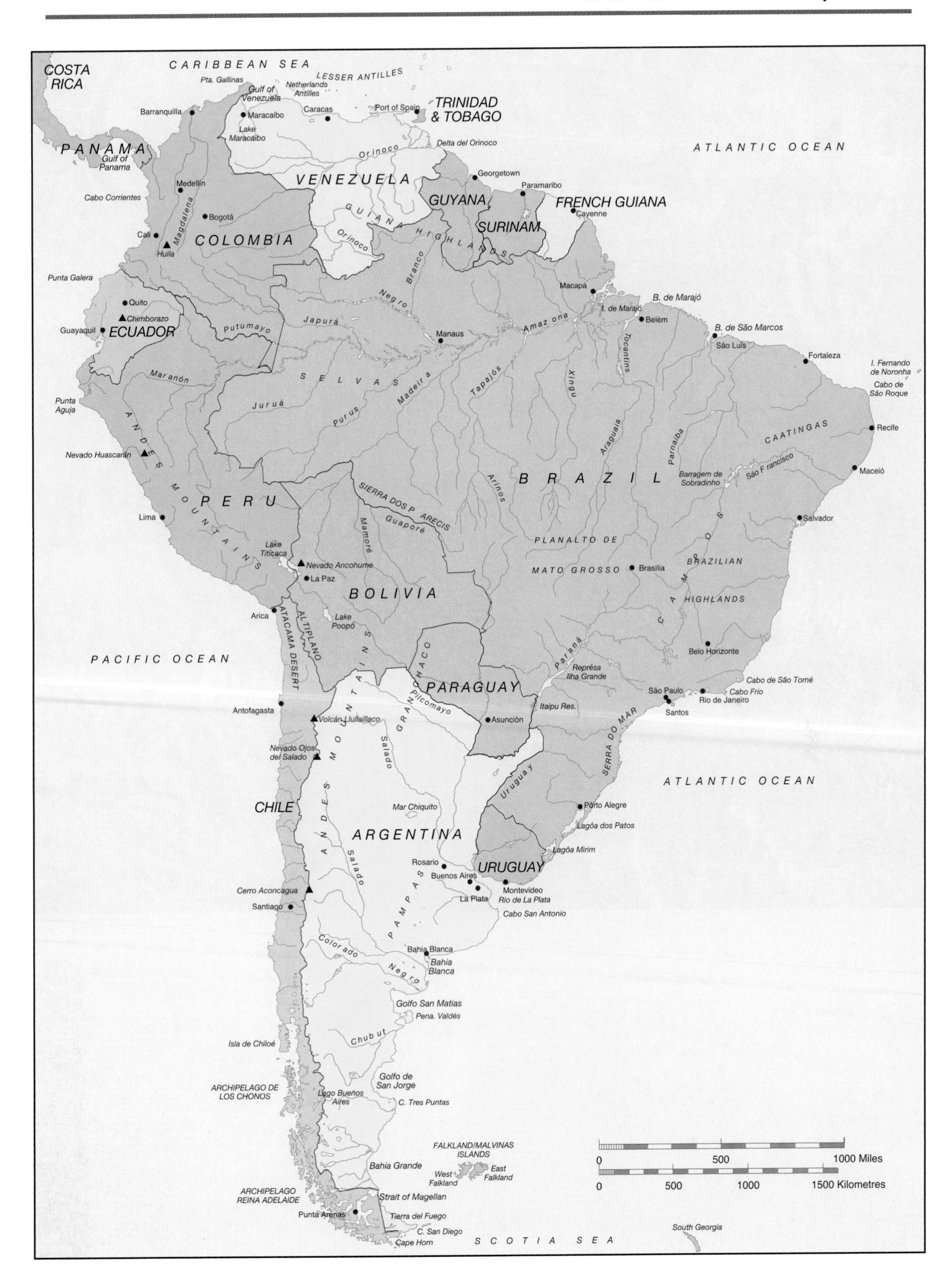
COSTA RICA
CARIBBEAN SEA
LESSER ANTILLES
Pta. Gallinas
Gulf of Venezuela
Netherlands Antilles
Barranquilla
Maracaibo
Caracas
Port of Spain
TRINIDAD & TOBAGO
Lake Maracaibo
PANAMA
Gulf of Panama
Delta del Orinoco
Orinoco
ATLANTIC OCEAN
VENEZUELA
Georgetown
Paramaribo
Medellín
Cabo Corrientes
GUYANA
FRENCH GUIANA
Cayenne
Bogotá
GUIANA HIGHLANDS
SURINAM
Magdalena
Cali
Huila
COLOMBIA
Orinoco
Branco
Punta Galera
Macapá
Negro
B. de Marajó
Quito
Chimborazo
Japurá
I. de Marajó
Belém
Guayaquil
ECUADOR
Putumayo
Manaus
Amazona
B. de São Marcos
São Luís
Tocantins
Fortaleza
I. Fernando de Noronha
Marañón
SELVAS
Madeira
Tapajós
Xingu
Cabo de São Roque
Punta Aguja
Juruá
Purus
Recife
CAATINGAS
ANDES MOUNTAINS
Araguaia
Parnaíba
Nevado Huascarán
BRAZIL
Barragem de Sobradinho
São Francisco
Maceió
Arinos
PERU
SIERRA DOS PARECIS
Lima
Guaporé
Salvador
Mamoré
PLANALTO DE
Lake Titicaca
MATO GROSSO
Brasília
BRAZILIAN
Nevado Ancohume
CAMPOS
La Paz
BOLIVIA
HIGHLANDS
Arica
Lake Poopó
ALTIPLANO
ATACAMA DESERT
CHACO
PACIFIC OCEAN
Paraná
Belo Horizonte
Represa Ilha Grande
Cabo de São Tomé
PARAGUAY
São Paulo
Cabo Frio
Rio de Janeiro
Antofagasta
Pilcomayo
Itaipu Res.
Santos
Volcán Llullaillaco
GRAN CHACO
Asunción
SERRA DO MAR
Salado
Nevado Ojos del Salado
ATLANTIC OCEAN
Uruguay
CHILE
Mar Chiquito
Porto Alegre
ARGENTINA
Lagôa dos Patos
Lagôa Mirim
Salado
Rosario
URUGUAY
Buenos Aires
PAMPAS
Cerro Aconcagua
Montevideo
La Plata
Rio de La Plata
Santiago
Cabo San Antonio
Colorado
Bahia Blanca
Bahia Blanca
Negro
Golfo San Matias
Pena. Valdés
Isla de Chiloé
Chubut
Golfo de San Jorge
ARCHIPELAGO DE LOS CHONOS
Lago Buenos Aires
C. Tres Puntas
FALKLAND/MALVINAS ISLANDS
Bahia Grande
West Falkland
East Falkland
0
500
1000 Miles
0
500
1000
1500 Kilometres
ARCHIPELAGO REINA ADELAIDE
Strait of Magellan
Punta Arenas
Tierra del Fuego
C. San Diego
Cape Horn
SCOTIA SEA
South Georgia

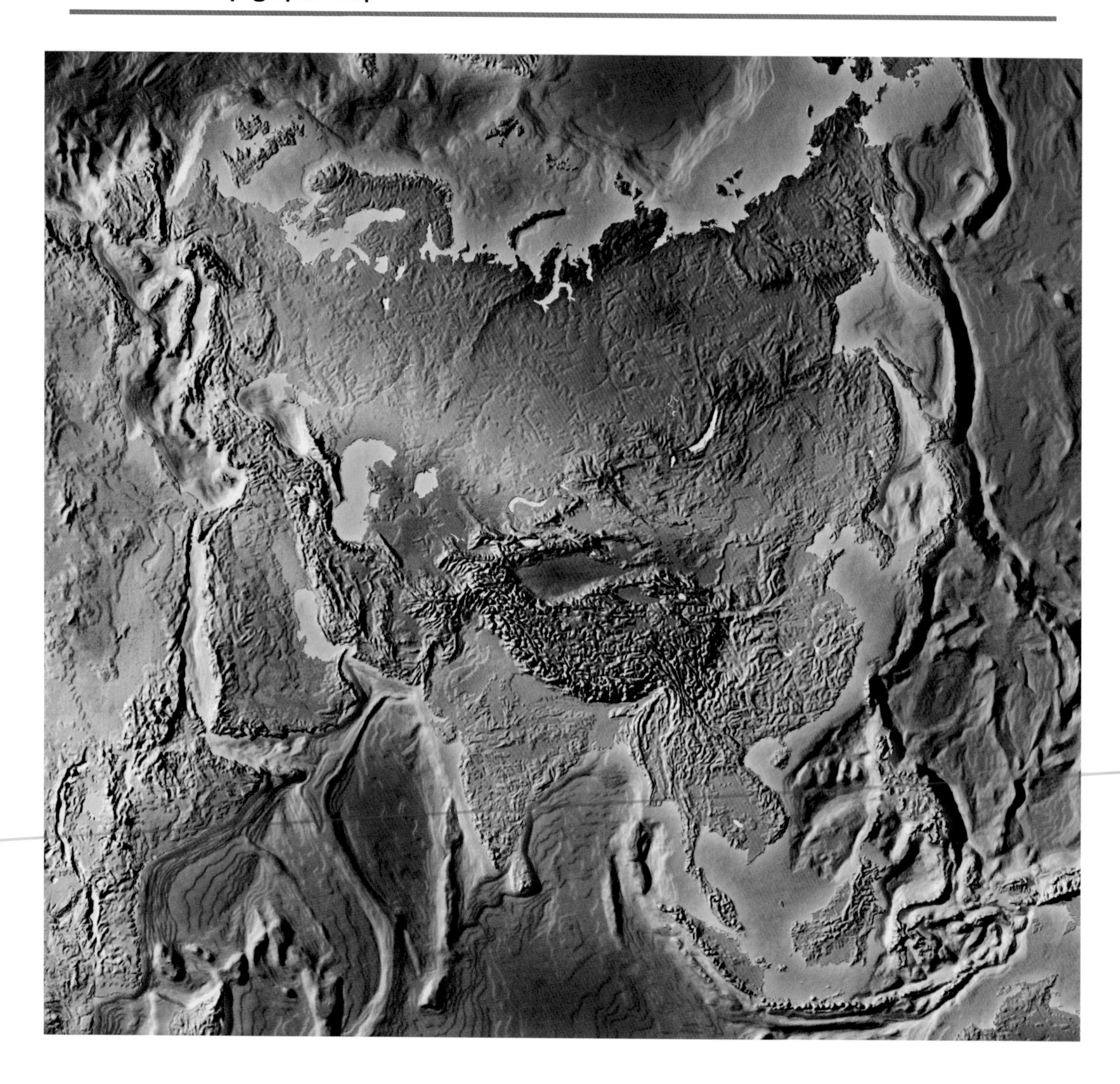

ARCTIC OCEAN
Chukchi Sea
Bering St.
ALEUTIAN ISLANDS
Bering Sea
Norwegian Sea
Svalbard
East Siberian Sea
Zemlya Frantsa Iosifa
Novosibirskiye Ostrova
North Sea
Barents Sea
Severnaya Zemlya
Laptev Sea
Kolyma
Indigirka
Kolyma Lowland
Loire
Lapland
Novaya Zemlya
Kara Sea
ALPS
ESTONIA
LITHUANIA
LATVIA
RUSSIA
CENTRAL SIBERIAN PLATEAU
Lena
Yenisey
BELARUS
Dnepr
Pechora
Sea of Okhotsk
Mediterranean Sea
Volga
URAL MOUNTAINS
Ob'
Siberian Lowland
Nizhnyaya Tunguska
UKRAINE
MOLDOVA
RUSSIA
Tobol
Angara
Black Sea
Ural
Amur
Ishim
Irtysh
Lake Baykal
CAUCASUS MTS.
Manchurian Plain
TURKEY
ANATOLIAN PLATEAU
Qattara Depression
Caspian Sea
Aral Sea
KAZAKHSTAN
SAYAN
Sea of Japan
ALTAI
MONGOLIA
Syr Dar'ya
Lake Balkhash
LEBANON
UZBEKISTAN
TURANIAN PLATEAU
NORTH KOREA
SYRIA
Tigris
Gobi Desert
JAPAN
ISRAEL
Euphrates
Amu Dar'ya
JORDAN
TURKMENISTAN
KYRGYZSTAN
TIEN SHAN
SOUTH KOREA
Yellow Sea
IRAQ
ZAGROS MTS.
TAJIKISTAN
Tarim Basin
Huang He
Nile
IRAN
AFGHANISTAN
HINDU KUSH
Great Basin
KUWAIT
KUNLAN SHAN
PACIFIC OCEAN
SAUDI ARABIA
Red Sea
QATAR
PLATEAU OF TIBET
CHINA
East China Sea
PAKISTAN
UNITED ARAB EMIRATES
Indus
HIMALAYA
Chang Jiang
NEPAL
Mt. Everest
BHUTAN
TAIWAN
ETHIOPIAN HIGHLANDS
YEMEN
OMAN
Arabian Sea
Deccan
Ganga
BANGLADESH
Irrawaddy
Philippine Sea
Gulf of Aden
INDIA
BURMA
LAOS
Salween
South China Sea
Mekong
Bay of Bengal
THAILAND
L. Victoria
CAMBODIA
VIETNAM
PHILIPPINES
L. Tanganyika
Kilimanjaro
Celebes Sea
BRUNEI
SRI LANKA
INDIAN OCEAN
MALAYSIA
SINGAPORE
L. Nyasa
INDONESIA
EAST INDIES
Mozambique Ch.

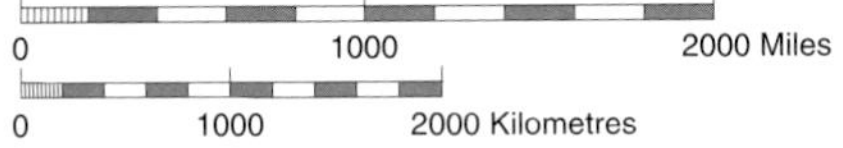
0
1000
2000 Miles
0
1000
2000 Kilometres

UZBEKISTAN
KAZAKHSTAN
MONGOLIA
GOBI
Dzungaria
Ürümqi
SHAN
Hami
Bishkek
Almaty
Oz. Issyk Kul'
Tashkent
Bukhara
Samarkand
KYRGYZSTAN
TIEN
Aksu
Turfan Depression
Beijing
Tianjin
Bo Hai
Baotou
ORDOS
Shijiazhuang
Yinchuan
Taiyuan
TURKMENISTAN
Dushanbe
TAJIKISTAN
Sheberghan
TARIM PENDI
Yumen
QILIAN SHAN
Qinghai Hu
ALTUN SHAN
HINDU KUSH
Hotan
Hwang Ho
Zhengzhou
Luoyang
AFGHANISTAN
Kabul
Khyber Pass
KARAKORAM
KUNLAN SHAN
Lanzhou
Xi'an
Huang He
Srinagar
Islamabad
Qandahar
SULAJMAN RANGE
Indus
HOH XILSHAN
BAYAN HAR SHAN
Tongtian He
HUA SHAN
DABA SHAN
Wuhan
QING ZANG
CHINA
Chang Jiang
Quetta
Lahore
Multan
Sutlej
TANGGULA SHAN
GANGDISE SHAN
HIMALAYA
Chengdu
Dongting Hu
PAKISTAN
Bahawalpur
Qamdo
RED BASIN
Chongqing
Changsha
KIRTHAR RA.
Sukkur
Lhasa
HENGDUAN SHAN
DALOU SHAN
Hengyang
New Delhi
Annapurna
Xigaze
Hyderabad
Mt Everest
NEPAL
Thimphu
Guiyang
Jodhpur
Jaipur
Yamuna
Katmandu
BHUTAN
Ajmer
Lucknow
Ghagara
Kanpur
Brahmaputra
NAGA HILLS
Kunming
Liuzhou
Allahabad
Varanasi
Gauhati
Udaipur
Ganges
Patna
Gulf of Kachch
Ahmadabad
Son
BANGLADESH
Imphal
Nanning
Jamnagar
Dhaka
Bhavnagar
Vadodara
Narmada
Zhanjiang
Jamshedpur
Hanoi
Surat
Chittagong
Haiphong
Haikou
INDIA
Kolkata (Calcutta)
BURMA
Mandalay
Gulf of Khambhat
Raipur
Gulf of Tongkin
Nagpur
Mouths of the Ganges
Mahanadi
Arakan Yoma
Irrawaddy
PEGU YOMA
Louang Prabang
Hainan
Mumbai (Bombay)
DECCAN
Cuttack
Akyab
TANEN R.
Vinh
Pune
Godavari
Chiang Mai
LAOS
Prome
M.Lampang
Vientiane
Mekong
Solapur
WESTERN GHATS
EASTERN GHATS
DAWNA RANGE
Hue
Kolhapur
Hyderabad
Vishakhapatnam
Da Nang
Krishna
Henzada
Hubli-Dharwar
Vijayawada
BAY OF BENGAL
Bassein
Moulmein
THAILAND
INDIAN OCEAN
Kurnool
Yangon
Nakhon Ratchasima
VIETNAM
Qui Nhon
Mouths of the Irrawaddy
Gulf of Martaban
BILAUKTAUNG R.
PHANOM DANG
Nellore
Chu Yang Sin
Tavoy
Bangkok
CAMBODIA
Nha Trang
Mangalore
Koramangala (Bangalore)
Battambang
Kampong Cham
Chennai (Madras)
Mysore
ANDAMAN IS.
Mergui
Phnom Penh
LACCADIVE IS.
Kozhikode
Ho Chi Minh City
Coimbatore
Can Tho
Tiruchchirappalli
Andaman Sea
MERGUI ARCHIP.
Gulf of Thailand
Mouths of the Mekong
INDIAN OCEAN
Madurai
Jaffna
Ten Degree Channel
Mui Bai Bung
Thiruvananthapuram
Trincomalee
Nakhon Si Thammarat
C. Comorin
Gulf of Mannar
Phuket
SRI LANKA
Kandy
Songkhla
Colombo
Galle
Kota Baharu
Dondra Head
Georgetown

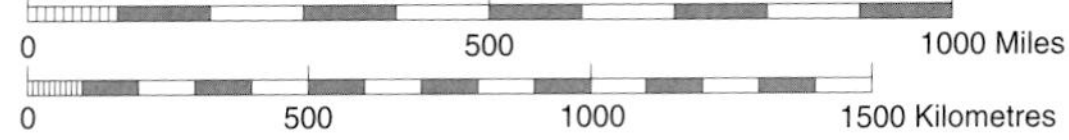
0
500
1000 Miles
0
500
1000
1500 Kilometres

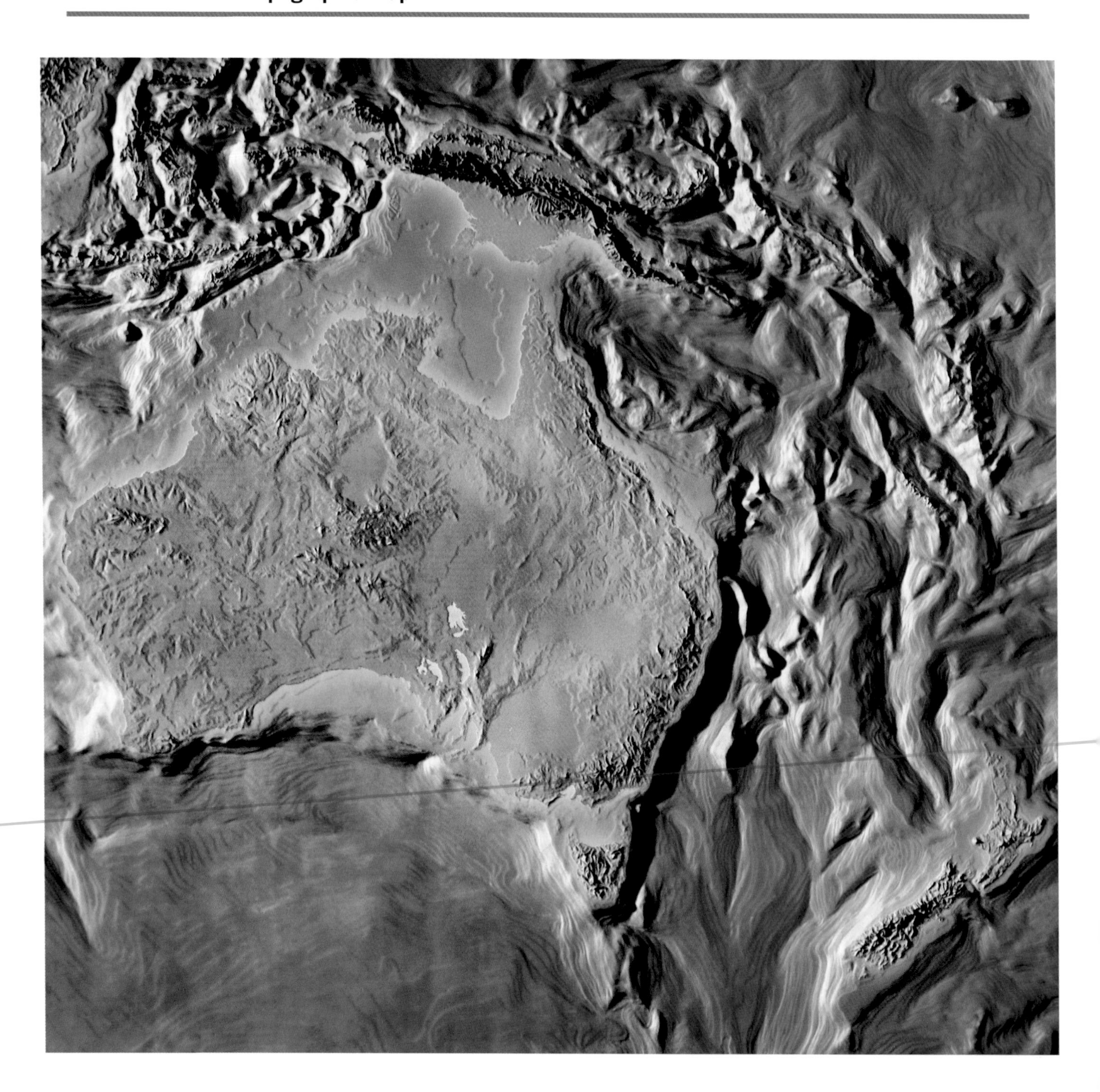

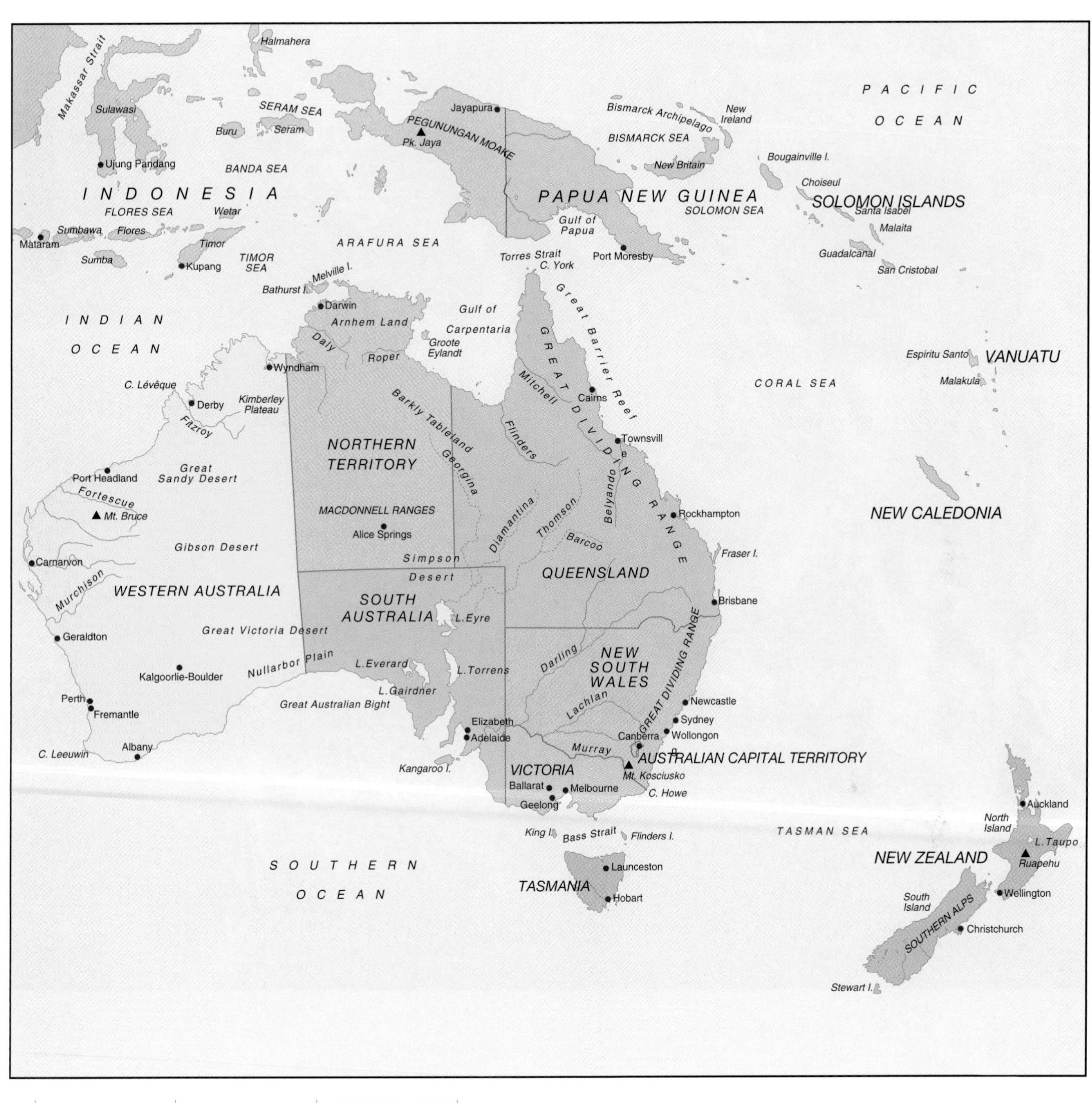

PACIFIC OCEAN
INDONESIA
PAPUA NEW GUINEA
SOLOMON ISLANDS
VANUATU
NEW CALEDONIA
NEW ZEALAND
INDIAN OCEAN
SOUTHERN OCEAN
CORAL SEA
TASMAN SEA
ARAFURA SEA
TIMOR SEA
BANDA SEA
SERAM SEA
FLORES SEA
BISMARCK SEA
SOLOMON SEA
NORTHERN TERRITORY
QUEENSLAND
WESTERN AUSTRALIA
SOUTH AUSTRALIA
NEW SOUTH WALES
VICTORIA
TASMANIA
AUSTRALIAN CAPITAL TERRITORY
GREAT DIVIDING RANGE
MACDONNELL RANGES
SOUTHERN ALPS
Great Barrier Reef
Darwin
Wyndham
Derby
Port Headland
Carnarvon
Geraldton
Kalgoorlie-Boulder
Perth
Fremantle
Albany
Alice Springs
Cairns
Townsvill
Rockhampton
Brisbane
Newcastle
Sydney
Wollongon
Canberra
Melbourne
Ballarat
Geelong
Elizabeth
Adelaide
Launceston
Hobart
Auckland
Wellington
Christchurch
Port Moresby
Jayapura
Ujung Pandang
Mataram
Kupang
Great Sandy Desert
Gibson Desert
Great Victoria Desert
Simpson Desert
Nullarbor Plain
Barkly Tableland
Arnhem Land
Kimberley Plateau
Great Australian Bight
Gulf of Carpentaria
Gulf of Papua
Torres Strait
Bass Strait
Makassar Strait
Mt. Bruce
Mt. Kosciusko
Pk. Jaya
PEGUNUNGAN MOAKE
Ruapehu
L. Eyre
L. Torrens
L. Everard
L. Gairdner
L. Taupo

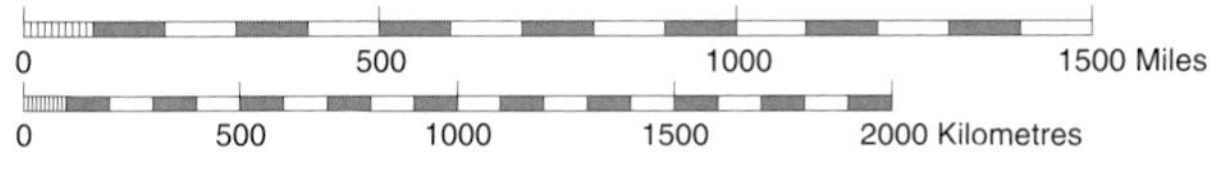

0 500 1000 1500 Miles
0 500 1000 1500 2000 Kilometres

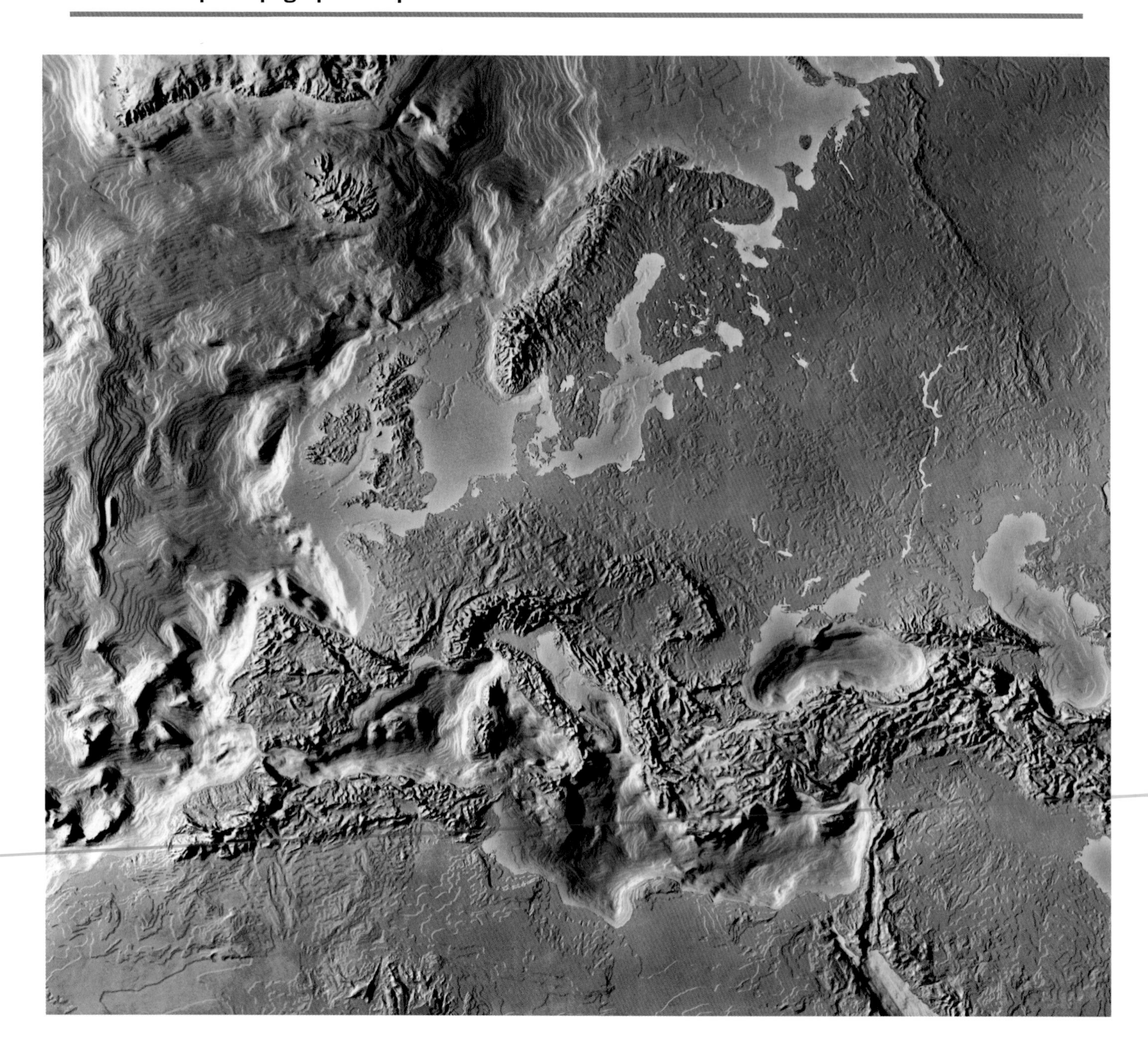

GREENLAND
GREENLAND SEA
ARCTIC OCEAN
BARENTS SEA
Denmark Strait
Ostrov Kolguyav
Reykjavik
ICELAND
Murmansk
KOLA PENINSULA
NORWEGIAN SEA
Arkhangel'sk
Severnaya Dvina
FAEROES
SWEDEN
Gulf of Bothnia
FINLAND
L.Onega
SHETLAND IS.
NORWAY
Sukhona
L.Ladoga
HEBRIDES
ORKNEY IS.
Oslo
Stockholm
Helsinki
Sankt-Peterburg
Tallinn
ESTONIA
SCOTLAND
NORTH SEA
Edinburgh
Skagerrak
Göteborg
LATVIA
Riga
Belfast
IRELAND
Dublin
DENMARK
Copenhagen
BALTIC SEA
LITHUANIA
Vilnius
RUSSIA
Moscow
ATLANTIC OCEAN
WALES
Cardiff
ENGLAND
London
NETHERLANDS
The Hague
Berlin
Minsk
BELARUS
Pripyat
Warsaw
POLAND
BELGIUM
Brussels
GERMANY
Paris
Luxembourg
LUXEMBOURG
Seine
Loire
Prague
CZECH REP.
Kiev
UKRAINE
Dnepr
Dnestr
Desna
Don
FRANCE
Donau
Vienna
Bratislava
SLOVAKIA
CARPATHIANS
MOLDOVA
Kishinev
Prut
Odessa
Bay of Biscay
Bern
SWITZERLAND
AUSTRIA
Budapest
HUNGARY
Ljubljana
SLOVENIA
CROATIA
ROMANIA
Bucharest
Bilbao
MASSIF CENTRAL
Rhône
ALPS
Po
Porto
Duero
Andorra la Vella
ANDORRA
Monaco
San Marino
BOSNIA-HERZEGOVINA
Sarajevo
Belgrade
YUGOSLAVIA
BULGARIA
Sofiya
PORTUGAL
Madrid
Tajo
Lisbon
Barcelona
Corsica
ITALY
Rome
Skopje
MACEDONIA
Tirane
ALBANIA
SPAIN
Sardinia
MADEIRA IS.
Sevilla
Malaga
Gibraltar
Strait of Gibraltar
BALEARIC IS.
GREECE
Athens
Algiers
Sicily
MEDITERRANEAN SEA
Casablanca
Rabat
Tunis
MOROCCO
Marrakech
CANARY IS.
Las Palmas
ATLAS SAHARIEN
HAUT ATLAS
TUNISIA
MALTA
Tripoli
El Aaiun
ALGERIA
LIBYA
Gulf of Sirte
Benghazi
WESTERN SAHARA
MAURITANIA
Crete
Istanbul
Izmir
Ankara
TURKEY
TOROS DAGLARI
Sevastopol'
BLACK SEA
CAUCASUS
GEORGIA
Tbilisi
Yerevan
ARMENIA
AZERBAIJAN
Baku
CASPIAN SEA
Nicosia
CYPRUS
Beirut
LEBANON
Damascus
SYRIA
Jerusalem
ISRAEL
Amman
JORDAN
Alexandria
Cairo
Suez
EGYPT
Nile
RED SEA
L.Van
Tehran
IRAN
ZAGROS MOUNTAINS
Tigris
Euphrates
Baghdad
IRAQ
Al Basrah
Al Kuwayt
KUWAIT
SAUDI ARABIA
Riyadh
URAL MOUNTAINS
WEST SIBERIAN PLAIN
Ob'
Irtysh
Tobol
Yekaterinburg
Kazan
Volga
Magnitogorsk
Samara
KAZAKHSTAN
Ural
0 500 1000 1500 Miles
0 500 1000 1500 2000 Kilometres

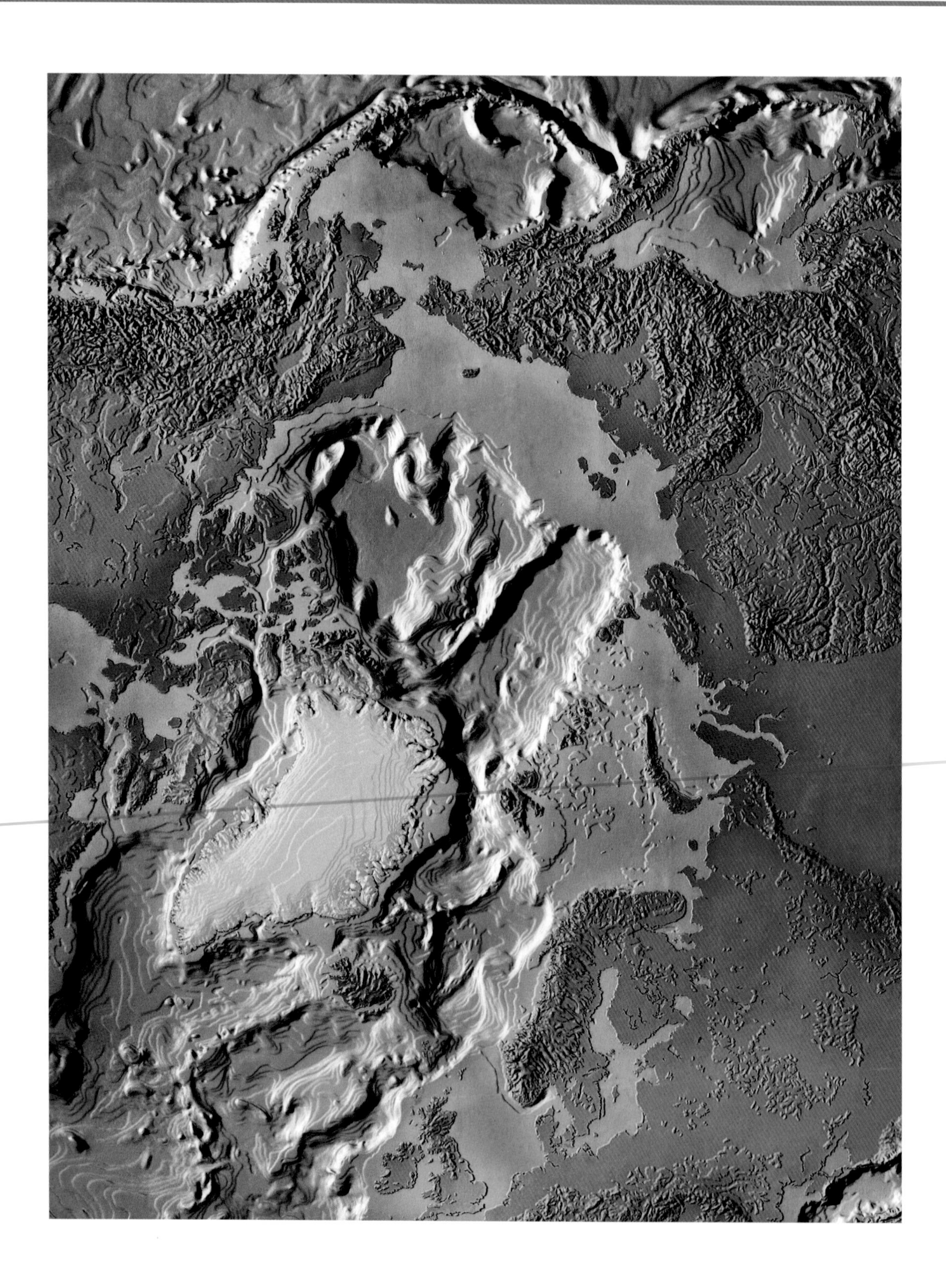

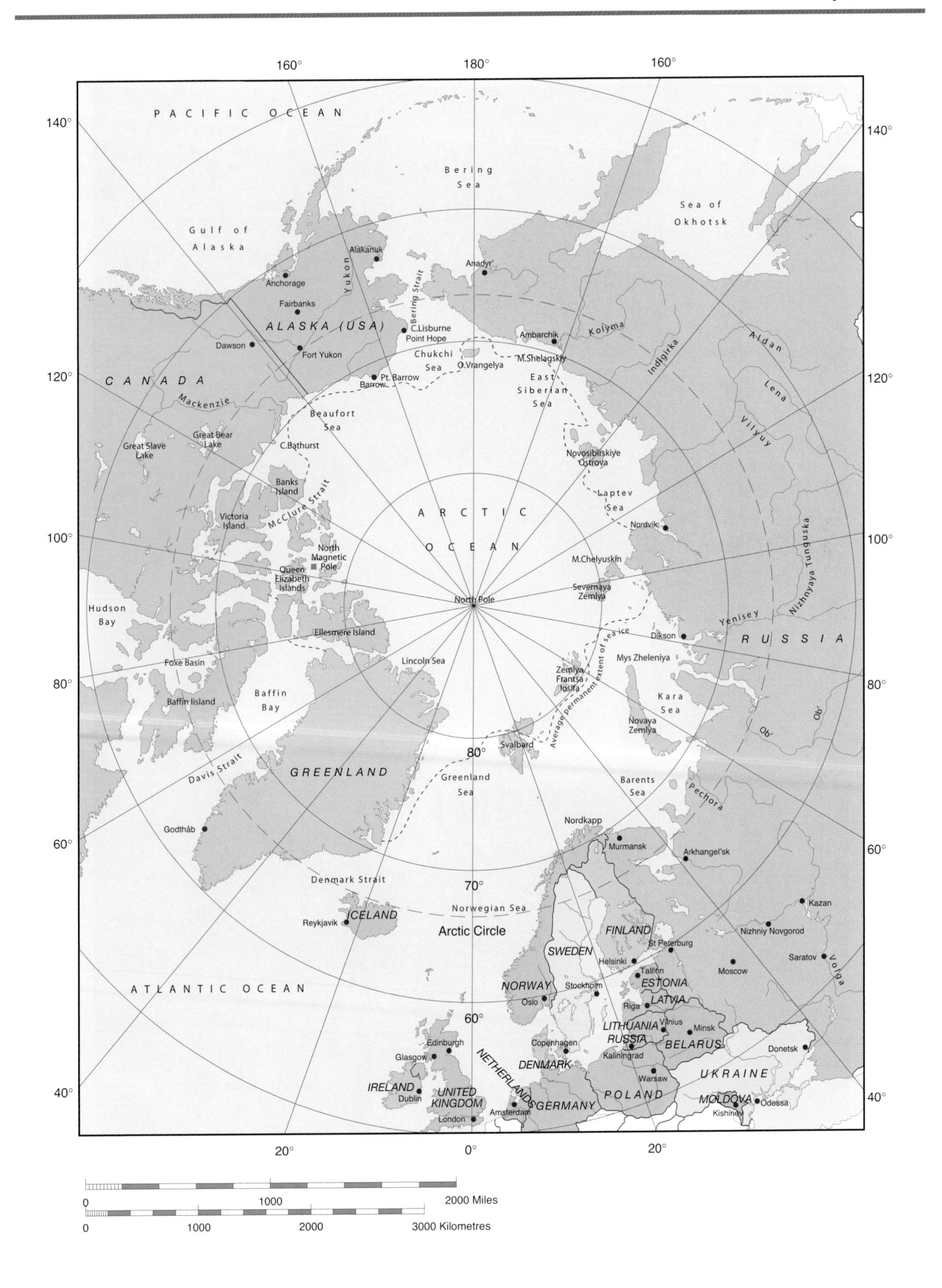
PACIFIC OCEAN
Bering Sea
Sea of Okhotsk
Gulf of Alaska
Alakanuk
Anchorage
Anadyr'
Yukon
Bering Strait
Fairbanks
ALASKA (USA)
C.Lisburne
Point Hope
Ambarchik
Kolyma
Dawson
Fort Yukon
Chukchi Sea
O.Vrangelya
M.Shelagskiy
Aldan
Indigirka
CANADA
Pt. Barrow
Barrow
East Siberian Sea
Lena
Mackenzie
Beaufort Sea
Vilyuy
Great Slave Lake
Great Bear Lake
C.Bathurst
Novosibirskiye Ostrova
Banks Island
Laptev Sea
Victoria Island
McClure Strait
ARCTIC OCEAN
Nordvik
Nizhnyaya Tunguska
North Magnetic Pole
M.Chelyuskin
Queen Elizabeth Islands
Severnaya Zemlya
Hudson Bay
North Pole
Yenisey
Ellesmere Island
Dikson
RUSSIA
Average permanent extent of sea ice
Foxe Basin
Lincoln Sea
Mys Zheleniya
Zemlya Frantsa Iosifa
Baffin Island
Baffin Bay
Kara Sea
Ob'
Novaya Zemlya
Svalbard
Davis Strait
GREENLAND
Greenland Sea
Barents Sea
Pechora
Nordkapp
Godthåb
Murmansk
Arkhangel'sk
Denmark Strait
Kazan
Norwegian Sea
Reykjavik
ICELAND
Arctic Circle
FINLAND
Nizhniy Novgorod
St Peterburg
SWEDEN
Helsinki
Saratov
Volga
Tallinn
Moscow
ESTONIA
ATLANTIC OCEAN
NORWAY
Stockholm
Oslo
Riga
LATVIA
LITHUANIA
Vilnius
Minsk
RUSSIA
BELARUS
Edinburgh
Copenhagen
Donetsk
Glasgow
Kaliningrad
NETHERLANDS
DENMARK
Warsaw
UKRAINE
IRELAND
Dublin
UNITED KINGDOM
POLAND
MOLDOVA
Odessa
Kishinev
GERMANY
Amsterdam
London
140°
160°
180°
120°
100°
80°
60°
70°
40°
20°
0°
0
1000
2000 Miles
2000
3000 Kilometres

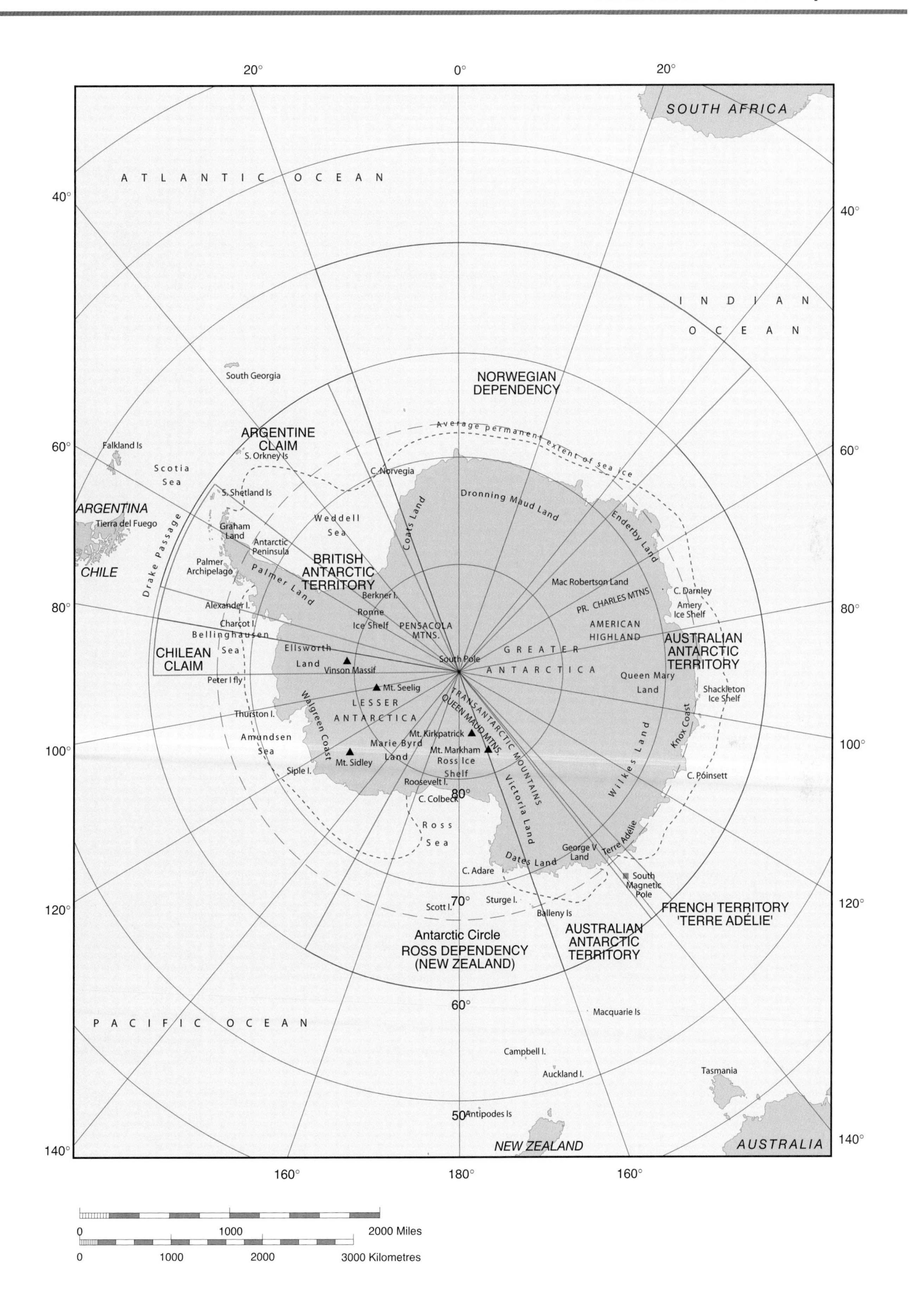
SOUTH AFRICA
ATLANTIC OCEAN
INDIAN OCEAN
NORWEGIAN DEPENDENCY
South Georgia
ARGENTINE CLAIM
S. Orkney Is
Falkland Is
Scotia Sea
S. Shetland Is
ARGENTINA
Tierra del Fuego
CHILE
Drake Passage
Graham Land
Antarctic Peninsula
Palmer Archipelago
Weddell Sea
C. Norvegia
Coats Land
Dronning Maud Land
Enderby Land
Average permanent extent of sea ice
BRITISH ANTARCTIC TERRITORY
Palmer Land
Berkner I.
Ronne Ice Shelf
PENSACOLA MTNS.
Alexander I.
Charcot I.
Bellinghausen Sea
CHILEAN CLAIM
Ellsworth Land
Vinson Massif
Peter I fly
South Pole
Mac Robertson Land
PR. CHARLES MTNS
C. Darnley
Amery Ice Shelf
AMERICAN HIGHLAND
AUSTRALIAN ANTARCTIC TERRITORY
GREATER ANTARCTICA
Queen Mary Land
Shackleton Ice Shelf
Mt. Seelig
LESSER ANTARCTICA
Thurston I.
Walgreen Coast
Amundsen Sea
Marie Byrd Land
Mt. Kirkpatrick
Mt. Markham
Mt. Sidley
Ross Ice Shelf
TRANSANTARCTIC MOUNTAINS
QUEEN MAUD MTNS.
Wilkes Land
Knox Coast
C. Poinsett
Siple I.
Roosevelt I.
C. Colbeck
Victoria Land
Ross Sea
Terre Adélie
George V Land
Dates Land
C. Adare
South Magnetic Pole
Scott I.
Sturge I.
Balleny Is
FRENCH TERRITORY 'TERRE ADÉLIE'
Antarctic Circle
ROSS DEPENDENCY (NEW ZEALAND)
AUSTRALIAN ANTARCTIC TERRITORY
PACIFIC OCEAN
Macquarie Is
Campbell I.
Auckland I.
Tasmania
Antipodes Is
NEW ZEALAND
AUSTRALIA
20° 0° 20°
40° 60° 80° 100° 120° 140°
80° 70° 60° 50°
160° 180° 160°
0 1000 2000 Miles
0 1000 2000 3000 Kilometres

2,4-D

2,4-D (2,4-DICHLOROPHENOXYACETIC ACID, CAS No. 94-75-7) was first synthesized by Robert Pokorny of the C. Dolge Company of Westport, Connecticut, and patented during World War II by the American Paint Company as one form of the halogenated phenoxy monocarboylic aliphatic acids. 2,4-D saw its first widespread production and use in the late 1940s, coinciding with the introduction of weed science as a scholarly discipline. 2,4-D is used as a selective, systemic, hormone-type herbicide to control sedge and broadleaf weeds in cereal crops such as rice, wheat, corn, sorghum, and barley; as well as range/pasture lands, turf areas, lawns, rights-of-way, aquatic sites and forestry applications; and as a growth regulator for citrus and potatoes. Trade names include Aqua Kleen, Demise, Esteron, Weed-B-Gone, and Weedone.

The annual global market is estimated to be over $300 million and the main producers are Agrolinz (Austria), Atanor (Argentina), Dow (United States), AH Marks (United Kingdom), Nufarm (Australia), Polikemia, Rhone-Poulenc (France), Sanachem (South Africa), Sinochem (China), Ufa (Russia), and four other producers in Turkey. Dow is the largest producer at 20,000 tons. Rhone-Poulenc is the largest European producer at 7,000 tons, followed by Agrolinz at 4,000 tons. The United States, South America, Europe, and the former Soviet Union are major markets for 2,4-D, and global use is predicted to grow over the next decade. Annual U.S. 2,4-D usage is approximately 21,000 tons, with 14,000 tons (66 percent) used by agriculture and 7,300 tons (34 percent) used in nonagricultural settings. Most nonagricultural use is due to homeowners, who apply it to their lawns (18 percent), and landscape maintenance contractors (seven percent). 2,4-D is also used widely in developing countries; for example, India used 1,300 tons in 1994–95.

Arguably the most infamous use of 2,4-D occurred during the Vietnam War (1962–71), when U.S. military forces sprayed a 50:50 mixture of 2,4-D and 2,4,5-T, dubbed "Agent Orange," over Vietnam to destroy crops, strip the thick jungle canopy that helped conceal opposition forces, and clear vegetation from the perimeters of U.S. base camps. Approximately 30 years later, the Department of Veterans Affairs published a notice in the Federal Register (January 1994) that stated, "there is a positive association between exposure to herbicides used in the Republic of Vietnam and the subsequent development of respiratory cancers." During the time that Agent Orange was being produced,

2,3,7,8-tetrachlorodibenzo-p-dioxin (TCDD), now classified as a Group 1 Human Carcinogen by the International Agency for Research on Cancer, was an unintended contaminant formed in the production of 2,4,5-T.

Human exposure data exist to estimate 2,4-D exposures from dietary ingestion of residues, nondietary ingestion of contaminates in soil and dust, inhalation of contaminated indoor and outdoor air, and dermal penetration. Little or no data exist for exposures of young children, prenatal, and neonatal infants. Several human studies have suggested an association between exposure to 2,4-D (and other herbicides) and an increased incidence of tumor formation. However, it is not clear whether this represents a true association, and, if so, whether it is specifically related to 2,4-D. On August 8, 2005, the U.S. Environmental Protection Agency (EPA) released its comprehensive assessment of 2,4-D under the agency's reregistration program. The EPA's document concluded that 2,4-D does not present risks of concern to human health when users follow 2,4-D product instructions as outlined in the EPA's 2,4-D Reregistration Eligibility Decision (RED) document.

SEE ALSO: Agent Orange; Herbicides; Vietnam War.

BIBLIOGRAPHY. Environmental Protection Agency, www.epa.gov (cited September 2006); *Generic Pesticides—the Markets*, (PJB Publications, 1994); Robert Pokorny, "New Compounds: Some Chlorophenoxyacetic Acids," *Journal of the American Chemical Society* (v.63/1768, 1941); *Veterans and Agent Orange: Length of Presumptive Period for Association Between Exposure and Respiratory Cancer* (Board on Health Promotion and Disease Prevention, Institute of Medicine, 2004).

Marielle C. Brinkman
Battelle Memorial Institute

Taiwan

OFFICIALLY KNOWN AS the Republic of China, and located on the island of Formosa, Taiwan has long struggled with the People's Republic of China for recognition as an independent nation. Some two million Chinese sought refuge on the island after the Communist takeover of China in 1949. The issue of whether or not Taiwan will eventually be unified with China continues to dominate Taiwan's politics. Bordering on the East China Sea, the Philippine Sea, the South China Sea, and the Taiwan Strait, Taiwan has a coastline of 971 miles (1,566 kilometers). Mountains cover the eastern two-thirds of the islands, giving way to plains in the west. Tropical and marine climates produce a distinct monsoon season in the southwest from June to August, and extensive clouds cover Taiwan for much of the year. Home to 23,036,087 people, the islands are prone to earthquakes and typhoons.

With a per capita income of $26,700 and healthy foreign investments throughout southeast Asia, Taiwan is the 34th-richest nation in the world. In recent years, many banks have been transferred from government to private hands as private industries have grown in response to the growth of the export industry. Taiwan has the third-largest trade surplus in the world, and China and the United States are the major trading partners. Natural resources include small deposits of coal as well as natural gas, limestone, marble, and asbestos. Roughly a fourth of the land is arable, but only 6 percent of the workforce are engaged in agriculture.

Extensive industrialization has led to major problems with air and water pollution, and supplies of drinking water have been threatened by pollution that includes the release of raw sewage into fresh water sources. Improper disposal of radioactive waste has created low levels of radiation. A study by scientists at Yale University in 2006 ranked Taiwan 24th in the world in environmental performance, well above the relevant geographic group and slightly below the relevant income group. The lowest rating was received in the category of air quality. The high overall ranking is due to government efforts to promote sustainable development and conserve natural resources that have received priority in Taiwan for several decades.

While a number of agencies bear responsibility for protecting the environment, the Environmental Protection Agency (EPA) is the only agency that is solely dedicated to environmentalism. Departments that work under the EPA include Comprehensive

Planning, Air Quality Protection and Noise Control, Water Quality Protection, Waste Management, and Environmental Sanitation and Toxic Substance Management. In order to combat the extensive air pollution that has accompanied industrialization, the legislature enacted the Air Pollution Control Act of 1975. The law was revised in 2002 to increase the power of the EPA and the Taiwan Area Air Quality Monitoring Network to strictly enforce environmental laws. Because Taiwan's independent status is not universally recognized, the country does not participate in international agreements on the environment.

Current land development policies in Taiwan have been formulated under the Challenge 2008 National Development Plan that promotes the upgrading of agriculture by encouraging farmers to employ improved technologies and land use, including switching from land cultivation to agricultural tourism. In 2004, Taiwan was hit by several typhoons. Extensive land development in the mountains led to heavy flooding and massive landslides during the typhoons. This devastation subsequently provided the momentum for a new program of conservation under the National Land Planning Act of 2004, combining public education with strict enforcement. Additionally, the Soil and Water Conservation Bureau was charged with overseeing the construction of a number of water catchment facilities designed to improve access to safe drinking water.

Taiwan has one of the richest ecosystems in the world. Constituting 1.5 percent of all identified species, 150,000 separate life forms have been identified on the islands. Around 30 percent of these life forms are endemic to Taiwan. The government first began a conscious effort to protect this biodiversity with the passage of the Cultural Heritage Preservation Act of 1981 and followed it up with the Wildlife Conservation Act of 1989. Taiwan is home to 70 species of mammals, 500 species of birds, 90 species of reptiles, 30 species of amphibians, 18,000 species of insects (including 400 butterfly species), and 2,700 species of fish. Some 1,955 species of rare fauna have also been identified. Among the 3.9 million acres (1.57 million hectares) of Taiwan's forests, 72 percent are nationally protected, and Taiwan has 16 wildlife refuges, 19 nature reserves, and an extensive national park system.

SEE ALSO: Biodiversity; Ecosystems; Industrialization; Pollution, Air; Pollution, Water; Radioactivity; Sewage and Sewer Systems.

BIBLIOGRAPHY. Central Intelligence Agency, "Taiwan," *The World Factbook*, www.cia.gov (cited April 2006); Timothy Doyle, *Environmental Movements in Minority and Majority Worlds: A Global Perspective* (Rutgers University Press, 2005); Kevin Hillstrom and Laurie Collier Hillstrom, *Asia: A Continental Overview of Environmental Issues* (ABC-CLIO, 2003); Michael Howard, *Asia's Environmental Crisis* (Westview, 1993); *Taiwan Yearbook 2005* (Government Information Office, 2005); Yale University, "Pilot 2006 Environmental Performance Index," www.yale.edu/epi (cited April 2006).

Elizabeth Purdy, Ph.D.
Independent Scholar

Tajikistan

TAJIKISTAN IS A landlocked country located in Central Asia. It has an area of 55,251 square miles and a population of 6.8 million (2005 estimate). The capital of the country is Dushanbe, with an approximate population of 700,000. Tajikistan is bordered by China, Afghanistan, Uzbekistan, and to the north, both Uzbekistan and Kyrgyzstan. Over 90 percent of the land is mountainous, with most of the population located around river valleys that flow from glaciers in the Pamir and Fan mountain ranges westward into the Aral Sea Basin. The majority of the population is Tajik, a group strongly tied to the Tajiks of northern Afghanistan and to Iranians, with a sizeable Uzbek minority and small Russian and Kyrgyz minorities. Although the official language of the country is Tajiki—a language closely related to Farsi in Iran—the government also recognizes Uzbeki and Russian for educational and judicial purposes. The country is largely Sunni Muslim; however, there is a small minority of Shi'a Ismailis in the East.

Historically, the land now comprising the country of Tajikistan was lightly settled with the majority of the people living in the Hisor/Qaroteghin Valleys

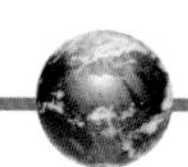

(now bisected by Dushanbe) and along the Zerafshan River, which flows to the historic cities of Samarqand and Bukhara. Prior to Russian advances in the region, the economy of the area was based on agriculture and transhumance pastoralism. This changed only slightly with the advent of the Soviet Union's hold over the region and most people remained either farmers or herders, though on collectivized farms. Industry was little developed outside of aluminum smelting (that took advantage of the hydropower in the region) and the necessary production around cotton, the republic's main agricultural crop.

With the collapse of the Soviet Union, Tajikistan obtained independence on September 9, 1991. Nascent tribal and Islamist feelings exploded in the southern Vakhsh Valley and triggered a debilitating civil war lasting until 1997. The Russian-backed winners, represented mostly from the cities of Kulob and Khojand, have governed ever since under the leadership of Emamoli Rahmanov. The Rahmanov regime has attempted to bring stability to the country by holding multi-party elections and having one of the few legislatures in the region with an opposition party represented in the parliament. Continued destabilization caused by opium-financed warlords who provide the transport of opium products from Afghanistan, however, threatens this still fragile country.

Like most former Soviet countries, Tajikistan's infrastructure is heavily tied to the other former Soviet republics. At independence, all of its roads, railroads, and pipelines led only toward Russia and the other former Soviet republics. Agriculture has remained the largest employer and cotton remains the most important economic indicator. The Tajik government, however, has maintained a monopoly on the cotton trade. Afghanistan's instability and the sheer height of the mountains within and around Tajikistan have hampered the country in its ability to increase economic activity with neighboring countries. Recently, however, the government has expanded trade with a new bridge now linking Tajikistan to Afghanistan, and ultimately on to South Asia and their ports; and more importantly, a new road to China through the Pamir Mountains has already dramatically increased trade into the country.

SEE ALSO: Cotton; Russia (and Soviet Union); Wars.

BIBLIOGRAPHY. Shirin Akiner, *Tajikistan: Disintegration or Reconciliation?* (Royal Institute of International Affairs, 2001); Mohammad Reza Djalili, Frédéric Grare, and Sharin Akiner, *Tajikistan: The Trials of Independence* (St. Martin's Press, 1997); Richard N. Frye, *The Heritage of Central Asia: From Antiquity to the Turkish Expansion* (Markus Wiener Publishers, 1996); Colette Harris, *Gender Relations in Tajikistan* (Pluto Press, 2004); Martha Brill Olcott, *Central Asia's Second Chance* (Carnegie Endowment for International Peace, 2005).

William C. Rowe
Louisiana State University

Takings

THE "TAKINGS" CLAUSE of the Fifth Amendment of the U.S. Constitution requires the government to pay compensation to owners of private property when their land is appropriated for public uses such as building roads. In early cases, only the physical invasion of private property for public use was considered a just taking. For example, in 1871, in the case of *Pumpelly v. Green Bay Co.*, the court ruled that when dam construction flooded private property, the loss of property was considered a taking. Conversely, regulatory action that affected the economic value of the property, such as rezoning legislation that forced an operating business to close, was not considered a taking.

A major shift in takings jurisprudence occurred in 1987 when the Supreme Court had to rule on three cases in the same term. The 1992 case of *Lucas v. South Carolina Coastal Council* is also an important case study supporting the claim that loss of economic value on private property due to governmental regulations constitutes a taking that requires just compensation. In 1986, David Lucas paid $975,000 for two residential lots on the Isle of Palms in South Carolina with the intention to build single-family homes on the properties. In 1988, the South Carolina Legislature enacted the Beachfront Management Act, which barred Lucas from building on his two parcels. Because the legislation had a significant economic impact on Lucas, rendering the lots practically "valueless," the court ruled that

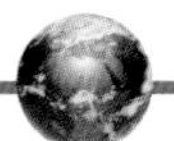

this accomplished a taking and required compensation. As a result of this and similar cases, regulatory takings have since been asserted whenever a regulation has reduced the value of private property.

"Takings" has received a secondary definition under the Endangered Species Act (ESA) of 1973, which has significantly challenged both the freedoms Americans attach to private property and the legal understanding of what constitutes a taking. An important provision of the ESA prevents the "taking" of endangered species on both public and private land. In the context of the ESA, the definition of taking is expanded to mean to "harass, harm, pursue, shoot, wound, kill, trap, capture, or collect" a listed species.

As a result of the ESA, a legal debate has emerged: If a take of an endangered species is illegal on private property, does the prohibition against taking endangered species constitute a taking of the private property under the Fifth Amendment? This question poses a major concern to both environmentalists and landowners. Even if most Americans wish to prevent the extinction of endangered species, that does not establish that individual landowners should be compelled to pay for (or bear the burden of) their preservation. Many have argued that if the government fails to compensate landowners for providing a habitat for endangered species on their land, then a landowner is better off killing and burying the plant or animal than potentially losing the rights generally associated with private property.

The shift from focus on species protection under the ESA to protection of their habitats has resulted in further broadening of the applications of the takings clause. In 1978, the Supreme Court's famous snail darter decision resulted in the temporary halting of the construction the nearly completed, multimillion-dollar Tellico Dam because it jeopardized the only known habitat of the endangered snail darters.

Another case involved logging in the Pacific Northwest, which was halted in the 1990s to protect the threatened spotted owl and its old-growth habitat, resulting in the loss of thousands of timber-related jobs in rural Oregon and Washington. The spotted-owl controversy has become an emblematic case in the takings debate, pitting the need for jobs against the protection of animals.

The application of the takings clause in respect to the ESA and private property has created much debate as Americans grapple with the conflicts between economic and ecological priorities in land use. Legal scholar Carol Rose has argued that in the United States environmental protection seems to be in conflict with the concept of private property when burdens are placed on particular individuals who are asked to preserve wetlands or endangered species. However, both secure private property rights and effective environmental protection share a common goal—the enhancement of the overall social well-being, both private and public. Rose believes that the current friction exists as new understandings of environmental harms and what constitutes a taking confront preexisting property laws. Because the legal community is still defining the boundaries of this modern usage of taking under the ESA, further modifications of the law through court cases are expected.

SEE ALSO: Endangered Species Act (ESA); Habitat Protection; Northern Spotted Owl; Private Property; Property Rights; Snail Darter and Tellico Dam; Timber Industry.

BIBLIOGRAPHY. Lynn E. Dwyer, Dennis Murphy, and Paul Erlich, "Property Rights Case Law and the Challenge to the Endangered Species Act," *Conservation Biology* (v.9/4, August 1995); Robert Percival et al., *Environmental Regulation: Law, Science, and Policy* (Aspen Publishers, 2003); Carol M. Rose, "Property Rights and Responsibilities," in Marian Chertow and Daniel Esty, eds., *Thinking Ecologically: The Next Generation of Environmental Policy* (Yale University Press, 1997); Barton Thompson, Jr., "The Endangered Species Act: A Case Study in Takings and Incentives," *Stanford Law Review* (v.49, 1997).

Amity A. Doolittle
Yale School of Forestry and
Environmental Studies

Tanganyika, Lake

LAKE TANGANYIKA IS the largest of a chain of lakes in the Great Rift Valley in eastern Africa. Geologic processes have formed the extraordinary

lakebed. Lake Tanganyika measures 673 kilometers by 15–90 kilometers, covers 32,900 square kilometers, and is the longest lake in the world. With a maximum depth of 1,470 meters, Lake Tanganyika is the second deepest lake in the world next to Lake Baikal. The lake is, therefore, considered the largest freshwater reservoir in Africa. It is also one of the most diverse freshwater ecosystems in the world, with several hundred endemic species.

The lake catchment is relatively small at 220,000 square kilometers, due to the structure of the Great Rift Valley's steep mountainous ridges. Two main rivers flowing into the lake are the Rusizi, draining Lake Kivu and entering Lake Tanganyika from the north, and the Malagarasi, which enters the lake in the east, draining western Tanzania. In addition, many smaller rivers drain the slopes of the surrounding mountains. There is one major outflow, the Lukuga River, which leaves Lake Tanganyika as a tributary to the Congo River in the west. However, due to the tropical climate, Lake Tanganyika loses most water to evaporation.

The countries of Burundi, the Democratic Republic of Congo, Tanzania, and Zambia share Lake Tanganyika. Of the lake's shoreline perimeter, 43 percent is in the Democratic Republic of the Congo, 36 percent in Tanzania, 12 percent in Zambia, and 9 percent in Burundi. An estimated one million people live on the shores of Lake Tanganyika. The main settlements are Bujumbura (400,000 people), the capital of Burundi; Uvira (100,000) in the Democratic Republic of the Congo; Kigona (135,000) in Tanzania; and Mpulungu (70,000) in Zambia. Economic activity in these urban areas includes industrial plants (such as paint, brewery, textile, soap, and battery plants), cotton processing, sugar production, and industrial fishing. Kigoma is the largest transit point for goods and people entering and exiting the lake region. Lake Tanganyika, with its well-developed shipping routes, is one of the most important inland traffic and communications links in eastern Africa.

Crop and livestock production and processing, and mining (tin, copper, coal), are the main industries in the catchment of Lake Tanganyika. Commercial, as well as small-scale fisheries are particularly important to the local economy and local livelihoods. The fast rate of urbanization is creating severe problems of lake pollution through urban wastewaters. Moreover, overfishing in the coastal zone and habitat destruction are increasing concerns. High rates of soil erosion in the catchment have increased lake sedimentation rates.

The appropriate utilization of natural resources in the catchment has become a priority, particularly since it plays a major role in regional poverty reduction strategies. After centralized systems failed, decentralization of government structures and empowerment of local governments to manage local resources for improving livelihoods is a new policy change in recent years. Commercial water provision, for example, has become a major strategy to improve water coverage. However, partial privatization of urban water providers largely failed, since municipalities and private investors often do not agree on how to pre-finance the investments into a deficient system.

SEE ALSO: Baikal, Lake; Congo, Democratic Republic of the; Lakes; Rift Valley.

BIBLIOGRAPHY. G.W. Coulter, *Lake Tanganyika and its Life* (Oxford University Press, 1991); G. Hanek, E.J. Coenen, and P. Kotilainen, *Aerial Frame Survey of Lake Tanganyika Fisheries* (Food and Agriculture Organization, 1993); A. Mills, *Pollution Control and Other Measures to Protect Biodiversity in Lake Tanganyika* (United Nations Environment Programme, 2000).

Wiebke Foerch
University of Arizona
Ruger Winnegge
University of Siegen, Germany

Tanzania

AFTER GAINING INDEPENDENCE in 1964, Tanganyika and Zanzibar merged into the United Republic of Tanzania. The next three decades were characterized by one-party rule, which ended in 1995 with democratic elections. Zanzibar has retained a semi-autonomous status that has contributed to hotly contested elections and charges of voting irregularities. With a per capita income of only $700, Tanzania is the seventh poorest country in the world. Some 36 percent of the population live in abject pov-

erty. Only four percent of the land area is arable, yet 80 percent of Tanzanians are engaged in the agricultural sector, which provides around half of the Gross Domestic Product and 85 percent of all exports. The United Nations Development Programme's Human Development Reports rank Tanzania 164 of 232 countries on overall quality-of-life issues.

Industries are generally involved with processing agricultural productions or in producing light consumer goods. Natural resources include: Hydropower, tin, phosphates, iron ore, coal, diamonds, gemstones, gold, natural gas, and nickel; Tanzania has recently begun exploiting these resources. The World Bank and the International Monetary Fund are working with the government to reduce poverty and rehabilitate the economic infrastructure.

Bordering the Indian Ocean, Tanzania has a coastline of 1,424 kilometers and inland water resources of 59,050 square kilometers. The total area of 945,987 square kilometers includes the islands of Mafia, Pemba, and Zanzibar. Tanzania shares land borders with Burundi, the Democratic Republic of the Congo, Kenya, Malawi, Mozambique, Rwanda, Uganda, and Zambia. The coastal plains of Tanzania give way to a central plateau and to northern and southern highlands. While the coast enjoys a tropical climate, the highlands are temperate. Tanzania is subject to drought, and flooding may occur on the central plateau during the rainy season. Elevations range from sea level to 5,895 meters at Mount Kilimanjaro along the Kenyan border. Kilimanjaro, which is Tanzania's most distinctive geographic feature, is the highest point in all of Africa. The world-renowned mountain is bordered by Lake Victoria, the second largest freshwater lake in the world, Lake Tanganyika, the second deepest lake in the world, and Lake Nyasa.

Like many of Africa's poorest countries, Tanzania's population of 37,445,392 is vulnerable to a number of environmental health hazards. With an adult prevalence rate of 8.8 percent, the HIV/AIDS epidemic has claimed 160,000 lives since 2003. Another 1.6 million people are living with the disease. While 73 percent of the population have access to safe drinking water, only 46 percent have access to improved sanitation. Therefore, Tanzanians are at very high risk for contracting food and waterborne diseases that include bacterial diarrhea, hepatitis A, and typhoid fever and the water contact disease schistosomiasis. In some areas, the population is at high risk for contracting vectorborne diseases such as malaria, Rift Valley fever, and plague. Consequently, Tanzanians have a lower-than-normal life expectancy (45.64 years) and growth rate (1.83 percent), and higher-than-normal infant mortality (96.48 deaths per 1,000 live births) and death (16.39 deaths/1,000 population) rates. The fertility rate of five children each places women at great risk and taxes strained resources.

Tanzania is experiencing extensive soil degradation and desertification as a consequences of human mismanagement and natural disasters. Around 44 percent of the land is forested, but deforestation is occurring at a rate of 0.2 percent per year. The destruction of coral reefs is jeopardizing marine life. Marginal agriculture has been seriously threatened by prolonged droughts. Illegal hunting and trade, particularly the ivory trade, is posing major threats to Tanzanian wild life.

Because Mount Kilimanjaro is the highest freestanding mountain in the world, it has provided valuable information on climate change created by greenhouse gases and global warming. Scientists have observed a visible depletion of the snow in recent years. This factor is affecting the tourist industry and decreasing Tanzania's supply of fresh water. There is some indication that the situation on Kilimanjaro has also reduced fish supplies in Lake Tanganyika. A 2006 study by scientists at Yale University ranked Tanzania 83 of 132 countries on environmental performance, well above the relevant income and geographic groups. Only the low score in environmental health prevented Tanzania from a higher ranking.

Approximately 30 percent of the land has been claimed by the government to protect Tanzania's rich biodiversity. Protected lands include nature reserves, wilderness areas, national parks, species management areas, and wetlands. Wildlife has been seriously threatened, however, partially by rural Tanzanians who hunt wildlife for food because they cannot afford domestic meat. Bush meat provides around a fourth of all meat consumption for this segment of the population. Of 316 identified mammal species, 42 are endangered, as are 33 of 229 bird species.

In the 1960s and 1970s, Tanzania became involved in ecotourism as a means of preserving vulnerable ecosystems while helping villages meet their economic needs. Until that time, many villagers had seen wildlife only as hunting prey or pests that destroyed crops. In the 1980s, the Campfire programs were initiated in Africa, and villagers were given an economic stake in protecting the wildlife that attracted tourists to their areas by providing services and sharing in profits. Tanzania's Wildlife Division developed programs using ecotourism as an alternative to hunting, which wrecked the environment and destroyed wildlife populations. Some of the proceeds from the Campfire projects are used to supply water and sanitation systems to villages and to support the national park system. Critics claim, however, that resources are being directed toward the state treasury and away from conservation as originally planned.

In 1983, Tanzania passed the National Environment Council Act that created the National Environment Management Council, which advises the government on all environmental issues and formulates and recommends policy and standards. The Ministry for Tourism, Natural Resources, and Environment is the statutory body charged with implementing and enforcing environmental laws based on the framework provided by the National Conservation Strategy for Sustainable Development and the National Environmental Policy, which comprise national policy on land use, natural resources and conservation, pollution, and environmental management. Tanzania participates in the following international agreements on the environment: Biodiversity, Climate Change, Climate Change–Kyoto Protocol, Desertification, Endangered Species, Hazardous Wastes, Law of the Sea, Ozone Layer Protection, and Wetlands.

SEE ALSO: Colonialism; Coral Reefs; Ecotourism; National Parks; Poaching; Safaris; Tanganyika, Lake.

BIBLIOGRAPHY. Central Intelligence Agency, "Tanzania," *World Factbook,* www.cia.gov (cited April 2006); Timothy Doyle, *Environmental Movements in Minority and Majority Worlds: A Global Perspective* (Rutgers University Press, 2005); Kevin Hillstrom and Laurie Collier Hillstrom, *Africa and the Middle East: A Continental Overview of Environmental Issues* (ABC-CLIO, 2003); Valentine Udoh James, *Africa's Ecology: Sustaining the Biological and Environmental Diversity of a Continent* (McFarland, 1993); One World, "Tanzania: Environment," uk.oneworld.net (cited April 2006); United Nations Development Programme, "Human Development Report: Tanzania," hdr.undp.org (cited April 2006); World Bank, "Tanzania," lnweb18.worldbank.org (cited April 2006); Yale University, "Pilot 2006 Environmental Performance Index," www.yale.edu/epi (cited April 2006).

ELIZABETH PURDY, PH.D.
INDEPENDENT SCHOLAR

Taxidermy

TAXIDERMY IS THE practice of preserving the skins of animals and using them to create tableaux or dioramas of the animals in what are meant to be accurate representations of their shape, habits, and/or life environments. Taxidermical techniques have developed considerably in the modern age. The practice may be considered to be an offshoot of the mummification of corpses, human and otherwise, as found in ancient Egypt and Xinjiang. However, mummification was conducted primarily for religious purposes and the resultant mummies were not meant for public display. When taxidermy began on a significant scale in the 18th century, it served a partly scientific purpose to help to disseminate information about creatures that most people would have almost no chance of seeing live in the wild. It also became important for hunters to display the prey they had managed to kill.

Chemical processes were established to preserve skins and some other organic remains, although the methods were often somewhat crude and inconvenient. In the 19th century, processes improved to the extent that large tableaux could be created that were displayed in some of the better-regarded natural history museums of the world. As time progressed, taxidermists moved from the concept of stuffing animals with some kind of nonreactive filler to the concept of mounting them, which involves using a variety of structural materials and

Competitive Taxidermy

Taxidermy has become an activity of considerable scope, especially in the United States, where manufacture and retail of taxidermy supplies has grown into quite a significant industrial sector. Competitive taxidermy, in which contestants display their creations for the deliberation of judges, has become intense. The field has grown to include invertebrates and reptiles. Rogue taxidermists prefer to use their artistic talents to create scenes that have never been seen in the real world; they mount creatures that appear to be mermaids, unicorns, or some other fantastic hybrid. Such items can fetch high prices.

Modern techniques have led to more realistic taxidermy, but its scientific role is waning.

techniques to create the desired lifelike appearance. The American naturalist Carl Akeley is associated with the transformation of taxidermy into a systematic and well-ordered process. Recently, new processes have enabled the realistic recreation of a wider range of organic remains and a broader selection of the animal world.

While taxidermy is likely to continue as an artistic hobby, especially among common animals, its use in science and conservation is likely to decline as computer graphics improve and are better able to recreate the life conditions of animals.

SEE ALSO: Animal Rights; Animals; Hunting.

BIBLIOGRAPHY. Carl Ethan Akeley, *Taxidermy and Sculpture: The Work of Carl E. Akeley in Field Museum of Natural History* (Field Museum of Natural History, 1927); Mary L. Jobe Akeley, *The Wilderness Lives Again: Carl Akeley and the Great Adventure* (Dodd, Mead and Company, 1944); Gerald Grantz, *Home Book of Taxidermy and Tanning* (Stackpole Books, 1985).

JOHN WALSH
SHINAWATRA UNIVERSITY

Taylor Grazing Act (U.S. 1934)

THE TAYLOR GRAZING Act of 1934 authorized the U.S. Secretary of the Interior to establish and regulate grazing districts on vacant, unappropriated, and unreserved public domain lands deemed "chiefly valuable for grazing," to issue grazing permits for up to 10 years, and to collect grazing fees. Federal regulation of grazing first began on Forest Reserves in the early 1900s, but it took many years for legislation regulating grazing on the remaining public domain lands to pass. Although cattle ranchers were generally in favor of legislation that would bring order and stability to what was essentially an unregulated grazing commons, sheepherders, settlers, and farmers opposed it. In addition, the U.S. Department of Agriculture and the U.S. Department of the Interior battled over which one of them would administer the public rangelands. Legislation was also delayed because the future status of these lands was unclear. On the one hand, since the passage of the first Land Ordinances in the 1780s, Congress had adopted a policy of transferring the public domain lands to private ownership. On the other hand, President Herbert Hoover offered to give the public rangelands back to the states in 1929, but the states declined. Finally, in 1934, the Depression, coupled with an extended drought throughout the West, provided the necessary impetus for the passage of the Taylor Grazing Act.

The purpose of the Taylor Grazing Act was to "stop injury to the public grazing lands...to provide

for their orderly use, improvement and development ... [and] to stabilize the livestock industry dependent on the public range." The Act directed the Secretary to cooperate with "local associations of stockmen" in the administration of grazing districts, a feature that its sponsor, Congressman Edward Thomas Taylor from Colorado, referred to as "democracy on the range" or "home rule on the range." It also gave "preference" in issuing permits to local residents. A crucial clause, "pending its final disposal," indicated that Congress was still trying to decide what to do with the remaining public domain and that these arrangements might be temporary.

Secretary of the Interior Harold Ickes established a Grazing Division (which became the Grazing Service in 1939, and the Bureau of Land Management [BLM] in 1946) to guide these procedures and appointed Farrington Carpenter, a Colorado cattleman and attorney, as its first director. The Secretary delegated authority to set up boundaries for grazing allotments, decide who would receive permits, and determine initial stocking rates to grazing advisory boards composed of local ranchers. State advisory boards and the National Advisory Board Council took on more general policy issues.

This structure kept bureaucracy to a minimum and reduced the potential conflict involved in implementing the new program by taking historic use patterns into account, providing relatively secure tenure, and giving local grazing interests real decision-making authority. It has been heavily criticized for effectively instituting private grazing rights on public lands, allowing the BLM to be "captured" by local interests, and allowing range degradation to continue. Although the management structure set up by the Taylor Grazing Act has been significantly altered by subsequent legislation, it should be noted that, in its initial form, it bore a remarkable resemblance to collaborative and community-based approaches to resource management now being promoted throughout the world and embraced by federal and state agencies in the United States.

SEE ALSO: Bureau of Land Management (BLM); Livestock; Overgrazing; Ranchers.

BIBLIOGRAPHY. J.N. Clarke and D.C. McCool, *Staking out the Terrain: Power and Performance among Natural Resource Agencies* (State University of New York Press, 1996); Debra Donohue, *The Western Range Revisited: Removing Livestock from Public Lands To Conserve Native Biodiversity* (University of Oklahoma Press, 1999); Christopher McGrory Klyza, *Who Controls the Public Lands? Mining, Forestry, and Grazing Policies, 1870–1990* (The University of North Carolina Press, 1996); Karen R. Merrill, *Public Lands and Political Meaning: Ranchers, the Government, and the Property Between Them* (University of California Press, 2002).

JULIE BRUGGER
UNIVERSITY OF WASHINGTON

Technology

ETYMOLOGICALLY, TECHNOLOGY (BASED on ancient Greek *technè* and *logos*) means the study of crafts. But the word is usually understood in a much broader sense to include objects (i.e., a hammer, car, or microscope), techniques (i.e., writing, agriculture, or advertising) as well as processes influencing aspects of human life (i.e., when technology is seen as creating its own culture or determining economic cycles). The relationships between technology and the environment are very close, as technology is sometimes defined as the way by which humans manipulate their environment to fulfill their needs. Indeed, through the invention of tools and techniques, humans have been able to take advantage of resources found in the environment. They have been able to increase resources, for instance through agriculture.

Today, as people are often surrounded by devices such as computers or cell phones, use transportation means ranging from cars up to planes for traveling, and eat food grown in greenhouses, there is a general feeling—at least in industrialized countries—that humanity lives in a more technological society than ever before. On the one hand, this sensation makes one think that humanity has freed itself of environmental constraints and is no longer dependent on what can be found in nature in order to survive. On the other hand, humans seem to have become dependent on these technologies, which contribute to the degradation and depletion of the environment

and in some most extreme forms—such as nuclear weapons—can be a threat to the very existence of humanity. However, the relationship between technology and the environment is much more complex that either improvement or depletion.

Technology is most often referred to by the objects it includes. In this understanding, a spade, a plow, or a combine harvester are all technologies. From the simplest up to the most sophisticated, these devices allow humans to do things they could not do otherwise. In this sense, technological objects are considered as means to reach ends defined by their designers and users.

But technology also refers to techniques. For instance, agriculture can be considered as a technology to provide food, or writing as a technology to communicate through space and time. A technique does not refer directly to a specific object, but to a set of knowledge, skills, and routines that allows one to arrange and use objects in order to reach a specific goal. Most techniques can be put into practice only through the use of specific objects—agriculture in its simplest form implies the use of basic tools. Even observation activities may require the use of technological devices: The monitoring of a national park relies on data transferred by satellites or on rangers equipped with vehicles. Some authors consider that a technique can only be viewed as a technology when it is reflectively used to solve problems or satisfy needs. In this understanding, a technique for playing a musical instrument would not be considered as a technology. However, the distinction is not always easy to make.

A third understanding of the word *technology* considers it as a more abstract concept defining a process that influences the way humans deal with the world. This understanding relates to expressions like "technological society" or "technological culture" when they are used to qualify the state we live in. In these expressions, technology does not refer to specific devices or techniques, but is perceived as a force that brings changes to the way humans interact among themselves and with their environment compared to a situation with less or no technology. This understanding is driven by a feeling that the proliferation of technological objects is something ineluctable, almost independent from human will.

TECHNOLOGY AND SCIENCE

Current technological developments seem to rely strongly on scientific discoveries. The petrochemical industry would probably not have had the same development and impact without modern chemistry. The idea that scientific research is the main factor behind technological progress became common during the 19th century and culminated in the post–World War II period, influencing both scientific and technological policies. In the United States, such a view can be found in the Vannevar Bush model that drove federal scientific policy from the 1940's on and stated that technological innovations would trickle down from massive investments in fundamental research. It led to increased cooperation between scientific institutions such as universities and high technology firms and government branches. The word *technoscience* has been proposed by some authors in order to define this state in which science and technology are so strongly related that it is not possible to distinguish them.

Despite the strong links between science and technology, they are different activities and should not be confused. Science is defined as an activity that seeks to gain knowledge about the world, while technology enables human beings to complete specific tasks. A common misunderstanding stemming from this confusion is considering technology as applied science. When scientific results are applied, they usually lead to technological innovations, but not all results have applications. Moreover, not all technological innovations result from the application of scientific findings. They can also be the result of trial and error processes undertaken without an understanding of the mechanisms behind the innovation. For example, agricultural innovation relied for a long time on observations made by farmers before agronomists and botanists fostered rapid changes using scientific results. It is also important to note that most scientific activity would not be possible at all without the technological instruments used for observing or measuring.

ENVIRONMENTAL IMPACTS

One of the reasons why technology is sometimes perceived as an independent force is its continuous

growth over long historical periods. Retracing a history of technology implies retracing a history of humanity. The periodization of early human ages is based on technological criteria such as Paleolithic (early stone age with chipped-stone tools), Neolithic (new stone age, which saw the development of polished stone tools and pottery) up to the Bronze Age. The distinction between prehistoric and historic ages is also based on the technological criteria of the appearance of writing. French philosopher of technology Gilbert Simondon goes as far as to say that humanity comes from the use of technology—by becoming technological beings humans have distinguished themselves from other species.

Technology is also a constituent of the human-environment relationship as it is the means by which humans are able to free themselves from environmental constraints, but it is also dependent on resources found in the environment. However, it seems that the human impact on the environment due to technologies is growing larger.

The first technologies used by humans consisted of very simple tools that could be found in their surroundings, such as sticks or rocks with useful shapes for reaching or breaking other objects. As humans learned to shape rock, wood, and other materials into more complex tools, direct dependence on environmental conditions began to decrease. The invention of cutting stones, for instance, allowed humans to use animal skins or furs, and therefore become less dependent on climatic conditions and survive in colder environments.

The development of agriculture and the subsequent sedentarization of communities it implied was probably the technological development that had the strongest impact on the environment for a long period. Historians of technology, although they use different types of periodization, usually consider the whole range of time from the point after which agriculture and sedentarization established themselves as dominant forms of human dwelling around 10,000 B.C.E. up to the Industrial Revolution in the 18th century as a single period. Agriculture contributed to powerfully transforming the environment by several means. First, with agriculture came the sedentarization of communities, with the first permanent settlements growing into towns, occupying space in a very different way than in former lifestyles. Second, the first areas where agriculture developed being floodplains of large rivers (the valleys of the Euphrates and Tigris Rivers in Mesopotamia, the Nile Valley in Egypt, or the Yangtze in China), there was a need to develop complex irrigation systems in order to take advantage of the regular floods of these rivers. These systems were developed as early as 3,000 B.C.E. and considerably shaped their environment. Chinese rice growing, for example, had a significant impact on the landscape. Even in Western Europe, where there was no need for such technological irrigation systems, early agriculture had an impact on the environment through deforestation.

Although there were great spatial variations over this long period of time, the main technological evolutions improved ways of living without radically changing them. The Agricultural Revolution that took place during the 17th century was essentially a rationalization and an intensification of the technologies available at that time. However, these were made possible through a new attitude toward technology of trying to understand how things worked, not just how to make them work.

The drive for scientists to foster technological innovation increased during the Enlightenment period and is one of the factors that favored the Industrial Revolution that began in Western Europe in the 18th century and eventually spread over most of the world. It triggered a period of fast technological transformations that also brought major changes to the ways humans occupied and exploited the earth's surface.

One of the main technological changes driving the Industrial Revolution was the invention and diffusion of steam-powered machinery, first in factories (i.e., spinning machines) then in transportation (i.e., trains or steamboats). The generalization of this form of power in industries throughout the 19th century had various environmental consequences. Steam engines were powered through the combustion of wood or coal; these resources had to be exploited more intensively in order to meet growing needs. The opening of new coal pits and deeper digging in existing ones had considerable roles in destroying ecosystems. The development of train transportation led to deforestation, as there was a great need for timber for crossties for the tracks. In

the United States the situation became so dramatic by the end of the 19th century that a governmental Forest Service was created in order to manage timber resources and avoid depletion. Another consequence of the use of coal as the major fuel was much air pollution in industrial cities.

Just as the first Industrial Revolution centered around coal, it was sources of power that characterized the second Industrial Revolution and its environmental consequences. The exploitation of mineral oil resources that developed in the 1860s enabled their use as fuel, leading to considerable development of prototypes of gas engines. The invention of gasoline and diesel engines in the last two decades of 19th century led to the car becoming a significant means of transportation, with all its consequences in terms of urban organization and pollution later on in the 20th century. But petroleum has also had important impacts on the environment with the development of the petrochemical industry. Another important technological impact on the environment at the turn of the century was the use of electricity as a source of power for lighting and above all for urban transportation means such as subways and tramways. The development of this source of power during the 20th century had spectacular environmental consequences with the damming of rivers in order to produce hydroelectricity.

The two World Wars are often depicted as periods of important technological development; in order to have an advantage over their opponents belligerent governments invested in research for new technologies. But these periods also brought some important changes in the way people considered technologies. Technologies based on nuclear power are symptomatic for this, as their development for civil use was accompanied by constant opposition from more or less large parts of the population, depending on the country. If at first it was largely the destructive capacities of nuclear power used as a weapon that fed this opposition, accidents in the power plants of Three Mile Island in Pennsylvania in 1979 or at Chernobyl in the former Soviet Union in 1986 reinforced popular opposition

Public enthusiasm for technology in the postwar era was also altered by the growth of production in sectors that had emerged with the first two Industrial Revolutions. These had much more actual effects on the environment than the destructive potential of nuclear power. In the context of a competition economy, growth of consumption led to more technological innovation in order to differentiate everyday goods and renew the need for consumption. Therefore it induced more waste or pollution because of the constant growing production of goods.

During the last quarter of the 20th century, the development of biotechnologies has led to a new understanding of the consequences of human activity on the environment. Biotechnologies are technologies that use or modify living organisms in order to favor some of their features or even add new ones. One of the ways biotechnologies may have consequences on the environment is through the introduction of new breeds of cattle or crops for agricultural use through the modification of genetic information of existing species. These new breeds are called genetically modified organisms (GMO). By modifying the gene stock of living organisms it becomes possible to create new species such as pest-resistant corns.

Environmental impacts of biotechnologies were initially considered rather promising. It was said that biotechnological plants would make agriculture less dependent on environmental constraints such as weather or soil conditions. By creating crops that are pest-resistant or adapt to specific soil conditions, the need for fertilizers or pesticide could be drastically reduced. Despite the introduction of such crops in some countries, genetically modified organisms are very controversial as there is evidence that they are altering their environment by making some pests and weeds resistant to pesticides. There also may be a transfer from the modified genes to other plants, with all the unintended consequences this may have. Finally, as some agro-industrial firms are designing crops that are best suited for their respective environments, they may be a threat to biodiversity, especially in developing countries where there is an urgency to increasing agricultural production and where it can seem attractive to replace less productive local crops.

A consequence of the rising impact of technology on the environment during the second half of the 20th century has been to trigger even more innovation, but this time in order to minimize negative

technological effects. These are sometimes called green technologies and are oriented toward efficiency in energy use, waste recycling, or reparation of damaged ecosystems. The main push for these efforts comes from governmental financial incentives and stricter regulation. In a way, technology is again clearly dependent on the state of the environment.

THEORIES OF TECHNOLOGY

Whereas for a long time impacts of technologies were mostly considered positive, the consequences of the introduction of new technologies since the Industrial Revolution have been viewed in more ambivalent ways. Relating to this ambivalence, there are different theories that can be used in order to study technologies. Philosopher of technology Andrew Feenberg offers a framework in order to characterize theories of technology by looking at whether they consider technologies to be value neutral (they are just means) or value laden (they include ends) and at whether they are considered autonomous or human-controlled.

Some theories of technology are deterministic in the sense that they consider that technologies have a functional logic that is autonomous from the rest of the social sphere and in return determines how society must function in order to make use of these technologies. However, the determination is considered to be value neutral in the sense that technologies develop only in order to meet human ends. In such a view, it would be said that water regulation techniques for rice culture in ancient China required centralization of decisions and therefore determined the form Chinese imperial societies took. Such a view can for instance be found in traditional Marxist theories.

However, regarding the proliferation of negative effects on the environment occurring from technologically mediated human activity, technology is often perceived as a threat to the environment or even to humanity as a whole and developed beyond the human needs it is supposed to fulfill. Such a view is largely present in public opinion, in some parts of the environmental movement, and also among some philosophers who have worried about the negative effects of technology on human freedom. In these views, technology is also conceived as autonomous but value laden. Such theories have been developed by philosophers such as Martin Heidegger or Jacques Ellul who claimed that technologies have their own ends that are guided by an instrumental rationality based on efficiency, which influences all other domains of activities not only by determining how the technology has to be used, but also by making efficiency the core value of all social activities.

Instrumentalism qualifies theories that consider technologies neutral. But unlike determinism, human agency is considered as the major force shaping them. In this view technology is pure instrumentality and its use is determined by other aspects of social life. Instrumentalism holds that technology is neither good nor bad, but depends on social, political, or cultural values and its outcome will depend on how humans make use of it. This view is very common among governmental institutions, engineers, and scientists. In instrumental theories there is an implicit idea that the negative consequences of a technology can be reversed, since those are consequences of human action. However, this view does not acknowledge the complexity of human agency, especially the variety of sometimes competing uses that can be made of a single technology.

A fourth kind of theory of technology holds that there is no clear distinction between means and ends when speaking of technology. This view can be found in critical theories of technology (i.e., Herbert Marcuse, Andrew Feenberg) or in some versions of Actor-Network Theory such as the one developed by Bruno Latour. The conflation of ends and means happens because technology implies human practice. On the one hand, a technology can be used in ways very different than it was intended, showing that it is not an end for itself. On the other hand, the very existence of a specific technology can lead humans to aim at ends they would not have thought about without this technology. So technology is also not just a means in order to reach predefined ends. In such views, there is no technological determinism but a path of dependency can occur once one technology is chosen. Humans retain a certain level of control, but because of the interrelatedness of ends and means, each technological change will bring changes in the way a society is organized. It is therefore impossible to reduce the question of how humans should deal with technology to a matter of

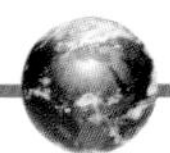

wise or bad use. It is rather a political matter and always implies questions of social organization such as: Who takes advantage of a particular technology? What does it imply for everyday life? What environmental consequences may derive from it?

TECHNOLOGY AND POLICY

Technology is an important issue in environmental policy since it is both a cause of environmental problems but also very often a solution to them. One of the main ways in which environmental policy affects technology is the adoption of environmental regulations that put constraints on the use of specific technologies. Because industrial development has had a strong negative impact on the environment through depletion of natural resources or pollution, legal measures have been taken in order to limit its negative effects. As in the past, technology allows humans to transform the environment to take advantage of its resources, but the possibility of irreversible negative impacts on the environment because of technology have become higher. Most industrialized countries now have sets of environmental regulations that try to limit the negative impacts of technology. These regulations can take different forms such as increased taxation for companies that use dangerous technologies, strict norms of how to use technologies or even the banning of some technologies or products. But there can also be incentives to promote or develop new technologies, as in the case of renewable energies such as wind or solar power.

The adoption of a regulation is often subject to resistance from business circles using the targeted technologies. Constraining environmental laws are often considered a threat to the competitiveness of companies using the technologies in question in their production process. However, strict environmental rules can also have positive effects on the competitiveness of firms in that they force them to be innovative in order to be able to match legal norms as with the high level of governmental taxation on gas in Western European countries, which has led car manufacturers to develop more fuel-efficient vehicles.

One of the main issues at stake in the formulation of environmental policy is the ability to diagnose the actual problems. Therefore, technologies for the measurement of various factors intervening in environmental problems are very important for environmental policies. For instance, satellites can track changes in wide land cover patterns, such as the progression of deserts. Some environmental matters, such as global warming, require use of a set of technologies before they can be grasped. The liability of those technologies and the results they produce are often contested and give way to controversies led by experts in order to determine if one technology or the other is appropriate to make the diagnosis.

Technology has shaped the environment even through repairing damage caused by humanity. In the second half of the 20th century, important efforts have been made in order to restore altered ecosystems. Even when restoration projects are able to recreate functioning ecosystems and enhance the environmental value, it is not a return to the original nature free from human intervention. Very often these ecosystems are recreated through what are called *green technologies*. They require much human monitoring or intervention in order to retain a high environmental value. In fact, most of the areas called natural today have been shaped by human activity. Most of resources, whether exploited in a sustainable way, depleted, or put under strict protection are included in technological networks that determine how humanity can or should deal with them. This has led some authors to talk about "technonatures" to qualify this state where nature is no longer untouched by human hands, but where all natural resources are somehow marked by technological activity, whether for exploitation, distribution, protection, or just observation.

SEE ALSO: Agriculture (including Agricultural Revolution); Biotechnology; Genetically Modified Organisms (GMOs); Genetics and Genetic Engineering; Green Chemistry; Green Production and Industry; Green Revolution; Industrial Revolution; Industrialization.

BIBLIOGRAPHY. Jesse H. Ausubel and Hedy E. Sladovich, eds., *Technology and Environment* (National Academy Press, 1989); Ruth Schwartz Cowan, *A Social History of American Technology* (Oxford University Press, 1997); Andrew Feenberg, *Transforming Technology:*

A Critical Theory Revisited (Oxford University Press, 2002); Bruno Latour, "Morality and Technology: The End of the Means," *Theory Culture and Society* (v.19/5-6, 2002); Bruno Latour, *Politics of Nature* (Harvard University Press, 2004); Isabelle Stengers, *Power and Invention* (University of Minnesota Press, 1997); Jeffrey K. Stine and Joel A. Tarr, "At the Intersection of Histories: Technology and the Environment," *Technology and Culture* (v.39/4, 1998); Erik Swyngedouw, *Social Power and the Urbanization of Water* (Oxford University Press, 2004).

Olivier Ejderyan
University of Zurich, Switzerland

Technology-Based Standards

TECHNOLOGY-BASED STANDARDS ARE the current legal basis for regulating pollution in American federal law. These new standards are more stringent and have replaced older demanding standards. They can be found in newer legislation affecting the removal of pollution from the environment. Until the 1970s, federal environmental policy focused on conservation as its approach to managing the environment. Pollution control was left to the states. In 1970 a dramatic shift occurred with the creation of the Environmental Protection Agency (EPA). Environmental policy shifted from management to regulatory control. Pollution became a national priority. By 2000, Congress had adopted at least 12 major environment acts and had amended them numerous times. This legislation created new standards for each kind of pollution or industry.

In the 1970s, the focus was on obvious forms of pollution such as raw sewage, industrial smoke stack pollution, water effluents, or automobile air pollution. The standards for handling pollution were those that were the best-known technologies of the day. At first, the emphasis was on health-based standards, but with a mandate to the EPA to engage in "technology forcing" standards-setting activities such as setting timetables for improvements. By 1972, the Federal Water Pollution and Control Act Amendments mandated federal permits for all future effluent discharges to use technology-based standards. Two standards in particular were mandated: "best practical technology" and the "best available technology." As knowledge increased about chemical pollution in potable water, and the dangers of carcinogens, public concerns increased. Environmental politics demanded the removal of numerous chemicals. During the 1980s and afterward, concerns about the cumulative effects of chemicals as carcinogens remained, but concern grew over the long-term effects on reproductive health and neurological toxicity.

Today, the Clean Water Act as amended employs technology-based standards. The act actually recognizes two other types of standards besides technology-based standards. These are the ambient or water-quality-based standards, and in a limited number of cases health-based effluent standards that deal with a small number of toxic compounds. The Clean Water Act uses four kinds of technology-based water pollution control standards. Industry-wide standards are set by the "best practical control technology currently available" (BPT) standard. This standard sets uniform effluent standards for operators in a particular industry, no matter where they are located. This standard sets the national ground floor for all existing sources of industrial water pollution for each industry.

The Clean Water Act also allows for the "best available technology economically achievable" (BAT). This standard applied to a limited number or toxic pollutants or to nonconventional pollutants such as thermal pollution. A third standard is the "best conventional pollutant control technology" (BCT). This standard is a modified form of the "best available technology" standard. The modified standard seeks benefits from pollution control where the costs are applied to "conventional pollutants" such as biochemical oxygen demand (BOD). The fourth standard is the "best available demonstrated control technology" (BACT) and is now the basis for judging the allowable pollution levels permitted in effluents. Technology-based standards have become more stringent.

In 1990, Congress amended the Clean Air Act. It gave the EPA the authority to go beyond health-based standards so that air quality regulations would be set by technology-based standards. The amendments allowed the EPA to relax standards in

some cases, with the responsibility to regulate 189 specific substances. The law permits the EPA to add new chemical pollutants to the list and anyone may propose a possibly polluting chemical. The EPA will then evaluate the proposal. Politically, the change to technology-based standards work more in the favor of industry than did the older health-at-any-cost-based standards. The newer standards bring a different kind of science to the policy issues. They reduce the political and economic penalties of regulation because the costs are considered.

SEE ALSO: Best Available Technology (BAT); Clean Water Act; Environmental Protection Agency (EPA).

BIBLIOGRAPHY. David Harrison, *Who Pays for Clean Air: The Cost and Benefit Distribution of Federal Automobile Emission Controls* (Harvard University Press, 1975); Noga Morag-Levine, *Chasing the Wind: Regulating Air Pollution in the Common Law State* (Princeton University Press, 2003); David P. Novello, *The Clean Air Act Handbook* (American Bar Association, 2005); Mark A. Ryan. *The Clean Water Act Handbook* (American Bar Association, 2004).

Andrew J. Waskey
Dalton State College

Tennessee Valley Authority (TVA)

DURING THE WORST days of the Great Depression in the 1930s, President Franklin D. Roosevelt initiated a series of new and innovative ideas to combat America's economic crisis. The programs, collectively known as the New Deal, offered relief and recovery to several groups and institutions that had been particularly hard-hit by the Depression, including farmers, youth, banks, industry, and workers. One of the most innovative accomplishments of the New Deal was the Tennessee Valley Authority (TVA), which became law on May 18, 1933, during the first 100 days of Roosevelt's initial term in office.

The TVA offered recovery and relief to an agriculturally devastated region in southern Appalachia, which incorporated seven southern states within the water tributaries of the Tennessee River system—an area encompassing approximately 40,000 square miles (103,600 square kilometers). This region, once identified by Roosevelt as the "nation's number one economic problem," included the states of Tennessee, Alabama, Georgia, Mississippi, North Carolina, Kentucky, and Virginia.

The Tennessee Valley had been prosperous in the past, but, according to Arthur E. Morgan, a Roosevelt intimate associated with the TVA, as quoted by Arthur M. Schlesinger Jr., by the 1930s, "only poverty remained—poverty, with thousands on thousands of families who never saw $100 cash income a year." Thus, the TVA sought to revitalize the South's economy but also to improve the lives of millions of families who lived in rural Appalachia. It was, Morgan suggested, "not primarily a dam-building program, a fertilizer job or power-transmission job," but, rather, a large idea for "a designed and planned social and economic order."

One of the primary goals of the TVA was to offer public power to a region devoid of electricity. The greatest champion of public power, Nebraska senator George W. Norris, had vigorously advocated the idea of public power during the 1920s. His defense of the government-owned dam at Muscle Shoals, Alabama, against private interests who hoped to purchase the complex led him to seek a broader federal program of dam construction along the Tennessee River. However, conservative Republican leaders in the House and Senate, as well as two Republican presidents, defeated six of his dam-building proposals before Roosevelt took office in 1933. Norris's efforts kept the dream of public power alive in the South and, in large measure, the TVA was actually his legacy.

Between 1933 and 1945, the TVA, in one of the largest construction efforts ever undertaken by the federal government, constructed 16 dams in the Tennessee River basin. According to Leuchtenburg, it was, in short, a "public corporation with the owners of government but the flexibility of a private corporation" that worked in conjunction with state and local agencies. TVA's benefits were immediately apparent. Millions of Americans found steady employment through the TVA. The dams also generated electricity to the rural South (only two out of every

David Lilienthal

David Eli Lilienthal (1899–1981) was a co-director of the Tennessee Valley Authority from 1933 until 1941, and then its chairman until 1946, and is one of the people most identified with the project.

Lilienthal was born at Morton, Illinois, and was educated at DePauw University at Greencastle, Illinois, and then at Harvard Law School. He then practiced as a lawyer, working on labor law and taking on public utilities. In one celebrated case he took on the telephone authorities in Chicago and managed to gain a refund of $20 million for the subscribers. This brought him to the attention of Philip La Follette, the governor of Wisconsin, who appointed him a member of the Wisconsin Public Service Commission in 1931. It was there that he reorganized the utilities statutes making them so much more efficient that six other states copied his scheme.

Lilienthal's genius had come to the attention of President Franklin Roosevelt and when Congress approved the TVA flood control project in 1933, the president appointed Lilienthal as one of the three directors of the power program; he became sole chairman in 1941. In 1944, he wrote *TVA: Tennessee Valley Authority—Democracy on the March* in which he defended the TVA from criticisms raised by detractors, many of whom were associated with private electric companies.

In 1946, Lilienthal had to resign from the TVA in order to become the first chairman of the Atomic Energy Commission. He assumed power over the U.S. nuclear-development program, which had, up until that point, been supervised by the U.S. Army. He was also involved in expanding nuclear power plants and building up stockpiles of nuclear weapons. He resigned in 1950, and for three years was chairman and chief executive officer of the Development and Research Corporation and was involved in major resource development projects. He died in 1981 in New York.

100 farms had electricity before the TVA) and manufactured fertilizer for the region's farmers. The lakes that it created offered recreational opportunities and a profusion of government-funded parks were built by the Civilian Conservation Corps (CCC) and other agencies along their shores. Conservationist goals, like soil conservation, the removal of depleted agricultural lands, flood control, and forestation, were also major components of the TVA program.

The TVA also pioneered in social experimentation. The altering of water levels at the system's dams, for example, helped to eliminate malaria as a serious health risk in the Tennessee Valley. Previously, about one-third of the region's inhabitants had been affected by malaria. Furthermore, the organization made major contributions to recreational lake management and architectural design.

TVA had its critics. The government's public power program, which infringed on rights of private utility companies, was highly controversial and led some opponents of the agency to label Roosevelt a Communist. Others leveled criticism at the TVA's bureaucracy, which they felt was riddled with fiscal mismanagement and poor planning.

Racial conflict also plagued the TVA. When the TVA compensated farmers for flooding their croplands, they paid landowners, but not sharecroppers and tenants who were forced out of work and off of the land. Hiring inequities and complaints of racial discrimination frequently surfaced within the TVA. Investigations during the mid-1930s by the National Association for the Advancement of Colored People (NAACP) substantiated evidence of racial discrimination and led to a congressional investigation. During the hearings, the TVA maintained that it had to uphold regional racial customs in order to maintain a good relationship with local authorities. Otherwise, the chances for a successful program that would benefit both blacks and whites would be jeopardized. The congressional committee only mildly chastised the TVA for its racial practices.

The conservation goals of the TVA were also compromised. The construction of dams disrupted the flow of freshwater in the Tennessee River valley and thus altered aquatic and riverine ecosystems. Several species of freshwater mollusks, for example, dependent on flowing water, became either endangered or extinct. In addition, thousands of acres of

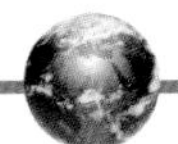

wetland areas were permanently lost under the waters of TVA lakes. The TVA also strip-mined coal fields in order to provide fuel to operate the power plants.

For the duration of the 20th century, the TVA continued to maintain its lakes and produce power. During the 1960s the TVA began producing nuclear power in several of its facilities. Although most of its nuclear plants have been phased out, the TVA has developed an energy plan for the Tennessee valley that will be phased in through 2020. In 1998, TVA announced a new plan for its facilities that would reduce pollutants that deplete ozone and cause smog. According to its own study, the agency will have spent $5.6 billion on clean air modifications to its coal-fired plants by 2010. As of 2005, the TVA operates 29 hydropower plants.

SEE ALSO: Coal; Dams; Electricity; Endangered Species; Floods and Flood Control; Habitat Protection; Hydropower; Justice; Malaria; Nuclear Power; Ozone and Ozone Depletion; Wetlands.

BIBLIOGRAPHY. Roger Biles, *The South and the New Deal* (University of Kentucky Press, 1994); William E. Leuchtenburg, *Franklin D. Roosevelt and the New Deal* (Harper Torchbooks, 1963); Arthur M. Schlesinger, Jr., *The Coming of the New Deal* (Houghton Mifflin Company, 1958); Tennessee Valley Authority, www.tva.gov (cited May 2006).

Clay Ouzts
Gainesville State College

Teratogenic Substances

TERATOLOGY IS THE study of the effects of teratogenic substances on the development of embryos, fetuses, or a pregnancy. Teratogenic substances can cause congenital malformations, or they can halt a pregnancy altogether. Clinical and experimental teratologists study how teratogenic agents that actively create birth defects can disrupt the normal development of a fetus or embryo. These include chemicals, drugs, maternal infections, and radiation. Exposure during pregnancy to teratogenic substances causes one or more structural abnormalities in the developing embryo or fetus. The discovery that a particular substance is a teratogen is usually the result of a significant increase in birth defects in an area or in a class of people.

The discovery of Minamata disease in Minamata Bay in postwar Japan is typical of the discovery of a teratogen. Minamata disease is a form of encephalopathy that closely resembles cerebral palsy. Eating fish contaminated with methyl mercury during pregnancy causes Minamata disease. The sudden appearance of over 3,000 cases of what appeared to be cerebral palsy puzzled health care officials. After an investigation it was discovered that Chisso Corporation, a chemical company, was dumping mercury into Minamata Bay where it had entered the food chain, including locally caught fish eaten by pregnant women. In the 1960s there was a surprising increase in the number of cases of phocomelia in Germany and Australia. An investigation identified thalidomide as a human teratogen. "Thalidomide babies" with severe birth defects were born to mothers who had been prescribed the drug thalidomide as a treatment for morning sickness. The exposure to the drug at the critical first trimester stage of development of the embryo produced severe birth defects.

The number of chemicals that are proven or suspected of being teratogens is growing as people are exposed to an increasing number of synthetic chemicals. In addition, attention to the problem has led to an increase in the number of defects identified. The identification of fetal warfarin syndrome, fetal hydantoin syndrome, fetal trimethadione syndrome, fetal alcohol syndrome, and low birth weight due to smoking during pregnancy, are fetal developmental problems that are given full recognition. Maternal infections during pregnancy from rubella or sexually transmitted diseases such as herpes simplex or syphilis can also cause teratogenicity. In addition, diabetes during pregnancy or other maternal health factors can interfere with fetal development.

Over-the-counter and prescription drugs are major sources of teratogenicity. Known teratogenic drugs are ACE inhibitors such as benazepril, captopril, and enalapril; antibiotics such as tetracycline and streptomycin; anti-depressants such as lithium; anticoagulants (blood-thinner) such as warfarin; or

anticonvulsants. In addition, illegal drugs such as cocaine and marijuana are teratogenic substances. There are no absolute teratogens. The damaging effects of teratogenic substances are due to size of the dose, the length of exposure, and the stage of fetal development.

SEE ALSO: Drugs; Infant Mortality Rate; Mercury; Minamata; Radioactivity; Sexually Transmitted Diseases.

BIBLIOGRAPHY. J.M. Friedman and Janine E. Polifka, *Teratogenic Effects of Drugs: A Resource for Clinicians (Teris)* (Johns Hopkins University Press, 2000); James L. Schardein, *Chemically Induced Birth Defects* (Marcel Dekker, 2000); Thomas H. Shepard and Ronald J. Lemire, *Catalog of Teratogenic Agents* (Johns Hopkins University Press, 2004).

Andrew J. Waskey
Dalton State College

Terraces and Raised Fields

TERRACES AND RAISED fields are earthworks created to improve cultivation conditions by manipulating slope, hydrology, soil fertility, erosion, and microclimate. While both require intense labor, they allow agriculture to flourish in potentially inhospitable topography and environments. Terraces carve out sections of hillsides to improve planting surfaces. Raised fields create dry platforms above wetlands.

Farmers around the world have used terraces since ancient times, from the Hanging Gardens of Babylon to Cajete agroecosystems of Mexico. Terraces are used for defense and aesthetics as well as agriculture. They are critical to growing crops in arid and semiarid climates. Dryland terraces in Yemen enhanced soil water retention, taking maximum advantage of local water sources. The Hopi of the American Southwest irrigated terraces to grow maize. Terraces are also used in moist climates. Wet rice production throughout southeast Asia uses terraces to regulate water flow and grow rice on mountainsides.

Check dams and cross-channel terraces are constructed in hillside hollows where water flows during rainy seasons. They control water runoff and prevent soil erosion. Sloping field or broad-base terraces are used on gently rolling hills or valley floor alluvial fans. Each terrace level is constructed around existing contours, and soil surface is not flattened. These terraces control surface water runoff, thus conserving soil moisture and nutrients and reducing soil erosion. In the United States, farmers use mechanized agriculture on broad-base terraces.

Bench terraces provide a horizontal planting surface and are common on steep hillsides. Examples include those found in Machu Picchu, Peru, and the Ifugao terraces in the Philippines. Bench terraces are constructed by cutting out soil and putting in retaining walls made of stone or other materials. Behind the wall, soil and rock are backfilled in to create a new planting surface. To aid drainage, bottom layers of the new soil profile consist of larger rocks, then smaller cobble, and finally soil. By controlling surface water runoff, terraces reduce erosion, nutrient loss, and flooding risk. Bench terraces modify microclimates and enhance sunlight capture. They often are integrated with irrigation channels. Cultivating on slopes avoids nighttime frosts that settle in valley bottoms. While cultivatable area may be smaller than an unterraced plot, agricultural yield increases significantly.

Raised fields are large platforms of soil constructed in areas with seasonal flooding or permanent wetlands. They are found in Mexican wetlands, coastal India, Indonesian tidal wetlands, New Guinea swamps, Venezuelan and Bolivian savannas, and surrounding Lake Titicaca. Raised fields take many forms, ranging from long ridges to mounds, with flat or ridged surfaces, scattered or in rows. Drainage canals often separate platforms.

Raised fields reduce crop failure risk by altering microclimate, soil nutrients, and hydrology. They allow cultivation in areas otherwise inundated with fresh or saline water, control flooding in rainy seasons, and conserve water during dry seasons. Water in drainage canals may be used for aquaculture. Cleaning muck from canals provides nutrients for raised beds. Around Lake Titicaca, where diurnal temperatures vary up to 30 degrees C, raised fields

By controlling surface water runoff, terraces reduce erosion, nutrient loss, and flooding risk.

reduce frost risk. Solar radiation heats canal water during the day. At night, this stored heat dissipates, heating raised bed soils and air.

Maintaining terraces and raised fields requires intensive labor, often in conjunction with social and religious rituals. Despite increases in agricultural productivity, thousands of acres of historic terraces and raised fields are abandoned today. The reasons why are complex, including indigenous population crashes after the Spanish conquest of Latin America, the high capital and labor investment required, low market prices for agricultural products, uncertain land tenure and water rights, shifts in land use and technology, and even climate change. Modern-day projects have reconstructed terraces and raised fields, demonstrating successful agricultural yields but uncertain longevity.

SEE ALSO: Agroecosystems; Farming Systems; Floods and Flood Control; Irrigation; Maize; Microclimates; Rice; Runoff; Soil Erosion; Titicaca, Lake.

BIBLIOGRAPHY. Lloyd E. Eastman, *Family, Fields, and Ancestors: Constancy and Change in China's Social and Economic History, 1550–1949* (Oxford University Press, 1988); S.K. Mickelson et al., "Effects of Soil Incorporation and Setbacks on Herbicide Runoff from a Tile-Outlet Terraced Field," *Journal of Soil and Water Conservation* (March 22, 1998); Arthur Stephen Morris, *Raised Field Technology: The Raised Fields Projects Around Lake Titicaca* (Ashgate Publishing, 2004).

KEELY MAXWELL
FRANKLIN AND MARSHALL COLLEGE

Thailand

NEVER FORMALLY COLONIZED, Thailand emerged as a modern nation-state in the mid- to late-19th century during the period of European colonial expansion in Southeast Asia. This country's elongated geography is a result of its own imperial expansion out from its central core in the valleys of the Chao Phraya River to the north, south, and northeast. Linguistically, the country is diverse, although Central Thai is the dominant, formally recognized language taught in schools and used in all public affairs.

This linguistic diversity parallels the ethnic and religious diversity in the country, which not only represents a number of Thai speakers (Northern Thai, Northeastern Thai, and Southern Thai), but also a number of minority ethnic groups, euphemistically called "hill tribes" or *chaaw khaaw* in Thai. While Buddhism is the dominant religion, represented in the image of the constitutional monarchy, animistic practices are common, as is Islam, which is dominant in the country's south.

Thailand is located in the tropics and is subject to a yearly monsoon, which brings significant amounts of precipitation through the rainy season from November to May. As a result, the country's main crop is rice. Thailand exports rice to other parts of the world and this crop remains one of its most significant economic sectors behind tourism. The extensive river system, which runs through much of the north and central regions of the country, fuels irrigated wet rice agricultural practices and a dynamic fisheries industry. Thailand is often portrayed as a rural country with extensive agricultural lands based around relatively small villages. Despite this popular image, the country's economy is diverse, including significant manufacturing and service sectors. This economic diversity parallels the growth of both urban and suburban areas, the latter of which includes the conversion of agricultural land into numerous housing subdivisions on the edge of the country's major cities, such as Khon Khaen in the northeast and Chiang Mai in the north.

This expanding economic diversity has had a number of environmental and social consequences. The expansion of the manufacturing and service sectors in cities, such as the capital of Bangkok, has drawn in massive numbers of people (it is estimated that one in six citizens of Thailand live in the capital for at least part of the year). This rapid expansion of the capital has fueled the growth of slums and other areas that are rife with environmental problems linked to the spread of illnesses, such as tuberculosis and cholera. Secondary cities are also witnessing similar patterns of urban growth, with a growing middle class occupying formal housing units in city centers and the poor living in the fringes often in illegal or temporary housing structures. The shifting nature of rural and urban life has also led to a number of social problems, including an expanding commercial sex work industry serving local and tourism communities as well as an increase in undocumented migrant populations from countries such as Burma, Cambodia, and Laos working in the rapidly expanding construction industry.

The Thai government, despite its commitment to its rapidly expanding economy and position as one of the new economic "Tigers" of Asia, has not made similar commitments to environmental regulation. There are numerous reports of environmentally related health problems linked to the pollution emanating from expanding rural and urban factories, as well as problems linked to the pesticides used in the agricultural sector. Thailand has also been a historical site for the export of wood products, and deforestation (and flooding) has been a growing concern that has been mitigated, to a certain degree, with the establishment of a national parks system. Despite this shifting environmentalism, which has been fueled by a growing nongovernmental sector and Buddhist environmental movement, Thais and their government often struggle over the ownership and use of common resource properties, including forest, river, and fishery resources, which have historically been critical to local economies.

SEE ALSO: Industrialization; Monsoon; Rice; Timber Industry; Tourism; Tropical Forests; Urbanization.

BIBLIOGRAPHY. Jim Glassman and Chris Sneddon, "Chiang Mai and Khon Khaen as Growth Poles: Regional Industrial Development in Thailand and its Implications for Urban Sustainability," *Annals of the Academy of Political and Social Science* (v.590/1, 2003); Clemens M. Grünbühel et al., "Socioeconomic Metabolism and Colonization of Natural Processes in SangSaeng Village: Material and Energy Flows, Land Use, and Cultural Change in Northeast Thailand," *Human Ecology* (v.31/1, 2003); Philip Hirsch, ed., *Seeing Forest for Trees: Environment and Environmentalism in Thailand* (Silkworm Books, 1997); Mary Beth Mills, *Thai Women in the Global Labor Force: Consuming Desires, Contested Selves* (Rutgers University Press, 1999).

Vincent J. Del Casino, Jr.
California State University, Long Beach

Thar Desert

LOCATED IN SOUTH Asia, the Thar Desert straddles the international border between western India and southeastern Pakistan covering an area of approximately 92,162 square miles (238,700 square kilometers). On the Indian side, most of this hot desert lies in the state of Rajasthan, but some parts extend into the states of Gujarat, Haryana, and

Punjab as well. In Pakistan, it is part of the Sindh and Punjab provinces. It is bounded by the Sutlej River in the northwest, the Aravalli mountains in the east, the Indus River in the west, and the Rann of Kutch (a salt marsh) in the south. Average annual rainfall in the Thar ranges between 6 to 31 cubic inches (100–500 millimeters) from the west eastward. It is an inhospitable desert with almost 10 percent consisting of shifting sand dunes and the other 90 percent of rocky outcrops, compacted salt-lake bottoms, and interdunal and fixed dune areas. Temperatures fluctuate from highs near 50 degrees C in May and June to as low as 5 degrees C in January.

With about 700 species of flora, of which 107 are grasses, the Thar is rich in biodiversity. However, there are few indigenous species of trees and strong winds facilitate desertification, which threatens the region's predominantly agricultural economy. According to the Central Arid Zone Research Institute in Jodhpur, India, wind erosion in the Thar can be checked through sand dune stabilization and shelterbelt plantation. Dune stabilization programs have involved plantation of naturally growing shrubs with extensive root networks such as Phog (*Calligonum polygonoides*), but have more commonly included a range of exotic species of eucalyptus, acacia, and cassia species, a few of which have proven to be pernicious invasives, causing unanticipated effects from well-meaning policy. Erecting shelterbelts around crop fields protects young seedlings from sand blasts and hot desiccating winds and reduces the loss of moisture from the fields.

The Thar's fauna includes threatened species like the blackbuck (*Antilope cervicapra*) and the great Indian bustard (*Ardeotis nigriceps*), as well as other species such as chinkara (*Gazelle bennettii*), caracal (*Felix caracal*), and the desert fox (*Vulpes bengalensis*). Recently, land use changes and agricultural intensification have brought problems to the region. As pasture has declined under the plow, livestock herds of the region have lost much of their traditional grazing land, leading to increased migration and some overgrazing. In India, almost 60 percent of the Thar is farmland and 30 percent is open pastureland. Intense grazing of livestock has altered the ecosystem by affecting soil fertility and destroying native vegetation.

The region relies on the Indira Gandhi Canal Project for irrigation and drinking water supply, but also increasingly upon direct tapping of groundwater through deep tubewells. As a result, groundwater levels are falling precipitously in some areas. Frequent droughts have also prompted the locals to devise ingenious water harvesting techniques such as *kunds* (covered underground tanks)—a traditional response to the scarcity of potable water exacerbated by the high salinity of groundwater, especially in western Rajasthan.

With a population density of about 263 persons per square kilometer for the entire Thar region, it is one of the most densely populated deserts. Multiple ethnic and religious groups make it culturally rich, with a variety of music, poetry, and architectural styles. It is home to Hindus, Muslims, Jains, and Sikhs on the Indian side while in Pakistan, Sindhis and Kohlis (both Hindus and Muslims) are the main ethnic groups.

Pokhran, in Rajasthan, has also served as the test site for India's nuclear programs. Underground testing of nuclear weapons was carried out in Pokhran in 1974 and 1998.

SEE ALSO: Biodiversity; Desert; Desertification; India; Overgrazing; Pakistan; Soil Erosion.

BIBLIOGRAPHY. Anil Agarwal and Sunita Narain, eds., *Dying Wisdom: Rise, Fall and Potential of India's Traditional Water Harvesting Systems* (Center for Science and Environment, 2003); Robyn Davidson, *Desert Places* (Viking, 1996); H.S. Mathur, *Arid Lands, People, and Resources* (RBSA Publishers, 1994); R.C. Sharma, *Thar, the Great Indian Desert* (Lustre Press, 1998).

Priyam Das
University of California, Los Angeles

Thatcher, Margaret (1925–)

MARGARET THATCHER WAS born in Grantham Lincolnshire, England, to a working-class family. She was elected as Member of Parliament for Finchley, London, in 1959 and held the posts of British Secretary of State for Education and Science, in the Ted

Heath government (1970–74); Leader of the United Kingdom (UK) Conservative Party (1975–90); Leader of the Government's Opposition (1975–79) and British Prime Minister (1979–90). She was the winner of three successive general elections in 1979, 1983, 1987 and is the longest-serving Prime Minister since Lord Liverpool. She is significant to British history because her policies represented a radical shift from the consensus politics that had characterized the previous postwar British political era.

Margaret Thatcher was a figurehead of a "new right" political philosophy and doctrine that became known as Thatcherism. Built on a belief in the free market and the private sector, this involved a range of actions including: Reductions in public/state spending, lower direct taxation, a tight monetarist policy, a program of privatization of government-owned industries, local and national deregulation of markets, and the curtailing of union powers and other perceived restrictions on the economy.

Thatcher's policies and style polarized public opinion of her, often along class lines. Her policies perpetuated the income gap between rich and poor, which were reflected geographically in a north/south divide. She resigned as Prime Minister in 1990 following an unconvincing victory in the first round ballot of a party leadership contest. In 1992, she was awarded the title Baroness Thatcher and became a member of The House of Lords. She was heavily engaged in public speaking, an activity that she was forced to curtail in 2002 due to ill health. Her legacy is that the new right has effectively become the current political center ground.

In terms of environmental policy, Thatcherism follows the broad outlines of free market environmentalism, which favors volunteerism, green consumption, and the trading of externalities over any form of regulation or imposition of systems or standards. As such, Thatcher was a reliable opponent of most forms of environmental regulation, which she viewed as extensions of socialist and left-leaning "big government" legacies.

Thatcher was throughout her career an opponent of multilateral and global governance systems, moreover, including in the realm of environmental issues. In her book *Statecraft*, she went so far as to say that global warming "provides a marvelous excuse for worldwide, supra-national socialism." While she was an early leader in raising and discussing the problem of global warming and the ultimate necessity of dealing with it, to the degree that such environmental issues might be addressed, Thatcher remained skeptical of international regulation and treaties or anything that might hinder economic development. The legacy of her approach to the environment in the UK remains somewhat unclear in the era of the Kyoto Protocol, to which the UK is a signatory, though there has yet to be a legislative framework to promote greenhouse gas reductions.

SEE ALSO: Kyoto Protocol; Markets; Trade, Free.

BIBLIOGRAPHY. M. Thatcher, *Statecraft* (Harper Collins, 2000); M. Thatcher, *The Downing Street Years* (Harper Collins, 2003).

Gavin J. Andrews
McMaster University
Denis Linehan
University College Cork

Thermodynamics

THE FIRST CRUDE thermometer was invented in the 1600s by Galileo. Accurate thermometers arrived some 200 years later. It wasn't until this time, in the mid-1800s, that the concept of energy was widely accepted. It is now clear that temperature is a measurable indicator of energy. In the last 150 years since the discovery of energy, fundamental laws governing the transport, conversion, and storage of energy have been developed. These laws are the basis for the science we call thermodynamics.

THE FIRST LAW OF THERMODYNAMICS

Simply stated, the first law of thermodynamics says that energy cannot be created nor destroyed. However, energy can be transformed and transferred. The main ways in which energy is manipulated are heat and work. It must be understood that heat and work are not properties of matter, but rather they are the routes by which energy is moved or converted.

When energy moves from areas of higher temperatures to areas of lower temperatures this is called heat transfer. If you touch a hot stove, large amounts of energy are transferred from the hot stove to your colder hand. Alternatively, if you make a snowball with your bare hands, energy moves from your warmer hands to the colder snow. The flow of energy in each of these situations is in the form of heat.

Work, as defined in physics, is force applied over a distance. In thermodynamics, work is a means of transferring and storing energy. If you pull a wagon up a hill, you are doing work, using your energy to move something over a distance. You have also stored some of your energy in the wagon. To see that energy, push the wagon off the top of the hill and it will race to the bottom, using the energy you gave it through work.

The first law says that no matter how energy is transferred, transformed, or stored through heat or work, the same total amount of energy is always present. For this reason, energy is said to be conserved. In other words, whenever energy decreases in one place, it must increase by an equal amount somewhere else.

THE SECOND LAW OF THERMODYNAMICS

The second law has been defined in many ways over the years. Heat cannot flow from areas of lower temperature to those of higher temperature. Creating order in one system must create equal or greater disorder in the surroundings. Perpetual motion machines are an impossibility. No process can convert heat completely to work. These are all valid statements of the second law.

The second law is concerned with the relationship between heat and work. Work can be completely converted into heat with no losses. For example, if you rub your hands together on a cold day, all the work you do is converted to heat. However, all of the energy in heat cannot be converted to work. Some of the heat will always be dissipated to the surroundings. In industry, hot steam is often used to drive work-producing turbines. This is a way of converting heat to work. However, some of the heat energy will be dissipated to the surroundings and not converted to work. The disparity between heat and work is filled by the concept of entropy.

Entropy is classically defined as disorder or randomness. It is also said that entropy tends to increase with time in natural environments. For example, pretend you have a large box with 100 rabbits in it. Fifty of the rabbits have black fur, and the other 50 have white fur. You put all of the rabbits with white fur on the far right hand side of the box, and all of the rabbits with black fur on the far left hand side.

At this point, the box of rabbits is very ordered and has very little randomness, hence low entropy. If you then leave and come back one minute later, it is likely that most of the white rabbits will still be on the right and most of the black on the left, with only a few of them mixing in the middle. So after one minute, there is a little more disorder in the box, therefore the entropy has increased slightly. If you then leave the box and come back several hours later, it is likely that the rabbits will be thoroughly mixed with black and white rabbits in all parts of the box. The box of rabbits now has a large degree of disorder and randomness, meaning very high entropy. To return to the original situation with all of the white rabbits on one side and all of the black rabbits on the other, you will have to do a significant amount of work to move and separate the rabbits.

Thermodynamic entropy works in a similar way. When heat is used to create work, some of the heat is "lost." This lost heat contributes to the entropy of the system. In other words, the extra heat makes the system more disordered and random. If you want to restore the original order to the system, you will have to add more work, similar to the rabbits in the box.

THE THIRD LAW OF THERMODYNAMICS

The third law also deals with entropy. It states that a system with a temperature of absolute zero (–273 degrees C or –459 degrees F or 0 degrees K) will have no entropy. Going back to our rabbit example, if the temperature in the box of rabbits is lowered, the rabbits will move around more slowly and the entropy of the box will increase much more slowly. However, if the temperature is made so low that the rabbits freeze in place, the entropy of the system will be at a minimum because the rabbits cannot mix, cannot increase their disorder.

It is the same with energy systems. As the temperature approaches absolute zero, the system becomes more and more sluggish, preventing any disorder from developing. The third law says that entropy will approach a limit of zero as the temperature approaches zero. These laws of thermodynamics are enough for a detailed energy analysis in most situations. Additional laws that deal with self-organizing and nonequilibrium systems may also exist.

USES OF THERMODYNAMICS

The practical uses of thermodynamics are limitless. Because energy is all around us, the laws of thermodynamics can be applied to almost anything. The practical application on which thermodynamics was founded is the engine. Studying and improving the engine were the real motivations for studying thermodynamics and developing the laws we have today. Throughout the last century, the engine has been the primary device for converting heat to work. Through thermodynamics, we have made engines more efficient and found ways that we can use engines to make our lives easier. Thermodynamics is also essential for understanding and designing air conditioning and heating systems. Understanding the flow of energy is pivotal to technology like refrigeration.

The transfer of energy in the body also follows the laws of thermodynamics. For this reason, thermodynamics is important for the medical field as well. Medical researchers use thermodynamics to develop medical equipment used for diagnosis and treatment of patients. Thermodynamics is also at the root of drug delivery systems, which govern how the medicine you take gets to the part of the body where it is needed.

SEE ALSO: Energy; Internal Combustion Engine.

BIBLIOGRAPHY. Roger Kinsky, *Heat Engineering: An Introduction to Thermodynamics* (McGraw-Hill Education, 1989); Anastasios Tsonis, *An Introduction to Atmospheric Thermodynamics* (Cambridge University Press, 2007).

Geoffrey Grubb and Bhavik R. Bakshi
Ohio State University

Think Tanks

THINK TANKS ARE nonprofit, research-oriented institutes whose primary objective is to influence public opinion and public policy. Think tanks have the objective of providing research and innovative policy solutions to legislators, the judiciary, and the public. Some scholars suggest that think tanks exist merely for the type of large-scale lobbying that aims to create a climate of opinion favorable to particular private interests. The term *think tank* was first used in the United States during World War II to refer to a secure room where defense scientists and army planners could meet to discuss war strategy, but the meaning has expanded to include any advice-giving institution, including public relations and marketing organizations.

Scholars who have studied the growth and development of American think tanks agree that the highly decentralized nature of the American political system, the lack of strict party discipline, and the large infusion of funds from philanthropic foundations have contributed to the expansion of think tanks in the United States. The first generation of think tanks includes the Carnegie Endowment for International Peace (1910), the Hoover Institution on War, Revolution and Peace (1919), and the Council on Foreign Relations (1921).

The second generation includes the Institute for Government Research (1916), renamed after a merger with other institutes into the Brookings Institution (1927), and the American Enterprise Institute for Public Policy Research (1943), a highly respected conservative think tank. This group was the first to focus on a foreign policy issues. After World War II, the RAND Corporation was created (1948) to promote and protect U.S. security interests during the nuclear age.

The third generation of think tanks were the "advocacy think tanks" such as the Center for Strategic and International Studies (1962), the Heritage Foundation (1973), and the CATO Institute (1977) that appeared in the 1970s. These think tanks combine policy research with marketing techniques.

Think tanks of the fourth and most recent generation, the Carter Center in Atlanta and the Washington, D.C.–based Nixon Center for Peace and Freedom, were created by former presidents with

the objective of leaving a lasting legacy on foreign and domestic policy. Scholars note that the influence of think tanks has shifted to the right since the 1970s. Of the 10 think tanks most often cited in the media, six are conservative or right-leaning, three are centrist, and one is left-leaning. More than half of all media citations of think tanks referred to conservative or right-leaning institutions, such as the Heritage Foundation, which states that its objective is to "formulate and promote conservative public policies based on the principles of free enterprise, limited government, individual freedom, traditional American values, and a strong national defense." Only 13 percent of media citations referred to progressive or left-leaning institutions.

Strategies employed by think-tanks to transmit their views to policymakers and the public include organizing public conferences, seminars, and public lectures; testifying before legislative committees; writing opinion pieces in the print media and giving expert comment on electronic media; and creating content on the internet.

Today, there are over 3,500 think tanks worldwide, half of which are in the United States, where they are distinguished from their counterparts in other countries by their ability to participate directly and indirectly in policymaking. Think tanks in Canada, Australia, Europe, Asia, the Middle East, and Africa have developed around the idea of promoting independent and objective research on relevant policy issues.

In Europe, think tanks are perceived as independent nonprofit associations, open and accountable providers of analysis and information to assist policymakers in research and evaluation. The European Policy Institutes Network (EPIN) is a network of dynamic think tanks and policy institutes that focus on current European Union (EU) and European political and policy debates. With 25 member think tanks in 21 countries, EPIN includes almost all the EU member states and accession and candidate countries. Think tanks in Brussels use regular conferences and seminars as platforms to network and discuss policy opinions with other EU actors, thus allowing participants from the private sector, media, academia, and civil society to meet EU institutional representatives in a neutral environment.

Transnational think tanks founded by philanthropic foundations, corporations, and international organizations such as the World Bank have become a global phenomenon. A new trend is collaboration between think tanks across continents, such as the World Economic Forum's Council of 100 Leaders on West–Islam relations.

The first think tank devoted exclusively to natural resource and environmental issues was Resources for the Future (RFF), a nonprofit and nonpartisan organization founded in 1952 to conduct independent research—primarily in economics and other social sciences—on environmental, energy, and natural resource issues. RFF has pioneered the application of economics as a tool to develop more effective policy for the use and conservation of natural resources by analyzing critical issues concerning pollution control, energy policy, land and water use, hazardous waste, climate change, biodiversity and the environmental challenges of developing countries.

Scholars note that conservative, corporate-funded think tanks contribute to confusion about the scientific basis of environmental problems such as global warming, species depletion, acid rain, and ozone depletion. Conservative think tanks oppose environmental regulations and promote free-market remedies for those problems. On the other hand, liberal think tanks have promoted the work of environmental economists, and many of the leading scholars in this area are associated with think tanks, including Robert Hahn, a resident scholar of the American Enterprise Institute; Terry Anderson, who has written for several think tanks in Australia and the United States; Robert Stavins and Bradley Whitehead, authors of a Progressive Policy Institute study; Alan Moran from the Tasman Institute, an Australian think tank; and Walter Block from the Fraser Institute, a Canadian think tank.

SEE ALSO: Expertise; Lobbyists; Policy, Environmental.

BIBLIOGRAPHY. Donald E. Abelson, *Do Think Tanks Matter? Assessing the Impact of Public Policy Institutes* (McGill-Queen's University Press, 2002); Sharon Beder, "Examining the Role of Think Tanks," *Engineers Australia* (November, 1999); Elizabeth T. Boris and C. Eugene Steuerle, eds., *Non-profits and Government: Collaboration and Conflict* (Urban Institute Press, 1998);

Michael Dolny, "New Survey of Think Tanks," *Extra* (July/August 1997), www.fair.org (cited May 2006); Diane Stone and Andrew Denham, eds., *Think Tank Traditions: Policy Research and the Politics of Ideas* (Manchester University Press, 2004).

Verica Rupar
Victoria University of Wellington

Thoreau, Henry David (1817–62)

BORN DAVID HENRY Thoreau, the naturalist, essayist, poet, and philosopher familiarly known as Henry David spent most of his life and dedicated much of his writing (published and private) to his native Concord, Massachusetts. Whatever the order of his given name (which he reordered shortly after his 1837 Harvard graduation), posterity simply refers to the celebrated sage of Walden Pond as "Thoreau," a surname often connoting things mystical and mythical to subsequent generations of Americans.

Indeed, the life and writings of Henry David Thoreau are continual sources of inspiration, specifically for environmentalists—in the United States in particular, and throughout the world more generally. Why this is so is quite simple. Thoreau was a highly sensitive observer of nature; he was deeply interested and forever curious about his natural surroundings; and he was greatly concerned for the present and future condition of his environs—local, regional, and national.

Those qualities, the latter especially, most endear Thoreau to present-day environmentalists. By the last decade of his life he had achieved a heightened awareness of nature shared by few of his generation: During the late 1850s, Thoreau entered the (then) rarefied realm of preservationism, an obscure place left to his successors to clarify and make better known.

Thoreau's recognition of nature's rights, however, was evolutionary—that (biocentric) viewpoint emerged over time and transformed him in the process. Similar to his predecessors and contemporaries (European Romantics and Romantically-inspired American Transcendentalists), Thoreau originally engaged the natural world aesthetically.

Yet, while poeticizing nature, a veritable Transcendentalist odyssey of discovering the "self" and the "divine" in the natural environment, he developed a profound respect for it. To be sure, Thoreau revered nature from the outset of what would be a lifetime of environmental sojourning. That reverence, however, ceased to serve his spiritual and philosophical needs only. Although nature remained a sacred space suffused with revelatory power, just as Thoreau's former mentor Ralph Waldo Emerson counseled, by the 1850s, Thoreau surpassed—without abandoning—purely anthropocentric concerns. His intimate wilderness relationship, eagerly sought and passionately cultivated, bore unexpected fruit: The acknowledgment that nature, like humanity, possessed inherent rights as well. That insight received forceful expression in 1857 when Thoreau vented his disgust at the destruction of something as seemingly insignificant as a favorite patch of bushes: "[I]f some are prosecuted for abusing children, others deserve to be prosecuted for maltreating the face of nature committed to their care."

Such a statement reveals that Thoreau was exceptionally attuned to his immediate environment (and by implication the natural environment more generally). Yet, Thoreau's environmental ethic was not only ahead of its time, but against it as well. Confronting a rapidly modernizing New England in which textile factories and railroads encroached upon pristine nature, and facing increasingly materialistic New Englanders who viewed nature as commercially exploitable, Thoreau leveled his discontent by retreating into what he called "the Wild."

From the moment that Thoreau commenced his now-legendary (26-month) experiment on the outskirts of Concord at Walden Pond on July 4, 1845 (a one-person utopia immortalized in print in 1854 as *Walden, or Life in the Woods*), his literary career and private life received inspiration and meaning in the nonhuman, the world of nature from which Thoreau criticized contemporaries for their careless and abusive treatment of it. And although Thoreau sought solace in the natural environment, his was not a permanent escape from society. True, as his writings emphatically indicate—in addition

to *Walden* consult, for example, the essay "Walking" (1850–62), the posthumously published *Maine Woods* (1864) and the voluminous private journals spanning his adult life—Thoreau expressed, directly and indirectly, his alienation from society. Among his primary artistic objectives, however, was his desire to reform American values concerning nature, not simply for humanity's own benefit (an initial concern), but for the welfare of the environment as well (a later concern).

SEE ALSO: Nature Writing; Preservation; Trancendentalism.

BIBLIOGRAPHY. Lawrence Buell, ed., *The American Transcendentalists: Essential Writings* (Modern Library, 2006); Robert L. Dorman, *A Word for Nature: Four Pioneering Environmental Advocates, 1845–1913* (The University of North Carolina Press, 1998); Richard J. Schneider, ed., *Thoreau's Sense of Place: Essays in American Environmental Writing* (University of Iowa Press, 2000).

RAYMOND JAMES KROHN
PURDUE UNIVERSITY

Three Gorges Dam

WHEN CONSTRUCTION IS completed in 2009, the Three Gorges Dam will be the largest hydroelectric dam in the world. Spanning Asia's longest river, the Yangtze, at Sandouping, Yichang, Hubei Province, the Three Gorges Dam will be 1.45 miles wide, 607 feet in height, and have 26 generating units with a combined capacity of 18.2 million kilowatts, producing 84.7 billion kilowatt-hours of electricity per year. When the reservoir created by the dam is filled, its waters will rise to be 574 feet deep and stretch for some 360 miles, flooding thousands of villages over 243 square miles of land, and displacing roughly 1.2 million peasants. A system of ship locks are intended to bring ocean liners from Shanghai 1,500 miles inland to the city of Chongqing, which was promoted to a provincial-level municipality under direct central control in 1997 in part to coordinate the resettlement of refugees from the dam. Official cost estimates for the project are roughly U.S. $25 billion.

Construction of the Three Gorges Dam was first proposed in 1919 by the father of modern China, Sun Yatsen. Serious planning began in the 1930s, and toward the end of World War II, the U.S. Bureau of Reclamation's chief design engineer conducted a major study. It was hoped that the dam would provide both electricity and relief from the long history of devastating summer floods along the Yangtze. After the founding of the People's Republic of China, Mao Zedong pushed for the building of the monumental dam as a symbol of national pride and human mastery over nature. A sharp debate emerged in the mid-1950s, however, between leaders who opposed the project on technical grounds and favored a series of smaller dams instead, and those who favored the project. Soon after, the economic depression of the Great Leap Forward and political upheavals of the Cultural Revolution put the plans on hold. Debates were revived after the death of Mao and the beginning of economic reform. Momentum picked up as Deng Xiaoping became an enthusiastic supporter of the dam, though there continued to be bitter disagreement on whether, when, and at what height the dam should be built, as well as how the surrounding area should be administered.

In 1986 a study commissioned by the government and funded by the Canadian International Development Agency concluded that the dam was feasible. This moved the project closer to implementation but also sparked a vocal debate within China, coinciding with China's democracy movement and growing international opposition to large dams. The State Council agreed in 1989 to suspend construction plans for five years, but this changed after the crackdown on Tiananmen Square, which led to the arrest of journalist Dai Qing and other critics, and silenced opposition to the dam. With a strong push from Premier Li Peng, the State Council and Politburo approved the project in 1992. Three months later, the National People's Congress (NPC) approved the project with a vote of 1,767 yes, 177 no, and 664 abstentions. This was an unprecedented level of dissent for the NPC, which generally rubber stamps leaders' proposals. Construction has proceeded in three stages over 17 years. From

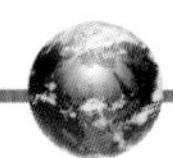

1993–97, the river was diverted; at the end of the second phase, 1998–2003, the first group of generators began to produce power, and a permanent ship lock opened for navigation; and in 2004–09, the entire project is to be completed. Corruption scandals and poor construction have plagued the project. In 1999 a bridge collapsed and a crack developed in the dam; in 2000, officials were arrested for extortion, kickbacks, and embezzling resettlement program money.

The major rationales for the dam are flood control, navigation improvement, and clean power generation to substitute for coal burning. By 2009, it should provide 10 percent of China's total power supply, but most of the electricity will be sent to the prosperous coastal region rather than used in the area around the dam. Power was in short supply when the dam first generated electricity in 2003, but in 2006, the Three Gorges Power Company was concerned about a power glut and how it should offload its supply. Nevertheless, the company was also already planning to build four more dams upstream in the Yangtze's longest tributary. The power generated by the Three Gorges Dam is eventually to pay for about 7.5 percent of its total cost. The rest has been financed by the China Development Bank, export credits, corporate bonds, and some taxes; the World Bank declined funding because of environmental concerns.

The gargantuan reservoir created by the dam threatens the habitat of many rare and endangered species including 36 endemic plants, the now-endangered Chinese sturgeon, and the Yangtze River dolphin. Though fish ladders were built, they have not been very successful. The dam also holds back sediment, which formerly carried nutrients downstream, and lowers water temperature; this further affects habitat for fisheries. Decomposing organic material in the reservoir will produce significant methane emissions. More importantly, the reservoir may alter the local climate, and worsen problems with schistosomiasis, a parasitic, snail-borne disease. Current patterns of dumping untreated garbage, sewage, industrial chemicals, and heavy metals into the river could lead to a public health disaster and affect normal dam operations if unchecked, because the reservoir will concentrate the pollution rather than flushing it out to sea. The government has laid plans to build numerous pollution control and treatment facilities, but critics remain skeptical of effective implementation.

The role of sedimentation is also disputed. While the reservoir is supposed to lessen the frequency of large downstream floods, and sluice gates are in place to flush out silt, critics warn that the technology is unproven. If ineffective, the build-up of sediment behind the dam could shorten the life-span of the dam, cause the reservoir to lose flood storage capacity, accelerate coastal erosion, and cause dam failure. Officials, however, claim that the technical issues have been resolved, that upstream erosion is being reduced through massive afforestation, and that after 100 years, the dam should still have 92 percent of its effective storage capacity.

Critics also warn that the dam is built on a fault and that water held by the dam could trigger landslides or an earthquake, but the government emphasizes the geological suitability of the chosen site, and claims that the dam could withstand even a class seven earthquake and a nuclear attack. Also of concern is the fact that the reservoir will flood thousands of graves and more than 1,000 recognized cultural and archeological relics.

Finally, the reservoir will inundate very fertile farmland and has necessitated the involuntary resettlement of some 1.2 million people, who have been moved either to higher land, to live with relatives in nearby urban areas, or to more distant provinces. Although government officials state that the involuntary migrants are "satisfied with their new lives, enjoying better living conditions," independent surveys have found a majority saying they are worse off. Among other problems they face in their new homes are inadequate or missing compensation, discrimination, and reduced standards of living because of the poorer quality of the land they received, and difficulty finding jobs.

SEE ALSO: China; Dams; Fish Ladders; Floods and Flood Control; Hydropower; Narmada Dam.

BIBLIOGRAPHY. Deirdre Chetham, *Before the Deluge: the Vanishing World of the Upper Yangtze River* (Palgrave, 2002); Li Heming, Paul Waley, and Phil Rees, "Reservoir Resettlement in China: Past Experience and the Three Gorges Dam," *The Geographical*

Journal (v.167, 2001); Kenneth Lieberthal and Michael Oksenberg, *Policy Making in China: Leaders Structures and Processes* (Princeton University Press, 1988); Dai Qing, *Yangtze! Yangtze!: Debate over the Three Gorges Project* (Earthscan Publications, 1994); Dai Qing, *The River Dragon Has Come: The Three Gorges Dam and the Fate of China's Yangtze River and Its People* (Armonk, 1998).

Emily T. Yeh
University of Colorado, Boulder

Three Mile Island Accident

THE THREE MILE Island (TMI) plant, located on the Susquehanna River, about 10 miles from Harrisburg, Pennsylvania, was the site of what the Nuclear Regulatory Commission (NRC) calls the most serious nuclear accident in American history. Metropolitan Edison then owned the TMI facility, consisting of two reactors. The accident began at about 4:00 A.M. on March 28, 1979, in the nonnuclear section of the power plant, when the pumps that feed water to the system to create steam (which propelled the power turbines) failed. Because these pipes carry away part of the heat from the reactor, the cooling system would be required to carry the extra load. The water and steam pressure increased in the cooling system, which caused a relief valve to open. The valve was supposed to close when pressure reached a safe level, but it did not, and, unknown to controllers, pressure and water in the cooling system was lost, thereby leading to a partial meltdown of the reactor core. The NRC summarizes the situation:

> As coolant flowed from the core through the [cooling water] pressurizer, the instruments available to reactor operators provided confusing information. There was no instrument that showed the level of coolant in the core. Instead, the operators judged the level of water in the core by the level in the pressurizer, and since it was high, they assumed that the core was properly covered with coolant. In addition, there was no clear signal that the pilot-operated relief valve was open. As a result, as alarms rang and warning lights flashed, the operators did not realize that the plant was experiencing a loss-of-coolant accident. They took a series of actions that made conditions worse by simply reducing the flow of coolant through the core.

This sequence of events describes what Charles Perrow calls a "normal accident" that results when redundant safety systems interact with human actors to result in unpredictable system accidents. By mid-day, the NRC, Environmental Protection Agency, and Department of Energy inspectors and scientists were at the scene. The utility's and the agencies' efforts appeared to be successful, but, on March 30, there was a small release of radiation due to attempts to release pressure on the coolant system. The radiation had come from an auxiliary building, not the containment, but the radiation was sufficiently worrisome to induce the governor of Pennsylvania to urge an evacuation of school-aged children and pregnant women within five miles of the reactor.

Another concern soon arose when it became evident that a bubble of hydrogen gas had appeared at the top of the containment structure; the highly flammable gas could explode and cause a small breach of the containment, thereby releasing dangerous radiation. By April 1, engineers and scientists had determined that the lack of oxygen in the containment would minimize the chance of an explosion, and, in any case, the size of the bubble had diminished.

What remained for the utility and the regulators was to secure the reactor, assess the damage, and figure out what went wrong. The accident was sufficiently serious that President Jimmy Carter created a commission, popularly known as the Kemeny Commission after its chair, to investigate what happened. The basic conclusion they reached is that the accident was partially caused by mechanical failure, but that failure was greatly compounded by human error. The commission noted that the control room technicians were poorly trained, that they failed to properly interpret the information their instruments provided, and that they did not suspect a loss of coolant accident (LOCA) until quite late in the day. It was not until late in the chain of events that the operators realized that the core was not covered by cooling water, and that a partial meltdown had

taken place. The LOCA was ultimately discovered, and by 3:30, the immediate crisis had passed.

In October 1979, the NRC fined the utility $155,000 (about $440,000 in 2005 dollars), a rather small amount considering the seriousness of the accident. In 1982 during a remotely controlled television inspection of the TMI-2 reactor, engineers found that the damage to the core was much greater than anyone had expected. Since the late 1980s, the TMI-2 has been in "monitored storage." Between 1979 and 1985, TMI unit 1 was shut down, but in 1985, the NRC granted permission to restart that unit. Many people believe that the TMI accident was the event that stopped nuclear power from becoming a more important source of energy in the United States. Under this logic, nuclear power had a bright future until TMI proved that nuclear power was too dangerous to be relied upon as an energy source.

There are other reasons for the decline in nuclear power plant construction in the United States. The first of these reasons is the reordering of the politics of nuclear power. By the early 1970s, the system of promoting and regulating nuclear power policy was beginning to break down. Congress's Joint Committee on Atomic Energy (JCAE) was seen as too closed and too powerful. Members of Congress pressed to break up the JCAE's responsibilities and ultimately distributed its responsibilities among several committees. The Atomic Energy Commission (AEC) was broken up, and its regulatory role was transferred to the new Nuclear Regulatory Commission in 1976, and its research and promotion function was transferred to the Department of Energy in 1979. Second, there was the increasing cost involved in building and getting nuclear power plants approved. This was a function of more aggressive regulation by the NRC even before the TMI accident, and of the lack of one or two industry-standard designs. Unique power plants often underwent costly design changes as knowledge of nuclear technology changed.

The third reason was the increasing strength and visibility of the antinuclear power movement in the United States and in Europe. The existence of an active and well-informed antinuclear movement was made clear in the aftermath of a 1975 accident at the Tennessee Valley Authority's (TVA) Browns Ferry, Alabama, nuclear plant, in which an accidental fire cut off communications between the reactor and the control room. This incident motivated the Union of Concerned Scientists (UCS) to argue that government estimates of the safety of nuclear power plants had been inaccurate. As a result, by 1979, the NRC had fully repudiated an earlier AEC report on the very low likelihood of an accident.

In a seemingly prescient coincidence, a Hollywood movie, *The China Syndrome*, about a potential LOCA, was released a mere 12 days before the TMI accident. The combination of the TMI accident, interest group opposition to nuclear power, and the dramatic power of the movie helped turn public opinion against nuclear power. After TMI, the rate of reactor licensing slowed considerably, peaking at 112 units in 1990 and declining to 104 units in 2006, although actual power output has been level since 1990 at abut 99,000 million kilowatts. No new nuclear plants have been ordered in the United States since the late 1970s, and while TMI did not trigger this downturn, it most likely accelerated it. Even today, with calls for alternative, less-polluting power generation, no new nuclear plants are on the horizon, in large part because of TMI.

SEE ALSO: Electrical Utilities; Nuclear Power; Nuclear Regulatory Commission (NRC) (U.S.); Nuclear Weapons; Tennessee Valley Authority (TVA).

BIBLIOGRAPHY. F. Baumgartner and B.D. Jones, *Agendas and Instability in American Politics* (University of Chicago Press, 1993); T.A. Birkland, *After Disaster* (Georgetown University Press, 1987); D.J. Chasan, *The Fall of the House of WPPSS* (Sasquatch Publishing, 1985); D. Nelkin, "Some Social and Political Dimensions of Nuclear Power: Examples from Three Mile Island," *American Political Science Review* (v.75/1, 1981); C. Perrow, *Normal Accidents: Living with High-Risk Technologies* (Princeton University Press, 1999); *Report of the President's Commission on the Accident at Three Mile Island* (Kemeny Commission, 1989); L. Stephenson and G.R. Zachar, eds., *Accidents Will Happen: The Case Against Nuclear Power* (Perennial Library/Harper and Row, 1979).

Thomas A. Birkland
State University of New York, Albany

Throughput

THROUGHPUT IS A term used in a number of fields. In industry, it denotes the amount of raw material that a processing plant handles in a set amount of time. In the computer world, it is the amount of work a computer system or a component does in a specified amount of time. In the pharmaceutical industry, it signifies chemical screening technology used to manufacture drugs.

Throughput, when denoting a processing plant, can refer to any kind of processing plant such as an oil refinery that can handle a given number of barrels of oil per day. It may be a gas pipeline or an oil pipeline. Or throughput may refer to the loading facilities in a port and depot or some other point of debarkation or embarkation.

In the case of a gas pipeline the volume of natural gas that can be pumped through it in a given period of time such as a 24-hour day is very important information to planners. As long as demand for natural gas is steady the throughput will be a supply that is sufficient to meet demand. However, if a severe and prolonged period of freezing weather occurs it may well be that the carrying capacity of the pipeline is inadequate. It means that the volume of gas that can be delivered will not meet demand. If this should occur then the natural gas available for heating, for industrial use, or for other purposes may not be sufficient. As a result, people may freeze to death, or contract serious illnesses, or suffer injuries due to frostbite or other effects from the loss of heat. Planners are dependent upon accurate statements of the throughput of many resources.

Computer technology uses the term *throughput* to measure the effectiveness of large computers that run a number of different programs at the same time. Early in the history of computers their power was assessed by the number of jobs they could handle when bundled into batches of different jobs for a single day. Later computer operations used the term to measure the productivity of a computer in terms of its performance. The performance was the time it took from input to output. The time was reported as the computer's response time. The term has since come to mean the speed at which data is transferred between receptors. Since there are a number of factors that can interfere with data transfers it is a measure of its final communication time.

The notion of throughput is now being using in the pharmaceutical industry. Advances in several sciences have led to the development of many new drugs. The human genome project has greatly increased knowledge of human genes. Because there are thousands of proteins in human DNA, the knowledge is now available for seeking new receptors for drug testing as possible medicines. Since all drugs are chemicals that have an effect on some target such as an organ, a bacterium, a nerve cell, or even a plant cell, the development of new drugs is now using high throughput technology.

The development of robot chemistry toward the end of the 20th century has made it possible to assembly in combinations and permutations vast numbers of chemicals that can then be tested on the thousands of receptor sites in proteins that represent human organs or body parts, or that represent those of animals in veterinary medicine. The chemistry techniques that have been developed allow a single chemist to develop thousands of new compounds each week. Previously, a skilled chemist would have done well to develop one or two chemicals in a week.

Many new companies have arisen that synthesize chemical libraries. These libraries contain thousands or millions of chemicals. The data for these is stored in computers allowing them to be tested at different receptors in a test tube or cell culture. Pharmaceutical companies can buy high throughput technology systems in order to test for some target receptor that is of interest to it. Using the computer's database of chemical formulas it is possible for a pharmaceutical company to screen as many as a million chemicals against a chosen receptor site in only a month.

It is possible to design throughput libraries so that the natural characteristic of a receptor is presented with a library of chemicals. The chemicals can be reduced to those that are viewed as most likely to be matched and lock into the site. The goal is to use the throughput screening in a search for responses at receptor sites to chemicals being tested. With the vast number of new synthesized chemicals that can now be made throughput technology allows testing on a large scale at high speeds. This reduces the cost and time for developing new drugs significantly.

From the process of mass screening those compounds that show some effects are those that warrant further investigation. Those chemicals that do not show results are not abandoned because they may prove useful in future trials. Rather, they are returned to the library for another day's testing against another receptor site. The number of receptor sites is growing rapidly as knowledge about cells, genes, and the use of gene splicing is increasing.

SEE ALSO: Analytical Chemistry; Human Genome Project; Industry; Measurement and Assessment; Natural Gas.

BIBLIOGRAPHY. C. K. Atterwill, W. Purcell, and P. Goldfarb, eds., *Approaches to High Throughput Toxicity Screening* (CRC Press, 1999); Thomas Corbett, *Throughput Accounting* (North River Press Publishing Corporation, 1999); John P. Devlin, *High Throughput Screening: The Discovery of Bioactive Substances* (Marcel Dekker, 1997); James Kyranos, ed., *High Throughput Analysis for Early Drug Discovery* (Elsevier Science & Technology Books, 2004); David A. Wells, *High Throughput Bioanalytical Sample Preparation: Methods and Automation Strategies* (Elsevier Science & Technology Books, 2003).

Andrew J. Waskey
Dalton State College

Thunderstorms

THUNDERSTORMS ARE DRAMATIC weather phenomena that may deliver much-needed rain or cooling relief from oppressive heat and humidity, but they are also the makers of dangerous, violent winds, flash floods, damaging hail, and deadly air-to-ground lightning. Hundreds of people are killed every year by thunderstorm winds, floods, and lightning. Thunderstorms also destroy property, crops, and significant numbers of livestock every year.

Thunderstorms are formed by rising air that is warm and moist. The warm air currents become strong updrafts that can reach 7–11 miles high. Some storms have updrafts so powerful that the warm moist air reaches the troposphere where the updraft circulation produces hail. The hail may be the size of a marble, an orange, or even a grapefruit. These hail storms may kill people caught in the open and do significant property damage.

Thunderstorms may occur as a single storm; these are usually products of local conditions and may be called air mass or convective thunderstorms. In contrast, frontal thunderstorms are associated with rapidly moving moist air masses that are colliding with cold air masses. When these conditions occur the thunderstorms may occur in clusters or along frontal lines between advancing cold and slow-moving warm air masses.

Most thunderstorms occur in the late afternoon and early evening hours when the heating of the earth's surface produces the most updrafts. They have three stages to their development. The first is the rising of unstable warm moist air. As the mass of warm air reaches heights of 50,000–80,000 feet the cumulonimbus stage is reached with its anvil-shaped head. This is the mature stage, however, as cooling moisture moves out of the updraft and creates downward currents that strike the ground as precipitation. In the final stage the warm upward air drafts are shut off by the descending cold air and the storm ends.

Most single thunderstorms last from a few minutes to over an hour. If a single thunderstorm explodes over a small area and lasts a significant amount of time it may dump great quantities of rain in amounts greater than six inches (18 centimeters or more) in an hour. Thunderstorms producing large amounts of rain in a short period of time can cause flash floods. The western United States is very prone to flash floods caused by thunderstorms. Many people have been killed by flash floods that caught them in a canyon or arroyo that was normally dry. In the Middle East thunderstorms can easily flood the *wadis* of an area, sweeping away anything in the path of the storm water.

Thunder is caused by electrical discharges between clouds or between clouds and the ground. The electrical charge is visible as lightning. The sound is caused by the lightning breaking the sound barrier as the discharging current superheats the air between the positive and negative contact points for the electrical discharge. Within milliseconds, the air is heated to temperatures of about

18,000 degrees F (10,000 degrees C). Expanding violently, the superheated air forms pressure waves that are audible for up to 15 miles (24 kilometers). The rumbling sound is due to the variations in the sound waves caused by the various parts of the lightning channel.

Some regions of the world, like the western United States, experience dry thunderstorms that do not produce rain because the rain evaporates before it hits the ground. This type of thunderstorm produces lightning that often starts forest fires.

There are about 50,000 thunderstorms occurring on the earth every day. The tropics, where heating of the land and sea occur steadily all year long, are prone to the most thunderstorms. In the Northern and Southern Hemispheres the spring and summer months are the seasons with the greatest numbers of thunderstorms.

SEE ALSO: Fire; Floods and Flood Control; Livestock; Tropics; Weather.

BIBLIOGRAPHY. John A. Day and Vincent J. Schaefer, *Peterson First Guide to Clouds and Weather* (Houghton Mifflin Company, 1998); Ron Holle et al., *National Audubon Society Pocket Guide to Clouds and Storms* (Knopf Publishing Group, 1995); Thomas D. Potter and Bradley R. Colman, eds., *Handbook of Weather, Climate and Water: Dynamics, Climate, Physical Meteorology, Weather Systems, and Measurements* (John Wiley & Sons, 2003).

Andrew J. Waskey
Dalton State College

Tides

Tides provide natural cleansing of estuaries and bays and renew nutrient levels in marshes.

TIDES ARE THE twice-daily rising and falling of the ocean at the shores of continents and islands. Over an approximately 12-hour cycle the tide in a locality flows out to low tide (ebb tide) and flows in to high tide (flood tide). The cycle is repeated again over the next 12 hours so that each day and each night there is a high and a low tide.

Tides also occur in all bodies of water on the earth. In some freshwater lakes, such as the Great Lakes, the tide may be only an inch or two, as most tides are too small to be noticed without instruments. In some places, tides are relatively shallow; in other places, the turning of the tide creates tidal rushes that are many feet in depth. Tidal highs and lows vary between localities because of local topological conditions, the strength of winds blowing on or off shore, and the position of the earth relative to the sun and the moon.

Tides in all places display the enormous amounts of energy that it takes to create them. The forces creating tides are the gravitational pull of the moon and the sun, the rotation of the earth, and the drag of the earth's uneven surface upon the movement of the oceans' waters.

The envelope of water on the surface of the earth is attracted by the moon's gravitational pull toward the moon. As the moon rises and sets relative to an observer on the earth, the earth rotates on its axis as well. High tide will be on the side of the earth facing the moon and on the side of the earth opposite to the moon. Low tide will be on the two sides of the earth at 90-degree angles to the moon. The moon's gravity pulls the water closest to it toward itself causing high tide. In addition, it causes high tide on the side of the earth opposite to the moon by also pulling the solid earth toward itself.

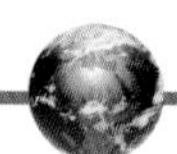

The earth rotates on its axis once every 24 hours. However, the moon always faces the earth with the same side as it rotates around the earth in every 29 and a half days. When the moon and the sun are aligned with the earth, the sun's gravity combines with the moon's to create the highest tides (spring tides) of the month. The tides occur at the full of the moon and at new moons when the pull of the moon and sun combined are at their greatest. The gravitational pull of the sun is 46 percent less than the moon's gravitational pull. The weaker gravitational pull of the sun is due to its distance from the earth, though its mass is many times greater.

Twice each month, the moon's orbit takes it to a position that is at a 90-degree angle to the earth and the sun. This is the time of neap tides, which are the lowest tides of the month. These occur at the times of the first and third quarter of the moon when the moon's gravitational pull is most out of line with the gravitational pull of the sun.

Tides are constantly changing as the sun, moon, and earth change. Tides are also affected by the winds of the seasons and by the local topography of the land next to the seaside. In estuaries and bays that are broad and open, the tide will be less than in those places where the entrance to the land is narrow and confined. In some places, such as the Bay of Fundy, the tidal range may be as much as 50 feet (15 meters) between tides.

The tides in the Atlantic and Pacific are different because of the size of the oceans. Pacific tides can be so-called "mixed" tides. Some islands have mixed tides where the tidal flows are such that one ebb tide is slight and then after the next high tide the following ebb tide is great. In a few localities in the Pacific, there is only a daily tide of a high and a low tide.

Tides provide natural cleansing of the estuaries and bays and renew the nutrient levels in the marshes. People who live beside the sea will usually regulate their activities according to the tides. Ships come and go with the tide using high tide in harbors to avoid shallows or underwater obstacles. Digging clams, crabbing, or some kinds of fishing (flounder) are done at low tide. There are also tides in rock and tides in the atmosphere, but these are too slight to detect without instruments.

SEE ALSO: Beaches; Estuaries; Fisheries; Oceans.

BIBLIOGRAPHY. John D. Boon, *Secrets of the Tide: Tide and Tidal Current Analysis and Predictions, Storm Surges, and Sea Level Trends* (Horwood Publishing Limited, 2004); James Greig McCully, *Beyond the Moon: A Conversational, Common Sense Guide to Understanding the Tides* (World Scientific Publishing Company, Inc., 2006); David T. Pugh, *Changing Sea Levels: Effects of Tides, Weather, and Climate* (Cambridge University Press, 2004); John Wright, ed., *Waves, Tides, and Shallow-Water Processes* (Elsevier Science & Technology Books, 2000).

Andrew J. Waskey
Dalton State College

Tigers

THE LARGEST OF the big cats, the tiger *Panthera tigris* has become the global face of wildlife conservation. Tigers are carnivorous mammals classified in the biological family Felidae, characterized by territorial behavior and specialized hunting skills. The tiger is listed in the International Union for the Conservation of Nature (IUCN) Red List of Threatened Species as a critically endangered species. One of the commonly voiced benefits of protecting tigers is that tiger conservation requires the protection of entire terrestrial ecosystems, essentially large areas of land, which in turn helps protect myriad other plants and animals that live in those ecosystems. Thus, tiger conservation efforts should ideally lead to larger gains in terms of the conservation of biodiversity and genetic diversity in the wild.

The recent history of tigers, however, continues to concern conservationists, biologists, wildlife managers, and others. Despite a complete ban under the Convention on International Trade in Endangered Species (CITES) since 1976 of the sale and use of tiger skin, bones, or any other body parts, tiger-derived products continue to be used in traditional Chinese medicines, for which the United States is the main market outside Asia. The plight of the tiger and the dramatic decline in the populations of its subspecies has gained worldwide attention since the 1960s. By then, the tiger was already on a dangerous path toward becoming an endangered spe-

cies. It was systematically hunted either as a pest or as a trophy by Indian royalty and the colonial British elite in the 19th and early 20th century in undivided India, the epicenter of its historic range. These days, despite the protection that tigers receive within parks, sanctuaries, reserves, world heritage sites, protected forests, and occasionally as a result of community action to protect their habitat, they are coming in contact with humans as never before. Agricultural expansion; loss of forest cover to mining, dams, and other developmental projects; conversion of natural forest to plantations; poaching of prey species; and destructive activities due to human migration and population growth in areas bordering protected tiger habitat are some of the reasons contributing to the decline in tiger populations.

Historically distributed from the Caspian Sea in the west through south and southeast Asia and up to Siberia and northern China, the tiger is often depicted in Asian mythology representing both good and evil. Durga, a female Hindu deity, is depicted in temples throughout India riding a tiger. Buddha is believed to have offered his body to a starving tigress. Regenerative powers have been attributed to tigers and people believe they are protectors, guardians, and the harbingers of peace. Tigers have inspired ecologists, such as George Schaller of the Wildlife Conservation Society, and continue to inspire artists specializing in tiger paintings at the Ranthambhore School of Art.

India is home to the world's largest population of tigers, estimated at between 3,000 and 4,700 individuals, while an optimistic estimate for the entire world pegs the number at about 7,500. Traditionally, in countries like Nepal and India, tiger numbers have been estimated by counting their pug marks; plaster casts or paper tracings of pug impressions are taken from the ground. This method is in some cases supplemented by the use of radio telemetry and DNA-based scat (tiger droppings) and hair

Tasmanian Tigers

The thylacine on the Australian island of Tasmania is more commonly known as the Tasmanian tiger, though it is not actually related to the tigers of Asia. It was well-known by the Aboriginal peoples, and engravings of the thylacine going back to 1000 B.C.E. have been found. It was not known to Europeans until it was encountered by French explorers in 1792. George P.R. Harris, a surveyor, wrote the first description of the animal in 1808, and it seems likely that it was extremely rare, even then—five years after the British settlement of Tasmania (then called Van Diemen's Land).

The thylacine was similar in size to a large short-haired dog, being about 100–130 centimeters long. It had a stiff tail, and because it was striped, it gained its name of "tiger." The thylacine was actually a marsupial, with the females holding their young in a pouch. It was nocturnal and is believed to have had a highly developed sense of smell allowing it to track prey easily. Living in the woodlands and heath of the midlands and coastal regions of Tasmania, areas that were turned into agricultural land by early British settlers, the thylacine was hunted by them. The thylacine lived on kangaroos, wallabies, small animals and birds, and possibly on the Tasmanian emu, a large flightless bird. The Tasmanian emu was hunted to extinction in 1850; this might also have contributed to the fall in the number of thylacines.

By the early 1920s, the number of thylacines in the wild had become extremely low; plans were advanced to reintroduce them into the Australian mainland. The last known wild thylacine was shot dead by a farmer in 1930 in northeast Tasmania; the last captive thylacine died in the Hobart Zoo in 1936. Known as Benjamin, in spite of its probably being a female, it may have died of neglect and exposure, as it did not have a sheltered place to sleep. Thought to be the last of its species, it was much photographed. A naturalist shot a short film of it—and was bitten in the process. Since its death, there have been many reported sightings of the thylacine, and rewards offered for evidence that any have survived. However, in spite of the discovery of some droppings and paw prints similar to that of a thylacine, none has ever been photographed or found.

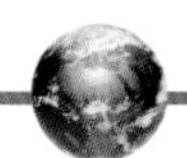

analysis. In recent years, field trials of camera traps to photograph and record individual tigers have been conducted. Human error and technical issues mean that all methods have their limitations and tiger numbers are best viewed as estimates.

Only about five percent of the tigers alive at the beginning of the 20th century now roam the forests, grasslands, and swamps such as those of the Sundarbans Tiger Reserve and World Heritage Site, a mangrove forest straddling the India-Bangladesh border and one of the largest protected areas for the Bengal tiger as well as 260 bird species. Tigers can easily weigh up to 225 kilograms and consume one-sixth of their body weight in food at a time. A good prey-base is essential to maintain tigers in the wild and some studies indeed suggest that tiger densities can be predicted if the approximate number of prey is known. Cameras triggered when an infrared beam is broken by prowling tigers have been used by ecologists to determine that these largely solitary animals that thrive in dense undisturbed vegetation can reach densities as high as 16 tigers in a 100-kilometers-square area at Kaziranga National Park in India.

But such high tiger densities are uncommon, and three subspecies of the tiger have gone extinct in just the past 70 years. These were the Bali, Caspian, and Javan tigers. The five subspecies that remain are all threatened by poaching and loss of habitat. Of these five, the Amur Tiger (*Panthera tigris altaica*, found largely in the easternmost provinces of Russia, China, and the Korean peninsula), the Sumatran Tiger (*Panthera tigris sumatrae*, found only on the Indonesian island of Sumatra), and the South China Tiger (*Panthera tigris amoyensis*, found in four Chinese provinces) have very low or declining populations, and are classified as critically endangered by the IUCN. The Indochinese Tiger (*Panthera tigris corbetti*, found primarily in southeast Asia from Bangladesh to Vietnam) was recognized as a distinct subspecies as recently as 1968, and the Bengal Tiger (*Panthera tigris tigris*, found largely in India with some in Nepal, Bhutan, and Myanmar), are both endangered. All tiger subspecies are listed in Appendix I of CITES and are protected in most of their range under CITES and national laws. The level and extent of enforcement of these laws, however, varies widely from country to country.

SEE ALSO: Convention on International Trade in Species of Wild Flora and Fauna (CITES); Poaching; World Conservation Union (IUCN); World Wildlife Fund (WWF).

BIBLIOGRAPHY. David Alderton, *Wild Cats of the World* (Facts on File, Inc., 1993); Simon Barnes, *Tiger!* (St. Martin Press, 1994); Government of India, "Project Tiger," projecttiger.nic.in (cited April 2006); George B. Schaller, *The Deer and the Tiger* (The University of Chicago Press, 1967); John Seidensticker, ed., *Riding the Tiger: Tiger Conservation in Human-Dominated Landscapes* (Cambridge University Press, 1999); John Seidensticker and Susan Lumpkin, *Cats: Smithsonian Answer Book* (Smithsonian Books, 2004); Valmik Thapar, *Land of the Tiger: A Natural History of the Indian Subcontinent* (University of California Press, 1997); Valmik Thapar, ed., *Saving Wild Tigers, 1900–2000: The Essential Writings* (Permanent Black, 2001); Alan Turner, *The Big Cats and their Fossil Relatives* (Columbia University Press, 1997); World Wildlife Fund, "Tigers," www.worldwildlife.org/tigers (cited April 2006).

Rahul J. Shrivastava
Florida International University

Tigris and Euphrates Rivers

WATERS FROM THE Tigris (in Arabic *Dijla*, in Turkish *Dicle*) and Euphrates (in Arabic *Furat*, in Turkish *Firat*) Rivers gave rise to some of the first known agricultural civilizations. These early societies of the Mesopotamian plain arose with irrigation techniques and water infrastructure dating back to 4000 B.C.E. and earlier. The Tigris and Euphrates Rivers are also well known for their high inter-annual and seasonal variability, leading to intense flooding with maximum flows as much as 10 times minimum flows. Sediment loading on the rivers is also high, resulting in raised river beds that historically have facilitated irrigation.

The "twin rivers" begin approximately 30 kilometers from each other in the highlands of eastern Turkey, travel through Syria and Iraq, and join together as the Shatt al Arab for 150 kilometers before draining into the Persian (or Arab) Gulf. The

Euphrates extends a total of 2,700 kilometers (40 percent in Turkey, 35 percent in Iraq, and 25 percent in Syria); the Tigris travels 1,900 kilometers (20 percent in Turkey, 78 percent in Iraq, and 2 percent in Syria). The mean flow of the Euphrates is approximately 30 billion cubic meters per year (BCM/Y); and the Tigris conveys considerably more after contributions from tributaries in Iraq, with approximately 50 BCM/Y.

Water diversion and damming of the twin rivers for irrigation and other uses continues today with major engineering works pursued by all three riparians (Turkey, Syria, and Iraq) for agriculture, hydroelectricity generation, and to overcome flooding and the intense seasonality of the rivers. In Turkey, there are approximately 20 dams completed or planned that comprise the GAP project (Southeastern Anatolia Project, with the Atatürk dam the largest among them, filled in 1991). In Syria, several large dams are in operation, including the Tabqa dam, completed in 1975 with the aid of Soviet engineering and financing during the Cold War. In Iraq, large reservoirs include Lake Abu-Dibbis, Lake Habbaniyah, and Lake Tharthar (serving flood control and facilitating water transfers from one river to another).

Planned future use of the rivers by all three countries is estimated to outstrip available freshwater supply by 148 percent on the Euphrates and 111 percent on the Tigris. Given this, the Tigris-Euphrates basin is often cited as a potential site of future water-related conflict, with troops already having been deployed to the Syria-Iraq border in the 1970s over water use conflicts; diplomatic hostilities that have led to decades-long stalemates and the absence of a tripartite water sharing agreement (there is, however, a 1987 agreement between Turkey and Syria guaranteeing 500 cubic meters per second of Euphrates water to flow over the border, and a 1990 agreement between Syria and Iraq, but none involving all three countries); as well as regional and international concern over Turkey's use of the river waters with continued implementation of the Southeastern Anatolia Project.

As Turkey's GAP project involves damming and water diversions on both rivers, it threatens to reduce the quantity and quality of freshwater for downstream Syria and Iraq—figures cited are as high as a reduction of 80 percent of Euphrates flow for Iraq, and 60 percent for Syria. Within Turkey, water diverted for irrigation has led to considerable agroecological and societal changes in the Southeastern Anatolia region and poses long-term sustainability concerns, such as issues related to pollution and salinization. There are also important implications of the project for the long-standing Kurdish conflict, as Kurdish populations occupy areas around the rivers in all three countries, and as former Turkish President Özal is said to have threatened Syria with cutting off flow of the Euphrates if it did not stop support for Kurdish insurgents.

However, there are also efforts underway to promote regional cooperation in the basin, for instance with recent efforts to establish ETIC, the Euphrates Tigris Initiative for Cooperation, involving technical experts from all three countries and led by the former president of Turkey's GAP project, Dr. Olcay Ünver.

Each of the three riparians invokes different rights claims to use of the rivers, for instance, with Turkey taking advantage of its upstream position, and also arguments related to territorial "contributions" to the rivers. Iraq similarly highlights "contributions" to make claims over Tigris waters, but also invokes historical use claims to the river waters. When examining much available data on the Tigris-Euphrates system, politics related to these different claims are often readily apparent. For instance, sources might insist on treating the rivers as a combined system, as doing so also gives Turkey a majority share in terms of contributions to the joined river system, while isolating the rivers might enhance Iraq's claim to the Tigris (Turkey contributes 60 percent to the joint river system, while separately, Iraq contributes a majority share to the Tigris system). Others similarly argue for concepts such as "economically irrigable lands" to argue that irrigation uses are more reasonable in one country over another, or analysts invoke the suitability of transferring water from the Tigris to the Euphrates to meet demand.

In terms of other notable political issues related to use of the rivers, historically, floodwaters would dissipate into the extensive marshes of the lower Mesopotamian plain. More recently, these marshes are the site of degradation and contentious engineering transformations. Under the regime of Saddam Hussein in Iraq, marshlands were drained to allow for agriculture and a canal was built to improve

Wilfred Thesiger and the Marsh Arabs

The travel writer Wilfred Thesiger has often been called the last of the Victorian travelers, born half a century after his time. He is still heavily identified with the Marsh Arabs of the Tigris River Delta with whom he lived, and about whom he wrote so evocatively.

Wilfred Thesiger was born in 1910 in Addis Ababa, Abyssinia (now Ethiopia), where his father was the British Minister. He was educated at Eton College and Oxford University, returning to Abyssinia at the age of 20. In 1935, he joined the Sudan Public Service and at the outbreak of World War II was seconded to the Sudan Defense Force. During the war, he served in Abyssinia and Syria, and then with the S.A.S. in the Western Desert.

After the war, he traveled to Arabia, and then went to Southern Iraq in 1951, planning to stay for a fortnight. He ended up staying for eight years, living with the Madan tribe, which became well-known through Gavin Maxwell's *A Reed Shaken by the Wind* (1957). The Marsh Arabs treated him well and in June 1958 Thesiger left Iraq and moved to Copenhagen, Denmark, where he wrote *Arabian Sands,* which was published in late 1959 and received excellent reviews. He then started work on his next book *The Marsh Arabs,* which was published in 1964.

Thesiger appeared to have enjoyed the eight years he spent with the Marsh Arabs and during that time with them he shared many experiences including pig hunts and fishing, and was even an eyewitness to many blood feuds. He prophesized in the introduction to *The Marsh Arabs* that he expected the marshes would soon be drained and the way of life of the Marsh Arabs, little changed over thousands of years, would disappear. In the 1970s, this came to pass, and many historians are thankful that Thesiger took the time to describe a culture and way of life that no longer exists. Wilfred Thesiger died in 2003.

transport from the Persian Gulf to Baghdad. While the project enjoyed international support in the decades before its completion, more recently it has been contested, with some suspecting the draining of 57 percent of the marshland was a move against Shiite dissidents opposed to the Ba'athist regime. The draining of the marshes is also widely cited for having posed significant threats to the livelihoods of Marsh Arabs, as well as to migrating birds and other wildlife dependent on the marsh ecosystem. While engineering works in Iraq clearly resulted in degradation and losses of the wetland ecosystems, the changes must also be understood in the context of withdrawals occurring throughout the river system, including withdrawals in neighboring Iran.

The Iran-Iraq war, and the more recent wars in Iraq in the past two decades have also had important implications for the rivers. For instance, the burning of oilfields and loss of infrastructure for production has resulted in considerable water pollution, with ongoing disposal of "black oil" in locations that threaten to degrade Tigris River waters. This is just one example of emergent insecurities and vulnerabilities with implications for water and conflict possibilities throughout the region.

SEE ALSO: Iraq; Persian Gulf; Persian Gulf Wars; Riparian Areas; Rivers; Syria; Turkey.

BIBLIOGRAPHY. D. Altinbilek, "Development and Management of the Euphrates-Tigris Basin," *Water Resources Development* (v.20/1, 2005); Center for Strategic and International Studies, "The Tigris and Euphrates River Basins: A New Look at Development," *Future Watch* (Global Strategy Institute, June 2, 2005); J. Glanz, "Waste Oil Dumps Threaten Towns in Northern Iraq: Seepage into Tigris River is Seen as Potential Disaster for Region," *New York Times* (A1, June 19, 2006); P. Gleick, *The World's Water: The Biennial Report on Freshwater Resources* (Island Press, 2005); L. Harris, "Water and Conflict Geographies of the Southeastern Anatolia Project," *Society and Natural Resources* (15, 2002); D. Hillel, *Rivers of Eden: The Struggle for Water and Quest for Peace in the Middle East* (Oxford University Press, 1994); J.F. Kolars and W.A. Mitchell, *The Euphrates River and the Southeast Anatolia Development Project* (Southern Illinois University Press, 1991); S. Postel, *Pillar of Sand: Can the Irrigation Miracle Last?*

(W.W. Norton, 1999); S.L. Postel, and A.T. Wolf, "Dehydrating Conflict," *Foreign Policy* (September/October 2001); V. Shiva, *Water Wars: Privatization, Pollution and Profit* (South End Press, 2002); A.T. Wolf, "A Hydropolitical History of the Nile, Jordan and Euphrates River Basins," in A. Biswas, ed., *International Waters of the Middle East: From Euphrates-Tigris to Nile* (Oxford University Press, 1994).

Leila M. Harris
University of Wisconsin, Madison

Timber Industry

WOOD AND WOOD fiber are found in thousands of different products that meet a variety of human needs. Accessible forests have been increasingly harvested around the globe, although relatively untouched areas still exist in isolated regions such as Siberia. Nevertheless, due to rising awareness of the problems associated with deforestation and biodiversity loss, there is increasing protection for primary forests, especially rare or endangered ecosystems. As much as 10 percent of the earth's forested areas are designated for conservation, yet the degree of protection varies by country and region.

Transition to plantation forestry is rapidly expanding, although less than 5 percent of global forests are currently within plantations. Large-scale, industrial production is used to produce low-cost lumber and paper products. Plantations cannot be considered a replacement for native forests as there is a considerable reduction in biodiversity. Forests managed for commercial production face other ecological risks, such as the use of herbicides as a treatment during reforestation in order to give timber species an advantage over vegetation that is not commercially valuable. There is also controversy over international research in genetic engineering to promote desirable traits for rapid lumber production, which is sometimes advocated as a means to take pressure off of natural forests, but there is concern that altered genes may drift into wild areas.

Productive timber lands are deemed a good economic investment, but a variety of natural disturbances can impact harvests: Fires, insect damage, floods, ice, wind, and hurricanes. Hurricane Katrina is believed to have damaged billions of board feet in the southern United States, and only a small percentage was recovered before the wood began to deteriorate.

Despite the large number of tree species existing around the globe, the timber industry has focused on just a few. On average, 5 percent of the tree species native to a country are either vulnerable, endangered, or critically endangered. In most regions as few as 10 tree species dominate commercial markets. Illegal logging contributes to the overextraction of favorite tree species, such as big-leaf mahogany. Timber companies increasingly support international campaigns against illegal logging because they realize that illegal practices keep global timber prices low. Illegal timber extraction also means a loss of revenue for state programs, including some targeting forest conservation.

Timber companies frequently harvest in areas of indigenous peoples, as their homelands are where the largest expanses of natural forest remain. In many countries the government maintains control over natural resources and can often grant concessions without local approval. While more than 80 percent of the world's forests are publicly owned, private ownership is increasing.

In spite of corporate mergers and acquisitions, as a result of a harsh operating environment for the industry post-2000, timber is often sold through a series of small- and medium-size intermediaries located in countries all over the world. When unable to compete to produce cheap roundwood, companies in industrial countries look to encourage additional processing with paneling or engineered products. While wood remains a popular construction material, there has been some transition away from traditional lumber boards to a variety of engineered wood products, such as particle board, oriented strand board, and composite board. These products make use of smaller trees or waste from milling, but there is concern over the environmental safety of some chemicals used as binding agents.

Timber extraction and lumber processing involves environmental regulation related to air emissions, wastewater discharge, solid and hazardous waste management, site remediation, and forestry operations. While there have been undisputed

improvements to many logging operations around the world in recent years, there are still widespread challenges to sustainability. Slickly marketed "green" forest products are sometimes only slightly improved from their conventional counterparts.

Since 1995 there has been a transition in the timber industry created by independent certification of sustainable ecological lumber harvested within socially just production systems. The original intent was to focus on tropical areas, but international certification grew most quickly in temperate and boreal forests. The world's largest certified forest product marketing companies include giant chains like Home Depot, IKEA, and B&Q. Such stores were previously criticized for contributing to tropical deforestation and thus have been able to improve their public image with this "eco-friendly" product line. However, there are many competing certification standards around the globe, and they are not equally stringent. Industry standards are often not as rigorous as independent regimes, such as the Forest Stewardship Council.

SEE ALSO: Biodiversity; Boreal Forests; Chipko Andolan Movement; Deforestation; Endangered Species; Forest Management; Forest Service (U.S); Habitat Protection; Indigenous Peoples; Plantation Forestry.

BIBLIOGRAPHY. American Forest and Paper Association, "'Illegal' Logging and Global Wood Markets" (November 2004); "Down in the Woods," *The Economist* (March 2006), www.economist.com (cited May 2006); Food and Agriculture Organization, *Global Forest Resources Assessment* (2005); PricewaterhouseCooper, *Global Forest and Paper Industry Survey* (2005); J.P. Siry, F.W. Cubbage, and M.R. Ahmed, "Sustainable Forest Management: Global Trends and Opportunities," *Forest Policy and Economics* (v.7, 2005).

Mary M. Brook
University of Richmond

Time

THE RELATIONS BETWEEN time and nature are so complex and so close that the two are often implicitly collapsed into one category. Whether or not "time" describes some movement in the world beyond the social (philosophers continue to debate the question), the experience of time is always socially determined. Still, time almost always seems to be the most "natural" phenomenon. Time appears to be the fabric of everything dynamic, from night and day, to the seasons, to something so fundamental to current thinking about "nature" as evolution. Without ideas of time, our current notions of change and causation are unthinkable. In much of the contemporary world, in which we understand the "passage" of time as "linear", and the future as a result of what has "come before"—i.e., the past cannot return, and the future is passive—the idea that time could be constituted otherwise seems not only culturally alien, but empirically false.

Conceptions of time as reversible, cyclical, or fate, are associated with "primitive" mythology. But there is nothing necessarily more false about these ways or understanding the order of experience. We cannot prove the existence of the pace or form of a "natural" time, by which others could be standardized. We are time's only fixed measure; Einstein's relativity shows that even the rate at which time ticks away depends on how fast we are moving through space. Indeed, the concept of space-time that falls out of physics is based on the idea that the two dimensions are in fact one: All space exists in time, all time ticks away in space, and, perhaps most importantly, all movement and change takes place in both time and space simultaneously.

Consequently, while assuming a constant linear rate of temporal change (years, seasons, days, milliseconds) is necessary for strictly biophysical analyses of the environment, in the investigation of the relation between nature and society, it is much more problematic to rely on a single time; for the problem of time is already posed by the idea of nature-society relations. Its most obvious manifestation is perhaps the diversity of "timescales" we use to understand change: We speak of different timescales depending upon the phenomena of interest. For example, think of the idea that human life has only existed for a brief instant on the "evolutionary" or "geological" clock, or of the charge that those who are not obviously concerned with a sustainable future are "myopic"—i.e., their personal timescale is too

short or too private. Even the vagueness of phrases like *short-term*, or *long-run* show how geographically, culturally, and historically specific any notion of time is.

The problem of time is also common in historical studies of more recent nature-society dynamics, and in theoretical attempts to deal with the direction and form of those dynamics. First, the patchiness of data often makes it difficult to establish past conditions, and even harder to guarantee the continuity across the times and places for which we do have data. Archeologists trying to piece together the fate of ancient societies usually find only a few points of reference across thousands of years, with no information directly pertaining to the centuries in between. Evidence that diets changed radically at a certain point, for instance, often offers no clue as to the how and why of the transition.

To fill in those blanks, we rely on assumptions about relations between humans and human relations to the non-human world to provide a priori narratives about the form and direction of change. These assumptions can be grossly inaccurate, as Fairhead and Leach show in a famous study of west African forests. They explain how the expectation that indigenous people's environmental practices are inevitably destructive led scientists to completely misread the forest history of some parts of the region, seeing deforestation where there has actually been active afforestation. Temporal assumptions like these also trouble dominant narratives of progress and development, which lead to nature-society analyses that frame environmental degradation as natural, or presume an inevitable if as yet unnamed technological fix. A good example of this is the so-called environmental Kuznets curve, which suggests that ecological damage is inevitable in the process of national economic growth, but will decrease after a certain development plateau has been reached.

Time is also central to the study of the difference nature's difference makes in human productive systems. Mann and Dickinson's seminal work framed this problem as the mismatch between periods of circulation or reproduction in biology and those in political economy. In other words, crops and money have different timescales over which they can grow or be reproduced. Corn will only grow at some times and in some places, and it takes a more or less specific duration to do so. Money can grow anywhere at any time, at least in theory. To the extent that nature does not instantly reward investment in nature-based sectors like agriculture or forestry, then, capital's circulation and accumulation is slowed while it must wait for the weather to get warmer, crops to ripen, or trees to reach merchantable dimensions. From this perspective, agricultural biotechnology can be seen as the efforts of business and the state to overcome this mismatch by, for example, accelerating crop growth rates to speed up production, or increasing cold or heat tolerance to extend growing season or geographic range. In addition, from this view it is the degree of the mismatch in circulation times that defines a natural resource as renewable or nonrenewable; coal, for example, despite the fact that it is produced over time, cannot attract capital willing to wait out the time of production.

Another way in which the problem of time affects the study of society-environment relations is crystallized in what might be called the ecology of the future. Many environmental narratives take what is called a *declensionist* form—it is presumed that humanity always harms the nonhuman world, and that, barring radical change, we are on a downward slide to an apocalypse that will significantly alter the biology of the planet, and possibly remove us from the picture altogether.

Other, more progressive narratives are founded upon the idea that we can weather this storm, or technology will allow us to avoid it. Either way, the future has become one of the principle frames through which environmental change is understood today, and our ecological expectations matter as much as, if not more than, present conditions in the planning and management of socio-environmental systems. All so-called "environmental policy" is thus a political statement about time. Whether it is the local and informal arrangements for the sustainability of a common property fishery, or the formal and state-enforced management of nuclear waste storage, it is always about how fast change is happening and in what direction, what temporal horizon should be meaningful to society, what form the future will take, and how we should care about it.

SEE ALSO: Anthropology; History, Environmental; Industrialization; Intergenerational Equity; Nature, Social Construction of.

BIBLIOGRAPHY. Robert Ayres, "Limits to the Growth Paradigm," *Ecological Economics* (v.19, 1996); Frederick Buell, *From Apocalypse to Way of Life* (Routledge, 2004); Gregg Easterbrook, *A Moment on Earth: The Coming Age of Environmental Optimism* (Viking, 1995); James Fairhead and Melissa Leach, "False Forest History, Complicit Social Analysis: Rethinking Some West African Environmental Narratives," *World Development* (v.23, 1998); David Harvey, *Justice, Nature and the Geography of Difference* (Blackwell, 1996); Claude Lévi-Strauss, *Structural Anthropology* (Basic Books, 1963); Susan Mann and James Dickinson, "Obstacles to the Development of a Capitalist Agriculture," *Journal of Peasant Studies* (v.5, 1978); Joseph Masco, "Mutant Ecologies: Radioactive Life in Post-Cold War New Mexico," *Cultural Anthropology* (v.19, 2004); Bill McKibben, *The End of Nature* (Random House, 1989); W. Scott Prudham, *Knock on Wood* (Routledge, 2004).

Geoff Mann
Simon Fraser University

Titicaca, Lake

LAKE TITICACA IS the highest commercially navigable lake on Earth located at 2.4 miles (3.8 kilometers) above sea level. The lake is located in the high Andean Altiplano on the border of Bolivia, with a port at Guaqui, and Peru, with ports at Puno and Huancane. It is the largest lake in South America and has a surface area of 3,205 square miles (8,300 square kilometers) with a length of about 121 miles (195 kilometers) and an average width of about 31 miles (50 kilometers). Lake Titicaca is a deep lake with a maximum depth of 922 feet (281 meters) and an average of 351 feet (107 meters). The mean water temperature of the lake remains about 51.8 degrees F (11 degrees C).

Lake Titicaca is fed by rainfall and water from numerous streams and rivers that originate in the snow-capped contiguous mountain ranges. Lake Titicaca is drained by the Desaguadero River, which flows south throughout Bolivia. The lake has more than 40 islands, including human made floating Uros islands, some of which are very densely populated. The extenuating effect of Lake Titicaca to the surrounding climate coupled with its water allows for irrigation of such crops as potatoes, barley, and maize. Trout farming and herding of alpacas and llamas are also common agriculture practices.

The basin of Lake Titicaca is one of the few intact and undisturbed areas in the Americas where indigenous societies and cultures developed. The Urus people, an indigenous ethnic group that appeared on earth about 8000 B.C.E. and today is extinct, originally settled the territories of Lake Titicaca basin. Later, Lake Titicaca was conquered by Aymara warlords, Quechuas of the Inca Empire who considered the lake a sacred place, and finally by the Spanish conquerors. The banks of the lake territories were dominated by the culture of Tiahuanaco (Tiwanaku) people whose descendants went north and founded the Inca kingdom after their kingdom was destroyed. As evidence of a flourishing ancient civilization, the Tiahuanaco culture left behind ruins of megalithic constructions, statues, and a temple to the sun. The territory of the Lake Titicaca basin is one of the very few places in the world where the craft of balsas (reed boats) building, which was practiced by the Urus people, still exists.

Poverty remains one the core causes of many social problems experienced by the population of the Lake Titicaca basin. The poor condition of education and health care systems in the Lake Titicaca region are the major socio-economic characteristics of living conditions. Major health problems are linked to the problems of malnutrition, lack of sanitation, and ecosystem fragility with regard to flooding. The major economic activities of the population inhabiting the lake basin are focused on food production activities, agriculture, and cattle herding. Only small-scale subsistence agriculture is possible due to rural property fragmentation land reforms, limited machinery and fertilizer supplies, natural drought, floods, and frosts. Providing very low crop yields, subsistence agriculture encourages over-harvesting and overexploitation of the fertile lands of the lake basin causing soil degradation and further environmental problems. Irrational use and mismanagement of natural resources has caused

serious organic and bacteriological contamination, particularly poor waste disposal, and mining of the important urban cores in the basin.

SEE ALSO: Basin; Bolivia; Indigenous Peoples; Lakes; Peru; Poverty.

BIBLIOGRAPHY: Clark L. Erickson, "*The Lake Titicaca Basin: A Pre-Columbian Built Landscape*," ccat .sas. upenn.edu (cited December 2006); Charles Stanish and Brian Bauer, *Archaeological Research on the Islands of the Sun and Moon, Lake Titicaca, Bolivia: Final Results of the Proyecto Tiksi Kjarka* (Cotsen Institute of Archaeology at UCLA, 2004); Charles Stanish, Amanda Cohen, and Mark Aldenderfer, *Advances in Titicaca Basin Archaeology* (Cotsen Institute of Archaeology at UCLA, 2005); United Nations Environmental, Scientific, and Cultural Organization (UNESCO), "Water: A Shared Responsibility, Lake Titicaca Basin," www.unesco.org (cited December 2006); UNESCO World Water Assessment Program, "Lake Titicaca Basin, Bolivia," www. unesco.org (cited December 2006).

Jahan Kariyeva
University of Arizona

Tobacco

TOBACCO IS INDIGENOUS to the Americas and was unknown to Europeans until the late 16th century. Its use among Native American peoples was widespread by this time; it was consumed largely for medicinal and religious purposes. Successfully grown by early settlers and exported to Europe, the plant (*Nicotiana spp., L.*) became a crucial crop for the pre-revolutionary American colonies, which relied heavily on slave labor in cultivation and processing.

Today, tobacco is grown in more than 100 countries and most tobacco is used for smoking. Tobacco is an essential ingredient for cigarettes, pipes, cigars, hand-rolling tobacco, bidis, and *kretek* cigarettes. Cigarettes account for the largest share of manufactured tobacco products in the world—96 percent of total sales. Except for chewing tobacco in India, and possibly *kreteks* in Indonesia, cigarettes are the most common method of consuming tobacco.

China is the world's leading producer. According to data provided by the World Health Organization (WHO), worldwide over 15 billion cigarettes are smoked every day. The global tobacco industry is dominated by three large multinationals: Altria Group (formerly Philip Morris) based in the United States, Japan Tobacco, which is government-owned and controls 75 percent of the Japanese market, and British American Tobacco (BAT) based in the United Kingdom. Tobacco is one of the United States's oldest and most profitable industries, but the tobacco market has been hit by price increases, higher state taxes, increased consumer awareness of health risks, and hefty litigation costs. Smoking has been linked to many types of cancer by medical research institutions. For years, the tobacco industry presented studies of its own in attempts to counter growing scientific knowledge about the additives and adverse health effects of cigarettes. Efforts to curtail tobacco use have increased throughout the world as many countries continue to tax tobacco heavily and restrict its use in public facilities.

Employment in the tobacco industry has been declining in developed countries as a result of the introduction of new technologies and national and international tobacco control policies. In developing countries, on the other hand, tobacco consumption and employment in the tobacco industry have been on the rise.

In some developed countries, consumers spend more on tobacco than they do on alcoholic beverages; however, the popularity of smoking is in decline. The main factors driving the long-term decline include: Concerns relating to the impact of smoking on health, the increasing view that smoking is an anti-social habit, growing restrictions governing where individuals can smoke and how companies can market their products, and the rising cost of legally bought tobacco. Consumers are increasingly turning to economy brands and smuggled tobacco—contraband products and those legally bought abroad account for 31 percent of sales—in response to taxation increases. Cigarettes are a legal, but controversial product.

Several Western European countries have increased taxes on cigarettes far more aggressively than the United States to discourage smoking, and they have imposed greater restrictions on cigarette

advertising, but have been less aggressive in prohibiting smoking from workplaces and restaurants.

Only in the United States has litigation against tobacco companies become an important feature of national tobacco control efforts. The U.S. Department of Justice is pursuing a case against the industry, citing 50 years of evidence it claims points to a cover-up of the health risks associated with smoking. Smokers stricken with cancer and other smoking-related health problems have also tried to pool their complaints together in large class-action lawsuits. Often, the courts frown upon such tactics; however, individuals have fared much better, but face lengthy appeals from the tobacco giants.

A $3 billion California award against Philip Morris in 2001 was among the top 10 jury verdicts in the country. However, the U.S. Department of Justice's case against the industry has weakened permanently, and awaits appeals (elimination of a $280 billion disgorgement claim). Other significant triumphs for big tobacco occurred in late 2005, when the Illinois Supreme Court dismissed the appeal of the Price "lights" class-action case. The third major problem, the review of the $145 billion Engle verdict, resulted in a dismissal by the Florida Supreme Court in July 2006.

Since late 1998, when cigarette manufacturers raised prices sharply as a consequence of the Master Settlement Agreement (MSA), deep-discount cigarette producers saw their market share increase from about two percent in 1998, to over 13 percent in 2003, with about a 45 percent price discount to premium brands. Increasing cigarette prices also encouraged purchase of cigarettes over the internet, sacrificing convenience for cost savings. Federal lawmakers contended that these internet stores were clear tax evasions: The Jenkins Act requires that both the retailer and consumer report online purchases to aid in tax collection. State governments aware of the loss in tax revenue, and retailers feeling the competition, have pushed for stricter regulation and greater enforcement.

The U.S. market is dominated by four main manufacturers known as Big Tobacco: Altria Group, which sells approximately half of the nearly 500 billion cigarettes sold in the United States, Reynolds American Inc., Loews subsidiary Lorillard Tobacco Company (a subsidiary of the Carolina Group), and Vector Group's Liggett unit. In the United States, people are quitting smoking in great numbers, while restrictions on advertising impede manufacturers' ability to attract new smokers.

China, with some 25 percent of the world's 1.2 billion smokers, is the big prize. Government-

Philip Morris

Philip Morris was a tobacconist who ran a business in Bond Street, London, selling Havana cigars (known in those days as "seegars") and pipe tobacco from Virginia. Many British gentlemen took snuff, and Philip Morris ran a discrete business until the Crimean War of 1854–56. During the war, many British soldiers had been based in Turkey and came across cigarettes there; they began asking Morris for them. Philip Morris started producing his own cigarettes, which he called Oxford and Cambridge Blues, and later called another brand the Oxford Ovals. However, the production of cigarettes was slow with no more than 1,500 or 2,000 produced by a single roller each day. Philip Morris stressed in his advertisements that he only used the best paper, the cleanest factory conditions, and the "purest aromatic tobacco," with a fine cork tip to prevent the cigarette from sticking to the mouth.

Philip Morris died in 1873 at the age of 37, and his widow Margaret continued running the business along with Philip's younger brother Leopold. The business grew with the patronage of Prince Albert, and Leopold bought out his sister-in-law in 1880, running it with Joseph Grunebaum. When the company was floated on the stock market soon afterwards for £60,000, the public offering was oversubscribed six times. In 1894, however, owing to Leopold Morris running up large debts, the company was in the hands of creditors and was then sold to another company that, in 1901, helped form Imperial Tobacco. Philip Morris, which in 2003 changed its name to Altria Group, Inc., is now one of the largest tobacco companies in the world.

owned China National Tobacco, the world's largest tobacco producer, primarily operates in the domestic market. The major tobacco companies have signed licensing agreements with Chinese partners to distribute their brands in the Chinese market. A country with a largely restricted market and laws against tobacco advertising, China still has 300 million smokers and four times the consumption rate of the number two world market, the United States, and is just about the only major international market that is growing. Tobacco contributes a tenth of all tax revenues in China.

Despite the health problems, lawsuits, and rising prices associated with cigarettes, tobacco companies still make profits. Altria Group, the U.S. and global tobacco leader, grew revenues by 17 percent in 2003 as economies in the United States and abroad grew. BAT, number two in the world, also held its own with about 15 percent growth in sales. Tobacco manufacturers are increasingly focusing activities on developing countries, which tend to have less stringent health and advertising regulations, and where the potential for brand development remains significant.

SEE ALSO: Cash Crop; China; Disease; Drugs; Smoking.

BIBLIOGRAPHY. *Economics of Tobacco Control, Curbing the Epidemic: Governments and the Economics of Tobacco Control* (World Bank, 1999); Judith Mackay and Michael Eriksen, *The Tobacco Atlas* (World Health Organization, 2002); G. Van Liemt, *The World Tobacco Industry: Trends and Prospects* (International Labor Office, 2002).

Alfredo Manuel Coelho
UMR MOISA Agro, Montpellier, France

Togo

FORMERLY FRENCH TOGOLAND, the Togolese Republic won its independence from France in 1960. Togo was governed by military rule for the next several decades. The government has repeatedly been accused of human rights violations, and the political situation remains unstable. Because of the accusations, most bilateral and multilateral aid to Togo is frozen, although the European Union has resumed some aid in exchange for promises of political reform. More than 46 percent of Togo's land area is arable, and the 65 percent of the labor force that is engaged in the agricultural sector is employed in both commercial and subsistence agriculture. Nevertheless, some basic foods are imported. Cocoa, coffee, and cotton are the chief export crops, generating around 40 percent of the Gross Domestic Product. Togo ranks fourth in the world in phosphate production. Other natural resources include limestone and marble. With a per capita income of $1,700, Togo ranks 191st in world incomes. Almost a third of the population lives below the national poverty line, and over a fourth of Togolese are seriously undernourished. The United Nations Development Programme's Human Development Reports rank Togo 143 of 232 countries on overall quality-of-life issues.

Bordering on the Bight of Benin in the Atlantic Ocean, Togo has a 56-kilometer coastline and 2,400 square kilometers of inland water resources. Togo shares land borders with Benin, Burkina Faso, and Ghana. Northern lands are comprised of gently rolling savanna that gives way to hills in central Togo and to plateau in the south. The coastal plain contains extensive marshes and lagoons. Elevations range from sea level to 986 meters at Mont Agou. The length of Togo extends for 317 miles, allowing it to stretch through six distinct geographic zones. The tropical climate is hot and humid in the south and semiarid in the north. Togo is prone to periodic droughts, and the north experiences the harmattan, a hot, dry, dust-laden wind that accelerates the pace of environmental damage and reduces visibility in the winter months.

Togo's population of 5,548,702 is at great risk for the environmental health hazards that go hand-in-hand with poverty and an unstable political system. One of the major threats comes from the 4.1 percent adult prevalence rate for HIV/AIDS. Some 110,000 Togolese have this disease, and another 10,000 have died with it since 2003. Only 35 percent of rural residents and 51 percent of all Togolese have sustained access to safe drinking water. In rural areas, only 17 percent have access to

improved sanitation, as compared to 34 percent of all Togolese. Consequently, the population has a very high risk of contracting food and waterborne disease that include bacterial and protozoal diarrhea, hepatitis A, and typhoid fever, the respiratory disease meningococcal meningitis, and the water contact disease schistosomiasis. In some areas, there is a high risk of contracting vectorborne diseases such as malaria and yellow fever.

Because of environmental health factors, the Togolese have a lower-than-expected life span (57.42 years) and growth rate (2.72 percent), and higher-than-expected infant mortality (60.63 deaths per 1,000 live births) and death (9.83 deaths/1,000 population) rates. The low literacy rate (60.9 percent), particularly among women (46.9), contributes to the high fertility rate (5.4 children per female) and adds to the difficulty of disseminating information on birth control and disease prevention.

At one time, much of Togo was covered with dense rain forests. The Togolese have engaged in slash-and-burn agricultural tactics, however, in addition to cutting down trees for fuel and selling woods such as acajo, sipo, and aybe for export, with the result that deforestation of the rain forest is occurring at a rate of 3.4 percent per year. Extensive water pollution is endangering health and threatening the fishing industry. Urban areas are experiencing elevated levels of air pollution, in large part because of the extensive use of so-called taximotos that ferry people around cities such as Lome, the capital of Togo. Solid waste management is a major issue in both rural and urban areas.

In 2006, scientists at Yale University ranked Togo 103 of 132 countries on environmental performance, in line with the relevant income and geographic groups. The overall ranking was reduced by the low score on environmental health. Existing rain forests have been reduced to river valleys and small sections of the Atakora Mountains, even though the government has protected nearly eight percent of land area. Of 196 identified mammal species, nine are endangered; however, none of the 117 bird species are threatened with extinction.

Although Togo established an environmental framework with the Environmental Code of 1988, environmentalism has not always been a priority with the Togolese government. The Minister of Environment and Tourism and the Minister of Rural Development bear the major responsibility for implementing and enforcing Togo's environmental laws and regulations, which are focused on: Sustainable development through reinforcement of legal and environmental institutions; enhancing environmental education, communication, training, and research; eradicating poverty; and checking pollution. Two of the major policy goals of the Togolese government are designed to provide 100 percent access to safe drinking water and improve sustained access to sanitation in the near future.

Togo participates in the following international agreements on the environment: Biodiversity, Climate Change, Climate Change–Kyoto Protocol, Desertification, Endangered Species, Law of the Sea, Ozone Layer Protection, Ship Pollution, Tropical Timber 83, Tropical Timber 94, and Wetlands.

SEE ALSO: Deforestation; Poverty; Rain Forests; Waste, Solid.

BIBLIOGRAPHY. Central Intelligence Agency, "Togo," *World Factbook*, www.cia.gov (cited April 2006); Timothy Doyle, *Environmental Movements in Minority and Majority Worlds: A Global Perspective* (Rutgers University Press, 2005); Kevin Hillstrom and Laurie Collier Hillstrom, *Africa and the Middle East: A Continental Overview of Environmental Issues* (ABC-CLIO, 2003); Valentine Udoh James, *Africa's Ecology: Sustaining the Biological and Environmental Diversity of a Continent* (McFarland, 1993); United Nations Development Programme, "Human Development Report: Togo," hdr.undp.org (cited April 2006); World Bank, "Togo," lnweb18.worldbank.org (cited April 2006); Yale University, "Pilot 2006 Environmental Performance Index," www.yale.edu/epi (cited April 2006).

Elizabeth Purdy, Ph.D.
Independent Scholar

Tomato

TOMATO (*LYCOPERSICON ESCULENTUM*) is a member of the genus *Lycopersicon* and was categorized thus by the Swedish botanist Linnaeus. Its

The tomato made history as the first crop plant to be modified with modern genetic engineering techniques.

botanical name is literally translated from ancient Greek as "wolf peach," which reflects the once widely held belief that the tomato was poisonous. In contrast, *esculentum* means "edible." The genus *Lycopersicon* is in the Solanaceae family of plants, which contains several species of plants of food or ornamental value, including the potato, tobacco, eggplant (aubergine), tamarillo, hot and sweet peppers (capsicum), Cape gooseberry, ground cherry, and various nightshades. The tomato is a vine that bears round or oval fleshy fruits with a high juice content. It is an annual plant whose stem grows between three to 10 feet (one to three meters). The stem is not self-supporting, so it climbs up or trails along neighboring plants in the wild or along supports when cultivated.

Precisely where the tomato was first selected as a crop plant is unknown, but it may have been the coastal Andes of Peru-Ecuador-Bolivia, a region characterized by a high diversity of tomato's wild relatives, as is also the case for tobacco and potato. It was sufficiently useful to be introduced into Mesoamerica, where domesticated tomato was a component of the food resource. Indeed, the word *tomato* derives from the Aztec word *xitomatl.* It was being cultivated in Mexico when first encountered by European colonists in the early 1500s and there is written reference from the 1530s to recipes that include tomatoes mixed with chiles. Spanish colonists probably introduced it to Florida in the mid-1500s, from where tomatoes spread along the eastern seaboard. Portuguese explorers introduced tomato to west Africa, and Spanish explorers brought seeds to Europe, where the plants flourished in Mediterranean environments, producing irregularly shaped and rough-skinned fruits.

The first known appearance in written records in Europe dates to 1554. Tomatoes became known as *pome dei Moro* (Moor's apple) and later as *poma Peruviana, pomme d'amour,* or in Italy as *pom d'oro.* They were not widely embraced outside southern Europe as suspicions about possible poison persisted, mainly due to recognized botanical links with the nightshades. The first tomato plants were grown in England in the 1590s, mainly for ornamental purposes, until the mid-1700s when tomato first began to appear in British cookbooks. Colonists from England reintroduced it to the United States, and although grown in the 1780s by estate owners such as Thomas Jefferson, it was not until the early 1800s that tomatoes were first consumed in the United States, beginning in the southern states and spreading to northern states by the 1850s. These fruits probably looked like the cherry tomatoes available today.

Today, the tomato is grown worldwide; almost 11.1 million acres (4.5 million hectares) are planted worldwide, generating an annual yield of more than 265 billion pounds (120 million metric tons). Numerous tomato types are now produced from a wide range of cultivars that have been bred for specific properties such as flavor and shape; red varieties dominate but yellow and orange varieties are also available. Nutritionally, tomatoes are about 94 percent water and are a low-calorie food; they are rich in vitamins A and C, calcium, and fiber and are a source of the antioxidant lycopene. Apart from being marketed and consumed as fresh salad or salsa ingredients, tomatoes are canned, processed to produce paste used in many prepared foods such as soups and pasta and meat sauces, and are also used to produce juice, jams, and chutneys.

The tomato made history by being the first crop plant to be modified using modern genetic engineering techniques involving the manipulation of plant deoxyribonucleic acid (DNA). It was modified to

enhance flavor, which gave rise to the name Flavr Savr. It was produced in 1994 and sold to the general public in the United States and United Kingdom as tomato paste. No adverse effects of its consumption have been recorded, but lack of interest prompted its withdrawal.

SEE ALSO: Biotechnology; Crop Plants; Gardens; Genetically Modified Organisms (GMOs); Parasites.

BIBLIOGRAPHY. Mark Harvey, Stephen Quilley, and Huw Beynon, *Exploring the Tomato: Transformations of Nature, Society and Economy* (Edward Elgar, 2004); Belinda Martineaux, *First Fruit: The Creation of the Flavr Savr Tomato and the Birth of Biotech Food* (McGraw-Hill, 2002); Judith Sumner, *American Household Botany. A History of Useful Plants 1620–1900* (Timber Press, 2004).

A.M. Mannion
University of Reading, England

Topographic Maps

A TOPOGRAPHIC MAP is a depiction of the earth's landscape, displaying elevation and selected natural and human features. The map portrays elevation as contour lines, lines that connect points of equal elevation, the natural features of hydrology and vegetation, and a variety of cultural characteristics such as roads, buildings, or cemeteries. Topographic maps are generally produced at different scales, with corresponding variations in the configuration of contours, natural and human features and labels to make the map legible.

Landscape portraits depicting relief have been used for over 2,000 years. Beginning in the Middle Ages and continuing into the 1800s, hachuring was used to illustrate slope, using lines drawn downhill to illustrate steep or low relief. In the early to mid-1800s, the French were the first to use contour lines for elevation. To begin, elevations at known locations are collected using surveying techniques. Contours lines are formed by connecting points of equal elevation, interpolating the elevation values between known point heights creates the equal elevation sites.

The topographic map depicts a specific portion of the earth's surface based on the map scale. For instance, the 1:24,000 scale map covers an area of 7.5 minutes of latitude by 7.5 minutes of longitude, an area of approximately 57 square miles (147 square kilometers), while the 1:100,000 scale map covers 30 minutes of latitude by 1 degree of longitude and an area of approximately 1,805 square miles (4,675 square kilometers). The 1:250,000 scale map represents an area of 1 degree of latitude by 2 degrees of longitude, an area of almost 7,845 square miles (20,320 square kilometers). The areal coverage will increase with latitudes closer to the equator or decrease with latitudes closer to the polar regions because of the convergence of longitude lines.

The topographic map presents the terrain as a series of lines depicting levels of constant elevation. Each line represents a set height above mean sea level (MSL). Most maps will have a standard contour interval—such as 10, 20, or 40 feet—by which the elevation will increase (upslope) or decrease (downslope) from one contour to another. Index contour lines are labeled with the elevation in either feet or meters, depending on the scale. In addition to the contour lines, the topographic map will display spot elevations, hydrologic features (streams, lakes, ponds), vegetation (green areas), select cultural features and several different coordinate systems (latitude/longitude, UTM, SPC, and PLSS). The amount of additional information illustrated on the map will depend on the map scale, with the primary purpose not to display everything but to present a visually-readable map of the terrain with additional information for locational reference.

To begin an interpretation and analysis of a topographic map, the map scale, location of the mapped area, and the contour interval has to be known. There are five basic rules in topographic map interpretation. First, any location on the same contour line will have the same elevation. Second, contour lines will never cross each other. Third, generally speaking, a move to an adjacent contour line is a change in elevation either an increase or a decrease. Fourth, the closer the contour lines are together the more rapid the change in elevation and the steeper the slope, conversely, the farther apart the contour lines the less slope or the flatter the terrain. Fifth, contour lines crossing a stream or drainage channel

will form a V-shape, with the apex pointing uphill or upstream.

Today in the United States, the U.S. Geological Survey produces over 54,000 topographic maps at a scale of 1:24,000 for the conterminous United States and Hawaii, and a scale of 1:63,360 for Alaska. In addition, the entire United States is covered at scales of 1:100,000 and 1:250,000. Currently, it requires the integration of field surveying for horizontal and vertical accuracy and control, the use of aerial photography and analysis for contour mapping, printing techniques to produce the topographic map, and computer analysis and databasing for storage and reproduction. Topographic maps are now produced as a paper copy or a computer-compatible file, to be downloaded into mapping software or into global positioning system units for visualization and location. The use of computers for computation and illustration also allows the cartographer to combine contour lines with different types of color-shading and light enhancement to create a 3-dimensional perspective, emphasizing the landscape relief.

SEE ALSO: Global Positioning Systems (GPS); Latitude; Longitude; Maps.

BIBLIOGRAPHY. A.J. Kimerling, P.C. Muehrcke, and J.O. Muehrcke, *Map Use: Reading, Analysis, and Interpretation* (JP Publications, 2005); V.C. Miller and M.E. Westerback, *Interpretation of Topographic* Maps (Merrill Publishing, 1989); M.M. Thompson, *Maps for America: Cartographic Products of the U.S. Geological Survey and Others* (U.S. Government Printing Office, 1979).

William J. Gribb
University of Wyoming

Tornadoes

A TORNADO—also popularly known as a twister—is an atmospheric phenomenon associated with a supercell thunderstorm or hurricane. It consists of a small rapidly and violently rotating column of air—or vortex—extending continuously from a convective cumuliform cloud to the ground. When the vortex is spinning but not touching the ground, it is called a funnel cloud, which eventually may extend to the ground evolving into a tornado. It becomes clearly visible in daylight as a funnel or tube cloud when it carries water vapor and debris lifted from the ground. It may be thin and rope-shaped in the case of weak tornadoes with speeds below 175 kilometers per hour (110 miles per hour). Sometimes the funnel is not visible except by signs such as whirling debris on the ground. A prolonged roar and hail or heavy rain happen during the event.

Most tornadoes are associated with rotating and long-lasting supercell thunderstorms. Tornado formation starts with a vortex in the base of the storm cloud, out of the wall cloud. Next, an organizing phase follows when wind intensity increases and the vortex extends to the ground. In its mature stage, the tornado reaches its maximum width and speed. After that it weakens, dimensions are reduced, and it adopts a rope-like form.

Tornadoes can last seconds or hours. Most tornadoes last from one to 20 minutes; however, some have been observed to last hours. The vortex may touch ground several times in different locations. The vortex has a diameter of 20–100 meters (20–100 yards) and travels at a translational average speed of 50–65 kilometers per hour (30–40 miles per hour), reaching maxima of 115 kilometers per hour (70 miles per hour), with a rotational speed of 480 kilometers per hour (300 miles per hour). Forward speed is not the only factor in damage; lifetime also contributes, as slow-moving tornadoes may be more dangerous than fast-moving ones. Maximum rotational speed is developed at the edge, decreasing to the center, so that major destruction takes place where rotational and translational speeds sum up. Indirect measurements indicate there is a pressure drop at the center of the tornado. Vortex rotation is commonly counterclockwise in the Northern Hemisphere. The average path length is 8 kilometers (5 miles) although some tornadoes have traveled for 100 miles, and the average path width is 300–400 meters (300–400 yards), and some have covered up to a mile.

Seventy-four percent of tornadoes are in the F0–F1 range of strength level, the weak class, while less frequent, violent tornadoes cause 68 percent of fatalities. They are most likely to occur in the afternoon and move from southwest to northeast.

A single tornado can develop various smaller vortices, known as subvortices or suction vortices with higher speeds. A sequence of continuous tornados along a line of storms is called a tornado outbreak. When the tornado happens over water it is called a waterspout.

Tornadoes represent a major local hazard causing notable destruction, loss of lives, and injuries. Houses collapse, structures are uprooted, and pieces of debris become projectiles. About 1,000 tornadoes are reported every year across the United States, versus 30–50 in the United Kingdom. The number of tornadoes registered in the United States has increased with the implementation of the Doppler Radar Network by the National Weather Service, particularly F0 tornadoes, many of which were not formerly detected.

Tornadoes are frequent in central North America, including: The Canadian central provinces of Alberta, Saskatchewan, and Manitoba; northern Argentina; western and central Europe, South Africa; and eastern and southwestern Australia. About one-fourth of all significant tornadoes occur in Tornado Alley in the Central Plains region of the United States, which includes parts of Texas, Oklahoma, Kansas, Colorado, Nebraska, Iowa, South Dakota, and Minnesota. In the southern states peak tornado season is spring, while in the northern states it is summer, with maximum frequency in May–June.

The most deadly tornadoes happen, however, in areas where they are less frequent, particularly the southeast. In the period of 1950–99, the national year average was 89 deaths. The three deadliest tornadoes were the Tri-State (MO/IL/IN) tornado outbreak on March 18, 1925, which killed 689 people; Natchez, Minnesota, on May 6, 1840, with 317 victims; and St. Louis, Missouri, on May 27, 1896, causing 255 deaths. The three costliest tornadoes happened in Omaha, Nebraska, on May 6, 1975, with an estimated damage of $1.132 billion; Wichita Falls, Texas, on April 10, 1979, which reached $840 million; and Lubbock, Texas, on May 11, 1970, which caused damage of $530 million.

The F-scale (or Fujita scale) was proposed by Tetsuya "Theodor" Fujita in 1971 to categorize the intensity of tornadoes based on the structural damage caused to man-made structures, estimated once the tornado has passed. It comprises six categories, from F0 to F5, although a theoretical maximum F12 tornado is possible. From F0 to F1 a tornado is considered to be weak, strong if in the range F2–F3, and violent from categories F4 to F5. Although subjective in damage assessment, which causes overestimation of wind speeds, the F-scale has widespread use in the United States after being accepted as the official classification system.

Despite the limitations, it was decided to maintain and improve the scale in order to provide continuity to historical tornado records. An enhanced Fujita scale (EF-Scale) was developed for use in the United States after February 2007. Twenty-eight indicators are used and the degree of damage estimated up to eight levels, associating them to upper and lower wind speeds.

As damage is not necessarily associated to wind speed, however, Terence Meaden proposed the Tornado Intensity Scale in 1972, relating the levels to the well-established Beaufort wind intensity scale. As the speed can be measured directly—or, better, remotely estimated—this allows the determination of intensity even if the tornado causes no damage. The scale ranges from T0 to a maximum T10, each level representing a range of windspeeds. From T0 to T3 tornadoes are considered to be weak, from T4 to T7 are strong tornadoes and from T8 to T10 are violent tornadoes; a degree higher than T10 is possible.

U.S. emergency administration issues two differentiated levels of risk to alert population. A tornado watch indicates there is a high probability of a tornado in the area and recommends remaining alert to future evolution. A tornado warning indicates a tornado has been sighted or detected by radar in the area and recommends taking shelter in pre-designated places of safety.

SEE ALSO: Hazards; Disasters; Thunderstorms; United States, Central South; Weather.

BIBLIOGRAPHY. Howard B. Bluestein, *Tornado Alley* (Oxford University Press, 1999); Thomas P. Grazulis, *The Tornado: Nature's Ultimate Windstorm* (University of Oklahoma Press, 2003); Long T. Phan and Emil Simiu, *The Fujita Tornado Intensity Scale: A Critique Based on Observations of the Jarrell Tornado of May 27, 1997* (National Institute of Standards and Technology, 1998);

Philip W. Suckling and Walker S. Ashley, "Spatial and Temporal Characteristics of Tornado Path Direction," *The Professional Geographer* (v.58, 2006).

URBANO FRA PALEO
UNIVERSITY OF EXTREMADURA

Totalitarianism

IN ALMOST ALL modern totalitarian governments, environmental policy is used as an instrument to control society and economic development for the benefit of those in power. Social theorists like Max Weber believe that the roots of totalitarian power can be found in specific environmental conditions of society. These conditions will have a direct impact on rule, necessitating a totalitarian state. Thus, according to this theory, nature itself determined the nature of government.

Alluvial societies (societies based on river irrigation like Egypt) were based on highly-centralized, and most often totalitarian systems of government since the dawn of civilization. The pharaohs ruled Egypt and the great Sumerian and Assyrian kings ruled Mesopotamia for millennia because only a strong ruler could guarantee the effective and efficient maintenance of canals and river irrigation. Recent genetic and archaeological research indicates that throughout most of human evolution, people lived in small bands and roving tribes; it was the growth of river agriculture that transformed this pattern of human existence into vast, centralized civilizations. The totalitarian ruler represented the maintenance of a predictable environmental order. Totalitarian rule is almost always supported by the ruler's actual or even mythical ability to manipulate nature and the environment.

The powers of even the most influential or charismatic rulers of the ancient past were pale in comparison to the potential power of rulers to harness the environment in modern totalitarian states. Yet, even as the methods are different and potentially far more devastating, the objectives of modern totalitarian environmental policy are almost identical to those of ancient regimes: To prove that the totalitarian ruler not only has power over people but power over nature itself, making resistance to the totalitarian system as futile as resistance to nature itself. By taming the Nile River with the Aswan Dam, Gamal Nasser of Egypt not only tamed a mighty river, but he tamed and controlled a society historically dependent on the Nile for its existence. By draining the marshes of Southern Iraq and fundamentally altering an entire ecosystem, Saddam Hussein eliminated much of the resistance from Marsh Arabs who were resisting his rule.

Often, however, modern totalitarian environmental policies have resulted in disastrous consequences. During the Cultural Revolution, Chairman Mao of China commanded peasants to kill all of the country's small birds as they were eating grain and crops. This mass culling of birds, however, only led to an even more massive infestation of insects. Gamal Nasser's Aswan dam has upset the natural balance of flooding, silted up portions of the Nile and has made much of the river undrinkable and dangerous even to touch. The construction of the enormous Three Gorges Dam in China shows that centralized and environmentally risky projects can still be pursued in China's hybrid command capitalist system. The taming of rivers was also one of the major objectives of Stalin's rule. Dams and other massive centralized projects not only allowed the efficient, domestic production of electricity and the centralized control of resources, but they also provided the totalitarian ruler with a great deal of prestige. Unlike the pharaohs and kings of the past who called on the gods to bring down the rains, the modern totalitarian need only flip a switch.

Unlike democratic societies, in which environmental policy is often shaped by popular movements to preserve human welfare, totalitarian systems have little regard for long-term environmental consequences. The primary concern of the totalitarian ruler is how environmental policy can enhance the regime's grip on power.

SEE ALSO: Aswan High Dam; China; Dams; Egypt; Movements, Environmental; Three Gorges Dam.

BIBLIOGRAPHY. Hannah Arendt, *The Origins of Totalitarianism* (London, 1986); Juan Linz, *Totalitarian and Authoritarian Regimes* (London, 2000); Max

Weber, *The Agrarian Sociology of Ancient Civilizations* (London, 1976).

Allen J. Fromherz, Ph.D.
University of St. Andrews

Total Maximum Daily Load (TMDL)

THE PASSAGE OF the federal Clean Water Act in 1972 established the water standard qualities and the total maximum daily load (TMDL) ceiling. The TMDL is intended to maintain the quality standards of water. To do so, TMDL is set to control the amount of pollutants that are allowed to flow into a given water source. The TMDL is calculated by adding the allowed pollutant loads for point sources, nonpoint sources, projected growth, and a margin of safety, resulting in a sum that is the TMDL.

Every state is responsible for its own water quality levels by establishing its own TMDL. However, if the state fails to do so, the U.S. Environmental Protection Agency (EPA) is then responsible to prevent pollution to the water. Because TMDL implications were not clearly defined in the federal regulations, its details continue to evolve and the EPA must adapt in many ways.

The TMDL process starts with identifying the water sources that do not meet the water quality standards. When water that does not meet the standard is identified, the cause of the pollutant for that particular area is investigated. Water sources containing more important or potentially dangerous pollutants are prioritized to be addressed earliest, so that the ones with more minor pollution issues, which often occur naturally, are at the bottom of the priority list. In order for certain areas to pass the state water standards, the TMDL staff must work hard to control the amount of the pollutants allowed into the water.

More than 40 percent of the water from the total U.S. watersheds did not meet the federal quality standards, leading the EPA to take actions to improve the TMDL programs. In the 1992 TMDL regulations established by the EPA demanded that the states and authorized groups publicly list waters that were polluted. Those water sources had to meet the standards in order to be removed from the list. There is a two-year listing cycle, and the authorities in charge of a particular water source are required to submit the list of polluted waters on the first of April of every even year. To avoid excluding certain polluted waters, the EPA required that the authorities provide a good reason to not include certain waters or remove waters from the list of polluted waters. The authorities in charge have 30 days to provide the list of polluted waters to the EPA. If some of them are disapproved by the EPA, the EPA has to come up with the list within 30 days and have the approval of the public or the list they disapproved from the authorities will be approved.

After an approval for a TMDL, all the authorities and responsible organizations must regularly update the progress of the process. The evaluation of a TMDL is performed by monitoring the loading of pollutants, keeping track of the controls of the pollutants, assessing water qualities, and then reevaluating the TMDL for water standards.

The priority of cleaning a certain watershed is determined not by the percentage of pollution, but by the priorities set on how the water is being used. When the water directly affects people's health, the source is ranked higher. Some of these risks include water used for fishing, swimming, and drinking water.

In 1997 the EPA established updated guidelines for the TMDL program to address issues raised as the program itself developed. In the new program, there are some recommendations that were also included to help address these issues. After the authorities establish the list of polluted waters, they normally have to come up with a resolution within eight to 13 years. When the schedules are made for the water sources, there are factors that need to be considered, including the number of segments of the polluted water, the distance of the water that needs cleanup, the number of similarities and differences among these waters, and significance of the threat of the pollutant in the water.

SEE ALSO: Clean Water Act; Drinking Water; Nonpoint Source Pollution; Point Source Pollution; Pollution, Water; Water Quality.

BIBLIOGRAPHY. Center for TMDL and Watershed Studies, tmdl.net (cited May 2006); North Carolina Division of Water Quality, "Modeling and TMDL Unit: The N.C. TMDL Program," h2o.enr.state.nc.us (cited May 2006); State of Maryland Department of Environment, "Total Maximum Daily Loads," www.mde.state.md.us (cited May 2006); U.S. Environmental Protection Agency, "Overview of Current Total Maximum Daily Load-TMDL-Program and Regulations," www.epa.gov (cited May 2006).

Arthur Holst
Widener University

Tourism

TRAVELING TO DISTANT places and lands has been a human activity since people first began to spread over the earth. However, there has always been a difference between visiting, inspections, migrations, business trips, scientific expeditions, pilgrimages, and tourism.

The great improvements in transportation in the 19th century opened the way for enormous numbers of people to go on tours around Europe and beyond for recreation such as visiting spas, or other leisure purposes. At first, touring was affordable mainly to the gentry, but with rising levels of prosperity members of the emerging middle class went touring.

Before World War II, most touring was of the United States, Canada, or in Europe west of the Danube. In the United States and Canada a great deal of tourism was by private automobile. After the war, increasing numbers of people traveled from frigid winters in the north to winter in Florida with an inevitable ecological impact.

Much of the postwar tourism was to Europe by cruise ships, until the advent of trans-Atlantic and then global passenger air travel made almost any place in the world accessible in just a matter of hours. The boom in the mass tourism industry since 1945 has had a significant impact on tourist sites, both historic and natural. Today, from San Francisco to Sydney, Australia or from Alaska to Antarctica, masses of tourists travel over the globe in search of leisure, recreation, or educational experiences. Companies compete for tourist dollars by advertising travel to almost any place in the world at an affordable price for most people. Cruise ships ply the Mediterranean, Baltic, and Caribbean Seas, as well as other waters.

ECOTOURISM

So voluminous has the tourist trade become that "ecotourism" has developed as a form of tourism. Ecotourism or ecological tourism seeks to give travelers on nature tours experiences of nature that do not harm the environment. The goal is to create a benign, sustainable tourism.

Ecotourism may seek volunteers to be part of scientific research on natural areas. It usually takes tourists to places where the cultural heritage or fauna and flora are the main attraction. This may mean being paddled by expert boatmen in *bancas* (traditional dugout canoes) up the Bumbungan River to the Pagsanjan (Magdapio) Falls on southern Luzon island, the Philippines. Or it may mean touring Palawan Island in the Philippines for the rich diversity of species that can be found there.

Ecotourism to Costa Rica features tours that present the extremely rich environment of Costa Rica, which can include tours of active volcanoes. Tourism of the volcanoes in Hawaii and well as of some of the numerous ecological areas in the islands is oriented toward preserving the unique ecosystem.

In the case of wilderness adventures, ecotourism may mean hiking with backpacks or riding horseback into remote areas of the Rocky Mountains or other wild areas of the world. The number of people visiting such areas has grown tremendously and shows no signs of leveling off.

Many of these wilderness adventures may stress personal growth; others may teach new ways to live in harmony with nature. Or they may focus on local cultures or volunteering to preserve areas of cultural or natural interest. Always these programs seek to minimize the impact of traditional tourism. They also seek to protect or encourage the preservation of local cultural heritage areas. To minimize adverse effects on the environment or the traditional culture, the touring program is designed to minimize the impact of the visitors.

To design an ecologically friendly touring program requires an evaluation of the natural environment and the cultural heritage area of the local people. The goal is to ensure hospitality providers have means for recycling and efficient use of water and energy, while creating economic opportunities for local people. Conservation practices that preserve both biological and cultural diversity must be implemented. Sustainability must be sought to prevent heritage or habitat destruction. The jobs created must include jobs for indigenous people; their input is also absolutely necessary and their participation in the management of tourism is essential.

To achieve these goals, the focus is put upon sustainable activities. For example, ecotourism is an issue in the Carpathian Mountains in Romania which are little changed since the Middle Ages. The region is still filled with bears, bison, lynx, wolves, and a variety of other wildlife. A program backed by the United Nations is seeking to promote sustainable tourism there. Balea Lac, Romania, is the site of a wintertime "ice hotel." High in the mountains, it can be reached only by cable car. The cost of building it is low and it melts away in the spring; yet it attracts those willing to pay for a sustainable adventure.

Globally, there are efforts underway to define and describe ways to create environmental tourism for the sake of the planet and for future generations. The use of environmental certificates is probably not sufficient because some tours are to extremely sensitive areas. Some tour companies treat ecotourism as a marketing tool, or as some critics call it, "green-washing." Other critics have pointed out that putting a magnificent hotel in a beautiful landscape does not qualify as environmental tourism; in fact it is just the opposite.

Some suggest the aim should be to make all *tourism more environmentally, economically, and culturally sustainable.*

ENVIRONMENTAL IMPACTS OF TOURISM

Humans can have an enormous environmental impact even outside of extremely sensitive areas. In some places the ecological impact of great numbers of people can be very serious. The influx of tourists to Zion Canyon in Utah's Zion National Park frightened away its population of mountain lions. This allowed the deer population to explode; deer browsing on a great number of plants led to the destruction of cottonwood seedlings. This affected a great many species, including toads and butterflies, in a "trophic cascade" in which most species disappeared. Comparison with nearby areas where humans normally do not go and where mountain lions still prowl showed a balanced ecology.

Tourism has been economically profitable to many areas of the world. The income earned from tourist visits has in many areas provided incentives for developing, managing, and preserving tourist sites. Tourism to environmentally sensitive areas such as the coral reefs in the Florida Keys grows, but so does local concern for protecting such vital resources.

Many nations are now seeing environmental tourism as essential for the preservation of tourist income. In many places tourism is first, second, or third in income generation for a nation's gross domestic product.

SEE ALSO: Beaches; Coral Reefs; Costa Rica; Development; Ecotourism; Galapagos Islands; Globalization; Indigenous Peoples; National Parks; Poverty; Recreation and Recreationalists; Safaris; Sustainability; Transportation; Underdeveloped ("Third") World.

BIBLIOGRAPHY. Brian Garrod and Julie C. Wilson, *Marine Ecotourism: Issues and Experiences* (Multilingual Matters Ltd., 2003); Andrew Holden, *Environment and Tourism* (Taylor & Francis, 2000); Stephen Page and Ross K. Dowling, *Ecotourism* (Longman Group, 2001); Joseph L. Scarpaci, *Plazas and Barrios: Heritage Tour-*

ism and Globalization in the Latin American Centro (University of Arizona Press, 2004); Hellen Vriassoulis and Jan van der Straaten, eds., *Tourism and the Environment: Regional, Economic, Cultural, and Policy Issues* (Kluwer Academic Publishers, 1999).

ANDREW J. WASKEY
DALTON STATE COLLEGE

Toxaphene

TOXAPHENE IS A now-banned insecticide that was previously used extensively across the United States to control insect pests in cotton-growing areas in particular. It is generally encountered as a gas or as a waxy, yellowish solid with an aroma of turpentine. Toxaphene is made up of some 670 separate chemical substances and has an average chemical structure of $C_{10}H_{10}Cl_8$ (it is also known as chlorinated camphene and other names). Although effective in its role as an insecticide, it has also been found to have serious negative impacts on human health including damage to the kidney, lungs, and nervous system; severe exposure to the substance might lead to death. The substance is persistent in the environment and not susceptible to biodegrading. Consequently, although it has been banned completely since 1990 (and in most of the country since 1982), its presence is still regularly found in many parts of the country. It accumulates within the bodies of mammals or fish and so its effects can start to occur years after initial exposure.

Because toxaphene will evaporate from its solid state and only imperfectly dissolves in water, it can remain active in the atmosphere for extended periods. Research indicates its pervasive presence in the Great Lakes area as well as other locations. While in use, its presence in the air was measured at one part per billion and this level has presumably declined in subsequent years. Its presence in drinking water is very rare, although it is most prevalent in those bodies of water in which it was used to eliminate what were considered excess species or numbers of fish. Its presence is greater at the lower levels of such bodies of water, where it tends to collect. However, there is comparatively little information about the impact of mild exposures to the substance or the point at which mild exposure becomes dangerous. Nevertheless, it is classified as a probable carcinogen for humans. Exposure is possible through atmospheric interaction, drinking contaminated water, or eating fish or shellfish that are contaminated. Research suggests that while the substance is detectable over wide areas, few such areas are heavily contaminated to the extent of representing a serious menace to human life.

Toxaphene is an example of the almost indiscriminate use of a substance that subsequently turns out to be dangerous to people and animals and damaging to the environment. It demonstrates the need for a properly rigorous testing regimen prior to the licensing of new chemical substances and the need to monitor their effects in the light of new learning.

SEE ALSO: Carcinogens; Marine Pollution; Pesticides; Pollution, Air; Pollution, Water.

BIBLIOGRAPHY. Agency for Toxic Substances and Disease Registry (ATSDR), "ToxFAQs for Toxaphene" (ATSDR, 2006), www.atsdr.cdc.gov (cited December 2006); Environmental Protection Agency (EPA), "Toxaphene" (EPA, 2000), www.epa.gov (cited December 2006); Jianmin Ma et al., "Tracking Toxaphene in the North American Great Lakes Basin: 2. A Strong Episodic Long-Range Transport Event," *Environmental Science and Technology* (v.39/21, 2005); M.A. Saleh, "Toxaphene: Chemistry, Biochemistry, Toxicity and Environmental Fate," *Review of Environmental Contaminants and Toxicology* (v.118, 1991).

JOHN WALSH
SHINAWATRA UNIVERSITY

Toxics Release Inventory (TRI)

THE TOXICS RELEASE Inventory (TRI), established under the U.S. Emergency Planning and Community Right-to-know Act (EPCRA), is a database produced by the U.S. Environmental Protection Agency (EPA) that tracks chemical releases and waste reported by major industrial facilities.

The EPCRA established the TRI and it requires the industrial facilities to annually report releases of waste into the environment to the EPA.

Not all toxic releases are covered by the EPCRA Act; the categories that qualify to be on the list are determined by whether or not they may damage public health, such the possibility of causing cancer, reproductive defects, or anything harmful to the neurological functions. In addition, the possible effects of toxic releases on the environment and animal life are also addressed.

The EPA has never officially inventoried the number of chemicals produced in the United States that are actually able to meet the TRI's requirements. The number of substances that are currently covered by the TRI is 650, which is equivalent to only 1 percent out of over 75,000 substances that are manufactured in United States.

TRI is based on company self-reporting and companies are held responsible for the accuracies of the reports. There is currently no penalty mechanism for those who provide inaccurate reports. Unfortunately, because TRI has been lenient about the accuracy and exact measurement of waste releases, many companies estimate them rather than actually measure them. The Environmental Integrity Project (EIP) estimated after a study that the companies are inaccurate about their waste release by as much as 15 percent.

The EIP tries to pinpoint the sources of inaccurate data in company reports, and one of the most frequent causes is improper emission monitoring, which has been replaced by the method of estimation. Even though most of the companies use the method of estimation to monitor the waste emissions, there are as few as 4 percent of all the companies who do in fact use the proper accurate emission monitors.

Companies frequently underreport emissions, causing the public to remain unaware of exposure to waste emissions and toxic substances, especially those released into the air. Companies have also made unintentional, though no less egregious, reporting errors, such as putting down the incorrect geographic locations of their company. In demanding more accurate reports, environmental defense associations have worked to correct these errors.

The TRI has expanded significantly since 1987 dramatically increasing the number of substances it covers. Seven additional sectors have been added to broaden industry coverage. Being a crucial tool for community information and advocacy, the TRI has gained a great deal of support nationwide with the hope of holding companies more accountable for chemical waste emissions.

SEE ALSO: Environmental Protection Agency (EPA); Pollution, Air; Pollution, Water; Waste, Solid.

BIBLIOGRAPHY. Louisiana Department of Environmental Quality, www.deq.state.la.us (cited May 2006); U.S. Environmental Protection Agency Toxics Release Inventory Program, www.epa.gov/tri (cited May 2006); U.S. National Library of Medicine, National Institutes of Health, www.nlm.nih.gov (cited May 2006).

Arthur Holst
Widener University

Trade, Fair

IN THE RAPIDLY-EXPANDING alternative trade movement known as fair trade, consumers in more developed regions of the globe subsidize the livelihoods of small-scale, marginalized farmers in poor countries by paying a premium for their goods. This new and more direct ethical relationships between rich consumers and poor producers contributes to the social, economic, and community development of fair trade growers. Fair trade has been characterized as "working in the market but not of it." The multi-million dollar fair trade market has become serious business; it is now defined by its moves into mainstream retailers as much as by its alternative economic model, new markets, and novel development opportunities for third world farmers.

Begun in the 1970s with the importation of handicrafts, fair trade was developed by European and American aid organizations with social justice underpinnings. Early activist groups include Max Havelaar (Netherlands), Oxfam and Traidcraft (United Kingdom [UK]), Ten Thousand Villages (United States), and GEPA (Germany). The idea has long been to develop trading relationships that give poor farmers opportunities to enter markets under

favorable conditions and move out of poverty. Current arguments put forth fair trade as an antidote to the excesses of "free trade" globalization with its control of markets by multinational corporations and wildly fluctuating or plummeting prices for coffee, cocoa, sugar, bananas, and other tropical crops. For example, the world price for coffee is at its lowest in decades, and, of the average price paid at a café, less than 1 percent of the value of a cup of coffee is captured by growers.

As the market has grown, a formalized regulatory system has developed. Currently, the Fairtrade Labelling Organizations International (FLO), created as an umbrella organization of 17 Euro-American labeling groups, maintains the standards for the Fairtrade logo. International production standards work to stimulate demand through consumer trust in the fair trade market.

Generalizing across products, fair trade standards involve: (1) A guaranteed price floor for all commodities—the minimum price for a pound of coffee is $1.21, which covers the cost of production and producers' livelihood support; (2) a "social premium" of 10 percent or more ($.05 for coffee) is tacked on to pay for community development such as new schools; (3) transparent and long-term trading contracts so communities might invest in new production techniques; (4) access to credit to smooth income streams; and (5) shorter supply chains to reduce intermediaries and permit farmers to capture more of the value of commodities.

To participate in fair trade, producers—like the long-standing Mexican coffee cooperative UCIRI—must: (1) Be a democratically-run cooperative and use the fair trade premium to the benefit of members; (2) be committed to improving the environmental conditions of production; and (3) prohibit child and slave labor. New standards have been created for workers on fair trade estate-farms and plantations.

The FLO then certifies that these and other standards are being adhered to by performing audits at each stage in the commodity chain. After this certification, the Fairtrade logo can be used on the packaging of all commodities deemed fairly traded. This logo is important as it demonstrates to consumers the "fairer" conditions under which a particular commodity was produced and differentiates these products on supermarket shelves.

Fair trade contributes to environmental conservation in several ways. First, production standards dictate that all cooperatives must work to resource management plans to encourage environmental conservation. Second, fair trade supports small farmers actively involved in resource conservation through the use of traditional farming methods, such as growing coffee under a canopy of trees. Eighty percent of U.S. fair trade coffee is grown this way (i.e., shade-grown) as it maintains a greater biodiversity of trees, insects, and birds. Third, greater economic resources have permitted many cooperatives to become organically certified (80 percent of U.S. fair trade coffee is organic) and institute sustainable post-harvest techniques. Recent work by the anthropologist Mark Moberg, however, describes how fair trade's specified environmental standards can be difficult to comply with given the poverty and marginality of third world farmers.

Fair trade markets have boomed in the last few years. In 2004, the UK market total for all fair trade products was £140 million ($252 million), with a massive 70 percent increase from the year before. Also, in 2004, the U.S. market for fair trade coffee was $369 million, a 75 percent increase from 2003. Over one million producers in 48 countries worldwide are now growing for fair trade markets. Through the fair trade "difference"—minimum prices combined with the price premium—the U.S. fair trade coffee market alone provided $26 million in additional money to producers in 2004. Many attest to these real economic and allied social benefits as evidence of fair trade's impact beyond its tiny international market share.

While there is little doubt that fair trade markets will continue to expand, some new developments include: The movement of fair trade products into large supermarkets in the United States and the UK; the growing ability of fair trade markets to build-up small-producer capacities to enter into larger commodity markets and leave the niche of fair trade behind; the growth of certified "fair trade towns" and "fair trade universities" in the UK where fair trade products are supplied to citizens and students; and the expansion of fair trade into manufactured products like soccer balls and clothes.

SEE ALSO: Coffee; Ethics; Globalization; Green Consumerism; Markets; Moral Economy; Organic Agriculture; Poverty; Trade, Free; Underdeveloped ("Third") World.

BIBLIOGRAPHY. Fairtrade Labelling Organizations International, www.fairtrade.net (cited November 2006); Mark Moberg, "Fair Trade and Eastern Caribbean Banana Farmers: Rhetoric and Reality in the Anti-Globalization Movement," *Human Ecology* (v.64, 2005); Douglass Murray, Laura Raynolds, and Peter Taylor, *One Cup at a Time: Poverty Alleviation and Fair Trade Coffee in Latin America*, (Fair Trade Research Group, 2003), www.colostate.edu (cited November 2006); Alex Nicholls and Charlotte Opal, *Fair Trade: Market-Driven Ethical Consumption* (SAGE, 2005); Oxfam International, *Mugged: Poverty in Your Coffee Cup*, www.maketradefair.com (cited November 2006).

Michael K. Goodman
King's College, England

Trade, Free

MUCH OF CONTEMPORARY economic policy is based on the notion that free trade raises the standard of living of all countries. This is the position espoused by advocates of neoliberal economic policies such as those of the World Trade Organization (WTO), its precursor, the General Agreement on Treaties and Tariffs (GATT), and the North American Free Trade Agreement (NAFTA). Yet, many environmentalists and antiglobalization activists remain critical of that assertion that free trade benefits the environment, suggesting that free trade actually leads to a race to the bottom in which polluting industries migrate to countries with the lowest standards.

The argument about free trade begins with the assumption that tariffs, taxes, subsidies, and regulations across international boundaries distort the terms of trade. The purported consequences of this leads to economic inefficiencies and imposes artificial costs on the movement of goods. One of the central tenets of free trade is the principle of comparative advantage, a theoretical explanation for why two countries can both benefit from international exchange. The first mention of comparative advantage is in economist Robert Torrens's *Essay on the External Corn Trade*. But the idea is often attributed to the English economic philosopher David Ricardo who sought to explain why both Portugal and England benefitted from their exchange of wine and cloth. Ricardo argued that Portugal could produce wine more cheaply than England. Even though England and Portugal could both produce cloth at similar costs, it was still beneficial for Portugal to produce more wine for trade with England because of benefits from economies of scale.

Since the time of Ricardo, the theory of comparative advantage has come under fire for some of its primary assumptions. For example, Ricardo assumes that transportation costs are negligible. This assumption may hold true in today's economy of subsidized oil, but it certainly does not account for externalities. Ricardo also assumed negligible labor costs because he assumes that there would also be a free market in labor, something that does not hold up in today's economic and political circumstances.

Perhaps the biggest pitfall to following the comparative advantage policy prescription is the narrow, even risky, set of economic enterprises that a country relies on. A good example of this can be found in the Yucatan, Mexico, as told by historian Sterling Evans. In the early 1900s, the Yucatan exported extensive quantities of sisal fiber, derived from an agave plant (*spp. Agave sisalana*) similar to that used to produce tequila, to the United States where it was used to bind wheat. Successive years of bumper wheat crops and a free trade zone in New Orleans made sisal production so profitable that by the 1920s and 30s the Yucatan was the wealthiest Mexican state and the first with electricity. But soon after extensive sisal monoculture plantations were established throughout the region, the combine harvester, which could bind the wheat at harvest, was invented, rendering the fiber useless to the U.S. wheat industry. The sisal plantations were subsequently abandoned and soon the Yucatan was one of the poorest states in Mexico. The low diversity of economic engagements and the dependency on a single industry were to blame.

Before free trade became the dominant economic ideology, it was opposed by mercantilism. The

mercantilists argued that the state should protect national interests through policies that promote protectionism. The protectionism was in the form of tariffs and import restrictions. Since mercantilism rested on an economic base of gold and silver bullion, it provided much of the impetus for early European imperial ambitions. Mercantilism was the dominant economic ideology of the 16th–18th centuries until it was supplanted by the laissez-faire economic policies of free trade proponents such as the physiocrats and later by the teachings of Adam Smith and Ricardo.

Often, nations will enact trade restrictions in order to improve the health of its citizenry, yet they will still be accused of protectionism, creating a conflict between the autonomy to protect human health and the global goal of free and unrestricted trade. The most disconcerting of these conflicts culminated with the two 19th-century Opium Wars. The Chinese emperor had completely restricted the import of opium as the nation was suffering from an addiction epidemic. Subsequently, British gunboats surrounded Chinese ports demanding that they open their markets to British imports of opium because the British had a considerable trade imbalance with China.

Often environmental regulations are considered barriers to free trade. Some rulings at the WTO have made this explicit. For example, the United States banned the import of Mexican tuna caught with purse seine nets because it inadvertently killed significant numbers of dolphins; the ban was justified by the U.S. Marine Mammal Protection Act. However, a GATT tribunal ruled that all like products should receive like treatment and that how a product was harvested could not be considered in the determination of likeness. Another case involved an amendment to the Endangered Species Act to include a ban on shrimp imported from countries that do not require turtle excluder devices. After suit was brought by Malaysia, Thailand, Pakistan, and India, the WTO dispute settlement tribunal reasoned that the requirement was excessive and an illegal barrier to trade.

More disconcerting to environmentalists are the investor rights provisions of free trade agreements that offer an opportunity for private companies to sue nations for regulatory takings. Part of the neoliberal free trade policy paradigm is the belief that regulations that affect investment are also seen as barriers to trade. NAFTA's Chapter 11 provides protections to the rights of investors and has been considerably controversial. In one case the Ethyl Corporation of North Carolina sued the Canadian government for banning the fuel additive MMT, a manganese-based fuel additive already banned in the United States. Fearing a NAFTA dispute tribunal would rule that the Canadian government ban was made without enough scientific evidence to support its environmental and health consequences, the Canadian government paid $13 million to the Ethyl Corporation and reversed their ban on MMT. This was the first of several cases under NAFTA's Chapter 11 that had direct implications for environmental policies. A similar, but global, investor rights provision called the Multilateral Agreement on Investment was targeted by activists in 1997 and defeated.

Free trade in cross border capital transactions has trended toward short-term speculation, putting some countries at the mercy of international investment trends. This can greatly affect the relative strength and stability of some national currencies. Opposed to the idea of free trade is the proposed Tobin tax, a tax that aims to discourage short-term currency speculation by taxing individual cross border financial investments in currency. Because the proposed tax is small and done on a per volume basis, long term investments, necessary for maintaining the strength of some currencies, would not be affected. It has been proposed that the United Nations manage the tax fund, which is estimated to generate hundreds of billions of dollars per year and apply it to humanitarian and emergency situations. The Tobin tax is championed by many environmental nongovernmental organizations (NGOs), green movements, and the antiglobalization movement.

One of the significant questions debated is whether or not free trade is a race to the top or a race to the bottom regarding environmental regulations. Critics of free trade's impact on the environment argue that polluting industries will move to poor nations that have weak environmental regulations, creating pollution havens. Proponents of free trade argue that it will result in a ratcheting up of environmental policy, citing the famous Kuznets curve, which argues that environmental conditions improve with increased

national income. Cases have shown both to be true. For example, the regulation of genetically modified organisms in the European Union has led the United States to improve its environmental regulations. On the other hand, in the United States, some polluting industries have moved across the Mexican border out of the Los Angeles Basin in the years subsequent to NAFTA. It might be argued that if the nations involved are on equal footing, it is possible that free trade will lead to a situation where regulations improve. But if the countries are in disparity regarding wealth or environmental protections to begin with, it could lead to a situation where regulations improve.

The negotiation of global free trade agreements has proceeded in a series of rounds since the founding of the GATT after World War II. It was during the Uruguay Round that the WTO was founded as an organization to implement the principles previously held by the GATT. Perhaps the most widely covered trade negotiation session in the popular press was in Seattle where the WTO Millennial Ministerial was shut down by tens of thousands of activists and labor union organizers in 1999. Since then the WTO, while still sticking to the neoliberal orthodox of free and unhindered trade, has become enmeshed in the discourse of sustainable development.

Since the Uruguay Round, when many developed nations promised to decrease subsidies to their producers of agricultural goods, and many developing countries actually removed their own, agriculture has been a sticking point in free trade negotiations. The high level of protectionism and the entrenched subsidies provided to developed world agricultural producers has led to a disparity in the impact of free trade on agricultural producers where many farmers in developing countries are exposed to the whims of international competition, unlike their counterparts in the developed world. These debates and negotiations are played out through the WTO Agreement on Agriculture.

There are still debates about whether or not multilateral environmental agreements like the Convention on Biodiversity, the Montréal Protocol, and the Basel Convention have jurisdictional precedence over WTO rules, a question of considerable importance to international environmental law. For example, the WTO Agreement on Trade-Related Aspects of Intellectual Property Rights comes into conflict with the Convention on Biodiversity on the issue of benefit sharing with indigenous farmers regarding plant genetic resources. The Cartagena Biosafety Protocol of the Convention on Biodiversity also comes into conflict with the Agreement on Technical Barriers to Trade because it is often asserted that regulations and bans are surrogates for protectionism.

Whether by slip of tongue, or honest confusion, people often confuse *free trade* with *fair trade*. But the terms have very different meanings. Proponents of free trade often suggest that free trade is fair trade. But advocates of fair trade use the term to signify an attempt to link consumers more directly to producers, redistribute inequality in terms of trade, and provide a minimum price to low-income producers, notably in cocoa, bananas, and coffee.

SEE ALSO: Capitalism; Globalization; Markets; Movements, Environmental; North American Free Trade Agreement (NAFTA); Race-to-the-Bottom Hypothesis; Subsidies; Sustainable Development; Trade, Fair; World Trade Organization (WTO).

BIBLIOGRAPHY. Sterling Evans, *Bound in Twine: Transnational History and Environmental Change in the Henequen-Wheat Complex for Yucatan and the American and Canadian Plains: 1880–1950* (Texas A&M University Press, forthcoming 2007); Kevin P. Gallagher, *Free Trade and the Environment: Mexico, NAFTA, and Beyond* (Stanford University Press, 2004); Howard Mann, *Private Rights, Public Problems; A Guide the NAFTA's Controversial Chapter on Investor Rights* (International Institute for Sustainable Development, 2001); Gary Sampson, *The WTO and Sustainable Development* (United Nations University, 2005).

Dustin Mulvaney
University of California, Santa Cruz

Trade Winds

THE TRADE WINDS are a persistent band of easterly winds that blow toward the equator in both hemispheres, covering most of the earth between 25 degrees N and 25 degrees S latitude. These winds

originate on the equatorial sides of subtropical high-pressure systems that exist over the tropical and subtropical oceans and represent a major component of the general circulation of the atmosphere. The high-pressure areas force air to move toward a belt of low-pressure near the equator called the doldrums. The air converging at the doldrums rises high over the earth, recirculates toward the poles, and sinks back toward the earth's surface to about 30 degrees latitude, thus completing a cycle. The surface air that flows from the subtropical highs toward the equator is deflected toward the west in both hemispheres because of the earth's west-to-east rotation. This results in the northeast trade winds in the Northern Hemisphere and southeast trade winds in the Southern Hemisphere.

The most reliable winds on earth are unquestionably the trade winds. They are extremely consistent in both direction and speed throughout the year, averaging about 11 to 13 miles per hour (18 to 21 kilometers per hour). These steady winds are called trade winds due to their ability to quickly propel trading ships across the ocean. The trade winds were named by the crews of sailing ships that depended on these winds during ocean navigation. The name *trade winds* derives from the Old English "trade," meaning "path" or "track," because of the regular course of the winds. These winds helped carry Christopher Columbus on his voyage to the New World in 1492. Mariners of the 16th century recognized early that the quickest and most reliable route for their sailing vessels from Europe to America lay in the belt of the northeasterly winds in the tropical North Atlantic Ocean.

The trade winds are best developed on the eastern and equatorial sides of the subtropical high-pressure systems, especially across the Atlantic Ocean. The trade winds are stronger and more consistent over the oceans than over land due to increased friction on the continental surfaces. When the trade winds reach the western edge of an ocean basin, they turn toward the poles and then loop back east to become part of the prevailing westerlies. The trade winds are primarily a surface wind and move north and south about 5 degrees with the seasons.

The trade winds originate as warm, dry winds capable of holding a tremendous amount of moisture. As they blow across the tropical oceans, they evaporate huge quantities of moisture. The trade winds are overlain by warmer and drier air, creating a temperature inversion in which temperatures increase with height. The temperature inversion often limits the vertical development of clouds, producing clear skies that make trade wind islands a popular tourist attraction. As the trade winds blow against mountain ranges, they are forced to rise and cool. This allows the moisture to condense and fall as rain. These conditions create large differences in rainfall due to topographic variations. Low-lying islands usually experience desert-like conditions, while the windward slopes of some islands are among the wettest places in the world.

SEE ALSO: Atmosphere; Climatology; Orographic Effect; Precipitation; Weather.

BIBLIOGRAPHY. Edward Aguado and James E. Burt, *Understanding Weather and Climate* (Prentice Hall 2004); Frederick K. Lutgens and Edward J. Tarbuck, *The Atmosphere* (Prentice Hall, 2004); Tom L. McKnight and Darrel Hess, *Physical Geography: A Landscape Appreciation* (Prentice Hall, 2005); Brian J. Skinner and Stephen C. Porter, *The Blue Planet: An Introduction to Earth System Science* (John Wiley and Sons, 1994).

Darren B. Parnell
Salisbury University

Tragedy of the Commons

THE "TRAGEDY OF THE COMMONS," probably the most common framework through which environmental issues are understood today, was made famous by biologist Garrett Hardin in a 1968 essay in *Science*. Hardin was specifically concerned with population growth and invoked the notion made popular by Thomas Malthus in 1798 that because population grows exponentially while food supply grows only linearly, population growth will lead inevitably to starvation, war, and disease, and eventually to a collapse in population levels. Hardin argued that population growth is a tragedy of the commons, which he explained with this image: "Picture a pasture open to all. It is expected that

each herdsman will try to keep as many cattle as possible on the commons." Every animal that is added contributes to pasture degradation, but this negative effect is shared by all of the herders. Each herder enjoys full benefits, however, from adding an additional animal to his own herd. Because each herder acts to maximize his or her own gain, more and more animals will be added. In the end, this leads to overgrazing and a tragedy of pasture degradation: "Each man is locked into a system that compels him to increase his herd without limit—in a world that is limited...Freedom in the commons brings ruin for all." Hardin argued that there are only two solutions to this problem. The first option is coercion, or control of each individual's behavior by an outside agent, particularly the state. The second is to privatize the commons; only if the common pasture is divided up into privately owned parcels will individuals take care of the resource, and thus preserve it from overuse and destruction.

Though Hardin's essay was focused on overpopulation, its broader legacy has been the idea that environmental resources held in common are naturally and inevitably subject to overuse and degradation. The metaphor of the "tragedy of the commons" has become conventional wisdom for understanding all kinds of environmental problems today, including: the depletion of ocean fisheries, hunting that led to the extinction of the American passenger pigeon, rangeland degradation, the overuse of national parks, and a variety of pollution problems.

Despite its popularity, Hardin's tragedy of the commons is flawed in many respects. First, Hardin's idea was modeled after an inaccurate understanding of the medieval English commons. Far from being completely unregulated and free for the taking, common pastures were available for the use only of specific villagers or individuals; even for them, there were often regulations including limits on the numbers of animals each tenant could put on the pasture. In other words, the model assumes that "commons" are in fact what is more accurately termed an "open access" situation, in which there are no rules and limitations whatsoever on the use of a resource.

The model also assumes that users do not have or develop social or cultural norms that might cause them to regulate their resource use, and that users are inherently selfish, have perfect information, and always seek to maximize their short-term gains. In other words, the model of the tragedy of the commons fails to take into account the specific historical, social, and cultural contexts of resource users.

To further understand this model, it is helpful to distinguish between the characteristics of a resource itself, and the characteristics of the system that governs resource management. Two key characteristics of resources in general are first, whether it is easy or hard to control access to the resource (excludability); and whether one person's use of the resource takes away from another's (subtractability; sunlight is an example of a nonsubtractable resource). Common property resources (sometimes called common pool resources) are those that are subtractable but not easily excludable; these include forests, pastures, fisheries, and sinks for various types of environmental pollution. Resources can be classified into those that are private (subtractable and easily excludable); common; public or state (nonsubstractable and difficult to exclude); and club or toll resources (easily excludable and nonsubtractable).

There is no necessary or automatic correspondence, however, between the type of resource and the management regime under which it is governed. One type of management regime is open-access, where there are no regulations and anyone can easily gain access to the resource. Another type is private property, where the right to use and exclude others is vested in an individual or legal individual (such as a corporation). Third, common property regimes are those in which there is an identifiable set of users who can exclude outsiders and who have legal or informal rules governing use. Finally, in public or state property, the government makes decisions about access to and use of the resource.

The resource overexploitation and degradation predicted by Hardin in the "tragedy of the commons" is indeed often seen in cases where resources are managed through an open access regime. Examples include the depletion of unregulated ocean fisheries, and the current unregulated emissions of carbon dioxide leading to anthropogenic global climate change. Where Hardin went wrong, however, were the assumptions that common property resources are always governed by open access regimes and that common property regimes never work.

In fact, case studies from around the world have shown that common property regimes often work quite well. Well-known cases include Native American hunting and fishing lands in James Bay, cooperative-based coastal fisheries in Japan, communal meadows and forests in the Swiss Alps, medieval irrigation systems in Spain, and contemporary lobster fishing territories in Maine.

Research on successful cases around the world has suggested a number of institutional factors which contribute to the success of common property management. Favorable factors include having clearly defined boundaries around both the users and the resource; ability to monitor resource use; mechanisms for conflict resolution, congruence between local conditions and rules; graduated sanctions; and the legal right to devise institutions and sustain ownership of the common property resource. When these conditions exist, the tragedy of the commons is likely to be averted.

Another flaw often arising from use of the tragedy of the commons model is the assumption that if commons are the problem, then conversely other forms of ownership will not lead to resource degradation. In fact, no type of resource or type of management regime is completely guaranteed in advance to be either sustainable or subject to degradation through overuse. The particular context is important in determining the outcome. As a model of resource degradation, the tragedy of the commons works in some cases, but is too over-simplified to accurately predict or explain the sustainability of resource use in general.

SEE ALSO: Common Property Theory; Hardin, Garrett; Lifeboat Ethics; Malthusianism; Overfishing; Overgrazing; Overpopulation; Pastoralism; Prisoner's Dilemma (PD); Rational Choice Theory; Resources; Scarcity.

BIBLIOGRAPHY. Susan J. Buck Cox, "No Tragedy of the Commons," *Environmental Ethics* (v.7, 1985); Thomas Dietz, Elinor Ostrom, and Paul C. Stern, "The Struggle to Govern the Commons," *Science* (v.302, 1912); David Feeny, Fikret Berkes, Bonnie J. McCay, and James Acheson, "The Tragedy of the Commons: Twenty-Two Years Later," *Human Ecology* (v.18, 1990); Clark C. Gibson, Margaret A. McKean, and Elinor Ostrom, eds., *People and Forests: Communities, Institutions, and Governance* (MIT Press, 2000); Garrett Hardin, "The Tragedy of the Commons," *Science* (v.163, 1968); Thomas Malthus, *An Essay on the Principle of Population* (London, 1798); Bonnie McCay and James Acheson, eds., *The Question of the Commons: The Culture and Ecology of Communal Resources* (University of Arizona Press, 1987); Elinor Ostrom, *Governing the Commons: The Evolution of Institutions for Collective Action* (Cambridge University Press, 1990).

Emily T. Yeh
University of Colorado, Boulder

Transamazon Highway

THE TRANSAMAZON HIGHWAY was constructed as an east-west road corridor across the Brazilian Amazon during the 1970s. This was one of several interregional road projects promoted by Brazil's then military government as a means of impelling frontier expansion. The military not only sought to relieve agrarian tensions in other parts of Brazil by opening land to landless populations, but also to secure remote portions of the national territory against perceived geopolitical threats by other countries.

Highway construction comprised one component of an integrated model of colonization, which the government implemented by selecting colonists, surveying and demarcating agricultural lots, and otherwise supporting frontier land settlement. The state land agency, INCRA, oversaw design and construction of the road network, which formed a "fishbone" pattern with feeder roads running perpendicular to the Transamazon itself. Colonist families settled along the highway corridor, first in the east, and increasingly toward the west, during the early 1970s. Integral to land settlement was the clearing of upland primary (old-growth) forest, which was not only necessary for colonists to establish land claims but also to plant food crops to feed their families.

In the early 1980s, the legitimacy of the military declined with Brazil's worsening economic situation, forcing the withdrawal of state support for colonization along the Transamazon. This left colonists on their own in very precarious circumstances.

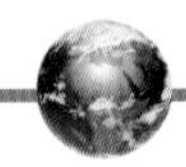

However, in the mid-1980s, prices rose for two key perennial crops, cocoa and black pepper, and colonists who were producing these commodities earned rising incomes. This stimulated a second wave of in-migration to the Transamazon corridor, expanding the population as well as forest clearing for agricultural land use.

In the late 1980s, as Brazil underwent democratization, municipal governments and social movement organizations emerged to support colonists along the Transamazon. Soon after, cocoa and pepper prices declined, as did crop production due to pests, leading again to difficulties. However, this circumstance did not halt deforestation, as colonists shifted their land use, this time to pasture for cattle, which requires much larger clearings than do crops. The availability of new credit lines for small-scale farms, including for colonists along the Transamazon, meant that many colonists had a new source of funds to expand pasture for cattle ranching, something that has continued into the new millennium.

Beginning in the late 1990s, fiscal decentralization in Brazil's government system provided greater funds and responsibilities to municipalities. This intensified local politics over roads as a means of ensuring access by rural populations to urban services and markets. Road building by local groups along the Transamazon has encouraged continued deforestation and forest fragmentation, even in indigenous reserves. The paving of the Transamazon highway west to the town of Altamira, in anticipation of construction of the Belo Monte dam on the Xingú river there, is bringing new changes to the Transamazon corridor. Capitalized interests are increasingly arriving in the area, speculating in land and timber and seeking to expel colonists. The onset of land conflicts between speculators and colonists led to the murder of Sister Dorothy Stang in February 2005, which has again called attention to the link between environmental damage and human rights abuses previously seen elsewhere in the Brazilian Amazon.

SEE ALSO: Amazon River Basin; Brazil; Deforestation.

BIBLIOGRAPHY. John O. Browder and Brian J. Godfrey, *Rainforest Cities: Urbanization, Development, and Globalization of the Brazilian Amazon* (Columbia University Press, 1997); Douglas Ian Stewart, *After the Trees: Living on the Transamazon Highway* (University of Texas Press, 1994).

Stephen G. Perz
University of Florida

Transboundary Rivers

TRANSBOUNDARY RIVERS ARE natural freshwater systems with at least one perennial tributary crossing the political boundaries of two or more states. Also known as international rivers, the 1978 United Nations (UN) Register of International Rivers, updated in 1999, classifies them. According to this update, there are 261 transboundary rivers, which cover 45.3 percent of the earth's surface, excluding Antarctica. A transboundary river basin includes both surface water and groundwater, which contribute hydrologically to a first-order stream before finding an outlet to an ocean, a lake, or an inland sea.

As an important source of freshwater, transboundary rivers are a major environmental concern for the 21st century. This concern is not only related to the land surface included in these basins, but also to the flow generated within these basins: 87 percent of the flow of 25 of the world's largest rivers, representing almost half of the world's total runoff, is generated within transboundary river basins. Additionally, 145 nations, as small as Liechtenstein and as populous as Bangladesh, have territory within such basins.

Rivers have been historically important in determining geopolitical borders, an example is the Danube, which formed the border of the Roman Empire and now forms the boundary between Romania and Bulgaria. Rivers have also served multiple purposes, from providing drinking water, irrigation, hydroelectric power generation, and industrial uses. However, management of transboundary rivers has often proven complicated due to disagreements about river flows, each nation's contribution to that flow, historic uses, and future demand. Such disagreements also focus on the social, ecological, and economic needs of each nation. To visualize di-

lemmas of management and allocation of water resources in transboundary river basins, one can look at the number of countries which share a transboundary river basin: 19 of 261 basins are shared by five or more riparian countries. A total of 17 riparian countries share the Danube basin, whereas 11 share the Congo and Niger.

Such complexities, combined with increasing water stress due to increases in population, economic growth, and reductions in the quality and quantity of world's freshwater resources, have caused some studies to focus on the danger of violence and wars in the transboundary river basins. Although there has not been an incident of such violence in recent history, these environmental security studies link environmental degradation and scarcity to armed conflict.

Such studies do not usually focus on environmental problems in transboundary basins as the main factor leading to insecurity. Instead, they take environmental degradation, such as pollution, or scarcity as an accessory factory of insecurity. From this perspective, a negative change in the quality and quantity of the renewable, nonsubstitutable resources is but one factor leading to conflict. Environmental change acts as a variable of conflict, exacerbating fault lines between state and society and worsening existing political, economic, and social tensions. For example, studies have envisioned water-war scenarios in the Middle East due to tensions in transboundary rivers, such the Jordan River, which is shared by Jordan, Israel, Syria, Lebanon, and the Palestinian Authority. Water diversion plans by riparian states have been interpreted as aggravating tensions related to the region's aridity, population pressure, and political situation.

Other studies emphasize the possibility of transboundary cooperation based on historic evidence. For example, the Mekong Committee, established over the Mekong by Cambodia, Laos, Thailand, and Vietnam in 1957, continued to exchange data over river basin management throughout the Vietnam War. The Mekong, the seventh largest river in the world in terms of discharge and the 10th in length, is shared by six riparian nations (the Mekong Committee members, Myanmar, and Laos). Various international organizations, including UN institutions and the Global Environmental Facility, have supported the Mekong Committee in promoting sustainable management of the basin. The Mekong Committee has become a useful model for governance of transboundary rivers by providing an international and institutional framework for cooperation prior to the outbreak of a water crisis in the transboundary river basin. Recent studies have focused on river management and allocation of water resources by comparing different transboundary river basins in various disciplines, such as international relations, international law, geography, and environmental studies.

Nations and international organizations have formulated international conventions and established joint management committees to reduce the risk of water-related conflicts and ensure quality and quantity of river flow. Transboundary river management has evolved from the Harmon Doctrine, stating absolute sovereignty over the waters of a transboundary watercourse within the state's territory, to the doctrine of equitable utilization.

The latter doctrine has been applied since the International Law Association's 1966 Helsinki Rules. The UN Convention on the Law of the Non-Navigational Uses of International Watercourses also adopted the equitable utilization principle in 1997, which emphasized the prevention of significant harm and prior notification of planned measures related to the river basin. The convention has not been ratified by the 35 nation states necessary for it to enter into force. Although 103 countries have adopted the convention on a preliminary basis, these principles offer only general guidance and are not mandatory to riparian states.

In addition to the efforts of the international community, bilateral treaties have proven an effective mechanism in transboundary river management. Treaties such as the 1944 Colorado Treaty between the United States and Mexico, and the 1959 Nile River Treaty between Egypt and Sudan regulate the allocation of water for both the upstream and the downstream nations. Similarly, commissions among riparian nations also enable the transparent exchange of data over future development plans, and provide a forum for continuous dialogues. For example, the Indus River Commission has continued to function despite two major wars between India and Pakistan. Such mechanisms enable states to

overcome differences in legislation, economic and policy goals, administrative structures, and social and cultural perceptions.

Many transboundary rivers have been dammed or diverted for irrigation, hydroelectric power production, or similar purposes. Such projects come with ecological and social costs. Before the building of the Aswan Dam, the Nile River carried over 120 million tons of sediment to the Mediterranean Sea each year and nearly 10 million tones of this was deposited in the floodplain and Nile Delta. After the construction of the dam, almost all the sediment has remained behind the dam. This caused not only coastal erosion, but also serious reduction in agricultural productivity in the Nile Delta. Moreover, 100,000 people had to be resettled from 1963 to 1969. Dam construction itself may create temporary jobs and generate income for local people. However, loss of lands and livelihoods (such as loss of fisheries after dam construction) raises questions about the social costs of water development projects. Moreover, tensions between states may increase during the filling of a reservoir when the downstream flow is cut off temporarily, especially when there is no prior notification, agreement, or institutional arrangement regarding the flow of the river.

Although the precise nature of related ecological changes is still unclear, global warming is likely to affect the availability of freshwater resources and disrupt the global supply and demand of water and arable land. Increased productivity of water use, stronger policies to regulate the use of water within territories, and the establishment of constructive dialogue among riparian states can help to sustain transboundary river cooperation. The maintenance of such cooperation is crucial in ensuring water and food security, political stability, and the protection of biodiversity.

SEE ALSO: Aswan High Dam; Mekong River; Nile River (and White Nile); Riparian Areas; Riparian Rights; Rivers.

BIBLIOGRAPHY. Stephen C. McCaffrey, *The Law of International Watercourses: Non-Navigational Uses* (Oxford University Press, 2001); Sandra Postel and Brian Richter, *Rivers for Life: Managing Water for People and Nature* (Island Press, 2003); A.T. Wolf, ed., *Conflict Prevention and Resolution in Water Systems*, vol. 5 in *Management of Water Resources* (Edward Elgar Publishing, 2002).

Nurcan Atalan
Ohio State University

Transcendentalism

TRANSCENDENTALISM WAS A series of new ideas that flourished among writers and philosophers in New England during the 19th century. The concept of transcendentalism was a state of being that was beyond the reach or comprehension of experience. These ideas centered on a belief in the essential unity of all creation, an innate goodness of man, and the supremacy of insight over logic.

The original concepts of transcendentalism came from Europe, but were also influenced by old Indian texts such as the *Upanishads* and the *Bhagavad-Gita*, Chinese ideas of Confucius, the teachings of Buddha, and work by the Muslim Sufis. Henry David Thoreau (1817–62) paid much tribute to the ideas that came from Vedic thought.

These traditions were merged with Platonism and Neoplatonism by British writers such as Samuel Taylor Coleridge (1772–1834) and Thomas Carlyle (1795–1881), as well as the Swedish thinker and mystic Emanuel Swedenborg (1688–1772) and the German philosophical mystic Jakob Böhme (1575–1624). Some of the ideas came from Prussian philosopher Immanuel Kant (1724–1804) who coined the phrase in his *Critique of Practical Reason* (1788) when he wrote, "I call all knowledge transcendental which is concerned, not with objects, but with our mode of knowing objects so far as this is possible *a priori*." Other ideas were synthesized from a number of German philosophers.

Although transcendentalism in Europe had been an abstract philosophical concept, in New England, especially in Concord, Massachusetts, it was a liberating philosophy which drew from the Romantic movement. It encouraged the adaptation of ideas from Europe and elsewhere to form what became from about 1830 until 1855 as a battle between the younger and older generations in the United States over the emergence of a

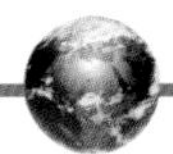

national cultural identity based around intrinsic American concepts.

The main people represented in the idea of transcendentalism were Ralph Waldo Emerson (1803–82), Orestes Augustus Brownson (1803–76), James Freeman Clarke (1810–88), Margaret Fuller (1810–50), Elizabeth Palmer Peabody (1804–94) and Henry David Thoreau. Others connected with the movement included Bronson Alcott (1799–1888), the younger W.E. Channing (1780–1842), W.H. Channing (1810–84), Christopher Pearse Cranch (1813–92), Theodore Parker (1810–60) and George Ripley (1802–80). Julia Ward was on the fringes of the group as were others like Jones Very (1813–80).

Although the ideas of the transcendentalists were being formed in the early 1830s, it was the publication of the essay "Nature," by Emerson that proved to be the catalyst for the movement. The Transcendentalist Club in Cambridge, Massachusetts, opened on September 8, 1836, with Emerson, George Putnam, and Frederick Henry Hedge as members. The height of transcendentalism saw Emerson and Fuller establish *The Dial* in 1840. It was a "little magazine" that was published until 1844 and contained many of the best writings by minor transcendentalists.

In 1841, Brook Farm was established as a cooperative community near West Roxbury, Massachusetts, nine miles from Boston. There George Ripley and others from the Transcendentalist Club tried to apply their social, religious, and political theories to a farm of 200 acres called the Brook Farm Institute of Agriculture. It came under the influence of Albert Brisbane, a prominent Fourierist, and was renamed the Brook Farm Phalanx. However, the idea finally collapsed in October 1847, and the group was dissolved.

Another cooperative experiment of the transcendentalists was Fruitlands, which was established by Alcott from 1842 until 1843 at Harvard, Massachusetts. It was planned as a place where members would labor on the land and conduct themselves in a simple manner, eating vegetarian meals and living in harmony with nature. It was even less successful than Brook Farm. It is best remembered through Louisa May Alcott's *Transcendental Wild Oats,* a fictional account of Fruitlands.

Bronson Alcott

Amos Bronson Alcott, to give his full name, was born in Connecticut in 1799 and initially worked as a salesman in the south before returning to New England as a schoolteacher. He taught at the Temple School in Boston from 1834 until 1839, becoming a school superintendent at Concord, Massachusetts, and elsewhere.

It was at the Temple School that Alcott began collaborating with Elizabeth Peabody, his assistant, who edited Alcott's *Record of a School, Exemplifying the General Principles of Spiritual Culture* (1835), which was followed by *Conversations with Children on the Gospels*, published in two volumes (1836–37). These books set forth Alcott's views on the theory and practice of education, but although he gained support from Ralph Waldo Emerson and others in the Transcendentalist Movement, some parents withdrew their children in protest against these radical ideas.

In 1842, Alcott went to England, where he met with Thomas Carlyle. The two exchanged ideas, although Carlyle found Alcott tiresome. Back in the United States, Alcott and his friends established the cooperative Fruitlands. It lasted only seven months, and the family returned to Concord in January 1845. Bronson Alcott continued to expound his political views and also managed to get singing, dancing, reading aloud, and even physiology introduced into the school curriculum.

In 1868, his daughter Louisa May Alcott (b.1832) published her book *Little Women*, which was about her childhood and included an account of Fruitlands. It gained the family the independence they needed, and allowed Bronson Alcott to push forward with the Concord School of Philosophy from 1879 until his death in 1888. Alcott has always been seen as impractical in his ideas, but with a genius for conservation. Louisa Alcott wrote many other books but none achieved the fame of *Little Women.* She died in 1888, two days after her father.

On the religious front, members of the transcendentalist movement were reacting against the 18th-century thought of Alexis de Tocqueville, Alexander Hamilton, and Thomas Jefferson. They were also reacting against New England Calvinism and the rationalist views of John Locke (1632–1704). They repudiated Unitarianism and the idea of an established order and argued in favor of major changes in school curriculum and teaching methods, the right for women to vote, better conditions for the working man, universal temperance, freedom of religious thought, and a change in fashion. These views, at their most extreme, drew many critics who saw the transcendentalists as supporters of anarchy, socialism, and even communism. Although they were certainly against slavery, their focus on New England meant that abolitionism, one of the major religious and social causes of the time, was not a central focus of the transcendentalist movement.

The transcendentalists were also the inspiration for many other writers such as Nathaniel Hawthorne, Herman Melville, and Walt Whitman, leading to the flowering of the American literary scene that critics believe to be the American Renaissance in literature. It has to be said, however, that Nathaniel Hawthorne did later parody the movement in his novel *The Blithedale Romance*, which was based on his time at Brook Farm. Still, it was also from the transcendentalists that ideas about environmental planning and architecture emerged. The environmental designs of Benton MacKaye and Lewis Mumford owe a huge debt to the transcendentalists, as do the architectural designs of Frank Lloyd Wright and Louis Sullivan. Others transcendentalist ideas can be seen in the works and thought of William James, John Dewey, and Alfred Stieglitz.

SEE ALSO: Communism; Locke, John; Religion; Socialism; Thoreau, Henry David; Vegetarianism; Wright, Frank Lloyd.

BIBLIOGRAPHY. Lawrence Buell, *Literary Transcendentalism: Style and Vision in the American Renaissance* (Cornell University Press, 1973); Rosalie Sandra Perry, *Charles Ives and the American Mind* (Kent State University Press, 1974); Joel Porte, *Emerson and Thoreau: Transcendentalists in Conflict* (Wesleyan University Press, 1966); Arthur Versluis, *The Esoteric Origins of the American Renaissance* (Oxford University Press, 2001).

Justin Corfield
Independent Scholar

Transmissible Spongiform Encephalopathies (TSFs)

TRANSMISSIBLE SPONGIFORM encephalopathies (TSEs) are a cluster of rare degenerative brain disorders. Also known as "prion" diseases, they leave their victims with tiny holes in the brain. The tiny holes give it a "spongy" appearance that can be seen under a microscope when a section of brain is dissected. Prion is short for proteinaceous infectious particle. The particle is a protein that occurs in a harmless normal form in the body's cells. However, deadly prion proteins are the proteins that can cause the disease. TSEs are transmissible because the deadly prion protein is acquired from an alien source. Researchers are examining the possibility that TSEs are also infectious. There are many forms of TSEs including Creutzfeldt-Jakob disease (CJD), Kuru ("laughing sickness"), fatal familial insomnia (FFI), and Gerstmann-Straussler-Scheinker disease (GSS).

In addition, a new type of CJD or a variant of it was first described in the United Kingdom and in several continental countries in 1996. The new TSE (designated as vCJD to mark it as a variant CJD) had symptoms that were different from classic CJD. Also, it afflicted much younger people. The cause of the disease may have been the consumption of beef that had a bovine form of spongiform encephalopathy (BSE). The name "mad cow disease" was used to label the new form of CJD.

Besides BSE, there are other forms of TSEs in animals. Since TSEs are transmissible, infected animals can spread the disease to other flocks if sold rather than being destroyed. In sheep and goats a fatal degenerative disease is known as scrapie. The disease causes loss of production and it prevents the sale of semen, embryos, or breeding stock to other countries. That elk and deer can be afflicted with TSE was discovered in the late 1960s. Chronic wasting

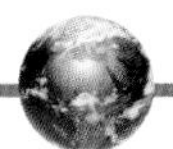

disease (CWD) is similar to mad cow diseases, but affects deer and elk populations. Herds in Colorado and Wyoming have been affected, and the disease is spreading. Feline spongiform encephalopathy (FSE) was found in cats in Great Britain in 1990. Since then it has been reported on the Continent as well. Transmissible mink encephalopathy (TME) was first diagnosed in 1947 in ranch-raised minks in the United States. TME affects the central nervous system. TME has been reported in Russia, Canada, Finland, and Germany.

Kuru was first identified in Papua New Guinea among the Fore people. With a Stone Age culture, they practiced ritualistic cannibalism. At funerary feasts they would eat the brains of deceased relatives, allowing infectious crystal protein to invade human cells. After the practice was outlawed in the 1970s, Kuru gradually disappeared. Fatal familial insomnia (FFI) is symptomatically similar to CJD. It is a hereditary prion disease that disrupts sleep, creating insomnia that lasts until death. The disease is due to a missense mutation at codon 178 of the prion protein gene on chromosome 20. The disease runs in families and is fatal in seven to 36 months after the onset of symptoms. Gerstmann-Straussler-Scheinker (GSS) is a very rare TSE; most cases have been inherited. It usually strikes between ages 35 to 55. As the disease progresses, symptoms resemble Parkinson's disease with dysarthria, nystagmus, spasticity, disturbances in vision, or deafness.

The appearance of cases of TSEs has produced political controversy over the safety of the food supply, blood supply, and medical procedures. The general public and the media have been fairly successful in pushing governments to exercise due regard for public safety from any threat connected with TSEs. There are no known effective treatments for TSEs. The National Institute of Neurological Disorders and Stroke (NINDS) and other organizations are working to find cures.

SEE ALSO: Bovine Spongiform Encephalopathy; Cattle; Disease; Livestock; Mad Cow Disease; Sheep.

BIBLIOGRAPHY. Harry F. Baker, ed., *Molecular Pathology of the Prions* (Humana Press, 2001); Rick Erdtmann and Laura B. Sivitz, *Advancing Prion Science: A Guidance for the National Prion Research Program* (National Academies Press, 2004); David A. Harris, ed., *Mad Cow Disease and Related Spongiform Encephalopathies* (Springer-Verlag, 2004); Robert Klitzman, *The Trembling Mountain: A Personal Account of Kuru, Cannibals, and Mad Cow Disease* (Perseus Publishing, 2001); Daniel T. Max, *The Family That Couldn't Sleep: A Social and Scientific History of Prion Disease* (Random House, 2006); Eve Sequin, ed., *Infectious Processes: Knowledge, Discourse, and the Politics of Prions* (St. Martin's Press, 2004); Barbara Sheen, *Mad Cow Disease* (Thomson Gale, 2004); Charles A. Spencer, *Mad Cows and Cannibals, A Guide to the Transmissible Spongiform Encephalopathies* (Pearson Education, 2003); Glen C. Telling, *Prions and Prion Diseases: Current Perspectives* (Taylor & Francis, Inc., 2004); Philip Yam, *The Pathological Protein: Mad Cow, Chronic Wasting, and other Deadly Prion Diseases* (Springer-Verlag, 2003).

Andrew J. Waskey
Dalton State College

Transportation

EFFICIENT, CONVENIENT, AND affordable transportation systems are viewed as an essential component for social and economic development. The United Nations (UN) has called for concerted action on transport issues in its 1992 Agenda 21 Program and the 2002 Johannesburg Plan of Implementation. However, because transport is a major source of atmospheric emissions, it must be carefully designed and managed so as to be sustainable—environmentally, socially, and economically—in the long term. Transport involves the movement of people and goods, and falls into three sectors: Air, sea, and land.

AIR TRANSPORT

Air transport has only existed for approximately 100 years; however, the consistent growth of this sector has raised concerns about its environmental impacts. Airplanes release pollutants associated with the combustion of fossil fuels, such as carbon dioxide, carbon monoxide, nitrogen oxides, sulfur oxides, volatile organic compounds, particulates, and other trace compounds. These pollutants are released into the upper troposphere and lower stratosphere, where they have an impact on atmospheric composition. A report issued in 1999 by the International Panel on Climate Change summarized the ecological impacts of aviation: Aircraft emit gases and particles that alter the atmospheric concentration of greenhouse gases, trigger the formation of condensation trails, and may increase cirrus cloudiness, all of which contribute to climate change; and aircraft are estimated to contribute about 3.5 percent of the total radiative forcing (a measure of change in climate) by all human activities and that this percentage, which excludes the effects of possible changes in cirrus clouds, is projected to grow. Emissions from aircraft are not included in targets set by the Kyoto Protocol of the UN Framework Convention on Climate Change (UNFCCC), though the Protocol does state that Annex I Parties do have the responsibility to limit or reduce greenhouse gas emissions from aviation fuels.

Urban light rail systems can now be found in such cities as Cairo, Manila, Geneva, and Bangkok.

SEA TRANSPORT

Sea transport is used primarily for long-distance shipping of goods. Two prime environmental concerns are the disposal of pollutants into the sea (marine or ocean dumping), and the introduction of alien invasive species through ocean transportation. Marine dumping occurs either accidentally, as is the case with oil spills, or intentionally through dredging spoil, nuclear waste disposal, sewage outfalls, and cruise ship waste. Various international agreements have been agreed to address intentional disposal of wastes at sea, such as the Convention on the Prevention of Marine Pollution by Dumping of Wastes and Other Matter (that is, the London Convention of 1972).

Introduction of alien invasive species is the second biggest threat to biodiversity after habitat loss. In terms of marine invasive species, a common method of transmission has been through ballast water, which "is now regarded as the most important vector for trans-oceanic and inter-oceanic movements of shallow-water coastal organisms," according to the World Conservation Union (IUCN). However, species can also attach themselves to the hulls of ships and be transported in that way. Examples of the introduction of alien invasive species by ballast water or hull fouling include the spread of zebra mussels into the North American Great Lakes water system.

LAND TRANSPORT

Land transport mainly refers to road and rail modes; road transport can be further divided into motorized and nonmotorized traffic. Nonmotorized transport is comprised primarily of walking and cycling, though other forms do occur, such as wheelbarrows, handcarts, pack donkeys, sledges, and animal-drawn carts. Nonmotorized transport has a number of environmental advantages, the most obvious of which is the minimal or nil production of emissions. Nonmotorized transport is also an integral element of transport for people in developing countries, who may not be able to afford private motorized vehicles or even public transport. Nonmotorized transport has been considered one of the most viable methods of address-

ing social and economic development issues in developing countries. Cycling is a common form of nonmotorized transport, and infrastructure for cyclists can also be integrated into public transport networks in order to improve accessibility and convenience.

Rail transport is generally more energy efficient than road transport, but despite this, in many regions of the world, rail freight and passenger numbers are decreasing. Light rail systems are gaining popularity as an urban transportation option in many countries, and can be found in: Cairo, Egypt; Kuala Lumpur, Malaysia; Christchurch, New Zealand; Manila, the Philippines; Geneva, Switzerland; and Bangkok, Thailand. These rail systems can be in the form of trains, tramways, or trolleys/streetcars. Other rail-based urban transport systems include subways/metros (underground) and monorail (elevated) systems.

Reduced environmental impacts from motorized road transport are possible through technological improvements in two main areas: Vehicle technology and fuel quality. Fuel quality has improved greatly over the past 30 years, and is continuing to improve. Unleaded petrol, first introduced in the 1970s, is now available in most countries around the world, and many countries are reducing the concentrations of sulfur and other substances in petrol and diesel fuels. Alternative fuels have also emerged, such as biodiesel and compressed natural gas.

However, many newer and cleaner fuels require more advanced vehicle technologies. Therefore, vehicle engines have consistently become more fuel-efficient, and many countries have been implementing stricter regulations to encourage further advances in vehicle efficiency. In addition, new vehicle components have been developed that reduce the volume of pollutants that are released from vehicles, such as particulate filters and catalytic converters. Most recently, through the development of new ways of powering vehicles, low or no emission vehicles are now available, such as hybrid-electric vehicles. As research and development in this area increases, the availability, affordability, and efficiency of these vehicles will grow. Innovations in fuel quality and vehicle technology can result in substantial improvements to air quality and human health.

RURAL AND URBAN ISSUES

Transportation issues vary between rural and urban areas. Transport in rural areas is critical in that these areas tend to be inhabited by poorer segments of the population. In South Africa, for example, half of the population is rural, but 72 percent of those living in the rural areas are poor, according to the South African National Department of Transport. Because rural transport links tend to service the poor, it is essential that they are cost-effective. Provision of affordable and well-maintained transport links for people and for goods is essential, especially in terms of strategies to increase economic development and reduce poverty in these areas.

In urban areas, where there is high population density and therefore a greater need for the movement of goods and people, congestion caused by traffic is a significant problem. The built environment of urban areas is often not designed for high volumes of vehicular traffic, and prevents the dispersion of emissions, and pollutants remain in the urban street canyons, resulting in poor air quality. Moreover, increasing urbanization and personal vehicle use can result in urban sprawl. Expansion of low-density metropolitan areas outside of urban centers can reduce the cost-effectiveness and efficiency of alternative transport options such as mass transit, and exacerbate the need for personal vehicle use. Therefore, transport planning is an essential part of any urban, regional, or national infrastructure strategy. Comprehensive and forward-looking plans are necessary for a rational transport infrastructure that supports current and future needs of society and is environmentally sustainable.

Public transport is viewed as a mobility option that has environmental and social benefits. From an environmental perspective, the higher density of people being transported in less vehicles results in lower energy inputs, lower emissions, and less space required for roads and parking areas. Social benefits are also gained from efficient public transport systems: enriched social contacts, increased time for activities such as reading, and lower risk of being involved in traffic accidents. A number of cities around the world, including those in developing countries, have undertaken a range of highly innovative land transport policy and projects that

promote social and economic development and that address the environmental impacts from transport.

Bogotá, Colombia has redesigned some of its transportation infrastructure to support nonmotorized traffic. For example, it has built about 120 kilometers of bicycle routes throughout the city, and has instituted a policy by which all cars are banned from over 100 kilometers of the city's main roads on Sundays and holidays, providing a safe space for cyclists. With these innovations, the number of cyclists in Bogotá has increased dramatically: the number increased from 0.1 percent of the population in 1997, to five percent of the population in 2001.

In 2003, London instituted congestion charges for driving in central London during certain peak hours. In addition, the transport strategy also has provisions for extra buses and the introduction of new routes. By early 2004, congestion levels during weekdays had fallen by one-third of the previous amount. In 1998 the Indian Supreme Court handed down a judgment with a list of measures to be taken to address air pollution. In order to comply with that judgment, all buses had to be converted to compressed natural gas (CNG) by the end of March 2001. Delhi currently has over 80,000 CNG vehicles on the road, including 9,000 buses. Pressured with quickly growing urban populations, Curitiba, Brazil started implementing express bus lanes in 1974. As of 2001, 75 percent of Curitiba's commuters used the bus system (over 1.9 million passenger trips each day), and the city's transport network included 58 kilometers of express bus lanes, 270 kilometersof feeder routes, and 185 kilometers. of inter-district routes.

SEE ALSO: Automobiles; Bicycles; Cities; Fate and Transport of Contaminants; Flight; Highways; Hybrid Vehicles; Marine Pollution; Oil Spills; Pollution, Air.

BIBLIOGRAPHY. *Aviation and Emissions: A Primer* (Federal Aviation Administration, 2005); N.J. Bax et al., "The Control of Biological Invasions in the World's Oceans," *Conservation Biology* (v.15/5, 2001); R.G. Bell et al., "Clearing the Air: How Delhi Broke the Logjam on Air Quality Reforms," *Environment* (v.46/3, 2004); J.K. Brueckner, "Urban Sprawl: Diagnosis and Remedies," *International Regional Science Review* (v.23/2, 2000); Andres Duany, Elizabeth Plater-Zyberk, and Jeff Speck, *Suburban Nation: The Rise of Sprawl and the Decline of the American Dream* (North Point Press, 2000); *IUCN Guidelines for the Prevention of Biodiversity Loss Caused by Alien Invasive Species* (IUCN, 2000); M.A. Miller and S.M. Buckley, *Institutional Aspects of Bus Rapid Transit—A Macroscopic Examination* (University of California Press, 2000); *Significance of Non-Motorised Transport for Developing Countries: Strategies for Policy Development* (World Bank, 2000); A.P. Tsai and Annie Petsonk, *Tracking the Skies: An Airline-based System for Limiting Greenhouse Gas Emissions from International Civil Aviation* (Environmental Defense, 2000); M. Velasquez-Manoff, "Look at What the Cargo Ship Dragged In," *Christian Science Monitor* (October 16, 2006); D. Van de Walle, "Choosing Rural Road Investments To Help Reduce Poverty," *World Development* (v.30/4, 2002).

Nick Low
Independent Scholar

Trichloroethylene (TCE)

TRICHLOROETHYLENE (TCE) IS a clear, colorless liquid with a sweet, chloroform-like odor. Long used in the 20th century in many industries, TCE is now considered a hazardous substance. Its production, transport, storage, and use are strictly controlled because of growing recognition and evidence that TCE is dangerous and even carcinogenic.

TCE has many synonyms: Acetylene trichloride, ethylene trichloride, ethinyl trichloride, trichloroethene, and colloquially "trike" and just "tri." Its chemical formula is ClCH=CCl2. It was widely produced after Word War I and was used in numerous ways in the following decades: In the food industry, it helped extract vegetable oils from plants such as soy and coconut and prepare flavoring extracts from spices; in the mid-20th century, it was also used as a dry cleaning agent, an industrial solvent to remove grease from metal parts, a refrigerant, a fumigant, a basic component in the pharmaceutical industry, and even in hospitals as a mild gas anesthetic. These direct uses of TCE stopped in the 1970s and 1980s when awareness of its toxicity was raised, together

with concerns of its carcinogenic potential. Presently, TCE is only found as an ingredient in adhesives, paint removers, and typewriter correction fluids.

Although TCE does not occur naturally, because of past use and careless disposal it can be now found in the environment, especially in the soil and groundwater. A third of the drinking water supply sources tested in the United States are said to have some slight TCE contamination. Direct exposure to TCE is unsafe and possibly lethal; its effects on health include skin irritation, headache, and dizziness, lack of coordination, hypotension, nausea, stupor, coma, and even death. Ingestion, inhalation, and skin contact require immediate medical treatment. Hazards include breathing air contaminated through household products containing TCE or swimming in contaminated water.

TCE is also regarded as particularly dangerous because of growing medical evidence, from both animal research and human population studies, that it can cause cancer. In its 11th report on carcinogens in 2005, the National Toxicology Program determined that TCE is "reasonably anticipated to be a human carcinogen." Other agencies concur: The International Agency for Research on Cancer has also concluded that TCE is "probably carcinogenic to humans." The Agency for Toxic Substances and Disease Registry has laid out particular guidelines to deal with exposure to TCE. In 2006, a report from the National Academies' National Research Council stressed that enough information exists for the Environmental Protection Agency to complete a credible human health risk assessment, although more research is still needed to improve understanding of precisely how TCE causes cancer and other adverse health effects.

SEE ALSO: Carcinogens; Environmental Protection Agency (EPA); Pollution, Water.

BIBLIOGRAPHY. Agency for Toxic Substances and Disease Registry, *Managing Hazardous Materials Incidents. Volume III—Medical Management Guidelines for Acute Chemical Exposures: Trichloroethylene TCE* (U.S. Department of Health and Human Services, 2003); Ginger L. Gist and Jeanne R. Burg, "Trichloroethylene: A Review of the Literature From a Health Effects Perspective," *Toxicology and Industrial Health* (v.11/3, 1995); National Academy of Sciences, *Assessing Human Health Risks of Trichloroethylene: Key Scientific Issues* (National Academies Press, 2006); National Toxicology Program "Trichloroethylene," *Report on Carcinogens, Eleventh Edition* (U.S. Department of Health and Human Services, 2005).

Loykie L. Lominé
Independent Scholar

Tropical Forests

TROPICAL FORESTS (COMMONLY referred to as *tropical rain forests*) are found in a narrow band around the equatorial belt. This belt experiences a huge amount of rainfall, which averages about 80 inches and varies from 50 to 260 inches (125 to 660 centimeters) a year. The vegetation is always lush and green, with dense growth and tall trees with giant buttresses. Vines and epiphytes such as orchids and bromeliads grow in the upper canopy on larger trees to reach sunlight. On average, temperatures within the rain forest range from 68 degrees F (20 degrees C) to 93 degrees F (34 degrees C) and humidity is between 77 and 88 percent. Although rain falls throughout the whole year, there is usually a brief season with less rain, and in some monsoonal regions a substantial dry season is often experienced.

It is estimated that rain forests cover less than seven percent of the earth's surface, about the size of the United States. The forests are scattered in a few geographical regions of the world along the equatorial belt, including: The Amazon River basin in South America, by far the largest portion of the rain forest accounting for two-thirds of tropical forests; the Congo basin in Africa, with a small area in west Africa and also eastern Madagascar; in Indo-Malaysia—the west coast of India, Assam, and portions of southeast Asia; New Guinea; and Queensland in Australia.

Scientists believe that these areas contain half of the world's plant and animal species (estimated at five to 10 million). To illustrate the richness of the tropical rain forest biome, in one tree in Brazil there may be as many as 40–50 species of ants. Scientists have counted anywhere from 100 to 300 species of trees in one hectare in South American rain forests. While on the (mainland) continent of Africa there is only one species of the majestic Baobab tree, in the tropical rain forest found on the island of Madagascar, there are seven different species of this tree. These forests also contain well over 95 percent of primates that are found nowhere else in the world. Scientists continue to discover new species of fauna and flora in the tropical rain forests of southeast Asia and Latin America. No one knows how many species of plants and animals are actually in the tropical rain forests. Some estimates indicate that there may be over 50 million species. Of the five to 10 million species that are suspected to be in the tropical rain forest, only six percent have been discovered, and of the six percent only a tiny proportion (about one-sixth) have been intensively studied.

Tropical rain forests are fast disappearing, cut or burned for short-term profit. Tropical rain forests in South America, Africa, and southeast Asia are felled at ever-increasing rates, with thousands of hectares of pristine forests lost every year. This loss is largely blamed on conflicting economic interests for control over forests and land that has made it hard to use the forests on a sustainable basis. As a local issue, the wholesale destruction of the tropical rain forest implies the removal of an important protective cover for the soil that results in severe soil erosion, impeding forest regeneration. Soil erosion also implies the reduction in the life expectancy of the many dams that have been built throughout the tropical areas (for example, in India, the Philippines, Ecuador, Colombia, and Brazil) to generate electricity for industrialization. In places where there has been massive deforestation in upland areas, silting has reduced the life expectancy of dams downstream by half, from an estimated 50 years to only 25 years.

Another local consequence of deforestation of tropical forests is flash flooding downstream. The tropical rain forest canopy absorbs rainfall like a sponge, and without this sponge flooding of farms and built-up areas downstream becomes a common occurrence. About 40 percent of farms are in river valleys in tropical environments. Tropical forests also act as an engine of rain locally through transpiration. Rain forest destruction in parts of Central America has led to a decline in amounts of rainfall. For example, Panama has experienced a decline of 17 inches of rainfall over the past 50 years, ushering in an era of ecological backlash.

Another important local consequence is the survival of indigenous peoples and their livelihoods. It is estimated that there were 230 ethnic groups in Amazonia in 1900 with a population of about one million people whose livelihoods depended on what the forests provided. It is estimated that this number has declined to about 140 ethnic groups with about 50,000 people. Many have lost their lives to newly introduced diseases such as flu and measles for which they had no immunity. Aggressive settlers that have moved into their habitats have wiped others out. The loss of the indigenous peoples is tragic for humanity, as their knowledge of the treasures the tropical forests hold, such as medicines, is also lost.

At the global level, tropical forest deforestation does not auger well for humanity. There will be loss of useful genetic materials, for example. About 25 percent of all medications are derived from rain forest plants. Curare comes from a tropical vine, and is used as an anesthetic and to relax muscles during surgery. Quinine, from the cinchona tree, is used to treat malaria. A person with lymphocytic leukemia has a 99 percent chance that the disease will go into remission because of vinblastine that is made from the rosy periwinkle from Madagascar. More than 1,400 varieties of tropical plants are thought to be potential cures for cancer. For example, a magic bullet for HIV might be hiding in the forests of Borneo or Central Africa. In 1991 University of Illinois at Chicago researchers brought a sample of a smooth barked gum tree (*Calophyllum lanigerum*) from Borneo. The National Cancer Institute determined the sample was effective against HIV, including strains resistant to AZT and Nevirapine. When the researchers went back to get more samples of the tree, the forest stand where the tree came from had disappeared due to logging.

Other global consequences of the destruction of the tropical rain forest include the release of huge amounts of carbon dioxide (CO_2) into the atmosphere due to widespread burning. CO_2 is a major contributor of global warming. Photos from the U.S. Space Shuttle have indicated that at any one given time there are over 5,000 fires burning in Amazonia as forests are cleared for plantation crops and pasture for beef cattle. The debate rages on as

to who is responsible for the massive destruction tropical forests are experiencing. The major agents of deforestation are commercial logging, plantation agriculture, cattle ranching, charcoal and fuel wood production, open pit mining of mineral ores, dams, and the growing of narcotics such as coca. Tropical hardwoods such as teak and mahogany are prized for furniture in highly developed countries.

Plantation agriculture has resulted in the clearing of thousands of acres of pristine tropical forests to give way to the growing of cash crops such as coffee, cocoa, oil palm, coconuts, coca, and rubber as well as to create pasture for the export beef industry. During the economic crisis of the early 1990s, land speculation in Brazil was seen as a way to hedge against runaway inflation. The government encouraged people to develop and improve the forestlands they had acquired by turning them into pasturelands, a phenomenon that has been termed the "cheeseburgarization" or "hamburgarization" of the rain forest in Central America and South America. To the ranchers, this is improved land, but in so doing the ecological balance is upset through the widespread use of herbicides and burning before pasture grass is sown.

Although it has been shown that shifting cultivation is a scientifically sound system of cultivation that makes sure that the forests are preserved, many governments in South America, Africa, and Southeast Asia continue to place blame on rain forest farmers as the culprits to the destruction of the tropical forests. Government departments responsible for forest management ignore the knowledge of indigenous forest farmers about the forest's ecology. In countries of southeast Asia, governments have actively encouraged resettlement of converted forestland for farming by landless settlers who then often cross the unmarked boundaries into permanent forested areas.

It is in this context that conflicting economic interests have resulted in large-scale deforestation of tropical forests. It has been suggested that the reserve solution would be an excellent way of preserving the tropical rain forests. Areas could be set aside for exploitation of forest products by indigenous peoples on a sustainable basis, such as rubber tapping, collection of Brazil and cashew nuts, and other forest products. These could then be processed and packaged for export to international markets. This approach was fervently campaigned for by Chico Mendes, a brave and persistent rubber tapper who challenged the people and institutions responsible for the devastation of the forest in Brazil. He galvanized local and international support for his vision of a self-sustained economy of the Amazon and was subsequently assassinated in December 1988 by those opposed to his vision. A New York Botanical Garden study found that money earned from collecting nuts on a hectare of forest land was more profitable, yielding $6,000 over a 50 year period, versus $3,000 if the trees were cut down for pasture, which would be productive for only three or four years given the infertile nature of tropical rain forest soils.

SEE ALSO: Climate, Tropical; Cloud Forests; Deforestation; Forests; Mendes, Chico; Nontimber Forest Products (NTFPs); Plantation Forestry; Rain Forests; Shifting Cultivation; Tropical Medicine; Tropics.

BIBLIOGRAPHY. E.F. Bruenig, *Conservation and Management of Tropical Rainforests: An Integrated Approach to Sustainability* (CAB International, 1996); Eldredge Bermingham and C.W. Dick, *Tropical Rainforests: Past, Present and Future* (University of Chicago Press, 2005); David Carr, Laurel Suter, and Alisson Barbieri, "Population Dynamics and Tropical Deforestation: State of the Debate and Conceptual Challenges," *Population and Environment* (v.27/1, 2005); Marcus Colchester and Larry Lohmann, *The Struggle for Land and the Fate of the Forests* (ZED Books, 1993); Amy Ickowitz, "Shifting Cultivation and Deforestation in Tropical Africa: Critical Reflections," *Development and Change* (v.37/3, 2006); C.C. Park, *Tropical Rainforests* (Routledge, 1992); S.E. Place, *Tropical Rainforests: Latin American Nature and Society in Transition* (Scholarly Resources, 1993); Andrew Revkin, *The Burning Season: The Murder of Chico Mendes and the Fight for the Amazon Rain Forest* (Island Press, 2004); H. Thompson and D. Kennedy, "Ecological-Economics of Biodiversity and Tropical Rainforest Deforestation," *Journal of Interdisciplinary Economics* (v.7/3, 1996); Michael Williams, *Planet Management* (Oxford University Press, 1993).

Ezekiel Kalipeni
University of Illinois, Urbana–Champaign

(v.75, 1985); Marcus Power and James D. Sidaway, "The Degeneration of Tropical Geography," *Annals of the Association of American Geographers* (v.94, 2004); Karl S. Zimmerer and Kenneth R. Young, eds., *Nature's Geography: New Lessons for Conservation in Developing Countries* (The University of Wisconsin Press, 1998).

W. Stuart Kirkham
University of Maryland

Tsunamis

A TSUNAMI IS a wave (sometimes improperly called a *tidal wave*) generated by an earthquake that can do substantial damage to property and create considerable loss of life when the wave crashes ashore. Tsunamis are created when an earthquake creates a pressure wave that moves ocean water upward, creating a generally small (sometimes as small as a few centimeters), but highly energetic wave that moves rapidly from the epicenter of the earthquake. The wave can be generated along the entire rupture zone of the earthquake, which, in the case of the 2004 Sumatra earthquake and tsunami, was over 100 kilometers long.

The size of the waves that hit the shore depends on the bathymetry of the coastal area—that is, the contours of the undersea topography. Deeper waters result in shorter waves, while a gently rising shoreline will see very large waves as the wave slows. Much of the bathymetry in less developed areas, and even in some developed nations, is not well known, which makes risk assessment difficult.

In 1946, a tsunami originating in the Aleutian Islands killed over 170 people on Maui. Because of this tsunami, the Pacific Tsunami Warning center was established in Hawaii in 1949. In 1960, 61 people were killed on Hawaii and over 160 on Honshu, Japan, from a tsunami that came from the coast of Chile. Many of the fatalities in the 1964 earthquake in Alaska were caused by tsunamis at Valdez, Alaska; near the epicenter in Kodiak, Alaska; and as far away as Hawaii and in Crescent City, California, where 11 people were killed. In 1998, a huge tsunami killed 10,000 in Papua New Guinea. Another destructive tsunami was generated by the 2004 magnitude 9.0 Sumatra earthquake, which struck Indonesia, Thailand, Burma, India, and Sri Lanka. Wave heights were highest in Indonesia, at 40 meters (about 130 feet); in other nations, the waves were smaller, but even a three-meter (10-foot) wave can damage property and drown people. The total death toll was over 200,000 people. Another tsunami hit Indonesia in 2006.

Keeping people safe from tsunami waves is a challenge, but some progress has been made. Perhaps the most effective method is warning systems. Such systems would need to integrate individual nations' tsunami monitoring networks and must ensure that information reaches where it is needed at the coastal areas, which may not be anywhere near a national capital.

Other ways to reduce the hazard to people is through disaster planning and hazard mitigation. Such efforts include hazard signage, informing people about the tsunami hazard and what to do if a tsunami happens, warnings (through the news media and sirens), and evacuation plans and drills (such as those conducted in communities in Oregon and Alaska). Public education is important for informing people of the natural precursors of a possible tsunami, such as a local earthquake that would cue people to move to higher ground, or the rapid outflow of ocean waters that often precedes a major tsunami wave. Mitigation efforts include buildings designed to be strong enough to withstand tsunamis, and tall enough to allow people to "vertically evacuate" into the upper floors of buildings. Such measures would eliminate the need to run inland if a warning is received or if one detects natural tsunami precursors. Some communities may choose to adopt set-back requirements to keep at least some buildings, like schools or hospitals, far away from the possible inundation zone.

Protecting the public from tsunamis is challenging because of the relatively low probability of such an event, and the widespread lack of knowledge about them. In the United States, for example, southeastern states have jointed the National Tsunami Hazard Reduction Program (NTHRP), because of the remote chance of a distant earthquake triggering a tsunami that would traverse the Atlantic. Historic evidence suggests that such events have happened in the past.

SEE ALSO: Disasters; Earthquakes; Floods and Flood Control; Hazards; Indonesia; Ring of Fire; Sri Lanka.

BIBLIOGRAPHY: E.N. Bernard, *Developing Tsunami-Resilient Communities: The National Tsunami Hazard Mitigation Program* (Springer, 2005); Thomas Aaron Green, "Tsunamis: How Safe Is the United States?" *Focus on Geography* (v.48/4, 2006); J.F. Lander, P.A. Lockridge, and M.J. Kozuch, *Tsunamis Affecting the West Coast of the United States 1806–1992* (National Oceanic and Atmospheric Administration, 1993).

Thomas A. Birkland
State University of New York, Albany

Tuna Fishing

TUNA FISHING AS a major industry did not exist before the 20th century, although there were some catches of "giant" mackerel off areas such as the British Isles that may have been bluefin tuna. Industrial tuna fishing began in the United States in 1903, when tuna was seen as a possible replacement for dwindling sardine catches, but the industry really took off during World War II, when canned tuna became an important source of protein. Tuna have been caught with driftnets, purse seine nets, long lines, and traps. The United States and Japan are currently the two largest consumers of tuna, accounting for 31 percent and 36 percent of the world catch, respectively. In 2000, an estimated 3.6 million tons of tuna were caught, 66 percent from the Pacific Ocean, 20.7 percent from the Indian Ocean, 12.5 percent from the Atlantic, and 0.8 percent from the Mediterranean and Black Seas.

Tuna belong to the teleost family Scombridae. Major target species include bluefin, albacore, skipjack, bigeye, and yellowfin. The largest of these are bluefin tuna (*Thunnus thynnus*), which typically grow to 20 feet (6 meters) and 1,102 pounds (500 kilograms), but some specimens have weighed over 1,499 pounds (680 kilograms). Bluefin in the south Pacific are estimated to be at 15–20 percent of their historic stock size due to overfishing. Moreover, recent research suggests that bluefin quotas in the Atlantic have been set too high, and due to the wide-ranging migration patterns of these fish, individuals from depleted populations may move to areas of high fishing pressure. Bluefin tuna are primarily traded with Japan and sold for consumption as sushi and sashimi.

Albacore tuna (*Thunnus alalunga*) are subtropical and have distinctive long pectoral fins but are much smaller than bluefin, at two feet (0.7 meter) in length and 22–44 pounds (10–20 kilograms), although they can grow up to five feet (1.4 meters) long and weigh 132 pounds (60 kilograms). This species was depleted by driftnet fisheries but may now be recovering with the introduction of large-scale driftnet bans. This species is frequently targeted for the canned tuna industry.

The main target species for canned tuna are the smaller skipjack (*Katsuwonus pelamis*), which are typically one foot (35 centimeters) long and about seven pounds (three kilograms) but can grow up to 40 pounds (18 kilograms), and the bigeye (*Thunnus obesus*). Bigeye tuna typically grow to about three feet (0.9 meter) and 33–44 pounds (15–20 kilograms) but have been caught at weights of up to 734 pounds (333 kilograms). This species dwells in deep, cool water and has a thick fat layer for insulation, making it a favorite of the sashimi market.

Yellowfin tuna (*Thunnus albacares)* are found in tropical and subtropical waters, growing up to eight feet (2.4 meters) and 441 pounds (200 kilograms). Yellowfin in the eastern tropical Pacific (ETP) are typically found swimming underneath schools of spinner (*Stenella longirostris*) and spotted dolphins (*S. attenuata*).

DANGERS TO DOLPHINS

It is not certain why yellowfin tuna swim under schools of dolphins, but it may be that the dolphins provide the tuna some protection from predators (the dolphins appear to gain no benefits from the tuna). This association is so strong that purse seine fishing operations targeting yellowfin set their nets around dolphin schools to catch the associated tuna, but also catch and kill many dolphins. Indeed, this method of tuna fishing has killed more dolphins than any other human activity.

It is estimated that six million spinner and pantropical spotted dolphins alone were killed as bycatch before conservation measures were introduced,

with up to half a million dolphins killed a year in the 1960s and 1970s. This high level of dolphin mortality was one of the key factors behind the introduction of the U.S. Marine Mammal Protection Act in 1972 (MMPA). The MMPA helped to reduce the amount of dolphin by-catch, and due to strict limits on mortality, fewer and fewer U.S. fishing vessels participated in the ETP fishery, although vessels from other nations remained in the region.

In 1990 the concept of "dolphin safe" tuna was introduced, which had a major effect on purchases of tuna worldwide, through consumer choice. Dolphin safe tuna was initially defined as tuna caught by methods other than setting purse seine nets on dolphins—it was an effort to eliminate a fishing method considered inherently harmful to dolphins. After use of the "dolphin safe" label became widespread, first voluntarily by tuna companies and later by law in the United States, fishers in the ETP made efforts to reduce dolphin mortalities, achieving a 95 percent decrease, but nonetheless did not cease setting purse seine nets on dolphins.

Furthermore, to reduce dolphin mortality, the Inter-American Tropical Tuna Commission (IATTC), the body governing tuna fishing in the ETP, adopted the International Dolphin Conservation Program (IDCP) in 1992, a voluntary by-catch reduction scheme codified in the so-called La Jolla Agreement. Nations fishing in the ETP agreed to voluntary limits on the number of dolphin deaths (the dolphin mortality limit, or DML) that could be inflicted by each vessel in the ETP tuna fishery. The aim was to reduce these limits each year until a zero mortality level was reached. Measures to monitor and reduce dolphin mortality included an onboard observer program to record dolphin kills, and the "backing down" of purse seine vessels, that is, sinking the top of the net to allow encircled dolphins to swim over the top and escape. In 1995, the Panama Declaration made these measures mandatory.

The Panama Declaration also introduced a new definition of "dolphin safe" that used an "observed mortality" standard rather than a "fishing method" standard. If observers did not see any dead dolphins in a net set, then the tuna from that set would be considered "dolphin safe." The U.S. Dolphin Conservation Act of 1997 stated that the United States would adopt this new definition, if the government made a final finding that the chase and encirclement of dolphins were not having a significant adverse impact on depleted stocks. In 1999, and again in 2002, the government tried to make such a finding, but both times the finding was challenged in court. In both cases, the courts ruled against the government, which means the original definition of "dolphin safe" has so far been retained.

The court challenges were based on the fact that, although dolphin mortality levels have been greatly reduced in this fishery, the dolphin populations have not recovered as expected. Scientific evidence strongly suggests that chasing and herding dolphins prior to setting nets separates mothers from calves and leads to debilitating stress, even when dolphins are released alive during the "back-down." Also there was evidence of underreporting of dolphin mortalities. The scientific data and underreporting evidence were vir-

In 2000, an estimated 3.6 million tons of tuna were caught; 66 percent were from the Pacific Ocean.

tually ignored when the U.S. government made its findings, a decision heavily criticized by the courts.

MERCURY CONTAMINATION

Another controversial environmental issue related to tuna fisheries is a high level of mercury contamination in some species. A quarter of canned albacore tuna examined in a recent study exceeded U.S. health regulation limits. Other species found to be highly contaminated include blackfin tuna (*Thunnus atlanticus*) and little tunny (*Euthynnus alletterus*) off the coast of Florida, of which 81 percent and 75 percent, respectively, contained more mercury than U.S. health regulation limits. As a result of possible health risks from mercury contamination in large predatory fish such as tuna, in March 2004 the U.S. Food and Drug Administration (FDA) issued guidelines recommending pregnant women, nursing mothers, and children limit their weekly intake of some tuna products. Other countries, such as those of the United Kingdom, have issued similar warnings.

SEE ALSO: Animal Rights; Dolphins; Fisheries; Habitat Protection; Mercury; Minamata; Overfishing.

BIBLIOGRAPHY. D.H. Adams, "Total Mercury Levels in Tunas from Offshore Waters of the Florida Atlantic Coast," *Marine Pollution Bulletin* (v.49, 2004); F. Archer et al., "Annual Estimates of the Unobserved Incidental Kill of Pantropical Spotted Dolphin (*Stenella attenuata*) Calves in the Purse-Seine Fishery of the Eastern Tropical Pacific," *Fishery Bulletin* (v.102, 2004); B.A. Block et al., "Electronic Tagging and Population Structure of Atlantic Bluefin Tuna," *Nature* (v.434, 2005); J. Burger and M. Gochfeld, "Mercury in Canned Tuna: White versus Light and Temporal Variation," *Environmental Research* (v.96, 2004); V. Gewin, "Pacific Dolphins Make Waves for U.S. Policy on Mexican Tuna," *Nature* (v.427, 2004); R.A. Myers and B. Worm, "Rapid Worldwide Depletion of Predatory Fish Communities," *Nature* (v.424, 2003); John R. Twiss, Jr. and Randall R. Reeves, *Conservation and Management of Marine Mammals* (Smithsonian Institution Press, 1999); U.S. Tuna Foundation, www.tunafacts.com (cited January 2006); U.S. Department of Health and Human Services and U.S. Environmental Protection Agency, "What You Need to Know about Mercury in Fish and Shellfish" (March 2004), www.cfsan.fda.gov (cited May 2006).

E.C.M. Parsons
George Mason University
Naomi A. Rose
Humane Society International

Tundra

TUNDRA, A TERM derived from the Finnish word *tunturia* and/or the Saami word, the genitive of *tundar*, both meaning "treeless plain," is the coldest and least species-rich ecosystem on earth. It is found both in areas of high latitude (Arctic and Antarctic tundra) and altitude (alpine tundra).

Arctic tundra technically refers to the areas of high latitude permafrost in the Northern Hemisphere. This includes large areas of Russia and Canada. Permafrost is permanently frozen ground, an ecosystem characteristic that—in addition to low water and temperatures—contributes to the high stress of plant and animal survival. The predominant flora is mosses, heath, and lichen. Mammals include wolf, fox, musk ox, polar bear, rabbit, vole, and caribou. In many of these permafrost areas the winter low temperatures dip to negative 60 degrees F. This does not, however, result in a frozen and unproductive ecosystem. The short window of summer—perhaps two months in length—provides ample warmth and extended daylight for the tundra to be highly productive for the native plants and animals and also for swarms of migratory birds and other fauna that flock there to reproduce and gorge on the tundra's ecological bounty.

Most of Antarctica is ice field except for a few areas of the continent, particularly the Antarctic Peninsula, whose rocky soil supports tundra. Its flora is made up of mosses, lichens, liverworts, and aquatic and terrestrial algae. Fauna are restricted to sea mammals and birds, most notably the penguin. Alpine tundra occurs wherever the altitude reaches above the tree line. It is devoid of permafrost and includes animals such as elk, marmot, mountain goat, pika, and sheep.

The tundra ecosystem—a biome of extreme temperature, daylight and moisture regimes—has

historically been considered a barren wasteland. These regions, however, contain many diverse ecological and physical environments. For example, the high arctic deserts are home to colonizer flora (e.g., lichens and mosses) and highly adapted fauna (e.g., caribou and spiders). The finite bounds of these diverse yet fragile ecosystems are limited because of the physical environment (i.e., low temperatures) for reasons such as: Lack of nutrients cycling, light availability, limited freshwater availability, and slowed biological process. These limiting factors make arctic ecosystems slow to recover from disruptions and impacts caused by invasive human development and resource exploitation practices. Furthermore, the faunal adaptive reliance on stored energy (i.e., fat) and long lifespan strategies makes food chains more susceptible to contaminants, such as persistent organic pollutants and radionuclides.

The adaptations and diversity of the tundra are not limited to flora and fauna. Arctic indigenous peoples represent an integrated part of the ecosystem who have developed modes of subsistence that include reindeer herding, sea mammal hunting, taiga terrestrial animal hunting, and settled river fishing.

In the 21st century the processes of global warming are being felt most acutely in tundra regions of the world. This process is most poignant in the Arctic tundra due to the melting of the permafrost and its effects on the plants, animals, and humans who have adapted to an ecosystem based on water in its solid state. Melting permafrost also involves the potential release of high amounts of methane, a potent greenhouse gas, from the bog that results. Although warming is already in progress, humans must take responsibility to forestall global warming before irreversible change occurs.

SEE ALSO: Antarctica; Arctic; Biome; Global Warming.

BIBLIOGRAPHY. David A. Wharton, *Life at the Limits: Organisms in Extreme Environments* (Cambridge University Press, 2002); Sarah J. Woodin and Mick Marquiss, eds., *Ecology of Arctic Environments* (Cambridge University Press, 1997).

Susan A. Crate
George Mason University

Tunisia

TUNISIA IS A country in the center of North Africa. It has an area of 63,170 square miles and a population of approximately 10 million. The capital of the country is Tunis, a city located near the once powerful city of Carthage. Tunisia is bordered by Libya and Algeria and has a long Mediterranean shoreline to the north and east. Physically, the country is dominated by the Sahara to the south and low mountains surrounded by large agricultural areas in the north. Most of the population lives either near the coast or in the major northern agricultural areas. The south, though intermittently dotted with oases, is considerably less inhabited. More than half of Tunisia is covered by pastoral land forests that are used for agriculture, supporting such crops as olives, dates, oranges, almonds, grain, sugar beets, wine grapes, poultry, beef, and dairy. Its main profits, however, are derived from industries such as the mining of phosphates and iron ore. The petroleum and textile industries of Tunisia are also significant.

The main environmental and conservation issues throughout North Africa is desertification, and Tunisia is not exempt from this problem. Poor farming techniques, such as overgrazing, along with deforestation, soil erosion, and a limited supply of natural sources of freshwater, are all contributing to the problem of desertification. Agricultural land is decreasing, not only because of desertification, but also because of increased salinization and siltation due to the increased erosion of the soil. Ineffective disposal of toxic and hazardous materials and water pollution from raw sewage (a problem common in North Africa) are posing risks to the well-being of Tunisia's citizens. Like many countries, Tunisia has fallen far short of the ideal global total of 10 percent of its land under some sort of environmental protection, protecting less than half a percent of its land. The people of Tunisia are mostly of Arab and Berber ethnicity. The official language of the republic is Arabic; however, there is a strong French language influence from the days of French colonization. Religiously, 98% of Tunisians are Muslims.

Unlike its neighbors, Libya and Algeria, Tunisia does not have abundant oil resources, thus Tunisia has a fairly diversified economy. The service industry, including the important European tourist sector, ac-

counts for over half of the economy and agriculture counts for less than 15 percent, with industrial jobs constituting the remainder. That said, agriculture still employs the greatest number of people and remains an essential part of the economy especially with high value products such as olives and dates. However, like other North African countries, unemployment is a continuing problem as the mean average of the population continues to get younger. This has fueled a migration of young people to Europe, both legal and illegal, in search of jobs. Politically, Tunisia is considered stable in relation to most other Middle Eastern countries. This in spite of President Ben Ali having not allowed any real democratization for fear of an Islamist party taking hold in the country and having effectively becoming a "president for life."

SEE ALSO: Desertification; Pollution, Water.

BIBLIOGRAPHY: Mounira Charrad, *States and Women's Rights: The Making of Postcolonial Tunisia, Algeria, and Morocco* (University of California Press, 2001); Leonard Cottrell, *Hannibal: Enemy of Rome* (Da Capo, 1992); Dwight L. Ling, *Tunisia: From Protectorate to Republic* (Indiana University Press, 1967); Kenneth J. Perkins, *A History of Modern Tunisia* (Cambridge University Press, 2004); Dirk Vandewalle, ed., *North Africa: Development and Reform in a Changing Global Economy* (St. Martin's Press, 1996).

William C. Rowe
Louisiana State University

Turkey

GEOGRAPHICALLY, TURKEY IS a Euro-Asian country. About three percent of Turkey is in the Thracian area of southeastern Europe. Ninety-seven percent is in the western areas of Asia called the Near East (Middle East). About the size of Texas and Louisiana combined, it has an area of 301,384 square miles (780,580 square kilometers) and a population of over 68 million people in 2004.

Turkey has seven major regions: The Black Sea region, the Marmara region, the Aegean, the Mediterranean, Central Anatolia, East Anatolia and Southeast Anatolia. The Black Sea coast extends from the Bosporus Strait to Georgia. Along the coast are the Northern Mountains running east until they meet the Pontic Mountains about mid-way along the Black Sea coast. The Northern and Pontic Mountains receive enough moisture for them to be heavily wooded and rugged.The Thracian part of the Marmara region is covered by the Northern Plains, an area of gentle rolling grass lands and farms. Historic Istanbul (Constantinople) is located in European Turkey on a peninsula at the intersection of the Bosporus and the Sea of Marmara. An estuary called the Golden Horn separates it from newer areas.

European Turkey is separated from Asiatic Turkey by three connected water ways: The Bosporus Strait, a narrow outlet from the Black Sea to the Sea of Marmara, and the Dardanelles. The Sea of Marmara is a saltwater sea, almost completely surrounded by land. It opens onto the Dardanelles strait formed by the Gallipoli Peninsula and the Northern Plains.

The Aegean Coast has many bays, peninsulas, coves, islands, and sandy beaches. Its narrow coastal plains rise through the broad fertile western valleys to the Anatolian plateau. The Aegean Sea changes off the Island of Rhodes to the Mediterranean Sea. The southern Mediterranean coast of Turkey has a narrow belt of plains that run to the border with Syria. The southern mountains include the Taurus Mountains (Toros Daglari) parallel the Mediterranean coast.

The Central Anatolian Plateau is a region of small rivers fed by occasional rainfall. There are several salt lakes in central Turkey. The Cappadocian volcanic tuff region is in the south. The Eastern Anatolian Plateau region is an area of rugged towering mountains. The area lies east of the Euphrates River and extends to Mount Ararat and the borders with Armenia and Iran. It also contains the large freshwater body, Lake Van, which is part of the original homeland of the Armenians.

The Southeastern Anatolian region is part of Mesopotamia and has fertile plains and river valleys that lie between the Euphrates and Tigris Rivers. The Euphrates rises near Erzurum in the Eastern Plateau not far from Mount Ararat. The Tigris rises

in the Taurus Mountains near Lake Hazer in the Eastern Plateau. Its eastern border is shared with Iran, Azerbaijan, Armenia, and Georgia.

Turkey's environment is varied and rich in natural beauty. Significant efforts at reforestation and land improvements have been made. However, Turkey's industrialization especially since the 1990s has caused a number of environmental problems. Expanding industry has consumed higher levels of energy and the damming of the Tigris and Euphrates Rivers and their tributaries has caused ecological concerns. Currently coal fired electrical plants are being replaced with natural gas fired plants. Maritime pollution is of grave concern to environmentalists as is the greatly increased flow of oil tanker traffic through the Bosporus Straits and the Dardanelles. In addition overfishing and pollution runoff have fouled the Black Sea.

With prosperity has come smog from factories and automobiles. Under its comprehensive environmental laws promulgated first in the 1980s, attention is being paid to environmental issues. However, Turkey's pollution control efforts have been severely criticized as inadequate by European agencies.

SEE ALSO: Armenia; Azerbaijan; Black Sea; Iran; Iraq; Syria; Tigris and Euphrates Rivers.

BIBLIOGRAPHY. Rashid Ergener and Resit Ergener, *About Turkey: Geography, Economy, Politics, Religion, and Culture* (Pilgrims' Process, 2002); William Spencer, *The Land and People of Turkey* (HarperCollins, 1990).

Andrew J. Waskey
Dalton State College

Turkmenistan

TURKMENISTAN DECLARED INDEPENDENCE from the Soviet Union on October 27, 1991. The president of Turkmenistan, the former communist insider Saparmurat Niyazov, used the Turkmen title Turkmenbashi: Leader of the Turkmen. Remarkably, however, Turkmenistan was until recently a collection of fragmented, interwarring tribes divided by dialect, ancestry and geography.

The environment of Turkmenistan, mainly a vast, untamed steppe in the heart of Central Asia, encouraged semi nomadic lifestyle that was not territorially defined. Trained and disciplined in war and eager to seize booty, it was the Turkmen, along with other steppe peoples to the East like the Mongols, who rained down from the East, conquering large swaths of territory in the Middle East and Eastern Europe during the Middle Ages. In this sense, the environment of Turkmenistan, the vast, dramatic and forbidding steppe and the need for open pasture and freedom of movement, has had a very significant impact on the social and historical identity of Turkmen. Today in contrast, most Turkmen today are settled in cities or on cultivated land. They participate in the growing industrial and gas sector, or cultivate cotton. Nevertheless, there remains a significant minority who live an almost exclusively nomadic life. They are often held with the highest respect and are considered "more Turkmen" by the rest of the population.

The forced cultivation of cotton, a single cash crop, was to benefit of the ruling elite in Moscow and their representatives in Ashgabat, the main city in Turkmenistan. Cotton, of course, was impossible to eat. This made Turkmen farmers and former nomads dependent on the central power for their food supply, leading to periods of famine and starvation. Moreover, the Soviet authorities often set the price for cotton artificially low, preventing the Turkmen from taking advantage of cotton as a source of economic growth.

Although it could be argued that the nomadic lifestyle of the Turkmen and the people of the steppe in Central Asia have had a long-term, millennial impact on the environment, preventing reforestation, and causing possible damages from long-term overgrazing, major economic projects with immediate environmental consequences have only been implemented very recently. Turkmenistan is in the process of signing contracts for Caspian Oil lines that will link the Oil of the Caspian Sea on the Western side of Turkmenistan, with China, a vast country thirsty for cheaper oil to fuel economic expansion. New damns are being built on fragile and usually temperamental rivers to fuel a plan of industrial and economic expansion orchestrated by the central authority. As Turkmenistan transitions rapidly into a settled, industrial economy, the traditional relationship between Turkmen society and the environment will

need to be reconsidered. Although the political situation in Turkmenistan is fairly totalitarian and centralized, it is possible that the nationalist character of Turkmenistan's steppe tradition, the closeness to the land and the interest in preserving identity, will help prevent an over-zealous exploitation of resources.

Turkmenistan and Uzbekistan sit on large reserves of oil and natural gas reserves yet both countries face challenges in getting those reserves to world markets. Neither country prefers to export their resources through Russian-controlled pipelines, and so each must seek to obtain capital and political support for pipelines either through Iran or through Turkey.

SEE ALSO: Caspian Sea; Natural Gas; Uzbekistan.

BIBLIOGRAPHY: Jacob Black-Michaud, *Sheep and Land: The Economics of Power in a Tribal Society* (Cambridge University Press, 1986); Adrienne L. Edgar, *Tribal Nation: The Making of Soviet Turkmenistan* (Princeton University Press, 2004). Robert Lewis, ed., *Geographic Perspectives on Soviet Central Asia* (Routledge, 1992); Oliver Roy, *The New Central Asia: The Creation of Nations* (I.B. Tauris, 2000).

Allen J. Fromherz
Independent Scholar

Tyler vs. Wilkinson

EBENEZER TYLER VS. Abraham Wilkinson was a court case decided in 1826 that continues to have important implications for the management of water resources in the United States. The case is also widely known as the Sargent's Trench Trial, as this was the name of the stretch of water above the water wheel over which the case was fought.

Tyler versus Wilkinson centers on the ownership of riverine water resources and the legal power to use them when they act as a finite resource. During the 19th century, water wheels had become an important source of power and were employed both in family farms and, increasingly, in larger-scale industrial enterprises. Development had already had a negative impact on the Sargent's Trench as it had previously been so blocked by construction waste as to obstruct the migration of fish traveling upstream to spawn. Even so, economic growth in the period meant that more people were attempting to draw power from the rivers and pressure was being placed on the extant legal regulations. Because access to such power also had numerous implications for economic and social opportunities, the precedent drawn in this case had significance for the whole country. However, it was also rooted in specific local circumstances and antagonism among the protagonists and among various coalitions of local interests. By exploring whether water resources should be made freely available to all or whether certain individuals or organizations should have privileged access, Judge Story was in part helping to determine the fate of traditional users of the river—fishers, farmers, and urban residents dependent on fresh river water for hygiene and sustenance. His decision was to allocate water resources among either upstream or downstream users and any excess or surplus. Only the surplus water resources could be sequestered by any subsequent claim to use of resources, although this principle was modified by the presence of any pre-existing claims or usages of the water, which were in turn to be reevaluated.

The consequence is that no individual was to be considered in possession of a permanent or inalienable right to water that would deny the use of that water to any other person. Such a principle has underlined a great deal of legal attitude toward the environment and its management in the United States.

Recognizing public interest in this case, Judge Story made sure the facts of the case and of his decision entered into the public domain. He wrote the decision for the newspapers personally to avoid possible distortion or bias by reporters or other intermediaries. The precedent came to be widely regarded and followed.

SEE ALSO: Riparian Rights; Rivers; Water Law.

BIBLIOGRAPHY. *Opinion Pronounced by the Hon. Judge Story in the Case of Ebenezer Tyler and Others vs. Abraham Wilkinson and Others: at the Last June Term of the Circuit Court, for the Rhode-Island District* (Randall Meacham, 1827).

John Walsh
Shinawatra University

Typhus

TYPHUS IS A disease caused by two different species of *Rickettsia* bacilli and associated with human crowding and poor sanitary conditions. Humans may contract two different types of typhus: Endemic or murine typhus, caused by *R. typhi,* which is transmitted by the rat flea; and epidemic typhus, caused by *R. prowazekii,* which is transmitted by the human body louse. Typhus causes very high fever, severe aches, delirium, vomiting, and a characteristic rash. Untreated, epidemic typhus presents a death rate of up to 60 percent; murine typhus is typically milder and less fatal.

Epidemic typhus has long been common in refugee camps, in times of war, in prisons, and after environmental disasters. Then called the "Hungarian disease," typhus swept through central and western Europe during the Thirty Years' War (1618–48); tens of thousands of civilians died in several German and French cities. During some wars in Europe, most famously the Napoleonic wars, typhus killed more soldiers and civilians than did warfare itself, leading some to claim for typhus a key role in history. Weakened from the potato famine and displaced from their communities, many Irish died of typhus in the 1840s. Between two and three million people died of typhus during World War I, primarily Polish, Romanian, and Russian soldiers and civilians; many more would have died had the louse vector not been discovered and delousing practices implemented. Thousands of prisoners in German concentration camps during World War II died of typhus.

Campaigns to prevent or eradicate typhus helped cement the role of DDT as a public health tool. Allied forces used DDT to kill lice during typhus outbreaks in Naples, Italy, in 1943. The exposure of residents to DDT on their clothing and skin has been cited as evidence of the chemical's safety. The threat of murine typhus has also spurred public health agencies to initiate rodent control campaigns. Laborers in Depression-era work-relief programs in the United States killed millions of rats in Georgia, Mississippi, and Texas in an effort to stop the spread of murine typhus and prevent more infected rats from entering the country via seaports. Murine typhus persists in low levels throughout much of the southern United States.

Today, typhus may be treated with antibiotics, most effectively in murine typhus but also with considerable success in epidemic typhus. A vaccine for typhus has also been used to protect vulnerable populations. In recent years, however, epidemic typhus has appeared in refugee camps following wars or disasters when public health aid is unavailable or limited. International public health observers and disease surveillance agencies suspect that recent civil unrest, refugee movements, and environmental disasters have set the stage for increasingly severe typhus epidemics to reemerge among vulnerable populations. Some epidemiologists believe that louse infestations are increasing worldwide as a result of a web of deteriorating social, political, economic, and environmental conditions.

Poor hygiene in refugee camps presents the greatest concern, and high percentages of lice found there carry the *Rickettsia* that causes epidemic typhus. Since the 1970s, refugee camps in central and eastern Africa have suffered the most severe outbreaks of epidemic typhus. During the 1990s thousands of refugees in Rwanda, Burundi, and what was then Zaire suffered typhus epidemics. In 1997 Burundi experienced the largest typhus epidemic since World War II, with 24,000 cases. The World Health Organization and other public health groups controlled the outbreak by treating patients with the antibiotic doxycycline. Since 1995 smaller outbreaks have also been reported in Russia, Peru, and Algeria.

SEE ALSO: DDT; Disasters; Disease; Poverty; World Health Organization (WHO).

BIBLIOGRAPHY. Pierre-Edouard Fournier et al., "Human Pathogens in Body and Head Lice," *Emerging Infectious Diseases* (v.12, December 2002); William H. McNeill, *Plagues and Peoples* (Random House, 1976); D. Raoult and V. Roux, "The Body Louse as a Vector of Reemerging Human Diseases," *Clinical Infectious Diseases* (v.29, October 1999); World Health Organization, "1997—Louse-borne Typhus in Burundi" (May 1997), www.who.int (cited May 2006); Hans Zinsser, *Rats, Lice, and History* (Little, Brown, 1934).

Dawn Day Biehler
University of Wisconsin, Madison

Udall, Morris King (1922–98)

BORN IN ST. Johns, Arizona, on June 15, 1922, Morris K. Udall would go on to be one of the foremost political leaders in environmental protection over a long and illustrious career. At an early age, he showed great leadership potential; he was not only the student body president of his high school, but also the school's valedictorian and basketball team co-captain. Udall graduated high school in 1940 and attended the University of Arizona in Tucson, Arizona, where he studied law. He left college two years later to serve in World War II from 1942 to 1945 in the U.S. Army Air Corps, where he was honorably discharged as a captain. After graduating college in 1949, he played professional basketball for the Denver Nuggets and also achieved the highest score on the state bar exam. He practiced private law with his brother Stewart and eventually worked as the county attorney in Pima County, Arizona, from 1953 to 1954.

Udall was elected as a Democrat to the 87th Congress and was subsequently reelected to the 15 succeeding Congresses until he retired in 1991. In 1976 he vied for the Democratic Party presidential nomination, only to lose to Jimmy Carter. From 1977 to 1991, he served as the Chairman of the Committee on Interior and Insular Affairs (95th through 102nd Congresses), where he worked on many issues related to the environment and public land policy. The Committee on Interior and Insular Affairs is now referred to as the Committee on Resources, and works on a diversity of environmental issues that include but are not limited to energy, forests, public lands, fish and wildlife, Native Americans, and water and power. Significant legislation passed with the help of Morris Udall includes:

- The Strip Mining Reclamation Act: requires coal companies to reclaim their strip-mined land;
- Archaeological Research Protection Act: secures protection of archeological resources on public lands;
- Southern Arizona Water Rights Settlement Act: outlines Indian water rights claims;
- Arizona Wilderness Act of 1984: designates 1.5 million acres of wilderness lands in Arizona;
- Arizona Desert Wilderness Act of 1990: designates 2.4 million acres of wilderness land in Arizona;
- Tongass Timber Reform Act: revokes the artificially high timber targets and protects over one million acres of watersheds.

Both Morris and Stewart Udall had a great appreciation for the natural environment that shone through in their political careers. Morris Udall, John Seiberling (D-Ohio), and 75 cosponsors successfully introduced the Alaska Lands Act of 1980 into Congress. The bill was heavily fought by mining, timber, and oil interests as it ultimately designated 55 million acres of new protected wilderness, expanded the national park system in Alaska by about 43 million acres (22.3 million hectares), creating 10 new national parks, and greatly expanded and created National Wildlife Refuge lands, Wild and Scenic River designations, and National Forest System lands.

Morris Udall spent much of his career promoting the environmentally detrimental $4.4 billion Central Arizona Project (CAP), the most expensive water project in U.S. history. Originally supporting the damming of two areas of the Colorado River for the project, Udall eventually sided with environmentalists after a massive public outcry; the dams were not built.

Later in his career (after the CAP was constructed) Udall made statements of regret over the water project and worried that he wrongly supported it: "Now we have cotton farms selling out and taking their money to enjoy in La Jolla [California]—and cities building lakes so people will have lakefront homes in the desert ... If I had to do it over, I think I'd say, 'Leave the water in the river.'" Udall died December 12, 1998, due to complications from Parkinson's disease.

SEE ALSO: Dams; Forest Service; National Parks; National Wild and Scenic Rivers Act; Native Americans; Public Land Management; United States, Alaska; Water Law.

BIBLIOGRAPHY. Donald Carson and James Johnson, *Mo: The Life and Times of Morris K. Udall* (University of Arizona Press, 2001); Morris K. Udall Foundation, "About Morris K. Udall," www.udall.gov, February 2006 (cited May 2006); National Parks Conservation Association, "ANILCA Fact Sheet," www.npca.org (cited February 2006); U.S. Congress Biographical Directory, bioguide.congress.gov (cited February 2006); U.S. House of Representatives Committee on Resources, resourcescommittee.house.gov (cited February 2006); "Udalls Prod Environmentalists to Tackle Growth Water Issues," *Arizona Daily Star* (April 14, 1987).

ANDREW J. SCHNELLER
INDEPENDENT SCHOLAR

Uganda

A DECADE AFTER achieving independence from Britain in 1962, the Republic of Uganda began a 14-year period marked by dictatorial governance, civil war, mass murders, atrocities, and extensive human rights abuses that sapped the country of both human and physical resources. By the end of that period, some 400,000 Ugandans had lost their lives. In 1987, many young Ugandans came under the influence of Joseph Kony, who further drained the country of its resources by recruiting soldiers for what he called The Lord's Resistance Army. Kony's tactics involved kidnapping children between the ages of eight and 12 and coercing them to be soldiers by threatening their lives and the lives of their families. Even after Kony was expelled, he continued to reinforce this children's army from neighboring Sudan. International organizations are currently involved in a massive effort to rescue and rehabilitate these children and bring an end to Kony's influence in Uganda. By the 1990s Ugandans had begun to recover politically and economically, dispensing with political parties to elect a new president and legislature.

Uganda's abundant natural resources include sizable deposits of copper and cobalt, hydropower, limestone, and salt. Nearly 26 percent of Uganda's land area is arable, and 82 percent of the work force is engaged in some form of agriculture. The abundant rainfall and fertile soils make it easy to grow a variety of products. Coffee is the major export crop, accounting for the lion's share of export revenue. Since 1986, international agencies have been assisting the Ugandan government in economic reform. In 2000, Uganda qualified for debt relief through funding from the Heavily Indebted Poor Countries initiative and the Paris Club. The return of exiled Indian-Ugandan entrepreneurs has also had a positive affect on the Ugandan economy.

With a per capita income of $1,700, Uganda is ranked 190th in world incomes. Income disparity exits, with the richest 10 percent holding more than one-fifth of the country's wealth and the poorest 10 percent sharing only four percent of resources. Over a third of Ugandans live in poverty, and nearly one-fifth of the population is seriously undernourished. The United Nations Development Programme's Human Development Reports rank Uganda 144 out of 232 countries on general quality-of-life issues.

While landlocked, Uganda has an abundance of lakes and rivers that provide 36,330 square miles of water resources. The largest body of water is Lake Victoria, which Uganda shares with Tanzania. Uganda also borders the Democratic Republic of the Congo (DROC), Kenya, Rwanda, and the Sudan. The terrain of Uganda is comprised of an alluvial plateau rimmed by mountains. Elevations range from 621 meters at Lake Albert along the border of the DROC to 5,110 meters at Mount Stanley. Most of Uganda experiences a typically tropical climate that is marked by two dry seasons from December to February and from June to August. The climate of northeastern Uganda is semiarid.

The population of 28,195,754 faces major environmental health hazards, including an HIV/AIDS adult prevalence rate of 4.1 percent. Some 530,000 Ugandans live with this disease, and another 78,000 have died from it since 2003. Forty-four percent of the population do not have sustained access to safe drinking water, and 59 percent do not have access to improved sanitation. Therefore, Ugandans have a very high risk of contracting food and waterborne disease that include bacterial diarrhea, hepatitis A, and typhoid fever and the waterborne disease schistosomiasis. In some areas, the population has a high risk of contracting African trypanosomiasis, popularly known as the "sleeping sickness." Ugandans have a lower-than-normal life expectancy (52.67 years) and growth rate (3.37 percent), and higher-than-normal infant mortality (66.15 deaths per 1,000 live births) and death (12.24 deaths per 1,000 population) rates. Ugandan women produce an average of 7.1 children each. The female literacy rate of 60.4 percent makes disseminating health information somewhat difficult.

The wetlands of Uganda have been repeatedly drained to gain land for agricultural use. Agricultural mismanagement has also led to overgrazing with extensive loss of vegetation, which has produced soil erosion. The process of eradicating tsetse flies has led to toxic pollutants being released into the environment. Water supplies have been contaminated by industrial effluents, including mercury released in mining operations. Even though over one-fifth of land area is still forested, deforestation is occurring at a rate of 2 percent per year. Lake Victoria is experiencing water hyacinth infestation that interferes with marine life and the fishing industry.

Uganda is rich in wildlife, and the government has protected almost a fourth of land area. These areas include a vast network of national parks, wildlife reserves, wildlife sanctuaries, and community wildlife areas. Nevertheless, biodiversity and habitats are threatened by extensive poaching. Of 345 identified mammal species, 20 are endangered, as are 12 of 243 bird species. In 2006, scientists at Yale University ranked Uganda 78 of 132 countries on environmental performance, above the comparable income and geographic groups. The overall ranking was decreased by the poor showing in environmental health.

The 1995, Ugandan Constitution established the right to a clean environment and created the Ministry of Water, Land, and Environment, which is the governing body charged with promoting sustainable development and protecting the environment. Specifically, under the framework of the National Environment Plan, the ministry implements and enforces laws and polices relating to the management of land, water, forestry, and wetlands and to weather and climate, and atmospheric pollution. The ministry also provides oversight for three statutory bodies: The National Environment Authority, the National Water and Sewerage Corporation, and the Uganda Land Commission.

Uganda participates in the following international agreements on the environment: Biodiversity, Climate Change, Climate Change–Kyoto Protocol, Desertification, Endangered Species, Hazardous Wastes, Law of the Sea, Marine Life Conservation, Ozone Layer Protection, and Wetlands.

SEE ALSO: Colonialism; National Parks; Pesticides; Poaching; Victoria, Lake; Water Hyacinth.

BIBLIOGRAPHY. Central Intelligence Agency, "Uganda," *World Factbook*, www.cia.gov (cited April 2006); Timothy Doyle, *Environmental Movements in Minority and Majority Worlds: A Global Perspective* (Rutgers University Press, 2005); Kevin Hillstrom and Laurie Collier Hillstrom, *Africa and the Middle East: a Continental Overview of Environmental Issues* (ABC-CLIO, 2003); Valentine Udoh James, Africa's *Ecology: Sustaining the Biological and Environmental Diversity of A Continent* (McFarland, 1993); Uganda Ministry of Water, Land, and Environment, www.mwle.go.ug (cited April 2006); United Nations Development Programme, "Human Development Report: Uganda," hdr.undp.org (cited April 2006); World Bank, "Uganda," *Little Green Data Book*, lnweb18.worldbank.org (cited April 2006).

Elizabeth Purdy, Ph.D.
Independent Scholar

Ukraine

AFTER A BRIEF period of independence from 1917 to 1920, the Ukraine was brought under repressive Soviet domination. After achieving independence in 1991 following the dissolution of the Soviet Union, the Ukraine continued to struggle with massive corruption that stymied efforts at economic and political reform. The "Orange Revolution" of 2004 precipitated a reform movement whose effects are still unclear. The struggle over the position of Ukraine in global politics, oriented either toward Russia or toward Western Europe, remains unresolved. The second-largest country in Europe, the Ukraine has a population of 47,425,336. With a per capita income of $6,800, the Ukraine is ranked 114th in world incomes. Some 29 percent of the population live below the poverty line. The United Nations Development Programme Human Development Reports rank the Ukraine 78th among all nations in overall quality-of-life issues.

Bordering on the Black Sea, the Ukraine has 1,725 miles (2,782 kilometers) of coastline. The climate is Mediterranean along the southern coast and temperate continental elsewhere. Precipitation is most frequent in the west and north. In the east and southeast, winters are cool around the Black Sea but inland temperatures are colder. Summers are generally warm, although it is hotter in the south. The Carpathian Mountains in the west and the Crimean Peninsula in southernmost Ukraine are major geographical features. The rest of the country is composed of fertile plains and plateaus.

The Ukraine is rich in natural resources that include iron ore, coal, manganese, natural gas, oil, salt, sulfur, graphite, titanium, magnesium, kaolin, nickel, mercury, and timber. Over 56 percent of the Ukraine is arable, and Ukrainian farmers export milk, grain, vegetables, and meat to neighboring countries. Agriculture generates almost one-fifth of the Gross Domestic Product (GDP). Despite the high level of agricultural activity, 67.3 percent of the population live in urban areas. With only 108 cars per 1,000 people, the Ukraine produces 1.5 percent of the world's carbon dioxide.

The Ukraine suffers from a lack of potable water. Air and water pollution are common in industrial areas, and deforestation is widespread. Residue from the explosion at the Chernobyl Nuclear Power Plant in 1986 continues to contaminate areas in the northeast. The past haunts the Ukraine in other ways. Like most former Soviet republics, the Ukraine was exploited with little care for the environment. Long-lasting environmental damage was ubiquitous after the Soviet withdrawal. For instance, elevated levels of dioxin-like polychlorinated biphenyls (PCBs) were identified in samples of human milk. Likewise, high concentrations of pesticide residues were found in water samples of the Black Sea. Emission experts have identified the Ukraine as one of the heaviest contributors to European pollution because half of the pollution generated in the Ukraine has been ultimately deposited in other European countries.

In 2006, a study conducted at Yale University ranked the Ukraine 51st among 132 nations in environmental performance, slightly higher than the relevant income and geographic groups. Ratings were particularly low in sustainable energy, biodiversity and habitat, and air quality. Only 3.9 percent of land area is protected, but plans to increase such areas are under way. Sixteen of 108 mammal species endemic to the Ukraine are endangered, and eight of 215 endemic bird species are in a similar situation.

The Minister for Environmental Protection and Nuclear Safety is in charge of implementing environmental policy in the Ukraine. Operating under the National Action Plan on Environmental Protection, the Ukraine has developed policies that target all levels of government by seeking to integrate sustainable development with economic growth. Since 1998 environmental policy has focused on exacting payments for nature resources and environmental pollution. With the Chernobyl accident always in mind, preventing future accidents is a priority in Ukrainian planning, and particular attention is paid to licensing procedures for hazardous activities.

The Ukraine has joined the environmental efforts of the global community by participating in the following agreements: Air Pollution, Air Pollution–Nitrogen Oxides, Air Pollution–Sulfur 85, Antarctic–Environmental Protocol, Antarctic–Marine Living Resources, Antarctic Treaty, Biodiversity, Climate Change, Endangered Species, Environmental Modification, Hazardous Wastes, Kyoto Protocol, Law of the Sea, Marine Dumping, Ozone Layer Protection, Ship Pollution, and Wetlands. The Ukrainian government has signed but not ratified the Air Pollution–Persistent Organic Pollutants, Air Pollution–Sulfur 94, and Air Pollution–Volatile Organic Compounds agreements.

SEE ALSO: Black Sea; Chernobyl Accident; Drinking Water; Pesticides; Pollution, Air; Polychlorinated Biphenyls (PCBs); Urbanization.

BIBLIOGRAPHY. Central Intelligence Agency, "Ukraine," *The World Factbook*, www.cia.gov (cited April 2006); Winston Harrington et al., *Choosing Environmental Policy: Comparing Instruments and Outcomes in the United States and Europe* (RFF Press, 2004); Kevin Hillstrom and Laurie Collier Hillstrom, *Europe: A Continental Overview of Environmental Issues* (ABC-CLIO, 2003); Minister for Environmental Protection and Nuclear Safety, "Ukraine: Implementing Environmental Reforms," www.mem.dk/aarhus-conference (cited April 2006); United Nations (UN) Development Programme, "Human Development Reports: Ukraine," hdr.undp.org (cited April 2006); UN Environment Programme, *Europe Regional Report: Chemicals* (Global Environment Facility, 2002); World Bank, "Ukraine," *Little Green Data Book*, lnweb18.worldbank.org (cited April 2006); Yale University, "Pilot 2006 Environmental Performance Index," www.yale.edu/epi (cited April 2006).

Elizabeth Purdy, Ph.D.
Independent Scholar

Uncertainty

THE NOTION OF uncertainty is used to characterize how well future events or scientific truths can be predicted or known. It is used in both social and natural science disciplines from mathematics to philosophy to risk assessment and public policy. If probability is a measure of likelihood, then uncertainty is a measure of how well the probability is known. Uncertainty can be classified into known and unknown probabilities. Events with known probabilities are referred to as events with statistical uncertainties. Events with unknown probabilities are often called events with true uncertainty.

In chemistry and quantum physics, the Heisenberg uncertainty principle is used to characterize the wave/particle duality of electrons. It postulates that the position and momentum of a particle cannot be simultaneously predicted. This has nothing to do with the experimental design as it is often suggested, but because as the scale gets smaller, it becomes less useful to model particles as spheres. The multiple wavelengths associated with the wave model of these particles add inherent uncertainty to the questions of position and momentum.

Uncertainty in the context of the environment mainly refers to scientific uncertainty. Here, science generates truths through the testing of hypotheses. But often the affirmation of hypotheses involves a certain degree of uncertainty due to the method or research design. Scientists often use the benchmark of 95 percent certainty when deciding whether or not cause and effect have been correctly identified. Scientists often report confidence limits based on research design and sampling error in their studies to account for uncertainty.

The significance of uncertainty for environmental policy makers is quite different. For example, the precautionary principle is often invoked under

conditions of uncertainty, particularly when the consequences are irreversible or permanent. This differs from the choice that scientists make when deciding what to do under conditions of uncertainty. Typically scientists are interested in avoiding false negatives because science is epistemologically conservative. Scientists do not want to suggest something as truth when in fact it may not be. In public or environmental policy, however, because the consequences are not epistemological but ethical, there is desire to avoid false positives and be ethically conservative.

In public and environmental policy it is important to understand how to make decisions in the absence of perfect information. Knowing the degree of uncertainty is particularly important when questions about risk arise. Risk assessment is a policy approach that deals with uncertainty. Risk assessment is widely used by the Environmental Protection Agency (EPA), but mainly focuses on known probabilities. Because of difficulties with codifying the precautionary principle into policy, the EPA has yet to include true uncertainty in environmental policy.

Schrader-Frechette describes four classes of scientific uncertainty dealt with by scientists and policy makers: Farming uncertainty, modeling uncertainty, statistical uncertainty, and decision-theoretic uncertainty. In framing uncertainty, scientists often use a two-value frame to accept or reject a hypothesis. Frechette argues that in public policy it is more appropriate to adopt a three-value frame that creates a category to deal with situations where significant uncertainty and serious consequences suggest adopting the precautionary principle. Modeling uncertainties involve those involved in the prediction of future scenarios. These are highly speculative despite claims of verified and validated models. In public and environmental policy, statistical uncertainty should be dealt with in such a way that highlights the difference between epistemological consequences and ethical ones. When faced with decision-theoretic uncertainty, scientists are forced to distinguish between using expected value rules and the minimax rule. The former argues that a decision should be based on the expected value, while the latter seeks to prevent the worse case scenario. More recently Bayesian statistics has been used to help evaluate data under conditions of uncertainty by adding updating the probabilities as new data come in to view. A Bayesian approach involves the introduction of prior knowledge into statistical models.

There are many environmental policy debates where questions about uncertainty are raised. In debates about global climate change, for example, scientists typically agree that there is significant uncertainty in the projection of future climate change models. Climate change skeptics often highlight uncertainty to discredit climate change science. In debates about genetic engineering, uncertainty about the prediction of how transgenic organisms will behave in the environment, or uncertainty about how markets will react to the adoption of transgenic organisms, is cited as a reason to invoke the precautionary principle. In debates about nuclear waste disposal at Yucca Mountain, uncertainty about how the storage facility will perform in the long term is cited as reason to question the suitability of the nuclear waste repository.

SEE ALSO: Genetically Modified Organisms (GMOs); Global Warming; Precautionary Principle; Risk, Perception, Assessment, and Communication; Yucca Mountain.

BIBLIOGRAPHY. Daniel Bodansky, "Scientific Uncertainty and the Precautionary Principle," *Environment* (v.33/7, 1991); John Lemons, ed., *Scientific Uncertainty and Environmental Problem Solving* (Blackwell Science Press, 1996); Allison MacFarlane and Rodney C. Ewing, *Uncertainty Underground: Yucca Mountain and the Nation's High-Level Nuclear Waste* (MIT Press, 2006).

Dustin Mulvaney
University of California, Santa Cruz

Underdeveloped (Third) World

THE "UNDERDEVELOPED WORLD" is a term used to describe the "third world." The *third world* was a common term used to differentiate between countries that aligned with neither the West nor the

East during the cold war. In academic literature, several terms such as the South, underdeveloped, less developed, and developing have since been used interchangeably to describe these countries.

While it should be recognized that there is a great deal of diversity among underdeveloped countries in relation to geographical location, climatic conditions, religion, population size, resource endowments and the extent of dual economy, in broad terms they share several common characteristics. These are: Low standards of living, low levels of productivity, high rates of population growth, high levels of unemployment, strong economic dependency on agricultural production and export of primary products, and high foreign debt. Most underdeveloped nations are unable to provide for the developmental and economic needs of their citizens. This has left underdeveloped countries dependent and vulnerable in the turbulent arena of international relations.

MEASURING SOCIOECONOMIC STATUS

Policy makers and academic experts measure the socioeconomic characteristics of a nation using the Human Development Index (HDI) and the Human Poverty Index (HPI). The HDI index was developed by the United Nations (UN) to differentiate and compare the relative economic and social well-being of nations. Annual HDI reports are published to establish the comparative economic status of a nation by comparing and contrasting the value of significant social and economic indicators including, mortality rates, education levels, health statistics, and income levels.

Countries with an HDI measurement of over 0.8 are categorized as developed world nations, while those with a measurement below 0.8 are described as being underdeveloped. Low HDI scores indicate low per capita income, a relatively undeveloped infrastructure (including transport and telecommunications), high mortality rates, and low levels of education and employment. The 2005 HDI Report shows that the large majority of countries with a low HDI are currently in Africa.

The Human Poverty Index (HPI), also developed by the UN, measures the extent of poverty in a country. The magnitude of poverty is used not only to reflect the level of social welfare attained by the country over time but also to rank the country on a development scale. The HPI measurement indicates the levels of material need, social need, and financial resources of a country.

The World Bank also studies global poverty rates and they have shown that poverty levels are highest in the developing world. The World Bank reported in 1996 that Asia accounted for over two-thirds of the world's 1.3 billion poorest people, and the World Development Report 2000–2001 states that 29.1 percent of people in selected developing countries live in poverty.

FOREIGN AID

It is the question of how to reduce poverty that dominates discussion between the countries of the developed and underdeveloped worlds. The allocation of foreign aid to the underdeveloped world has been the most common strategy employed by developed world nations to try to reduce poverty levels in underdeveloped countries.

Foreign aid refers to any money or resources that are transferred from developed to developing countries without expecting full repayment. The 1971 UN Conference on Trade and Development (UNCTAD) promoted the notion that one percent of the national income of developed countries should be allocated to easing third world poverty. A subsequent UNCTAD meeting in Chile, in 1972, also set growth targets for developing countries at 6 percent during the 1970s. Another initiative, the Lome I and II pacts of 1975 and 1979 between the European Economic Community (EEC) and 46 African, Caribbean, and Pacific (ACP) countries, exempted ACP exports from certain tariffs, and guaranteed income from agricultural exports.

While foreign aid can play an important complementary and catalytic role in promoting economic growth, generally the distribution of foreign aid has further strengthened the reliance of the underdeveloped world upon developed nations. Many experts suggest that conditional (foreign) aid has resulted in the delivery of overpriced technical assistance, the counting of debt relief as development aid, and the inclusion of immigration-related costs in aid figures, which has tended

to serve the interests of international donor countries rather than eradicating poverty.

Conditional foreign aid has also had a profound impact upon the environments of the developing world. Costa Rica, for example, received nine International Monetary Fund (IMF) and World Bank structural adjustment loans during the 1980s. These loans were designed to encourage the competitive entry of Costa Rica into the international markets for bananas and cattle. The developmental expansion, however, came at the cost of increased use of pesticides, intense deforestation, and species extinction, resulting ultimately in a decrease of 31 percent of Costa Rica's forest cover by 1987.

There is a growing consensus in the aid community that a considerable portion of the international aid budget should be restructured to focus on human development concerns in developing countries. Experts recommend that aid be practical, targeted, science-based, and measurable. Foreign aid policies must reflect local priorities, incorporate stronger commitment and partnerships, incorporate routine monitoring and evaluation and be consistent with other developmental policies of the donor countries.

GLOBALIZATION

As demonstrated by the Costa Rican example above, underdeveloped countries do not exist in isolation, but are part of a globalized economy. Globalization refers to the expansion of local concerns—such as markets, information technology, social, cultural and political systems—into the global arena. Significant financial power is required by a country to expand local concerns and take an active and competitive role in the global economy because the international economy is dominated by rich developed countries.

For the underdeveloped world, globalization has thus necessitated the realignment of national policies to be consistent with their global counterparts in order to allow economic integration. Unfortunately, most underdeveloped countries do not possess the financial power needed to fully exploit global opportunities. Consequently, sub-economies have been established within underdeveloped countries that essentially serve industrialized markets because market control over international goods and services is primarily dominated by large multinational companies based in the developed world. This means that underdeveloped countries are controlled by the price-setting measure of those companies; this has had the effect of limiting economic activity in developing nations to a few niche sectors and is preventing them from fully exploiting resources that could help reduce poverty levels.

RESOURCE INEQUITY

The economic dominance of the developed world over underdeveloped nations in globalized markets is best demonstrated by the major inequalities in the consumption and ownership of the world's resources. Resource consumption is defined as the exploitation or use of all resources that we extract from the environment (often termed environmental resources or inputs); including minerals, fossil fuels, fish, wood, water, land, and other forms of energy. These resources (inputs) are extracted to produce the goods and services that are manufactured and consumed in the market place.

The 1998 Human Development Report highlights the starkness of the inequity of resource use between the developed and developing worlds, noting that globally, the richest fifth of the world's population has 85 percent of its income, while the poorest fifth has just 1.4 percent. The same report also shows that 20 percent of the world's people in the highest-income countries account for 86 percent of total private expenditure in consumption of goods and services, while the poorest 20 percent account for just 1.3 percent. Further, the richest fifth of global nations consume 45 percent of all meat and fish, consume 58 percent of total energy, and own 87 percent of the world's vehicle fleet.

GLOBAL POPULATION GROWTH

The inequitable distribution and ownership of the world's resources can be demonstrated by an examination of projected global population statistics. The world population is estimated to reach 9.1 billion by 2050. Ninety-five percent of the world's population growth will be within the world's least-developed countries. Fertility rates in the underde-

veloped countries will remain high, while developed countries are estimated to reach "below replacement" level by 2050. Mortality rates in developing countries remain high, however, especially in those countries within Africa and some parts of Asia that have high HIV/AIDS and other contagious disease infection rates.

Theses projected trends in population growth will have significant implications for both human populations and the environment in underdeveloped countries. First, the growth of population in these countries will place added pressure upon the agricultural industries. As the population increases the demand for food will grow. It is estimated that the total demand for agricultural products in 2030 will be approximately 60 percent higher than today. More than 85 percent of the additional demand will be from the underdeveloped world.

A second implication of population growth is a predicted corresponding demographic shift in populations from rural to urban centers. Presently, there are 20 global cities of more than 10 million people. Fifteen of the 20 are in underdeveloped countries, containing four percent of the global population. By 2015 it is predicted that there will be 22 such mega-cities, 16 of which will be in developing countries, accounting for five percent of the global population.

The growth in mega cities will be due to the displacement of people from rural communities who will be forced from their land by the application of new technologies that displace local producers in favor of large commercial farms that can cater to the increased demand for food.

As is already being witnessed in China, the displacement of people from rural centers into urban areas is having a tremendous impact upon those forced to move. Unemployment rates are high, local cultural identities are being eroded and China's entry into the World Trade Organization (WTO) is lowering rural livelihoods as small producers compete with imports from other WTO countries.

POPULATION AND THE ENVIRONMENT

Another significant impact of population growth in the underdeveloped world is upon the environment. The drive for developing countries to achieve economic equity and status with the developed world, and meet the challenge of providing for rising populations, puts intense pressure on the environment, causing environmental problems such as degradation, erosion, salinity, and conversion of natural ecosystems.

In a report by the World Commission on Environment and Development, poverty is identified as a major cause and effect of global environmental problems. For example, China's population is estimated to increase by 25 percent by 2012. Excessive erosion rates resulting from this will have significant impact on over one-third of China's fields, while the burning of crop residues to cook and provide heating will denude the soil of important organic matter.

The use of pesticides in agriculture is of major concern. The World Health Organization (WHO) has estimated that over one million cases of pesticide poisoning occur annually, with most of these instances occurring within developing countries.

Dependence on and pressure to access the world's water resources will increase, as will the environmental impacts upon them. In the past century alone, growing populations have increased demand for freshwater six-fold. Moreover, there is an increasing demand for freshwater with industrialization, irrigated agriculture, massive urbanization, and growing populations. More than one-half of available freshwater supplies are now used for domestic purposes and the world's water demand is doubling every 20 years.

Deforestation is another major environmental issue facing many developing countries. Clearing for subsistence and commercial agriculture, fuelwood, logging, and mineral extraction pay economic dividends but the impact on ecosystems is dramatic. For example, since 1960, over a quarter of the rain forest in Chile has been cleared to provide land for cattle grazing, which has been encouraged to meet the demand for beef in developed countries.

Deforestation is having a global impact causing species extinction, ecosystem service loss, and the reduction of carbon sinks resulting in increased emissions of climate changing gases. Deforestation in the Amazon is affecting weather patterns and rainfall from Mexico to Texas, while deforestation in southeast Asia impacts rainfall in China and the

Balkan Peninsula. Twenty to 25 percent of global carbon emissions come from changes to land use, primarily the degradation of forests.

Population growth in underdeveloped countries is also having a negative impact upon the marine commons. The resources of the marine environment are a vital source of food and often the sole source of nutrition for over one billion people. A vast majority of those dependent on marine resources for nutritional needs are in underdeveloped countries. This resource is threatened both by unsustainable exploitation and contamination. Algal blooms—or eutrophication—are caused by the excessive input of nutrients from agricultural runoff into marine environments. These can cause health problems, poisoning, and sometimes death. In 1987 there were 200 cases of algal poisoning in Guatemala, 26 of which were fatal.

The United Nations Environment Program estimates that over 20 billion tons of waste—including sewage, agricultural waste, and industrial runoff—are discharged into the world's oceans annually. Ninety percent of this pollution remains near the coast, where 95 percent of fish are harvested. This is why implications of expanded agricultural areas and industrialization in developing countries will have a continuing detrimental impact upon the marine environment and health of local peoples.

DEVELOPMENT'S ENVIRONMENTAL COST

In order to cope with expanding populations and to participate in the global marketplace, major development projects are taking place across the underdeveloped world, including the construction of roads, dams, and railways. These often have immense environmental implications. The Projecto Grand Carajas in Brazil, an iron and other ores project, will occupy up to 10 percent of Brazil's land mass and will cut through 900 kilometers of Amazon rain forest. Impacts upon the rivers and hydrology of the Amazon will be significant because a series of hydroelectric dams will be constructed to provide energy to power the project; additional energy will be generated by burning charcoal products extracted from Amazonian timber.

The clearing of forest for Projecto Grand Carajas will release vast stores of carbon from the soil, plus the added effect of burning the timber to generate power will greatly increase Brazil's overall emissions of climate changing gases. Significant siltation of Amazonian rivers will occur as soil from cleared land is washed by monsoonal rains into waterways.

CONFLICTS AND REFUGEES

Conflicts due to demographic change, environmental pressures, population growth, resource use and the vagaries of natural climate are common across the underdeveloped world. Conflicts over resource use and exploitation have occurred, including disputes over the most precious of resources—water. Water scarcity the world over is causing significant tension. Forty percent of the world's population depends on 214 river basins for drinking water, irrigation, and power. In India, conflicts over water have occurred, such as the 1991 civil unrest between the states of Karantataka and Tamil Nadi over access rights to the Cauvery River.

Conflict can lead to refugees, and combined with natural environmental fluctuations, environmental degradation, and over-exploitation of natural resources, the underdeveloped world regularly faces significant challenges. At the end of 2000, the number of refugees in the world stood at 16 million people. The largest numbers of refugees were in underdeveloped parts of the world including nine million in Asia and four million in Africa. Three million refugees are displaced in developed countries.

Many refugees in the underdeveloped world are considered to be environmental refugees. This is mainly as a consequence of natural disasters such as drought. For example, in 1988, over one million Ethiopians were displaced due to a serious drought that also had major impacts in other developing countries, including drought-prone northeast Brazil.

People are also made refugees as a result of development projects such as dam or road constructions. For example, the construction of a dam at the Kaptai Lake in Bangladesh displaced 100,000 people without compensation and the Three Gorges Dam project in China has displaced 1.2 million people.

Human displacement due to environmental issues and social and economic development in the underdeveloped world has been a catalyst for conflict. The arrival of Mauritanians into Senegal led to armed clashes between the two groups.

The issue of refugees is only set to get worse as the underdeveloped world faces major environmental challenges as a result of long-term environmental degradation and human-induced climate change. Estimates indicate that sea level rise and a decline of food stock, due to climate change, has the potential to create up to 173 million environmental refugees, most of whom will be residents of the underdeveloped world.

EQUITY AND THE ENVIRONMENT

As discussed above, the developed world is the major consumer of the earth's resources, but population growth is centered in the underdeveloped world. Globalization is forcing the two global spheres to interact as never before. This presents the world with a question: Is it possible to support the rights of developing countries to enjoy the same levels of affluence and quality of life as the developed world, while minimizing environmental impacts?

Globalization means that human society now shares a global commons. Developing countries play a vital role in the global economy and provide the world with cheap resources and labor. However, current global consumption is increasingly happening at the expense of the environment and the earth's climate and is doing little to alleviate global poverty.

The establishment of effective legislative environmental controls in underdeveloped countries would help to minimize environmental impacts while creating an equitable world. At present, environmental laws in many developing countries are difficult to interpret, vague, or simply nonexistent. Where environmental legislation does exist in underdeveloped countries it is not accompanied by regulation or enforcement of those laws. In Central America and Mexico, surveys of air and water pollution laws show that while laws exist to regulate these issues, financial resources have not been allocated to legal enforcement. In the African countries of Zambia, Ethiopia, Ghana, the Sudan and Kenya, penalties for infringements of environmental law are so low it is cheaper to pollute than act within the law.

Ecologically sustainable development (ESD) has been suggested as a mechanism outside of legislative frameworks that can address these problems. The concept of sustainable development is seen as an essential measure of the impact of economic activity upon global survival. The World Commission on Environment and Development (UNCED) in 1992 defined ESD as "development that meets the needs of the present without compromising the ability of future generations to meet their own needs." The UNCED definition was based on the report *Our Common Future*, published in 1987 by the World Commission on Environment and Development, which called for strategies to strengthen global efforts to promote the concept of sustainability. The Commission highlighted that the political goals of socio-economic development must be treated as a crucial part of attaining sustainable development.

The Millennium Development Goals (MDGs) are another forum through which underdeveloped and developed countries can work together to minimize environmental impacts while countries develop their socio-economic structures. The MDG framework identifies eight major goals and 18 associated targets to evaluate the effectiveness of sustainable development. These goals include quantified aims in relation to achieving poverty alleviation by 2015. Furthermore, the forum acknowledges the importance of agriculture to rural development and how it may contribute toward meeting the MDGs.

Developing countries are also emerging with a collaborative voice to address some of these challenges. At the UN summit on climate change in Montreal in 2005, a coalition of 10 tropical developing countries called the Coalition for Rainforest Nations and led by Papua New Guinea and Costa Rica tabled a proposal for compensation for rain forest services that their nations provide for the rest of the world. UN figures show that the countries within this coalition collectively represent approximately 13 percent of the world's rain forest, reflecting $1.1 trillion in carbon storage. The coalition stressed the inequity of expecting developing nations, burdened by poverty, to give up income through deforestation while other countries benefit from the services derived from rain forest protection. These services include carbon storage, water filtration, biodiversity protection, climate regulation, and fisheries protection. They argued that if the developed world wished to save the rain forest, they should pay for rain forest services. Examples like these highlight the dilemmas

developing countries face when resource use and exploitation are the only way a nation is able to survive economically.

Underdeveloped countries, despite aid programs and their own efforts, largely remain in poverty. Poverty combined with the pressure of competing on the international stage, population increases, environmental degradation, and unsustainable resource use indicates the underdeveloped world still faces significant challenges. Cumulatively, these factors undermine the peace and stability of the entire world. While it is crucial that poverty is effectively addressed, development per se is not the panacea it was once thought to be. Pathways must be found to reconcile development needs, eradicate poverty, and decrease inequity between the underdeveloped and first worlds in ways that are environmentally sound, socially just, and economically viable.

SEE ALSO: Debt; Developed ("First") World; Globalization; Industrialization; International Monetary Fund; Land Degradation; Markets; Population; Poverty; Rain Forest; United Nations; World Trade Organization.

BIBLIOGRAPHY. R. Butler, *Developing Countries: Pay Us to Save the Rainforests* (Monganbay Press, 2005); Global Forest Resources Assessment, www.unfpa.org (cited April 2006); F. Lechner and J. Boli, eds., *The Globalisation Reader (*Blackwell Publishing Ltd., 2004); T. Oatley, ed., *The Global Economy: Contemporary Debates* (Pearson Education, 2005); B. Ogolla, "Water Pollution Control in Africa: A Comparative Legal Survey," *Journal of Africa* (v.33, 1989); R. Ramlogan, *The Developing World and the Environment, Making the Case for Effective Protection of the Global Environment* (University Press of America, 2004); E.S. Simpson, *The Developing World: An Introduction* (Longman Scientific and Technical, 1987); H. Steinfeld, "Economic Constraints on Production and Consumption of Animal Source Foods for Nutrition in Developing Countries," *The Journal of Nutrition* (v.133, 2003); E. Tang, "China and the WTO: A Trade Union View of Social Impacts and Workers' Responses," www.union-network.org (cited April 2006); T. Tietenberg, "The Poverty Connection of Environmental Policy," *Challenge* (v.33, 1990); United Nations (UN), "The State of Human Development," *United Nations Development Report, 1998* (UN 1998); United Nations, *World Population Prospects, The 2004 Revision* (Department of Economic and Social Affairs, 2005); J. Wheeler, "The Interwoven Strands of Development," *The IECD Observer* (v.167, 1990); Worldwatch Institute, *State of the World 2001: A Worldwatch Institute Report on Progress Toward a Sustainable Society* (Worldwatch, 2002).

Melissa Nursey-Bray
Australian Maritime College
Robert Palmer
Independent Scholar
Shekar Bose
Australian Maritime College

Underground Storage Tanks

UNDERGROUND STORAGE TANKS (USTs) are large containers that have at least 10 percent of their volume and associated piping underground. Underground storage tanks usually contain petroleum or other hazardous gaseous or liquid materials. In 2006, there were about 680,000 underground storage tanks in the United States. The petroleum or other materials that they contain are almost always hazardous to humans, animals, or to the general environment. Whenever an underground storage tank leaks it causes damage to the environment. Quite often water wells in the area have to be shut off because of contamination. In addition to the underground storage tanks in use, there are a large unknown number of old abandoned underground storage tanks. Until the middle of the 1980s, underground storage tanks were made from plates of bare steel that had been welded together. With a high potential for rusting and leaking, the life expectancy of these older tanks is only 30–50 years.

Thousands of underground storage tanks were installed in the United States after World War II in order to supply gasoline to the growing number of automobiles that Americans were driving. Since 1950 many of these hundreds of thousands of underground storage tanks have leaked. Usually leaking gasoline, some tanks have included petroleum distillates such as diesel, heating oil, kerosene, and jet fuel. Gasoline additives pose an even more important danger than leaking gasoline. These have

included lead, which can cause brain damage, and benzene, which is a known carcinogen. Just a small amount of benzene can pose a severe hazard because of its toxicity. Toluene, ethylbenzene, and xylenes are also toxic additives in gasoline that pose significant health risks when leaked into the environment. There are nearly 400,000 leaking underground storage sites in the United States that are being monitored by U.S. Environmental Protection Agency's (EPA) Office of Underground Storage Tanks.

Because over half of Americans get their drinking water from groundwater, the threat to health is very serious. In addition, leaking petroleum volatiles give off vapors that pose an explosive fire hazard that can accumulate in sewers and the basements of buildings. The EPA uses money from the Leaking Underground Storage Tank (LUST) Trust Fund to clean up the worst of the leaking underground storage tanks. Increasingly, the polluters are forced to pay for the cleanup. Since 1984, Congress has responded to the problem of leaking underground storage tanks with a range of laws. Besides the Comprehensive Environmental Response, Compensation and Liability Act ("Superfund") and the Resource Conservation and Recovery Act (RCRA), which list a large number of substances that are contained in USTs, Congress has provided cleanup funds.

In 1985 Congress banned the use of unprotected steel tanks and piping. It has also directed the EPA to publish regulations covering USTs that will require owners and operators of new tanks, as well as old tanks, to detect and clean up any releases from their tanks, and to establish financial resources to pay clean up costs in the event of a leak. The EPA works with state and local governments to manage current LUSTs and to prevent any new ones from occurring. The great number of tanks and their widespread distribution puts states and localities in the best position to supervise the regulation of USTs as a part of the powers of states to regulate the health and safety of their people. Many states have more stringent regulations than the federal government.

SEE ALSO: Drinking Water; Gasoline; Groundwater; Petroleum.

BIBLIOGRAPHY. Wayne B. Geyer, *Handbook of Storage Tank Systems: Codes, Regulations, and Designs*

Malta's Underground Grain Storage

When the Ottoman Turks attacked Malta in 1565, the Knights of Malta and the other Christians on the island were forced to retreat behind their fortifications. After bitter fighting for control of the island from May through August, the Ottoman soldiers were forced to withdraw, ending what has been called by some historians the "last battle of the Crusades."

Following the siege, the Knights of Malta decided to massively enlarge their fortifications. In particular, they were worried about having enough supplies of food should the Turks attack again and invest in a much longer siege. To that end the city of Valetta was built protecting the peninsula of Fort St. Elmo. Incorporated into the design were massive underground storage bins carved out of the rock. The view of the architect who designed the defenses, Francesco Laparelli, was that the defenses were strong enough that "if there are victuals and munitions, it will be impregnable."

Carving these underground storage bins was such arduous work that large numbers of Italian laborers had to be hired to carry out the work. Laparelli was so keen for labor that the terms he offered were very good. The contracts were for free passage to Malta and free rations at sea, and laborers were also paid a daily rate from the moment they signed up, whether they were needed for work or not. When the bins were completed, a large stone was carved and placed at the top of them to protect them from the elements and marauders. The Turks did not attack Malta again, but in World War II, with Malta being a British colony, it was attacked and bombed by the Germans and the Italians. Once again the storage bins were filled with food and withstood the German bombing raids. Laparelli's work was so thorough that even mid-20th century bombing techniques were unable to damage the underground storage bins.

(CRC Press, 2000); John P. Hartman, *Technology of Underground Liquid Storage Tank Systems* (John Wylie & Sons, 1997); U.S. Army Corps of Engineers, *Removal of Underground Storage Tanks (USTs)* (University Press of the Pacific, 2005).

Andrew J. Waskey
Dalton State College

UNESCO

THE UNITED NATIONS Educational, Scientific, and Cultural Organization (UNESCO) is an agency of the United Nations (UN). It was created in 1945 to promote world peace by focusing on culture and communication, education, natural sciences, and social and human sciences in order to further universal respect for justice, the rule of law, and the human rights and fundamental freedoms proclaimed in the UN Charter. UNESCO aims to create the conditions for genuine dialogue based upon respect for shared values and the dignity of each civilization and culture.

UNESCO's principal decision-making body is the General Conference, which is composed of representatives of the 191 member states (UNESCO also has six associate members). The General Conference elects the members of the executive board and appoints the director-general. The organization's headquarters is in Paris, and it has more than 50 offices around the world. Today, UNESCO serves as a center for the dissemination and sharing of information and knowledge in the fields of education, science, culture, and communication among its member states.

Although a founding member, the United States suspended its membership of UNESCO in 1984, believing that the organization had politicized subjects it dealt with and exhibited hostility toward the basic institutions of a free society, especially a free market and a free press; and has demonstrated unrestrained budgetary expansion.

The controversy was triggered by UNESCO's 1980 report on the state of the contemporary media, a document known as the MacBride Report, which criticized commercialization and unequal access to information and communication. The U.S. withdrawal was followed by that of the United Kingdom (UK) in 1985 and Singapore in 1986. The UK rejoined UNESCO in 1997 as did the United States in 2003.

UNESCO uses conventions, recommendations, and declarations as international instruments for establishing common rules. Conventions define rules that member states undertake to comply with and are subject to ratification, acceptance, or accession by these states. Recommendations are instruments and norms that are not subject to ratification but that member states are invited to apply. Declarations, another means of defining norms, are not subject to ratification either but, like recommendations, they set out universal principles to which the community of states wish to attribute the greatest possible authority and to afford the broadest possible support.

Noting that cultural and natural heritage were threatened with destruction by changing social and economic conditions, UNESCO adopted the Convention Concerning the Protection of the World's Cultural and Natural Heritage in 1972. The document defines natural heritage as:

> Natural features consisting of physical and biological formations or groups of such formations which are of outstanding universal value from the aesthetic or scientific point of view; geological and physiographical formations and precisely delineated areas which constitute the habitat of threatened species of animals and plants of outstanding universal value from the point of view of science or conservation; natural sites or precisely delineated natural areas of outstanding universal value from the point of view of science, conservation or natural beauty.

An Intergovernmental Committee for the Protection of the Cultural and Natural Heritage of Outstanding Universal Value, known as the World Heritage Committee, has been set up within UNESCO to establish, keep up-to-date, and publish, under the title of the World Heritage List, a list of properties forming part of the world's cultural heritage and natural heritage, in other words, properties considered as having outstanding universal value. The World Heritage List includes 812 places in 137 countries and regions.

In relation to the natural sciences, UNESCO adopted two documents relevant for the recognition of the interdependence of human beings and their environment: the Convention on Wetlands of International Importance Especially as Waterfowl Habitat (1971) and Recommendation on the Status of Scientific Researchers (1974).

Among UNESCO's millennium goals is the objective of helping countries to develop national strategies for sustainable development and reverse current trends in the loss of environmental resources by 2015. UNESCO states that the world urgently requires global visions of sustainable development based upon observance of human rights, mutual respect, and the alleviation of poverty.

UNESCO has developed several international programs to better assess and manage the earth's resources. The organization helps reinforce the capacities of developing countries in the sciences, engineering and technology. UNESCO's priorities in the field of the natural sciences are: water and associated ecosystems; oceans; capacity-building in the basic and engineering sciences; the formulation of science policies and the promotion of a culture of maintenance and promoting the application of science, engineering, and appropriate technologies for sustainable development; natural resource use and management; disaster preparedness and alleviation; and renewable sources of energy.

Under its Program on Man and Biosphere, UNESCO established in 1971 the world Network of Biosphere Reserves. In 2005 the network, which operated in 102 countries, included 482 biosphere reserves, places that promote and demonstrate a balanced relationship between humans and the biosphere.

SEE ALSO: Biosphere Reserves; Hazards; Man and the Biosphere Program (UNESCO); Renewable Energy; Sustainable Development; Underdeveloped (Third) World; United Nations; Wetlands; World Heritage Sites.

BIBLIOGRAPHY. Sagarika Dutt, *UNESCO and The Just World Order* (Hauppauge, 2002); UNESCO, www.unesco.org (cited April 2006); An Zi, "Preserving the Past," *Beijing Review* (2005).

Verica Rupar
Victoria University of Wellington

Union of Concerned Scientists (UCS)

THE UNION OF Concerned Scientists (UCS) is an independent nonprofit alliance of more than 100,000 concerned citizens and scientists advocating environmentally sound solutions to society's problems. The UCS was founded in 1969 at the Massachusetts Institute of Technology (MIT) by faculty and students protesting the misuse of science and technology. They put forth a Faculty Statement, the genesis of UCS, calling for greater emphasis on the application of scientific research to environmental and social problems, rather than military programs. The UCS now augments rigorous scientific analysis with citizen advocacy in order to build a safer, healthier environment.

The UCS's first report, *ABM ABC,* criticized President Nixon's proposed antiballistic missile (ABM) system. This opposition was part of a broad national movement that helped build public support for the ABM treaty, signed by the United States and Soviet Union in 1972. Similarly, the UCS mobilized opposition in the scientific community to President Reagan's Strategic Defense Initiative (SDI) popularly known as Star Wars. This stance culminated with more than 700 members of the National Academy of Sciences, including 57 Nobel laureates, signing UCS's *Appeal to Ban Space Weapons.* Most recently, the UCS's *Countermeasures* report, which demonstrated that the proposed national missile defense system could be defeated by missiles equipped with simple countermeasures, forced President Clinton to abandon the system.

After failures in government tests of emergency core-cooling systems at nuclear power plants, the UCS provided the principal technical expertise at national hearings, sparking the first public concern over nuclear power safety. In 1977, the UCS publication *The Risks of Nuclear Power Reactors* played a key role in the government's ultimate repudiation of its own faulty Reactor Safety Study. The UCS proposed alternatives to nuclear power and fossil fuels with their study *Energy Strategies: Toward a Solar Future,* starting the UCS's ongoing efforts to promote safe, renewable energy supplies for the United States.

In part, the UCS support of renewable energy stems from concerns over climate destabilization due to emissions from fossil-fuel combustion. More than 1,500 international senior scientists, including 105 science Nobel laureates, signed the UCS-sponsored *World Scientists' Call for Action at the Kyoto Climate Summit.* This document, as well as other UCS work with policymakers and scientists, set the stage for the Kyoto Protocol.

Many environmental trends such as climate change have caused the world's scientists to become "concerned." This is clear in the *World Scientists' Warning to Humanity,* which presented an unprecedented appeal from the world's leading scientists regarding the destruction of the earth's natural resources. It concludes:

> We the undersigned, senior members of the world's scientific community, hereby warn all humanity of what lies ahead. A great change in our stewardship of the earth and the life on it, is required, if vast human misery is to be avoided and our global home on this planet is not to be irretrievably mutilated.

The UCS has also tackled the public and environmental safety issues behind antibiotics in livestock feed and genetically engineered crops. In 2004, the union received a good deal of attention from the mass media by publishing a report titled *Scientific Integrity in Policymaking,* which criticized the George W. Bush administration for altering reports by the Enivironmental Protection Agency on global warming and West Virginia strip mining and for choosing members of scientific advisory panels based on their political views rather than scientific experience.

The UCS has become a powerful voice for change in U.S. policy. Its core groups of scientists and engineers collaborate with colleagues across the country to conduct technical studies on environmental topics. UCS experts work with citizens to disseminate their findings to influence local and national policy. In addition, the UCS Online Action Network gives citizens the means to stay informed on issues and help shape policy by expressing their views to government and corporate decision makers.

The UCS strives for a future that is free from the threats of global warming and nuclear war, and a planet that supports a rich diversity of life. Sound science guides its efforts to secure changes in government policy, corporate practices, and consumer choices that will protect and improve the health of our environment globally, nationally, and in communities throughout the United States.

SEE ALSO: Antibiotics; Bush (George W.) Administration; Clinton, William Administration; Genetically Modified Organisms (GMOs); Global Warming; Kyoto Protocol; Nixon, Richard Administration; Nuclear Power; Nuclear Weapons; Policy, Environmental; Reagan, Ronald Administration; Renewable Energy; Solar Energy.

BIBLIOGRAPHY. Henry W. Kendall, *A Distant Light: Scientists and Public Policy* (American Institute of Physics, 1999); Union of Concerned Scientists (UCS): Citizens and Scientists for Environmental Solutions, www.ucsusa.org (cited January 2006); UCS Online Action Network, www.ucsaction.org (cited January 2006).

Joshua M. Pearce
Clarion University of Pennsylvania

United Arab Emirates

IN THE ABU Dhabi of the 1950s, travelers described the capital of the United Arab Emirates (UAE) as a "small dilapidated town." Abu Dhabi—like most of the United Arab Emirates—was a completely different world just four decades ago. Modern Abu Dhabi and its sister emirate Dubai are glistening financial, commercial, transportation, and tourism centers. Dubai and Abu Dhabi have built massive, palatial malls, indoor ski rinks with artificial snow, artificial islands with seven star hotels, and one of the most modern and developed airports in the world.

Unlike most of the Middle East, which has stagnated economically since the 1960s, the United Arab Emirates—a loose confederation of different emirates, inherited "Princedoms" ruled by emirs or princes—has surpassed not only their Arab neighbors, but much of the Western world as well. The key source of this remarkable transformation seems obvious—oil. Vast amounts of easily-extracted oil continue to sell for massive

profits in the West. Yet other nations, such as Libya or Nigeria, have benefited from massive oil profits without nearly the same economic success as the Emirates.

The Emirates achieved this success through a careful and planned policy of economic diversification and an open attitude to free trade and investment. Far from relying only on the revenues from oil, the Emirates from the beginning planned industrial projects, encouraged agricultural development on once barren desert, and hosted the latest innovations in technology. The leaders of the Emirates often provide tax havens for the development of lucrative, major industries and technologies.

Often, however, the almost lightening speed of development in the Emirates has led to significant and troubling environmental consequences. One early example of the unforeseen environmental consequences of unbridled development was the creation of gas refineries. Before the mid-1970s, most natural gas was simply burned away or underutilized in the process of extracting oil. To make the most of all petroleum products, Abu Dhabi built the Umm al-Nar refinery deliberately near downtown as a symbol of industrialization and progress. This was an environmental blunder as the plant used corrosive materials and chemicals that had a harmful impact on the immediate area. Although there have been recent changes, a similar disregard for public health and safety has characterized some other development and diversification projects.

The Umm al-Nar refinery is only one example of how the prestige associated with economic development was often more important than possible long-term environmental and social concerns. Dubai and Abu Dhabi, and the rest of the Emirates to a lesser extent, have learned the importance of image, symbols, and marketing in a globalized economy. The environment is often relegated to a secondary concern. As skyscrapers fill the skylines of Dubai, and as pollution and congestion increase, it is possible that the UAE will overburden its social and environmental resources, leading to the potential for profound crises—not only crises of social identity, but crises caused by seemingly unstoppable economic and environmental overstretch.

SEE ALSO: Land Reclamation; Mining; Natural Gas; Organization of Petroleum Exporting Countries (OPEC); Persian Gulf; Petroleum.

BIBLIOGRAPHY. Ronald Codrai, *The Seven Skaykhdoms: Life in the Trucial States Before the Federation of the United Arab Emirates* (London, 1990); Christopher Davidson, *The United Arab Emirates: A Study in Survival* (Lynne Rienner, 2005); Sulayman Khalaf, "Gulf Societies and the Image of Unlimited Good," *Dialectical Anthropology* (v.17, 1992).

Allen J. Fromherz, Ph.D.
University of St. Andrews

United Church of Christ—Commission for Racial Justice (UCC-CRJ)

THE UNITED CHURCH of Christ's Commission for Racial Justice (UCC-CRJ) helped verify the concerns and claims of the emerging environmental justice movement in the United States with its 1987 study *Toxic Wastes and Race in the United States: A National Report on the Racial and Socioeconomic Characteristics of Communities with Hazardous Waste Sites*. Though the environmental legislation enacted by the U.S. Congress since the 1970s, such as the Clean Air Act and the Resource Conservation and Recovery Act, had addressed pollution problems that affect communities of color, these communities remained marginal in mainline national environmental organizations. Meanwhile, grassroots groups had begun to organize in the early 1980s around local toxics issues. Many asserted that communities of color faced disproportionate health hazards due to unfair practices of siting facilities for the treatment, storage, and disposal of toxics. *Toxic Wastes and Race* was the first study to test this assertion on a national scale.

The UCC-CRJ is the civil and human rights wing of the United Church of Christ. It began providing resources and research for grassroots antitoxics groups in 1982, particularly to rural communities in Warren County, North Carolina, who believed that

they had been targeted in the siting of PCB disposal facilities. The Warren County communities staged acts of civil disobedience to resist the siting of a new facility, and the UCC-CRJ intended its study to fuel further nonviolent protest.

The UCC-CRJ study sought to determine what demographic and economic variables correlated most strongly with the location of polluting facilities, especially uncontrolled hazardous waste sites, which the U.S. Environmental Protection Agency (EPA) describes as those that have been abandoned or closed by their operators and that pose serious health threats.

The study tested a variety of variables, including race, income, and housing age, and found that percentage of minority population most accurately predicted the location of these facilities. (The study counted African Americans, Latinos, Asian Americans, Native Americans, and Pacific Islanders as minorities.) In his preface to the report, UCC-CRJ Executive Director Benjamin Chavis called the disproportionate siting of these facilities "an insidious form of racism."

The report recommended, among other actions, an executive order by the president of the United States that would require that federal actions be evaluated on the basis of their environmental impact on communities of color. In 1994, seven years after the release of the report, President Clinton issued Executive Order 12989, which essentially added an environmental justice dimension to the National Environmental Policy Act.

Discussions of environmental justice since *Toxic Wastes and Race* have built on the UCC-CRJ study but have also raised questions about the processes behind the geographic distribution of toxic facilities. While it is clear in many cases that companies and governments have deliberately located facilities in communities of color because it was politically easier to do so, in many cases whites have been able to move away from older industrial areas into exclusive communities, in part because of discriminatory real estate practices, leaving behind more polluted areas for people of color. Thus, a historical examination is necessary to develop a causal explanation for these geographic patterns. Also, debates continue about whether income is a more important factor in determining exposure to toxic facilities, though many insist that issues economic and racial justice cannot be separated.

SEE ALSO: Bullard, Robert; Clean Air Act; Clinton, William Administration; Environmental Racism; Justice; Movements, Environmental; National Environmental Policy Act; Polychlorinated Biphenyls (PCBs).

BIBLIOGRAPHY. Robert Bullard, *Dumping in Dixie: Race, Class, and Environmental Quality* (Westview, 2000); Robert Gottlieb, *Forcing the Spring: The Transformation of the American Environmental Movement* (Island Press, 1993); Laura Pulido, "Rethinking Environmental Racism," *Annals of the Association of American Geographers* (2000); United Church of Christ—Commission for Racial Justice, *Toxic Wastes and Race in the United States: A National Report on the Racial and Socioeconomic Characteristics of Communities with Hazardous Waste Sites* (UCC-CRJ, 1987).

Dawn Day Biehler
University of Wisconsin, Madison

United Farm Workers (UFW)

THE UNITED FARM Workers (UFW) is an agricultural labor union that was founded in 1962 by Cesar Chavez and others, including Philip Vera Cruz, Larry Itliong, and Delores Huerta. Before 1962 Cesar Chavez had worked with Saul Alinsky and Fred Ross in California in community organization; together they formed the Community Service Organization (CSO). With the help of Roman Catholic leaders in southern California they organized a number of CSO units that benefited the Mexican-American population. In 1962 Chavez sought to create something more than a humanitarian organization to help unemployed workers to get unemployment insurance benefits.

The United Farm Workers Union arose from the merger of the National Farm Workers Association (FNWA) and the Agricultural Workers Organizing (AWO) committee. The farm workers who belonged to FNWA were mostly Filipinos. The members of the AWO were mostly Mexican. Both groups shared a Hispanic background. On Sep-

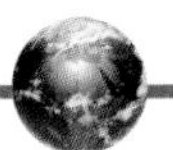

tember 8, 1965, the AWO led by Larry Itliong in Delano, California, began a strike with the grape growers. Both groups recognized their common goals and methods. Joining together they became the United Farm Workers. The strike over picking table grapes lasted five years. In the end the union won a contract with most of the grape growers in California. The UFW used the doctrine and methods of nonviolence promoted by Mahatma Gandhi and Martin Luther King, Jr. Chavez would engage in dramatic hunger strikes in order to draw attention to the union's issues. Chavez's fasting not only attracted public attention, it also gave him moral power with many of the members of the union who were often far from peaceful. Many sought to use violent methods employed in the California labor strife of the 1930s and illustrated in some of the novels of John Steinbeck.

In 1973 the Teamsters Union was able to sign a contract with many growers. The contract almost destroyed the UFW. For the next several years the UFW fought the Teamsters and the growers with strikes, lawsuits, and boycotts. Violence often occurred in the fields and a number of workers were killed. The State of California created the California Agricultural Labor Relations Board in 1975 in order to resolve labor union disputes. The new administrative agency was eventually able to move the UFW and its competitors in a more peaceful direction. In 2006 the UFW left the American Federation of Labor–Congress of Industrial Unions (AFL-CIO) and joined the Change to Win Federation, which is a coalition of unions.

The UFW has been active with a number of environmental issues. These include pesticide use and animal rights. A number of animal rights issues have been addressed by the UFW by petitioning responsible state officials about instances of abuse. In addition, boycotts of foods such as milk or meat from producers that are considered to be indifferent or irresponsible caretakers for farm animals have been instituted. Farm workers are exposed to pesticides at much higher levels than are most nonfarmers. Recent studies have suggested that Parkinson's disease may be linked to pesticide exposure. This is just one of a number of diseases that pose health hazards to farm workers and their families. The UFW battles for healthier farming conditions and for compensation for those whose health is injured by exposure to pesticides. It has in recent years found allies for its positions among scientists who work for the Environmental Protection Agency. The UFW also supports farm workers who speak out against alleged environmental hazards.

SEE ALSO: Animal Rights; Chavez, Cesar; Pesticides; United States, California.

BIBLIOGRAPHY. Richard W. Etulain, *Caesar Chavez: A Brief Biography with Documents* (St. Martins, 2002); Debra A. Miller, *Dolores Huerta, United Farm Workers Co-Founder* (Thompson Gale, 2006); Gary Soto, *Jessie de la Cruz: A Profile of a United Farm Worker* (Persea Books, 2002).

Andrew J. Waskey
Dalton State College

United Kingdom

THE UNITED KINGDOM (UK) of Great Britain and Northern Ireland is a sovereign state that occupies a number of large and small islands off the west coast of the continent of Europe. The state consists of a political union of four countries, three on the island of Great Britain (England, Scotland and Wales) and one that occupies the northern part of the island of Ireland (Northern Ireland).

The state is a constitutional monarchy, in which the monarch acts as head of state for a number of affiliated territories, including the Isle of Man and the Channel Islands, and for the 15 Commonwealth states that are the remnants of the world's largest empire. The UK was the home of the industrial revolution in the 18th century and remains one of the world's major industrial powers. Political influence is maintained through membership in the Group of Eight (G8) and a permanent seat on the United Nations (UN) Security Council.

HISTORY

The union of these four countries developed over a period of 700 years, beginning with the Statute

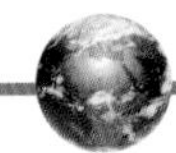

of Rhuddlan in 1284 which brought Wales under the control of the English monarchy. This relationship was formalized in 1535 with the Laws in Wales Act which made Wales subject to acts of the English parliament. In 1603, the failure of Elizabeth I to provide an heir for the English throne led to the accession of King James VI of Scotland and the union of the monarchies of England and Scotland. The Act of Union of 1707 suspended the Scottish parliament and led to the creation of a single unified parliament at Westminster. The Act of Union of 1801 achieved the same end for Ireland; however, this union lasted for only a little over 100 years and in 1922, the southern portion of the island of Ireland achieved independence as the Irish Free State and resulted in the change to the current name in 1927.

During the 16th and 17th centuries, the UK was one of the countries in Europe best positioned to take advantage of the innovations in navigation and exploration that opened up the New World to European exploitation. Unlike many European countries, colonization by the UK was not primarily driven by the crown but by independently financed merchant companies. This private entrepreneurship led to an explosion in private wealth that in turn was available as investment capital when the industrial revolution began in the 18th century. Colonization also provided raw materials and a critical mass of middle-class merchants who were accustomed to taking risks, and who formed the UK's entrepreneurial class.

The advantages of early industrialization and an expanding global empire to provide raw materials made the UK the first true world superpower in the 19th century. At its greatest extent the British Empire covered one quarter of the land surface of the earth and contained one third of the world's population. However, by the middle of the 20th century challenges from new powers such as Germany, Russia, and the United States, together with the physical and financial devastation of two World Wars led to the dismantling of the empire and the diminishing of the UK's industrial power. By the 1960s, the financial and political weakening was sufficient to persuade the UK to apply for membership in the newly formed European Economic Community. Membership was achieved in 1972 and while British membership in the European Union (EU) has never been overwhelmingly popular in the country, the current Labor government has created a much more positive working relationship with Brussels than the Conservative administration that was in power throughout the 1980s and much of the 1990s.

With new immigration, the UK now has the largest number of Punjabi and Hindi speakers outside of Asia.

GEOGRAPHY

The primary lowlands of the United Kingdom are in the midlands and south of England, with narrow lowland belts in central Scotland and along the south coast of Wales. The flattest land is in the Fens of eastern England, where land has been reclaimed from the marshes for agriculture since Roman times. The south coast and much of the midlands and southwest of England are covered with low hills that occasionally become a line of defined hills, such as the Cotswolds or the Chalk Downs. The primary highland areas of England are the limestone hills of the Peak District and the Cumbrian mountains of the Lake District, with the highest mountain in England (Scafell Peak at 978 meters) falling just short of 1,000 meters. Much Welsh terrain is very mountainous, especially in the north which contains Mount Snowden (1,085 meters). Scotland has the highest terrain in the UK, including the high-

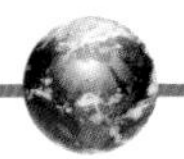

The Great Fire of London

The Great Fire of London swept through the English capital from Sunday, September 2, until Wednesday, September 5, 1666, destroying much of the old city, some of which dated from Roman times. Altogether 13,200 houses, 87 churches, and St. Paul's Cathedral were burned down, and 70,000 of the 80,000 people in London were made homeless. The fire followed a severe outbreak of the plague in London a year earlier, which had resulted in many deaths and also a large number of people (including the Royal Court) leaving London for nearby areas and other parts of the country—the population of London before the plague being about 300,000.

The fire started in a bakery in Pudding Lane after midnight on the morning of September 2. By Sunday night the fire was out of control and the Lord Mayor, Sir Thomas Bloodworth, held back from destroying houses to make firebreaks as he was concerned about destruction of private property. On Monday, September 3, the fire spread as rumors arose that it had been the result of Dutch arsonists—Britain was at war with the Netherlands at the time—and some foreigners were attacked. On Tuesday, St. Paul's Cathedral, which dominated the skyline of London, was on fire, and attempts to use the Fleet River (on the site of present-day Fleet Street) as a firebreak failed. Eventually, the army used gunpowder to blow up enough houses to create a firebreak and save the easternmost part of the city. There are many descriptions of the fire; the most famous are by the diarist Samuel Pepys.

As a result of the fire, much of the city of London was rebuilt. The architect Christopher Wren designed and built a new St. Paul's Cathedral, as well as many other churches in the city. It is not known how many people died; some books put the numbers in single figures. Others argue that many more must have died without leaving a trace, and furthermore, many others did not survive the impromptu refugee camps built after the fire.

est peak, Ben Nevis (1,344 meters) and is almost entirely mountainous to the north and west of the Highland Line.

The largest body of open water in the UK is Lough Neagh in Northern Ireland, and there are relatively few major rivers, with all major ports now located on estuaries. The primary rivers are the Thames, the Severn, and the Humber in England and the Clyde and Forth in Scotland. The UK's other notable geographic feature is the large number of islands that surround the coast particularly in the north and west, where Scotland's major island groups, the Shetlands, the Orkneys, and the Inner and Outer Hebrides, contain over 700 islands.

Located on the western edge of Europe, the UK has a marine west coast climate with high rainfall, particularly in the winter along the west coast. The moderating influence of the Atlantic and prevailing westerly winds leads to moderate temperatures, with winter highs averaging in the mid single digits (C) and summer highs around 15–20 degrees C.

POPULATION AND CULTURE

The current population of the UK is 60.6 million people, with the highest population densities in the south of England. England makes up the largest and fastest growing component of the UK population with a little over 50 million people. Scotland has a population of five million, which has remained static over the past 100 years, and the populations of Wales (3 million) and Northern Ireland (1.6 million) exhibit alternating patterns of very slow growth and stagnation depending on economic and political conditions.

While out-migration to the colonies drew off excess population during the industrial revolution, the UK has generally been subject to net immigration. From the 1950s through the 1980s, the primary origin of migrants was former colonies, particularly India, Pakistan, the West Indies and East Africa. Since the UK's admission to the EU, however, Europe has become the major source for immigrants (as well as becoming the main destination for British migrants).

The UK does not have an official language, although English is spoken by virtually the entire population and is the main language of government,

In 1968, with a recommendation from ECOSOC spearheaded by Sweden, the General Assembly decided to convene the UN Conference on Human Environment (UNCHE), which took place in Stockholm, Sweden, in 1972. At the same time, the GA also asked that the Secretary General of the UNCHE conference, Maurice Strong, prepare a comprehensive report on the problems facing developing and developed countries in regard to the human environment.

In preparing for the meeting, Strong worked with an advisory committee in which disagreements surfaced. Many developing countries were threatened by and resistant to the idea of integrating environmental protection as a part of their countries' policy agendas. The economic burden of dealing with environmental problems seemed daunting given that the path to economic and social development was already very difficult for the less-developed nations. But Strong was able to convince participants that without protecting the environment in the short term, the states cannot expect to grow in the long term.

This conference, attended by representatives of 113 nations, was a watershed event. These states came to an agreement on 26 principles on the human environment. The principles included the right to clean environment, the safeguarding of natural resources, the restoration of renewable resources, the protection of wildlife and endangered species, the prevention of the exhaustion of resources, the prevention of toxic pollution and global warming, and the prevention of pollution of the seas.

As a part of the negotiations, the developing countries received special consideration. Their particular needs were addressed under the 26 principles with a call for financial and technical aid to poorer regions and a clause stating that environmental policies should not hinder development. The principles also included a call for scientific research and education, a commitment to spend resources for making improvements on the environment possible, coordination among states to use resources more rationally, better urban planning, and recognition of the need for population control. There was also a call on states to create environmental agencies within their governments and additionally a call on states to start negotiations on the elimination and destruction of all nuclear weapons.

After the first comprehensive intergovernmental conference on the environment, the United Nations Environment Programme (UNEP) was created by the GA. UNEP is designed to coordinate UN agencies regarding environmental concerns and to generally serve as the lead agency on environmental affairs. The program is charged with assessing the state of the world's environment, developing mechanisms to deal with environmental problems, strengthening institutions, conducting research and raising global awareness on environmental issues.

THE 1980S: A NEW APPROACH

In the 1980s, the UN moved from a general discussion on the environment to more specific actions through treaty agreements on the ozone layer, toxic waste, and climate control. Inspired by a report by the UNEP governing body, the GA decided to create a special commission on Environment and Development to consider world "Environmental Perspective to the Year 2000 and Beyond." This commission is commonly referred to as the Brundtland Commission, named after its chairman, a former Prime Minister of Norway. The commission's 1987 report, *Our Common Future,* defined sustainable development, a concept that would serve as the guiding principle for global environmental policy. The statement read:

> The Governing Council believes that sustainable development is development that meets the needs of the present without compromising the ability of future generations to meet their own needs and does not imply in any way encroachment upon national sovereignty. The Governing Council considers that the achievement of sustainable development involves co-operation within and across national boundaries.

In many ways, the 1980s represented a new era for the UN in its approach toward the environment, which went beyond preservation and conservation. It explicitly recognized environmental degradation as a human-created problem that needed to be addressed along with its traditional development goals. New global treaties designed to address environmental problems were enacted. In an effort to address a growing hole in the earth's ozone layer, the 1985 Vienna Convention on the protection of the

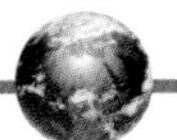

ozone layer was negotiated followed by the 1987 Montreal Protocol on Substances that Deplete the Ozone Layer. These treaties can be considered the most successfully implemented environmental treaties relative to other treaties negotiated. The Montreal Protocol called on state parties to take action to reduce and eliminate the production and consumption of ozone depleting substances (ODS) such as chlorofluorocarbons (CFCs) commonly used in refrigeration and air conditioning. Since going into effect, the treaty has been updated and renegotiated to further the goals of the Protocol.

One hundred and eighty-nine states made a commitment to the success of the Protocol and there has been tremendous progress in repairing the hole in the ozone layer due to these agreements. While the agreement is an international agreement among states, some have argued that the success of the Protocol would not have been possible without the commitment and leadership of the United States and the cooperation from major companies that produce CFCs, who aggressively sought alternatives to this harmful substance.

In another environmentally related development, in 1988 the Intergovernmental Panel on Climate Change (IPCC) was established in collaboration with World Meteorological Organization and UNEP. These entities were designed to investigate growing concerns about temperature increases associated with carbon dioxide buildup in the earth's atmosphere. Policies regarding climate control were seen as necessary after the first Assessment Report by the IPCC. This report led to the United Nations Framework Convention on Climate Change, which provides an overall policy framework for addressing climate change issues. A third major problem area addressed starting in 1989 is toxic waste disposal. This was first discussed under the Basel Convention on the Transboundary Movement of Hazardous Wastes. The agreement bans the export of hazardous waste from rich to poorer countries.

THE EARTH SUMMIT AND BEYOND

With the wider scope of environmental issues being discussed by the countries of the world, between 1989 and 1992 the governments prepared for a UN Conference on Environment and Development. This conference is also known as the Earth Summit, or the Río Summit, and took place in Río de Janeiro, Brazil, in 1992. The conference was another watershed event for the UN as it laid the groundwork for a number of agreements and proposed an ambitious environmental and social policy agenda detailed in the final conference document, Agenda 21. The framework laid out in Agenda 21 included issues of sustainable development such as combating poverty, protecting health, human settlements, population, as well as environmental concerns such as atmosphere, land management, deforestation, biological diversity, desertification, and sustainable agriculture.

While the participating nations were able to reach agreement, there were differences in emphasis sought by nations from different regions. The developing countries wanted to focus more on freshwater, deforestation, and pollution as opposed to the advanced industrialized countries, which sought to emphasize ozone depletion, hazardous wastes, and global warming. There were several conventions and commissions created to work on sustainable development during and following the Río Summit, including the 1992 Convention on Biological Diversity, the 1992 Framework Convention on Climate Change, the 1994 Convention to Combat Desertification, the 1995 Agreement Relating to the Conservation and Management of Straddling Fish Stock and Highly Migratory Fish Stocks, the 1998 Rotterdam Convention on the Prior Informed Consent Procedure for Certain Hazardous Chemicals and Pesticides in International Trade and the 2000 Stockholm Treaty on Persistent Organic Pollutants.

This gathering was also significant because nongovernmental organizations (NGOs) made inroads into the UN policymaking structure, a role that had traditionally been limited to nation states and in which NGOs were largely ignored or dealt with only informally. The Global Forum was a very effective parallel summit and it was hard for the states involved in the Earth Summit to ignore this important sector. Since that time civil society organizations have been more formally integrated into the UN's policymaking process.

Although the Earth Summit is still hailed as an important event, its significance lies more in the attention paid to environmental and development

issues than in the actual actions that resulted. While some agreements were reached and progress made, its goals have not been advanced nearly as rapidly as originally hoped. World leaders revisited the issues of sustainable development in Johannesburg in 2002, 10 years after the first Earth Summit, in order to discuss the lack of progress made toward meeting the goals of Agenda 21 and other agreements created under the Earth Summit.

The UN is the central organization where states of the world gather to make broad international agreements about many global issues including those related to the environment. Since its foundation, the UN and its subsidiary organizations have been working to create knowledge and awareness, set standards, and help carry out environmental policies accepted by the international community. So far these issues have been primarily addressed by the GA and ECOSOC. But, as resources become scarcer, states might start to consider environmental issues, especially those related to energy and water scarcity, as security issues and shift environmental policy responsibility to the Security Council. The goal of the UN has been to avoid environmental problems from becoming issues of international security and continue the work among member states to create a safer and cleaner environment. To this end, there have been more than 3,000 treaties signed and deposited with the UN and nearly 41 major multilateral treaties and conventions conducted under the auspices of the UN.

SEE ALSO: Agenda 21; Biosphere; Brundtland Report; Convention on Biological Diversity; Intergovernmental Panel on Climate Change; Montreal Protocol; Ozone and Ozone Depletion; Río Declaration on Environment and Development; Sustainable Development; Underdeveloped (Third) World; UNESCO; United Nations Conference on Environment and Development; United Nations Environment Programme; United Nations Framework Convention on Climate Change.

BIBLIOGRAPHY. Agenda 21, www.un.org (cited April 2006); Christopher Flavin, Hilary French, and Gary Gardner, *State of the World Report 2002: Special World Summit Edition* (Worldwatch Institute, 2002); Karen A. Mingst and Margaret P. Karns, *The United Nations in the Post–Cold War Era* (Westview, 2000); United Nations, *Basic Facts about the United Nations* (United Nations, 2004); United Nations, "Sustainable Development, Human Settlements and Energy," www.un.org (cited April 2006); United Nations Environment Programme, www.unep.org (cited April 2006); Thomas G. Weiss, David P. Forsythe, and Roger A. Coate, *The United Nations and Changing World Politics* (Westview 2004).

S. Ilgü Özler
State University of New York, New Paltz

United Nations Conference on Environment and Development (Earth Summit 1992)

THE UNITED NATIONS Conference on Environment and Development, also known as the "Earth Summit," was a major international conference held in 1992 at which representatives of almost every nation of the world participated in an effort to forge a plan for economic development and environmental protection. For many years prior to the meeting, conflicting views about the value of economic development versus environmental protection stood in the way of international progress on environmental issues in negotiations between states of the world. In 1987, a new understanding of the problem of environment versus development was introduced, and proposals for simultaneously dealing with problems of the environment and poverty were offered in a report called *Our Common Future*. The report was prepared by a United Nations (UN) special commission on Environment and Development headed by Prime Minister Gro Harlem Brundtland of Norway as a response to worsening environmental problems and the bleak environmental future of the world. According to this new understanding, sustainable development was defined as "development that meets the needs of the present without compromising the ability of future generation to meet their own needs."

After this report, between 1989 and 1992, member states of the United Nations started to prepare for a global summit to address sustainable devel-

opment comprehensively. There was a preparatory committee created under the General Assembly of the UN to gather and prepare documents to be discussed and agreed to during the 1992 conference. After a very thorough preparation process, the UN Conference on Environment and Development (the Earth Summit) took place June 3–14, 1992, in Río de Janeiro, Brazil. The conference was headed by General Maurice F. Strong of Canada, who also headed the UN Conference on Human Environment in 1972. There were 172 governments participating, of which 108 were represented by their highest government officials. There was also participation on the part of nearly 2,400 representatives from roughly 1,000 nongovernmental organizations (NGOs) who were among 17,000 people attending a parallel forum designed to coincide with the UN event. This injected perspectives from civil society into a policy forum at which only nation states had traditionally had input. This conference was the biggest and one of the most important environmental events in history.

A number of documents, prepared only by the official state participants, came out of the conference, including Agenda 21, a plan of action for advancing sustainable development. In addition, the Río Declaration on Environment and Development served as a set of guiding principles. Other official declarations included the Statement of Forest Principles, the UN Framework Convention on Climate Change and the UN Convention on Biological Diversity. The conference attendees also created institutions that would function as follow-up mechanisms for these documents, including the Commission on Sustainable Development, the Inter-agency Committee on Sustainable Development, and the High-level Advisory Board on Sustainable Development.

FIRST VERSUS THIRD WORLD

During the preparatory meetings and the conference, major compromises were made in order for all parties to reach a consensus. The central debate between the industrially developed "North" and the less developed "South" regarding the value of economic development versus environmental protection continued throughout the preparation process and during the conference. The developing countries were primarily concerned with issues such as the alleviation of poverty through financing of development from developed countries, access to technology for sustainable development as well as a focus on issue areas of freshwater, desertification, deforestation and pollution.

The issues were difficult to resolve even among the less-developed nations. For example, developing countries face a difficult dilemma in regard to deforestation. Forests are important to prevent landslides, which affect poor people disproportionately. Additionally, plants and animals in the forests can be a source of livelihood for the subsistence of the poor communities. Thus, the loss of these habitats affect the lives and livelihoods of people in the South. Yet, while protection of forests is an important issue, states are also hesitant to treat issues such as deforestation as global problems given sensitivity toward the protection of their sovereignty and the rights of state to exploit resources within their own borders. During the negotiations between and among developed and less-developed nations, these types of conflicts were difficult to manage. There were also pressures from NGO groups felt by governments before and during the conference.

In looking at the other major constituency, the advanced industrialized countries wanted to focus on the problems of ozone depletion, the production and disposal of hazardous wastes by industry, and global warming. These were issues that the less-developed countries blamed on the high levels of consumption by the wealthy nations of the North. Dozens of such topics had to be addressed throughout the four years and five rounds of negotiations between states before the conference.

RÍO DECLARATION

In Río, the final document was negotiated. Some, including conference president Maurice Strong, characterized the outcome of the Earth Summit as falling short of its originally envisioned goals. Yet many also felt it was a great accomplishment to have such a large meeting of world leaders and come to an agreement on a set of principles and an action plan that satisfied and reflected the concerns of such disparate groups. In the 27 principles of the Río Declaration on Environment and

Development, the states emphasized the importance of state sovereignty, the rights of states to equitable sustainable development, and the need for more aid for developing nations with a priority on the least-developed countries.

The advanced industrialized countries were acknowledged as responsible for global environmental problems and called for states to eliminate unsustainable patterns of production and consumption. The importance of technology, women, youth, and indigenous groups were recognized for their role in achieving sustainable development as well as citizens' right to know about their environment. The states were declared responsible for enacting legislation on environmental standards and protection of their citizens. The importance of transparency and cooperation between states on environmental issues was also emphasized.

AGENDA 21

The main document emerging from the conference, Agenda 21, lays out a comprehensive plan of action for global sustainable development. The document has four major subheading areas with a total of 40 chapters. The first area covers social and economic issues including poverty, consumption, population, health, human settlements, and the development of policies to be able to deal with these issues, especially in the developing nations. The second area on conservation and management of resources dealt with major environmental problems in the context of development. The areas of focus were atmosphere, land resources, deforestation, desertification, mountain development, sustainable agriculture and rural development, biological diversity, biotechnology, oceans, freshwater resources, toxic chemicals, and different kinds of hazardous wastes. The third section dealt with the strengthening of major groups necessary for the achievement of sustainable development. Women, children, indigenous people, NGOs, local authorities, workers and their trade unions, business and industry, and the scientific and technological community as well as farmers were recognized as important groups in bringing about sustainable development.

In the last section of the document, the means of implementation and financing sustainable development initiatives were discussed. The rich states reaffirmed their commitment to providing 0.7 percent of their Gross National Product (GNP) in aid to fund sustainable development in the less-developed nations. But there were no new commitments made by the rich countries to meet the goals set up by Agenda 21.

OTHER DOCUMENTS

Among the other major documents to come out of the conference was the Convention on Climate Change. This was negotiated and signed in Río with the goal of stabilizing greenhouse gas concentrations in the atmosphere. This was agreed upon without specific limitations set for nation states, but nevertheless this convention led to the Kyoto Protocol, which sets specific targets for different countries. Another agreement, the Convention on Biodiversity, was less successful in gaining acceptance due to disagreements on funding, but 158 signatories of the convention have agreed to conserve biological diversity. A final document created during the conference was the Forest Principles, which calls on states to further cooperate on forest issues.

SEE ALSO: Agenda 21; Brundtland Report; Kyoto Protocol; Río Declaration on Environment and Development; Sustainable Development; United Nations; United Nations Environment Programme; United Nations Framework Convention on Climate Change.

BIBLIOGRAPHY. Agenda 21, www.un.org (cited April 2006); S. Halpern, *United Nations Conference on Environment and Development: Process and Documentation* (Academic Council for the United Nations System, 1992); Karen A. Mingst and Margaret P. Karns, *The United Nations in the Post–Cold War Era* (Westview, 2000); E.A. Parson, P.M. Haas, and M.A. Levy, "A Summary of Major Documents Signed at the Earth Summit and the Global Forum," *Environment* (v.34/4, 1992); United Nations, *Basic Facts about the United Nations* (United Nations, 2004); Thomas G. Weiss, David P. Forsythe, and Roger A. Coate, *The United Nations and Changing World Politics* (Westview, 2004).

S. Ilgü Özler
State University of New York, New Paltz

United Nations Environment Programme (UNEP)

THE UNITED NATIONS Environment Programme (UNEP) was created in 1972 following the United Nations Conference on Human Environment (UNCHE) in Stockholm, Sweden. The General Assembly (GA) of the United Nations (UN) set the goals and mandate of the UNEP in its resolution 2997 (XXVII). The mission of the UNEP is "to provide leadership and encourage partnership in caring for the environment by inspiring, informing, and enabling nations and peoples to improve their quality of life without compromising that of future generations."

In general, the UNEP serves as a forum in which the member states of the UN can discuss environmental issues of international scope. For this, 58 member states are elected by the GA to serve as the governing body of the UNEP, which meets biennially and reports to the GA through the Economic and Social Council. The governing council has representation from different regions of the UN: 16 African, 13 Asian, 10 Latin American, six East European, and 13 Western European and other states.

The mandate of the organization evolved over time with amendments adopted at the United Nations Conference on Environment and Development in 1992 and during the 1997 session of the UNEP Governing Council meeting. Currently, the specific mandates of the organization include raising awareness about the environment globally and incorporating environmental considerations into all aspects of UN programs and activities from peacekeeping to disarmament and health to education.

The UNEP is charged with conducting scientific research and issuing reports about the human environmental problems and developments. When environmental threats are identified, UNEP monitors, gathers information, and calls for action to overcome these problems. The UNEP is the central coordinating body of all UN agencies on environmental issues. When the environment ministers from around the world met for the first time in Malmö, Sweden, they issued a declaration affirming the mandate of the organization and called for the institutional and financial strengthening of the UNEP in 2000.

One of UNEP's most important roles is to help states develop international environmental law by facilitating conventions among states and meetings among scientists to provide expert guidance. The first convention coordinated by the UNEP was in 1973 on International Trade in Endangered Species (CITES). Subsequently, there have been several conventions facilitated by the UNEP on issues ranging from migratory species, depletion of the ozone layer, climate change, transboundary movement of hazardous wastes, biological diversity, pollution of the seas, persistent organic pollutants, and biosafety and genetically modified organisms.

The UNEP has nearly 600 staff members who work to accomplish the mandate of the organization and the goals set by its governing body. Half of the staff works at the UNEP headquarters in Nairobi and the other half are stationed in various offices around the world to fulfill organization's global and regional goals. The UNEP has eight functional divisions for assessment of environmental threats, policy development, policy implementation, helping developing nations adopt environmental policies, regional cooperation, finding synergy between multilateral conventions, communication and education on the environment, and funding environmental programs.

Because UNEP does not have the field experience, institutional capacity, or the funding to carry out all of its goals, the organization cooperates and coordinates with other UN organs to accomplish its goals through these functional divisions. For example, the Division of Global Environment Facility (GEF) is a joint effort between UNEP and the World Bank. GEF is a trust fund established by the World Bank in 1991. An advisory panel of experts from the UNEP examines the GEF funded projects. These projects are established with the goal of making progress on one of the major treaty areas on the environment. Another functional division of the UNEP, Early Warning and Assessment, works in collaboration with several global networks of nongovernmental organizations (NGOs), private companies, and state agencies to monitor emerging environmental threats.

Since its foundation the UNEP faced several challenges. First it was difficult to build consensus on issues due to conflicts between the advanced industrialized

countries and the less-developed countries with competing opinions about environmental protection versus economic development. The UNEP has also been marginalized by donor countries whose financing is necessary for the success of the organization. Thus UNEP still lacks money and the organizational capacity to make significant progress toward meeting its mandate. The most important role of the organization still remains setting international norms on the environment and coordinating the overlapping environmental responsibilities of different UN agencies.

SEE ALSO: Convention on International Trade in Species of Wild Flora and Fauna (CITES); United Nations; United Nations Conference on Environment and Development.

BIBLIOGRAPHY. Christopher Flavin, Hilary French, and Gary Gardner, *State of the World Report 2002: Special World Summit Edition* (Worldwatch Institute, 2002); "Malmö Ministerial Declaration," www.unep.org (cited April 2006); "Nairobi Declaration of UNEP Governing Council," www.unep.org (cited April 2006); United Nations, *Basic Facts about the United Nations* (United Nations 2004); Thomas G. Weiss, David P. Forsythe, and Roger A. Coate, *The United Nations and Changing World Politics* (Westview, 2004); UNEP, www.unep.org (cited April 2006).

S. Ilgü Özler
State University of New York, New Paltz

United Nations Framework Convention on Climate Change (UN FCCC)

UNITED NATIONS FRAMEWORK Convention on Climate Change (UN FCCC) was signed during the 1992 Earth Summit in Río de Janeiro, Brazil. Climate change was introduced as an area that required special attention warranting a separate convention during the preparations to the Earth Summit. The UN FCCC was a first attempt to create a way to deal with climate change resulting from human activity. The states acknowledged that carbon emissions and greenhouse gasses do not recognize national borders, thus states need to cooperate to address climate change. Because the consensus on global warming due to human activity was not yet achieved among the expert communities and state parties, the convention was precautionary rather than proactive. The differential impact of advanced industrialized countries on bringing about this environmental problem was also recognized by the convention, giving more responsibility to the industrialized nations of the North. Eventually, 189 countries ratified the convention, which went into effect March 1994.

The Conference of Parties (COP) is the governing body of the UN FCCC. The COP meets every year to discuss climate change and develop new measures to deal with the evolving issue of global warming. The states in the convention have kept negotiations alive by working on the Kyoto Protocol of UN FCCC in 1997. This Protocol, using the same mechanisms included within the convention, assigns states legally binding targets to reduce or limit emissions.

Keeping in spirit with the differential responsibility of pollution, the convention separates countries into different categories. The Annex 1 countries include advanced industrialized countries, such as the United States and those of Western Europe, and countries in economic transition, mostly Eastern and central European countries. These are also the largest polluters in the world and are required to cut emissions on average at least 5.2 percent from their 1990 levels by 2008–12. Developing countries and the least-developed countries, on the other hand, are expected adopt environmental policies that will limit the increase in emission from a base of 1995 levels.

In accordance with the terms of the implementation of the agreement, the Protocol would take effect when more than 55 parties to the convention, who accounted for 55 percent of the total carbon dioxide emission based on 1990 figures, signed and ratified the treaty. This occurred in 2005 and the Protocol formally took effect at that time. The United States never ratified the Kyoto Protocol and withdrew from it in 2001. Since the United States was the largest carbon emitter in 1990 with 36.1

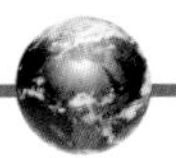

percent of the share of the world, every other large emitter had to ratify the Protocol in order for it take effect. The nations of the European Union (24.2 percent of total global emissions) and Japan (8.5 percent) had ratified the treaty in 2002 and with Russia (17.4 percent) joining the treaty in 2004, the treaty went into effect in February 2005.

The Protocol identifies the human sources of greenhouse gases such as energy consumption, fuel combustion, manufacturing, construction and other industries, transport, and production of minerals, metals, halocarbons, and sulfur hexafluoride. The Protocol allows for states to "trade" six greenhouse gasses: Carbon dioxide, methane, nitrous oxide, hydrofluorocarbons, perfluorocarbons, and sulfur hexafluoride, thus creating some flexibility in emissions.

The Convention is funded by financial mechanisms that ask the economically affluent countries to contribute to make compliance to the treaty possible by all parties. The Global Environmental Facility (GEF), a fund run by World Bank jointly with UN Environment Programme (UNEP) review, is also involved in funding for making progress toward meeting the treaty goals. Although the Kyoto Protocol represents progress in terms of strengthening international cooperation, most scientists believe that the limits set on greenhouse gasses by the Protocol are not sufficient to address global warming.

SEE ALSO: Carbon Trading; Global Warming; Greenhouse Gases; Kyoto Protocol; United Nations Conference on Environment and Development; United Nations Environment Programme.

BIBLIOGRAPHY. Seth Dunn, *Reading the Weathervane: Climate Policy from Rio to Johannesburg*, Worldwatch Paper 160 (State of the World Library, 2002); Seth Dunn and Christopher Flavin, *Moving Climate Change Agenda Forward* (State of the World Library, 2002); EarthSummit+5, "Special Session of the General Assembly to Review and Appraise the Implementation of Agenda 21, New York, June 23–27, 1997," www.un.org (cited April 2006).

S. Ilgü Özler
State University of New York, New Paltz

United States, Alaska

ALASKA IS THE northernmost, westernmost, and "easternmost" state in the United States (parts of the Aleutian Islands cross the 180th meridian). Because of its size and location, it encompasses many different geophysical areas, from rainy spruce and fir forests in southeast Alaska, to the desert of the Arctic. Alaska also contains many natural resources, and the exploitation of those resources has often been a point of controversy throughout its history.

Before its "discovery," the Inuit people, Eskimos of non-Inuit origin, Aleuts (of the Aleutian Islands), and Native Americans populated Alaska. These included the Athabascans of central Alaska and the Tlingit and Haida tribes of southeast Alaska. Unlike most of the United States, the indigenous peoples of Alaska were not forced into reservations. The Alaska Native Claims Settlement Act (ANCSA) of 1971 was an attempt to avoid the mistakes of past national policy toward indigenous peoples. ANCSA is also central to understanding the politics of land conservation in Alaska.

Russia, Spain, and Britain were the major powers in Alaska until 1876. Russia was most influential. Vitus Bering, a Dane working for the Russian Czar, "discovered" Alaska in about 1728, although there is evidence suggesting earlier Russian exploration. The Russian influence in Alaska is seen in numerous towns and villages throughout the Aleutians and coastal Alaska, many of which still have small and distinctive Russian Orthodox churches. The name of the state comes from the Russian interpretation of the Aleut *Alyeska* meaning "great land" or "mainland." Russian exploration extended to northern California, raising concern in Spain, which pressed its explorations north to Alaska. Spanish names in Alaska still exist, including the cities of Valdez (pronounced Val-DEEZ) and Cordova. British explorer Captain James Cook, seeking the inside passage, explored what later became known as Cook Inlet, and named Turnagain arm, near Anchorage, because, having failed to find the northwest passage, he had to "turn again."

Russians fur traders hunted sea otters to the brink of extinction. Timber and fish were also important, particularly as the otters became scarce. The Russian America company established its headquarters

Tourism, such as this cruise through Tracy Arm Fjord to view glaciers, has grown considerably since the early 1980s and is the second most important sector in the Alaskan economy after oil.

at Sitka (New Archangel), which was the center of Russian colonial government until 1867, when the United States, at the urging of Secretary of State William Seward, purchased Alaska from Russia for $7.2 million dollars, or about two cents per acre. This purchase was made well after the otter population had withered. Of course, there was widespread ridicule of "Seward's Folly," but most enlightened citizens and officials realized that Alaska contained considerable natural resources and provided strategic benefits for the United States in its competition with Great Britain over the Pacific Northwest.

AMERICAN DEVELOPMENT

After 1867, most Americans and the national government paid relatively little attention to the territory. The region remained under army control until the Organic Act created civil government in 1912. The formation of a territorial government was hastened by gold discoveries in the region, including the Klondike Gold Rush in Canada's Yukon. The most direct route to the Yukon was via Skagway, Alaska, and then over the treacherous Chilkoot Trail. Alaska had its own gold discoveries: At Juneau in 1880, along the beach in Nome in 1898, and near Fairbanks in 1902.

Gold was mined in Juneau until World War II, and gold panning is still a popular tourist activity. In the 1880s, John Muir visited Alaska (Muir Glacier bears his name), and became one of the earliest proponents of saving Alaska's natural treasures from development. The conservation movement of the early 1900s led to the creation of the Katmai National Monument in 1918, and Mt. McKinley National Park in 1917, both of which were expanded under the Alaska National Interest Lands Conservation Act (ANILCA).

Despite gold rushes, the population grew slowly, and was concentrated in southeast Alaska. The federal government completed the Alaska Railroad, which connects Seward, Whittier, Anchorage, Denali Park, and Fairbanks, in 1923. It is now owned by the state. The major supply port for the railroad, Anchorage, started as a tent city in 1914 and was incorporated in 1920. The first major population boom in Alaska came in the 1940s, during World War II. It continued through the cold war, as military spending increased Alaska's population and built key infrastructure. The Alaska Highway, linking the state to the lower 48, was completed as a military supply route in 1943, although it was not a reliable year-round road until the 1950s. The World War II and postwar period also saw the shift of population from southeast Alaska to Anchorage and the "railbelt" from Anchorage to Fairbanks.

In 1935, the Department of Interior and the Federal Emergency Relief Administration established Matanuska Colony about 45 miles north of Anchorage to encourage farmers from Wisconsin, Michigan, and Minnesota to develop agriculture in Alaska. The Matanuska valley soon became famous for its outsized vegetables—the long summer daylight makes crops grow quite rapidly—but the success of the colony was mixed, at best. Some dairy farming and relatively small-scale vegetable farming remains, but since the late 1980s, the Matanuska-Susitna area has become a major population center and, in essence, a suburb of Anchorage. Other industries that remain important in Alaska are fishing, timber, mining, and tourism. Tourism in particular has grown considerably since the early 1980s, and tourism is particularly important to the economies of southeast Alaska cities, and, to a lesser extent, Anchorage and Fairbanks.

The most important event in Alaska's economic history was the discovery of oil on the North Slope. In the late 1950s, some oil was discovered on the Kenai Peninsula and in Cook Inlet, but the potential of the North Slope discoveries was far greater. In 1969, the state sold oil leases for $900 million, a huge windfall in a poor state with a budget that in 1968 only exceeded $100 million. The ensuing construction of the Trans-Alaska pipeline created a boom in Anchorage and Fairbanks, and swelled state coffers. When oil started flowing through the pipeline in 1977, the importance of yearly royalty income became clear. Over 85 percent of the state's revenue comes from oil, and Alaskans pay no income tax or state sales tax. While the oil economy has paid substantial benefits to Alaska, it has also come at a cost. During the oil price crash of the late 1980s, state revenues plummeted, the economy suffered, and many people left the state. In 2006 state revenues (from oil royalties) were sharply higher due to increases in the world price for oil, although demands on state government remained quite high.

Increased oil wealth has increased expectations of state services to "rural" (that is, remote) Alaska, from airports, to state-subsidized television services (discontinued in 1996), to state-supported local schools, which have helped keep communities together because children no longer need to go to big cities or regional boarding schools. The erosion of native traditions is palpable in these communities, as modernity replaces, or is melded with, tradition. As snowmobiles ("snowmachines" in Alaska) and outboard motors replace dog sleds and oars, some might bemoan the loss of traditions; but many rural Alaskans appreciate how technology eases a very difficult way of life, even as modern conveniences take a cultural and environmental toll.

ENVIRONMENTAL ISSUES

Alaska has long experienced conflict over the exploitation of its rich natural resources. Like many other places in the western United States, Alaska's resource exploitation history involves periods of extreme overexploitation of species (otters), followed by near extinction and slow recovery. Improved management techniques avoided the worst environmental abuses, and Alaska's early economic potential was found in fish and timber, not gold and furs. These industries remain extremely important. Entire communities in Alaska, such as Cordova and Kodiak, are highly dependent on the fishing industry. At the same time, other industries have overtaken these traditional industries. Tourism is the second most important sector in the Alaska economy, after oil. This juxtaposition of industries is important, because Alaska has many industries that, for most of the time, are compatible with each other, but when an environmental crisis occurs, interests clash.

For example, some travel agents and other industry sources reported that bookings for Alaska tours dropped after the *Exxon Valdez* oil spill, which, while vast, still affected a relatively small part of Alaska, and was nowhere near the primary tourism markets in southeast Alaska and Denali Park. However, the exploitation of natural resources worries some in the tourism industry because they know that people visiting Alaska expect not to see clear-cut forests or oil spills. Oil is also incompatible with fishing if large oil spills occur. Fishing is an iconic Alaskan industry, but it is also a much smaller part of the economy than is oil.

The most heated environmental controversy in Alaska, or centered on Alaska, was the fight over what were called the "d-2" lands, so called after section 17(d)(2) of the ANCSA, This provision allowed the Secretary of the Interior to reserve 80 million acres of land for the "national interest." Many Alaskans were outraged by the very idea of allowing such designations of land, believing it interfered with the state's powers to select state lands—which then could be made available for development—under the Statehood Act. These Alaskans feared that the land would be "locked up" and forever unavailable for mineral or other development. The Congress did not take up an Alaska lands bill for years after ANILCA was enacted, so the issue came to a head when, in 1978, President Carter set aside over 100 million acres to protect them from development—40 million under ANCSA, and the balance under his powers to declare national monuments. This induced Congress to enact ANILCA, which created and expanded national parks, preserves, and monuments, including Denali, Katmai, Kenai Fjords, Glacier Bay, and Gates of the Arctic.

A continued environmental challenge in Alaska is one that is shared by many western states: The increased growth of the urban areas. Population growth is somewhat limited in Anchorage because there is not much available land. Much of the land to the north and east of the city is military or parkland. This has led to a form of suburban sprawl in the Matanuska-Susitna valley, as people move there to avoid expensive housing in Anchorage. They choose to commute as far as 100 miles round trip. Amidst these environmental challenges, Alaska faces the usual challenges of any economy based on natural resources: The commodity may run out, or prices may drop, or the environmental cost of extracting the commodity may exceed its benefits. Alaskans and Americans have met and addressed these challenges in the past, and have adapted to changes in their economy and to changes in perceptions of the value of the environment.

SEE ALSO: Alaska Pipeline; Arctic; Arctic National Wildlife Refuge; *Exxon Valdez*; Muir, John; Native Americans.

BIBLIOGRAPHY. P.A. Coates, *The Trans-Alaska Pipeline Controversy: Technology, Conservation and the Frontier* (University of Alaska Press, 1993); J.S. Hammond, *Tales of Alaska's Bush Rat Governor* (Epicenter Press, 1994); Anne Hanley and Carolyn Kremers, *The Alaska Reader: Voices from the North* (Fulcrum Publishers, 2005); S.W. Haycox, *Alaska: An American Colony* (University of Washington Press, 2002); S.W. Haycox, *Frigid Embrace: Politics, Economics, and Environment in Alaska* (Oregon State University Press, 2002); S.W. Haycox and M.C. Mangusso, *An Alaska Anthology: Interpreting the Past* (University of Washington Press, 1996); Larry Kaniut, *Tales from the Edge: True Adventures in Alaska* (St. Martin's Griffin, 2005); Orlando Miller, *The Frontier in Alaska and the Matanuska Colony* (Yale University Press, 1975).

Thomas A. Birkland
State University of New York, Albany

United States, California

CALIFORNIA IS THE third largest U.S. state in terms of area. In 2005, California's estimated population was 36,132,147, making it the most populous state in the country. Approximately 68 percent of Californians live in the greater Los Angeles and San Francisco Bay Areas.

The discovery of gold in 1848 led to the California Gold Rush, which marked the opening of the Sierra Nevada region to European-American occupation and development. What began as amateur mining with simple technologies in the early years transformed by the 1860s into large-scale mining,

including the excavation of mines with hydrologic operations. A federal injunction put an end to immense on-site and downstream environmental impacts of hydrologic mining during the 1890s. Mining in California, however, did not stop with the end of the Gold Rush. In 2004 California ranked first in the nation in nonfuel mineral production, accounting for nearly 8 percent of the national total.

California's diverse physical landscape is now managed through a mosaic of private, federal, and state landholdings. In 1890 Sequoia National Park became California's first national park. In 1902 Big Basin State Park became the first state park in California. Today, 47 percent of the land ownership in California is under the control of the federal government (20 percent Forest Service, 15 percent Bureau of Land Management, 8 percent National Park Service, and 4 percent military), while 5 percent of California's land ownership is under the jurisdiction of the state. The other 48 percent falls under private ownership.

Located at the interface of the Pacific and North American tectonic plates, California experiences frequent seismic activity. The San Andreas Fault runs the length of California and is one of the state's most active fault lines. One of the most destructive earthquakes along the fault was the 1906 San Francisco Earthquake. In 1989 the Loma Prieta Earthquake struck along the San Andreas Fault, causing widespread damage to the San Francisco Bay Area.

There are many other active fault lines in California. The 1994 Northridge Earthquake in Southern California, which is considered by many to be the most costly earthquake in U.S. history, occurred along the Santa Monica Mountains Thrust Fault. In reaction to California's vulnerable position, the California Legislature passed a landmark law in 1972 requiring the identification of seismic hazard zones. In these zones, special geologic studies are mandatory before structures can be assembled for human occupancy.

AGRICULTURE

California's Central Valley is the state's true agricultural breadbasket and one of the most productive agricultural regions in the country. In 2002, California ranked first in the nation in agricultural product sales, amassing a market value of $25,737,173,000. A significant portion of California's agricultural employment is composed of immigrants. Cesar Chavez, a Mexican American labor activist who founded the National Farm Workers Association (later the United Farm Workers), was central in the fight for immigrant worker rights throughout California and the United States. Chavez encouraged labor unions and fought for workers' rights, including reducing worker exposure to harmful pesticides. Through these efforts, Chavez brought the plight of immigrant workers to the public's attention and organized a number of important environmental justice agreements.

MEETING WATER DEMAND

Water has traditionally been the most contested natural resource in California. A number of factors complicate its use and management. First, California only receives precipitation nine months out of the year, leaving summers without a consistent water supply. Second, although the state contains a number of underground aquifers, many of them are over drafted, too deep to access economically, or contaminated by toxins such as MTBE. Third, the majority of precipitation in California falls in the northern portion of the state and in the Sierra Nevada Mountains, while a majority of the population resides in the southern portion of the state. Fourth, demand is high with a population of over 36 million and the nation's most productive agricultural sector.

All of these factors have necessitated the formation of a large-scale system for redistributing water across the state. The two key elements of this statewide system are the storage of runoff from mountain snow pack in reservoirs in order to control the timing of water delivery and the construction of waterways to control the geographic distribution of water.

In 1960 the California State Water Project commenced, enabling the production of a large water storage and transport system across the state. The Central Valley Project, containing 21 primary reservoirs and the 450-mile (725-kilometer) California Aqueduct, has enabled the delivery of water to Central Valley farmers and residents of Southern

Golden Gate

The strait connecting San Francisco Bay to the Pacific Ocean was first known as the Boca del Puerto de San Francisco ("Mouth of the Port of San Francisco") and was first recorded by José Francisco Ortega, the head of a surveying party that approached the bay by land. On August 5, 1775, Juan de Ayala steered his ship *San Carlos* through the strait, dropping anchor inside the bay; the *San Carlos* was the first recorded European ship to make the passage. A year later the Spanish, recognizing the potential military importance of the bay, established a military port at the tip of the San Francisco Peninsula as well as the small Dolores Mission.

San Francisco was a part of Spanish America until 1821 when it became a part of Mexico. It had one of the most isolated Spanish garrisons until it was withdrawn and replaced with a Mexican one. Most trade at the time took place in the more commercially successful settlement at Monterey. In 1848 the whole of California was ceded to Mexico, at which time San Francisco was a town of just 900 people.

The name "Golden Gate" appears to have been given to the strait just before the 1849 gold rush in California, which saw tens of thousands of prospectors heading through the strait in their search for gold. San Francisco boomed during the gold rush and the period that followed, developing from a frontier town to a thriving metropolis. It was a city of great wealth when much of it was destroyed by an earthquake on April 18, 1906. Nevertheless, the city recovered and in 1937 the Golden Gate Bridge was constructed across the strait. Until 1964, it was the longest main span bridge in the world.

California. Today, the system supplies water to over 23 million residents and 750,000 acres (303,514 hectares) of irrigated farmland.

Other significant water storage and transport systems include the Los Angeles Aqueduct and the O'Shaughnessy Dam at Hetch Hetchy. The Los Angeles Aqueduct was completed in 1913 under the jurisdiction of the Los Angeles Department of Water and Power. The controversial aqueduct devastated the Owens Valley but was instrumental to the growth of Los Angeles and the San Fernando Valley. Today, the aqueduct still carries water southwest from the Owens Valley to the city of Los Angeles. The controversial O'Shaughnessy Dam at Hetch Hetchy in Yosemite National Park was completed in 1923, much to the chagrin of preservationists like John Muir. Water from the Hetch Hetchy Reservoir serves the San Francisco Bay Area, which owes much of its early prosperity and economic growth to the reservoir.

LANDMARK POLICIES

California is home to a number of landmark policies including the California Environmental Quality Act (CEQA), which is a state-led version of the National Environmental Protection Act (NEPA). Enacted in 1970, CEQA requires that all projects undertaken or requiring approval by state and local governments be made public and reviewed for their potential environmental impacts. Projects deemed to have significant environmental effects are required to complete an Environmental Impact Report (EIR). CEQA requires these agencies prescribe ways of minimizing and/or mitigating deleterious environmental impacts.

The California Endangered Species Act (CESA) of 1970 (revised in 1984) remains a progressive piece of environmental legislation. The act was created to protect endangered and rare flora and fauna threatened by rapid statewide development. The CESA has resulted in the listing of numerous plants and animals that would not otherwise be protected by the Federal Endangered Species Act. As of mid 2006, California contained 155 listed endangered and threatened animal species with 31 only listed under the California Endangered Species Act. California lists 98 plant species under the sole protection of the CESA.

The passage of the California Clean Air Act in 1988 signaled a new era of statewide stringent air quality regulations. Under this act, California's am-

The San Francisco Bay Area owes much of its early growth to the Hetch Hetchy Reservoir in Yosemite National Park.

bient air quality standards are generally stricter than federal standards. Despite tough measures, Los Angeles was ranked in 2003 as the most polluted metropolitan area in the United States by the American Lung Association in terms of particulate matter and smog levels. Although other regions of California also experience elevated pollution on occasion, the Los Angeles Basin is particularly vulnerable due to low precipitation and a persistent inversion layer.

With over 28,000,000 registered vehicles as of 2000, transportation is responsible for roughly 35 percent of California's energy consumption and over 85 percent of total petroleum use. California is also one of the most progressive states in terms of setting strict motor vehicle efficiency standards. For example, California passed legislation in 2002, known as the Pavley Bill, requiring automakers to limit greenhouse gas emissions from motor vehicles in the state. The effects of California's progressive automobile legislation are far-reaching. By 2006, 10 other states adopted these strict California standards, which in turn put pressure on car manufacturers to fill this growing market.

In 2004 California produced 80 percent of its electricity with 10.6 percent from renewable sources. The other two major energy sources in California are petroleum and natural gas. In 2004, California produced 41 percent of its petroleum and 15.5 percent of its natural gas. The rest of California's energy needs are imported from other states. California deregulated its electricity market in 1996. A significant outcome was the 2000–01 California electricity crisis resulting in rolling blackouts, extremely high energy prices, numerous energy sector bankruptcies, a State of Emergency Declaration, and the eventual ousting of then governor Gray Davis.

SEE ALSO: Automobiles; Beaches; Earthquakes; Endangered Species; Gold; Hetch Hetchy Dam; Los Angeles River; Mining; Mulholland, William; National Parks; Owens Valley; Pollution, Air; Water Demand; Yosemite National Park.

BIBLIOGRAPHY. Marc Reisner, *Cadillac Desert: The American West and Its Disappearing Water* (Penguin Books, 1993); State of California, www.ca.gov (cited April 2006); University of California, *The Sierra Nevada Ecosystem Project: Final Report to Congress* (1996).

Gregory Simon
University of Washington

United States, Central South

THE STATES OF Arkansas, Kentucky, Missouri, and Tennessee are located in the central south region of the United States. They cover a vast area stretching from the Appalachian Mountains across the middle of the Mississippi River's course westward to the beginning of the Great Plains. Geologically, the central south region covers a diverse

topography of mountains, plateaus, hilly regions, and plains. Most of the region was submerged under shallow seas during several geological periods and is now covered in limestone rocks, which have eroded to make numerous caves in some areas and sweet soils for grass and crops.

TENNESSEE

Tennessee is named for the Over-the-Hills Cherokee Indian village of Tenasi. Tennessee's eastern borders are with the states of Virginia and North Carolina. In the north, it borders Kentucky; its southern border bounds Georgia, Alabama, and Mississippi. It stretches over 500 miles from east to west. The state is usually described as the areas of East Tennessee, Middle Tennessee, and West Tennessee, which are each marked by distinct land formations.

East Tennessee begins in the Blue Ridge Region which includes the Great Smoky Mountains and other ranges. The area from Franklin westward has numerous springs and forested areas. The boundary with Virginia runs along some of the highest peaks of the Appalachians. The area has numerous rivers and quiet mountain coves where small groups of people farm as the Cherokee did before them.

The Southern Appalachians that are in Tennessee include beautiful natural forests and wildlife that includes black bears. The upper altitudes are usually above 4,000 feet (1,220 meters) in height. The flora of the area includes numerous rhododendrons that flower profusely in the spring; there is striking fall foliage in the autumn. Because of the altitude, the flora is closer to that of Canada than to the flora of the Mississippi River Valley.

Among the rivers either beginning or flowing through East Tennessee are the Clinch, Holston, Ocoee, Hiawassee, and the French Broad. These rivers join the Tennessee River in a transition from the Appalachians to the Tennessee River Valley west of Knoxville. The transition is to the last of the valleys of the Ridge and Valley region of the Appalachians. Several ridges run in parallel lines from Georgia to Kentucky through the East Tennessee Valley. The area between Chattanooga and Knoxville is relatively flat and is excellent farmland.

The Tennessee River flows south from Knoxville to Chattanooga where it passes through a gorge and then continues in a looping circle through much of northern Alabama. At Florence, Alabama, it turns north again and flows to Tennessee where it separates the western part of the Highland Rim from West Tennessee. It continues flowing north to join the Ohio River in western Kentucky.

The third landform in the East Tennessee region is the Cumberland Plateau. It is a part of the vast Appalachian Plateau that runs from north of Birmingham and Gadsden, Alabama, into Kentucky. In Tennessee, the plateau in most places has rocky cliffs, which range from 1,500 to 1,800 feet (457 to 549 meters) in height. In the center are the Orchard Mountains, which are a range of peaks on top of the Cumberland, west of Knoxville. The area was extremely remote until the advent of modern automobiles and roads.

Middle Tennessee begins with the dramatic end of the escarpment that is the Cumberland Plateau. The area is part of the Highland Rim, which is a plain stretching to the Tennessee River at the boundary of West and Middle Tennessee. Nashville is located in a basin called the Nashville Basin.

The Cumberland River, flowing through the Highland Rim region, has left many eroded hills. The Cumberland River begins in the Appalachians near the junction of Tennessee and Kentucky. It flows through East Tennessee to Nashville where it turns to flow north. It joins the Tennessee River in Kentucky in the "Land-between-the-Lakes" a few miles from where the Tennessee River flows into the Ohio River.

The area of Middle Tennessee extending from north of Alabama north to Kentucky that lies between the Cumberland Plateau and the Nashville Basin is a garden area where enormous numbers of nursery farms are located.

West Tennessee is an extension of the Gulf Coastal Plain. It is an area of gentle undulations with rich farmland. In the northwestern corner of Tennessee lies Reelfoot Lake. The Lake was formed during the Mississippi River Valley earthquakes in late 1811 and early 1812. The quakes were centered along the New Madrid Fault. The severity of the earthquakes formed the shallow natural lake; it was filled by water from the Mississippi that for a while flowed north rather than south. Most of it is more like a swamp, however.

KENTUCKY

Eastern Kentucky is famous as an Appalachian region explored by frontiersman Daniel Boone. The area has also produced enormous quantities of coal. The eastern area is part of the Appalachian Plateau; in Kentucky, the area is triangle-shaped with many mountain ridges and deep canyons through which rivers flow. The area is a maze of narrow valleys caused by stream erosion. The two major mountain ridges in Kentucky are the Pine Mountains and the Cumberland Mountains. Between these ranges lies the Middleboro Basin. It is near Black Mountain, which is the highest in the state at 4,145 feet (1,263 meters).

The Bluegrass Region lies in the north-central area of the state and extends from the middle of the state to the Ohio River. In the northern part of the Bluegrass Region is a land that has a gentle roll and in which horses are raised for racing. Tobacco is also a major crop in the area. Surrounding the Bluegrass Region on the southern, western, and eastern boundaries are cone-like sandstone knobs. These are composed of light soils that easily erode, so much of the zone has been left wooded.

The southern and western areas of Kentucky are part of the Mississippian Embayment, or the Pennyroyal Region. The region has two arms that extend north to the Ohio River and west to the Mississippi River. The name Pennyroyal comes from an herb of the mint family that is widespread in the region. The southern part of the Pennyroyal region has very productive limestone soils. Further north in the center of the Pennyroyal region is an area called the Barrens. The name was given by the first pioneers who found the area treeless and barren. The northern part of the Pennyroyal rises in elevation and has rocky ridges and bluffs. It also has numerous limestone caves, the most famous of which is Mammoth Cave.

The western coal field region has rolling lands under which are beds of coal that are strip-mined. The region also is very productive farmland. The far western end of Kentucky is called the Jackson Purchase Region. It is part of the Gulf Coastal Plain. It is also an area of rich Mississippi River flood plains. It acquired its name in 1818 after Andrew Jackson participated in the purchase of the land from the Native Americans. The area has numerous ox bow lakes and swampy areas.

ARKANSAS

Arkansas lies across the Mississippi River from Tennessee and south of Missouri. In the south, it is bounded by Louisiana and Oklahoma and in the west by Texas. The Arkansas River flows from Oklahoma across the state to the Mississippi River. The eastern area of Arkansas is occupied by the Mississippi Alluvial Plan. The region is very flat and, in some areas, wooded and swampy. It is also very productive agriculturally. Rice and soybean fields stretch for vast distances during growing season. The area is also attractive to wildlife: Many migrating geese and ducks feed in the fields on their way south.

In the middle of Arkansas's eastern alluvial plain is a strange formation called Crowley's Ridge. It is a narrow ridge between half a mile and 12 miles wide that runs north and south for over 150 miles from the Missouri state line to near Helena. Its elevations reach nearly 550 feet at its northern end. The ridge was once covered with loess. Some geologists have explained Crowley's Ridge as a product of the Mississippi River's changes in its bed. This view contends that the ridge is the remains of a much larger formation eroded by the Mississippi before it shifted its course to the east. Other geologists believe that the ridge is the product of the earth's folding. Its elevation has increased over the decades in which measurements have been taken. Crowley's Ridge has flora that is different from the surrounding alluvial plain; its flora is much more like that of the Appalachians. It also is a source of garnet gemstones.

The southern part of Arkansas is a part of the West Gulf Coastal Plain. From Monticello west to the Texas state line, the area has reddish sandy soils and is forested with pines. Significant oil and gas discoveries have been worked at El Dorado and surrounding areas. In the area of the Red River at Hempstead County, the soil in prairies is a gumbo type of clay that expands and contracts significantly with increases and decreases in moisture. Across Hempstead and Howard Counties and other areas is a line of sandy hills that are believed to be the sand dunes of prehistoric times. At Murfeesboro

are several volcanic pipes, which form the Crater of Diamond State Park. Patrons can mine for diamonds and keep whatever they find.

The small area west of the Red River is in Miller County. The area was a swampy no-man's-land before the Mexican-American War. Today, very rich farmland produces cotton and other crops.

Between the Arkansas River Valley to the north and the Gulf Coastal Plain in the south are the Ouachita Mountains. The mountains are a series of ridges that are heavily forested. The highest peak is Blue Mountain at 2,623 feet (799 meters). Hot Springs lies at the eastern extremity. Quartz crystals are mined in the area, as is novaculite, which is a type of chert or flint used for sharpening whetstones. The Arkansas River Valley also produces rice and other crops and has deposits of gas and oil. Large vessels travel its waters to Tulsa, Oklahoma.

The Ozark Plateau covers the area of northern Arkansas from the western edge of the Mississippi alluvial plain to Oklahoma. It is an area of rugged ridges and eroded valleys. The Buffalo River, which flows through the area, is a favorite place for canoeing. The Boston Mountains form the southern boundary with the Arkansas River. The region is heavily forested and filled with man-made lakes.

MISSOURI

Missouri has four major land regions. First, in the southwest is the Missouri boot heel, which is a part of the Mississippi alluvial plain. It is also where the New Madrid Fault is centered. This seismic zone has historically had several major earthquakes. Second is the Ozark Plateau, which extends in southern Missouri from Arkansas to about the Missouri River. It is heavily wooded with narrow river valleys, numerous caves, and springs. Its elevation is about 500 to 1,700 feet (150 to 518 meters) above sea level.

Third are the Dissected Till Plains in northern Missouri, which was covered with glaciers during the ice ages. The soil is deep and rich, which makes it a major corn-growing region. Fourth are the Osage Plains, which lie roughly in the triangle formed by the Osage and Missouri Rivers. It is a plain with several areas of low hills.

Lead mining has played an important role in the economy of Missouri; the eastern Ozark region is the location of the Old Lead Belt.

SEE ALSO: Appalachian Mountains; Coal; Lead; Mississippi River; Mountains; Petroleum; Rivers; Wetlands.

BIBLIOGRAPHY. James Lane Allen, *Blue Grass Region of Kentucky and Other Kentucky Articles* (Reprint Services Company, 1989); Richard B. Drake, *A History of Appalachia* (University of Kentucky Press, 2003); Ann Heinrichs, *Arkansas* (Capstone Press, 2006); James Kavanagh, *Tennessee Birds* (Waterford Press Ltd, 2001); Charles P. Nicholson, *Atlas of the Breeding Birds of Tennessee* (University of Tennessee Press, 1998); John B. Rehder, *Appalachian Folkways* (Johns Hopkins University Press, 2004); Henry R. Schoolcraft, *View of the Lead Mines of Missouri: Including Some Observations on the Mineralogy, Geology, Geography, Antiquities, Soil, Climate, Population, and Productions of Missouri and Arkansas, and Other Sections of the Western Country* (Arno Press, 1972); Charles W. Schwartz and Elizabeth R. Schwartz, *Wild Mammals of Missouri* (University of Missouri Press, 2001).

Andrew J. Waskey
Dalton State College

United States, Great Plains

THE GREAT PLAINS region of the United States stretches in a broad band from Mexico in the south to Canada in the north and includes the states of Kansas, Nebraska, North Dakota, Oklahoma, and South Dakota. It is a region characterized by wide, flat lands with few trees similar to the steppes of Central Asia and the pampas of Argentina. The term *prairie* is often used, particularly in Canada. The Great Plains are part of a larger geographical area that extends to Texas and other more easterly states. It is possible to divide the Great Plains from the High Plains region approximately along the line of the 100th meridian. Land is more fertile to the eastern side, where the more generous rainfall supplements the ability of the land to sustain livestock, particularly cattle, as well as cereal crops and veg-

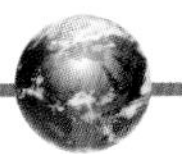

etables. The High Plains region to the west receives significantly less rainfall and the land is subject to drought and dustbowl effects, most notably during the 1920s and 1930s when, in combination with the effects of the Great Depression, depopulation of the area became a major trend as farm failures became endemic. In addition to the geographic features of the Great Plains, this region of land is characterized in modern culture by the perception of the people who have settled and lived there. Given the hardships of farming the often-difficult land, the remoteness, and the extremes of weather, the Great Plains has given rise to a notion of its people as being hardy, self-reliant, and stoic.

The Great Plains regions may be further divided into a number of different sub-regions, each with its own pattern of settlement and geography. These range from the Black Hills area of the Dakotas, which are composed of dark, igneous rock related to the Rocky Mountains, to the heavily eroded Red River Valley of Texas. The topography affects wind flow and this in turn affects rainfall and attendant plant life. For example, the warm air that flows up the Mississippi River Valley helps contribute to the moister climate of the east and bypasses the western region altogether. This makes for significant variation of conditions across the Great Plains, which in turn leads to some extreme weather conditions. Intense thunderstorms rage across the interior of the region and many settlements are threatened by seasonal tornadoes that erupt on an annual basis. The tornadoes are generally small but occur frequently. Other forms of extreme weather also mitigate against extensive population settlement and successful agriculture.

It is believed that the Great Plains were created as a byproduct of the geological processes that created the Rocky Mountains. Before recorded history, they were the home of many native animal species such as the buffalo (American bison) and other large mammals subsequently hunted to extinction. Tribes of indigenous peoples used the Great Plains on either a temporary or semi-permanent basis by building earth mounds or log houses of various types. Because their lifestyles relied on hunting and gathering more than sedentary agriculture, the indigenous tribes moved their residences on a rotating, seasonal basis to follow food sources. The peoples who were present in the Great Plains included the Sioux, Cherokee, Cheyenne, Arapaho, and many others. Low population densities meant that conflict between different tribes did not generally lead to high-intensity warfare, although raiding was a common occurrence. The horse was introduced by the Spanish and this revolutionized the lives of the native peoples whose long-range transportation abilities increased enormously; the ability to conduct hunting at long-range and to move further away from regular water supplies were significant.

With the arrival of European settlers, the lifestyles of the indigenous peoples became antithetical to new uses to which the land was being put. Settlers sequestered large parcels of land for private, household use and would not permit passage to any others. From Texas in the south, large cattle ranches were created with the animals driven north for fodder on a yearly basis. Both of these forms of agriculture yielded no space for indigenous tribes and warfare became inevitable.

The superior technology of the Europeans proved decisive, although not immediately or without great losses on either side. The remaining indigenous tribes were forcibly removed from the lands on which they had traditionally lived and required to live on circumscribed reservations where they were obliged to modify their lifestyles to the new situations in which they found themselves. Many have found it difficult to adjust and poverty and very low opportunity have been the result for many generations.

The new settlers were not automatically successful in their attempts at establishing farms; those who had experience with comparatively difficult situations tended to do better than others. This led to the tendency toward ethnic homogenization of the Great Plains as those with the skills to succeed in agriculture there typically came from similar geographical locations. The Homestead Act of 1862 in the United States and the Dominion Land Act of 1871 in Canada made provision for any settler to claim up to 160 acres of land (in America) on the basis that he and his family lived and worked the land for a specified period of time. Subsequent legislation strengthened the rights of farmers and regulated the rights of individuals and organizations to use and control water resources and other

significant inputs. Water resources were of particular importance in the western Great Plains because of the low rainfall levels and the persistent threat of drought. The *Tyler versus Wilkinson* case of 1826 had been instrumental in creating a regulatory regime in which the equitable and sustainable distribution of water resources became possible. Although the Civil War of 1861–65 was fought mainly in the eastern part of the United States, it had a significant impact on the Great Plains both in terms of actual military action and in the disruptions to patterns of supply and demand. Slave labor was not generally used on an intensive basis owing to the marginal profitability of many farms and their inability to feed many mouths. Nevertheless, slave labor built and developed the Great Plains just as it did the rest of the country.

As the railroad spread across the country, ranchers became accustomed to driving their cattle up from southern ranches to the railheads in the Great Plains where fodder was available for the livestock, which could then be transported east for slaughter and consumption in the eastern and coastal cities. The large-scale nature of this activity effectively prevented any change of land use in the intervening territory. However, drought and disease made the practice susceptible to external environmental shocks and it was more or less ended by the 1886 drought and famine, which witnessed thousands of over-crowded cattle starved and frozen to death. After this date, ranchers tended to move toward growing their own crops to support livestock throughout the winter.

While this had little positive effect on the Great Plains region, it did permit the wider spread of more intensive and scientific agricultural systems across the land. This was necessary since the hardy prairie grasses utilized deep root systems, which formed dense, interlocking sods that required multiple teams of oxen to plough for the first time. However, scientific advances ultimately contributed to the dustbowl conditions during the 1930s that led to the depopulation of the area. Decades of decline have followed, interspersed with some revivals as technology has suggested improvements in land use.

In recent years, since the 1950s in particular, the intensive use of the Ogallala aquifer has enabled widespread irrigation of much of the Great Plains area and this has made it possible for previously marginal or unusable land to be brought under agricultural land use. While this irrigation has revived many parts of the region, the use of the water may be unsustainable as presently organized since it is believed that more is being extracted than can be replaced by natural means. When aquifers are depleted in this way, the ultimate effects include subsidence, desertification, and the salinization of the remaining water as saltwater creeps in to replace the freshwater. However, steps have been taken since the 1970s to reduce the amount of water extracted and these have been partly successful. Even so, there has been a long-term tendency toward the depopulation of the region and the abandonment of its many small towns and villages.

Owing to the interaction of a number of different factors, including the failure of many farms, falling family sizes, the increased ability to use personal transportation to obtain paid income elsewhere, and the lack of enthusiasm for farming among younger generations, population density has gradually declined. When towns become insufficiently large to sustain their own infrastructure of schools, government agencies, and private sector firms, they are often amalgamated with neighbors for the purposes of providing such services and the cycle of depopulation intensifies as jobs disappear from the towns.

There are now many thousands of abandoned, or ghost, towns studding the Great Plains regions and many of those that have survived pose significant problems of rural poverty, isolation, and lack of basic services. However, research into agricultural practices suitable for the region have suggested a number of new options for agricultural practice that might bear fruit and help reinvigorate the region. One alternative approach is to consider the region effectively a failed development plan and reintroduce buffalo to roam freely over it.

SEE ALSO: Desertification; Dust Bowl, U.S.; Livestock; Native Americans; Ogallala Aquifer; Prairie.

BIBLIOGRAPHY. Geoff Cunfer, *On the Great Plains: Agriculture and Environment* (Texas A&M University Press, 2005); David Elstein and Don Cornis, "Moving away from Wheat/Fallow in the Great Plains," *Agricultural Research* (v.53/6, 2005); Tim Flannery, *The Eternal*

Frontier: An Ecological History of North America and Its Peoples (Atlantic Monthly Press, 2001); Ian Frazier, *Great Plains* (Picador, 2001); Ted Steinberg, *Down to Earth: Nature's Role in American History* (Oxford University Press, 2002).

John Walsh
Shinawatra University

United States, Gulf Coast South

THE UNITED STATES Gulf Coast states form one vernacular region, but several subregional areas can be delineated based on soils, characteristic vegetation, relief, and climate. These states include Alabama, Florida, Louisiana, and Mississippi. Beginning in the far southeast, subtropical South Florida encompasses the Everglades and a variety of mangrove, marsh, and forest zones, all resting on a foundation of limestone. To the west, the Gulf Coast comprises barrier islands and sounds protecting pine and palmetto forests. Coastal Louisiana, formed by deltaic sediments from the Mississippi River, consists of salt and fresh water marshes, cypress swamps, and areas of relatively high ground formed by distributary channels of the river. Heading north, the Mississippi Valley bisects the region with a broad alluvial plain of oxbow lakes, hardwood forest, and highly organic soils.

Bordering the Valley in north Louisiana and Mississippi, the Cotton Uplands, referring to their long-standing cultivation of the region's signature cash crop, are characterized by significant relief and forests of pine. This area, like much of the entire region, has red and yellow podzols which are generally less fertile than more northern soils which benefited from glacial deposits. The Black Belt, so named for its dark tertiary topsoil and for being the historic home of many African Americans, runs through eastern Mississippi and central Alabama. Northern Alabama's hills, the highest part of the region, are the southernmost extensions of those mountains.

The composition of the soils of the Gulf states is affected by the region's heat and high rainfall. These factors lead to high rates of oxidation of organic matter and a tendency toward erosion when soils are exposed. Regional soils formed under forested conditions therefore have low carbon content, especially after clearing and intensive cultivation.

The climate features hot summers, cool winters, and rainfall generally exceeding 40 inches per year. Places bordering the Gulf of Mexico have their climate mediated by the sea, but the entire region is susceptible to cold fronts from the north during the winter. The Gulf's limited circulation with the open Atlantic causes it to become very warm by late summer, making for high rates of hurricane activity along its northern coast.

This physical geography is of course much altered by human action. Native Americans lived in the region for thousands of years. Agricultural cultures extended down the river's valley, creating urban centers like Poverty Point in present-day Louisiana. Maize, originally from Mexico, formed the basis of these people's diet, as did the hunting of deer and bison. Widespread fire-setting by Native Americans to clear land and improve hunting is believed to have influenced the landscapes first witnessed by Europeans.

De Soto's expedition in the 1540s and the growing contacts with the Atlantic world that followed it brought Eurasian diseases like smallpox and malaria, which devastated native peoples and would pose barriers to European settlement in many parts of the region. Colonization moved very slowly, with the Spanish in Florida having only a minor presence and the French arriving in Louisiana in 1699, and remaining there by only the smallest margins. Unfamiliar with the territory, weakened by diseases, and unable to establish European crops to feed themselves, the French were only saved by their adoption of maize, their herds of cattle that ranged through the pastures opened up by burning the thick canebrakes that lined the Mississippi, and the introduction of rice and enslaved Africans knowledgeable in its cultivation. The French introduced the plantation system to the Gulf states, first focusing on indigo and tobacco, and then over time rice and sugar cane. Their export-oriented plantation economy, based on African slavery and leveeing the Mississippi to protect their settlements, set a precedent for the region's incorporation into global commerce.

Alongside a tiny but growing plantation sector, the trade in deerskins harvested by Native Americans grew to a great scale in the 18th century. The French and English waged a proxy war between the Choctaw and Chickasaw peoples in Mississippi and Alabama, fueled by European goods and firearms purchased with deerskins. The bison, present throughout much of the present U.S. southeast in 1700, disappeared by 1800, and other native animals like large hunting cats and parakeets declined greatly as European settlement advanced. English, then American, settlers began moving into the region from the Atlantic coast in the late 18th century. The incorporation of the Mississippi and Louisiana territories into the new United States combined with great demand for cotton made the expansion of slavery westward a compelling interest.

The Native American nations that occupied Mississippi and Alabama were an obstacle to this expansion. The successive removal of the Creek, Chickasaw, and Choctaw from the 1810s through 1830s opened the land to white settlement and plantations expended westward along the Black Belt as settlers moved into the Mississippi Valley from the north and south. A combination of inexpensive land, soil degradation, and the ambition of planters to chase new opportunities along the advancing frontier meant land was frequently abandoned, succeeding to scrub pine and extensive cattle raising.

Clearing fields contributed to erosion and their leaching from rainfall. The region's heat reduced dairy production, which worked against rotating pasture and farming and the corresponding benefits of manuring fields. The relations between antebellum land use, soil exhaustion, and the westward advance of slavery are not clear, but in the decades before the Civil War advocates for soil fertilization and conservation grew increasingly louder. Despite the massive acreages planted in cotton during this time, the staple food crop, maize, probably did more to deplete soil nutrients.

The clearing of snags in the region's rivers, levee improvements, and railroad development opened up new areas to plantation agriculture, but some areas remained isolated. The Mississippi Delta, south Florida, and much of coastal Louisiana remained inaccessible to planters due to the swampy nature of the terrain. The Swamp Land Acts in 1849 and 1850 attempted to stimulate the drainage of these places by promoting land sales that would finance reclamation.

The Civil War left many levees destroyed, farms abandoned, and investment capital in desperate scarcity. Emancipation and Reconstruction led to contestation over control of land. The counterrevolution of Redemption led to the reinforcement of monopoly over land and capital by planters and the prevention of black emigration from the region, which served to maintain a low cost rural labor force.

This period also saw increased Northern and European investment and control over land and infrastructure. The South became a principal source of lumber for the United States, as companies practiced a cut-and-get-out strategy that produced brief prosperity for lumber towns and left forests severely degraded. Market hunting decimated populations of waterfowl. The fad for feathered ladies' hats led to the slaughter of much of Florida's bird life and generated support for wildlife conservation that eventually banned market hunting and the plume harvest.

The turpentine industry spread to Florida and southern Alabama from the Carolinas and did much damage to pine forests and workers alike. The region around Birmingham became a coal and iron center and Florida became a source of phosphate fertilizers. Agriculture in the uplands became more dependent on cotton as cheaper provisions in stores undermined local produce. Diseases like malaria continued to drain the energies of many people in the countryside, while summer epidemics of cholera and yellow fever, especially in cities like New Orleans, killed thousands at a time.

The economic development of the region was marked by selective industrialization in extractive industries, dependence on outside capital, and a widening inequality relative to the United States as a whole and between localities within the region. A national depression in the 1890s contributed to the rise of the Populist movement, composed of small white farmers and black farmers, working separately or together, to break the monopoly of planter interests on state government and industrial trusts over the regional economy. The polarization of planters based in naturally rich regions like the newly drained Mississippi Delta from upland

smallholders facing economic ruin on land planters didn't want temporarily fractured the political alliance based on white supremacy that maintained the Southern social order.

The boll weevil infestation that swept the region destroying cotton crops beginning in 1894 added to a wave of rural emigration propelled by economic crisis. Pellagra, a nutritional deficiency disease caused by a maize-based diet, became a serious public health problem in the 1900s as cotton dependency reduced kitchen gardens.

Increased state and federal government intervention and rapid industrialization in the coming decades made for radical changes in the Gulf states. The massive Mississippi river flood of 1927 brought federal support for comprehensive flood control. New Deal–era programs brought rural electrification and reforestation. Endemic diseases finally began to be brought under control, then eradicated altogether through a combination of better nutrition and the elimination of breeding grounds for vectors.

The mechanization of agriculture was spurred by generous subsidies for price stabilization and soil conservation begun during the 1930s, programs influenced by the political power of large planters. This had the effect of accelerating the outmigration of rural farmworkers to regional and Northern cities. South Florida began its spectacular development from an isolated backwater to vacation destination and plans to drain the Everglades were implemented, creating the huge Everglades Agricultural Area for sugar cane farming. In Louisiana, exploitation of oil and gas wealth stimulated industrialization in the form of a petrochemical corridor along the Mississippi (known to environmental justice activists as "Cancer Alley") and did great harm to the state's wetlands by dredging thousands of canals to access drilling sites in the marsh.

The Civil Rights movement focused on issues like voting rights, but also had an important component of rural development and agrarian reform advocated by leaders like Fannie Lou Hamer, harkening back to historical African-American community development and earlier rural organizing efforts. During the mid-1960s peak of activism, plantations greatly reduced their workforces in favor of highly capitalized farming, resulting in a greater exodus of black farmworkers from rural areas and the dilution of the base of the movement.

Green Revolution technologies were incubated in the American South with public-private research partnerships, and as the old plantation order unraveled land companies transformed themselves into agribusinesses or even biotechnology firms. Delta and Pine Land, once a huge Mississippi Delta plantation, created the so-called terminator sterile seed technology and is now a part of Monsanto. After World War II pine plantations proliferated and the South became the United States's largest softwoods producer. Pulp and paper mills and new forms of agro-industry like off-soil chicken raising operations and catfish aquaculture were established.

All the Gulf states experienced the transformation of the Southern arc of the United States into the Sunbelt, a region of economic resurgence, net immigration, and suburbanization, but none more so than Florida. That state's population increased fivefold between 1950 and 2000, making it the fourth largest state in the United States. Air conditioning played a role in this growth, as did coastal tourism development across the Gulf Coast and the migration of many Northern retirees in search of warmer climates and a lower cost of living.

The ecological costs of regional development in the late 20th Century are particularly evident in coastal Louisiana and south Florida, the two largest areas of wetlands in the region. Louisiana's coast lost 1,900 square miles of land since the 1930s, due to a combination of oil development, reclamation, subsidence, and sediment deprivation caused by narrowly channeling the Mississippi river. This has compounded the vulnerability of south Louisiana to tropical storms.

Much of Florida's Everglades were drained for agriculture and urbanization. The remainder was affected by the hydrological engineering of South Florida that sometimes deprived the glades of seasonal water flows or flooded it with too much. A federal multibillion dollar restoration project for the Everglades using adaptive management was approved in 2000. The plan is driven by an interest in increasing fresh water availability for urban growth as much as a desire to preserve nature. Restoring the Louisiana coast has gained new urgency after Hurricanes Katrina and Rita in 2005, but has not

seen the same financial backing as Florida received for the Everglades.

SEE ALSO: Cancer Alley; Cotton; Everglades; Green Revolution; Hurricanes; Swamp Land Acts; United States, Central South; United States, Great Plains.

BIBLIOGRAPHY. E. Ayers, *The Promise of the New South* (Oxford University Press, 1992); A. Cowdrey, *This Land, This South: An Environmental History* (University Press of Kentucky, 1996); C. Woods, *Development Arrested: Race, Power, and the Blues in the Mississippi Delta* (Verso, 1998).

Thomas A. Birkland
State University of New York, Albany

United States, Hawaii

THE HAWAIIAN ARCHIPELAGO—LOCATED in the central Pacific—comprises eight larger and inhabited islands in the southernmost part, and other minor unpopulated islands, reefs and atolls, extended to the northwest along 1,000 miles (1,600 kilometers). The inhabited islands are Ni'ihau, Kaua'i, O'ahu, Moloka'i, Lana'i, Kaho'olawe, Maui, and the island of Hawaii. The northwestern islands are Kure Atoll, Midway Atoll, Pearl and Hermes Atoll, Lisianski Island, Laysan Island, Maro Reef, Gardner Pinnacles, French Frigate Shoals, Necker Island (Mokumanamana), and Nihoa. The islands belong to the Hawaiian Ridge, a volcanic chain of seamounts, residuals of old volcanoes, or active volcanoes with a northwest-southeast direction. It was formed in the last 30 million years from an active hot spot approximately located in the island of Hawaii, which is the youngest of the chain, as the result of the movement of the Pacific Plate.

Polynesian settlement occurred at least by 600 C.E. after long-distance voyaging across the Pacific, although first discovery might have happened three to four centuries before. Those explorers carried with them domesticated plants—at least 29 species—and animals to secure food for the journey. Once in Hawaii, their cultivation provided a source for carbohydrates that was not available in the islands. This was succeeded by the relatively late European discovery and settlement, compared to the other Pacific islands, in 1778. In his third voyage of Pacific exploration, the English Captain James Cook found the islands, which were named the Sandwich Islands. This last discovery initiated a rapid process of colonization because of their strategic position in the Pacific Ocean.

Population at the time of European contact, although still a matter of debate, was 300,000 native Hawaiians. A census completed by missionaries in 1831–32 gave an account of 130,000 and in 1876 a Hawaiian government census reported 54,000. This sharp decline of the native population was due to the period of wars and famine during King Kamehameha's mandate, the various diseases fatal to the local population introduced by the European navigators, and the migration of natives to work as sailors.

During the plantation era, a large labor force was needed and immigration was organized by the estates to supplement the diminished local force. Labor immigrants arrived from 1850 to 1950; the waves were first Chinese, then Portuguese, Japanese, Portuguese for a second time, and finally Filipino, which led to the present high ethnic diversity. About 180,000 Japanese and 168,000 Filipino departed to the islands. Thus, population begun to grow from 1876 onwards, as mortality rates dropped and birth rates increased. By 1940, the population exceeded 400,000 and rapidly doubled with the arrival of military and defense workers during World War II.

After the war population dropped with the departure of military and employees but increased again with the expansion of tourism by the mid-1950s. As of the 2000 Census, population was 1,211,537, mostly concentrated in O'ahu (73.3 percent), with population densities of 1,441 persons per square mile (547 per square kilometer); while the biggest island, Hawaii, only had 12.2 percent of the population, with a density of 36.8 persons per square mile (14.2 per square kilometer).

RESOURCE EXPLOITATION

Hawaii underwent various historical economic cycles based on different resources. Early phases

based on sandalwood, whales, sugarcane, and pineapple were followed by services, military, and tourism. During the first quarter of the 19th century, the principal commodity traded was sandalwood, harvested in great quantities and traded to China, which almost depleted the forests, until cutting restrictions were issued. The Forest Reserve system was created in 1903 to protect watersheds from deforestation and subsequent erosion; trees were planted including both native hardwood koa and short rotation nonnative tree species such as pine and eucalyptus. The abandonment of sugarcane plantations was an opportunity to gain new lands for forestry. Today forests cover 1.7 million acres (690,000 hectares), or 41 percent of the state's area.

The whaling industry in the North Pacific became a foremost activity for half a century until the 1870s. It turned into an opportunity for trade and facilitated port infrastructure development, which had already begun with sandalwood commerce. These two economies were not sustainable, but extractive and resource depleting. Eventually, whale stocks began to shrink. Changes in technology and petroleum use led to the decline of the industry.

The sugarcane industry started after this eclipse as the influence of American traders and landowners grew. This influence increased especially after the short-lived Republic of Hawaii, which succeeded the Kingdom of Hawaii, was annexed as a territory by the United States in 1898. This episode took place when Queen Lili'uokalani was overthrown in 1893 by those American traders and landowners willing to join the United States.

Cultivated area increased and the sugar industry expanded; production kept steady until the 1970s. At the outset of the 20th century, pineapple and other tropical fruit cultivation began, benefiting from a sustained year-round production which, although it declined in the 1970s, still has a local and mainland market. The number of related jobs in the 1990s fell to half of the jobs in the 1970s. Plantations shaped a relatively homogeneous landscape based on agriculture that turned into a highly diversified agriculture: Vegetables for the local market, fruits, flowers, and sugarcane for export.

Hawaii's central Pacific location provided an additional resource for at least half a century. The islands had a strategic military significance as a base

The Hawaiian Volcano Observatory

Thomas Augustus Jaggar, Jr., (1871–1953) was a petrographer and volcanologist, and was born in Philadelphia, the son of an Episcopalian Bishop. He originally studied geology at Harvard University and then went to Munich, Germany, to do research in mineralogy. When Mount Pelée erupted on the French Caribbean island of Martinique, Jaggar went to visit the site, and then also travelled to Japan to see Tarumai and Asama. He then went to Kilauea in Hawaii along with his colleague Reginald Daly (1871–1957), and founded the Hawaiian Volcano Observatory. In 1912 Jaggar resigned from a position he held at the Massachusetts Institute of Technology in order to become director of the observatory, a position he held until 1940.

At the Hawaiian Volcano Observatory, Jaggar saw to the designing and building of new machines to measure volcanoes, although he was always short of funds. In 1919, the observatory was placed under the control of the U.S. Weather Bureau, which ensured its survival. Five years later, it was placed under the U.S. Geological Survey, which in 1926 established a volcanology section with Jaggar as its head.

Jaggar used the observatory as a base to study volcanoes elsewhere in the United States. He studied the Aleutian volcanic chain in depth, leading a National Geographic Society expedition. When the observatory was transferred to the National Park Service in 1935, Jaggar saw that the volcanoes at Kilauea and Mauna Loa became part of a national park. Many of the ideas introduced by Jaggar had never been tried before. He designed ways of protecting cities from lava flows, including by aerial bombing, and wrote about safety precautions and dealing with eruptions. He remained in Hawaii after his retirement and died in Honolulu.

for operations in three major wars: World War II, Korea, and Vietnam. Interest rose with the Spanish-American war at the end of the 19th century, when the United States began to look to the Pacific as an area of territorial expansion. The Pearl Harbor Naval Shipyard was established in 1908. Military expenditures of $3.2 billion represent 9 percent of the Gross State Product and the second biggest resource after tourism. The military employs 8 percent of the state population, while installations occupy 5 percent of the land, mostly on Hawaii and O'ahu islands.

Mass tourism took off with the introduction of rapid air transportation in the 1960s. Although a principal domestic destination, it has followed a process of internationalization bringing visitors from Japan, Canada, Australia and New Zealand. As many as 7.4 million visitors a year spend $11.5 billion. Tourism makes up 23 percent of the state's employment, leaving the state highly dependent on the industry. Another problem results from the high numbers of visitors (hotel occupancy rates of 81.2 percent at times) and the resulting rapidly growing demand for new infrastructure.

ENVIRONMENTAL PROTECTION

The islands' tropical location, the exposure to the dominant northeast trade winds carrying moisture, and the high elevation of some islands result in an extraordinary diversity of environmental conditions and the emergence of a number of ecosystems. Windward slopes face the northeast trades and receive rainfall that decreases with altitude, whereas leeward sides are drier with less cloud cover. This orographic effect—particularly observable in the islands of Hawaii, O'ahu and Maui—produces rapid changes in short distances, such as the Ka'u desert next to the rain forest. The western slopes of the islands pose climatic conditions—milder temperatures and many sunny days—that favor the development of tourism.

There are about 15,000 native plant and animal species, which, added to the approximately 6,000 species intentionally or accidentally introduced by human colonization of the islands over the past 1,500 years, makes for a total of between 20,000 and 22,000 species, half of them insects. As species arrived to the islands from other areas, entire families and order of fauna and flora are missing, principally those without dispersal mechanisms, like mammals and amphibians. The long period of isolation and the slow environmental changes led species to acquire slow growth and low birth rates, or become flightless. Thus, rapid changes produced by the combined effect of habitat change and the introduction of predators makes many native species vulnerable to human impact.

One thousand species have become extinct—half of the avifauna—most of them plants and invertebrates. The main threats to native plants are both indirect factors such as alien weeds or introduced mammal predators and direct factors such as wildfires, species collecting, and urban development. Agriculture and livestock breeding brought the most changes. Wetlands were drained, pasture land displaced original species, and forest and shrubland ecosystems were cleared for cattle grazing and plantation agriculture. Coastal areas, valleys, and low altitudes are occupied by agriculture and residential areas, both leading to a loss of the lowland mesic and dry ecosystems.

A total of one million acres (404,685 hectares) of federal- and state-managed areas represent 25 percent of the state extension to protect fragile ecosystems and rare and endangered species. In 1916, the U.S. Congress established the Hawaii National Park, which was then split into the Hawaii Volcanoes National Park on Hawaii island and the Haleakala National Park on Maui in 1961. The Hawaiian Islands National Wildlife Refuge—administered by the U.S. Fish and Wildlife Service—comprises the northwestern Hawaiian islands, atolls, and reefs (with the exception of Midway and Kure Atolls) to protect seabirds and marine life. It was designated in 1909 to protect the endangered Hawaiian monk seal and the Hawaiian population of threatened green sea turtles.

Hawaii's 1961 State Land Use Law was the first enacted and implemented statewide land use zoning and planning system, establishing the State Land Use Commission, responsible for the classification of lands into the four districts: Urban, conservation, rural, and agricultural. The counties determine the zoning ordinances, and development or subdivision plans for urban use, the state

determines the use of conservation lands, and both share the administration of rural and agricultural lands. Ceded Lands, with an extension of 1.8 million acres (727,000 hectares), 43.8 percent of the territory, are the focus of an intensive and prolonged debate on their ownership and land use. After the Ceded Lands Act of 1963, these crown lands, which were transferred to the United States in 1898, are partly administered by the Federal and the State governments, but there has not been a final resolution to this prolonged conflict.

SEE ALSO: Biodiversity; National Parks; Orographic Effect; Plantation; Sugar; Tourism; Whales and Whaling.

BIBLIOGRAPHY. Ibrahim Aoude, ed., *Political Economy of Hawaii* (University of Hawai'i Press, 1994); David L. Callies, *Preserving Paradise* (University of Hawai'i Press, 1994); Sonia P. Juvik and James O. Juvik, *Atlas of Hawai'i* (University of Hawai'i Press, 1998); Ralph S. Kuykendall, *The Hawaiian Kingdom* (University of Hawai'i Press, 1969); Gordon A. Macdonald, Agatin T. Abbot, and Frank L. Peterson, *Volcanoes in the Sea. The Geology of Hawaii* (University of Hawai'i Press, 1983); Charles P. Stone and J. Michael Scott, eds., *Hawaii's Terrestrial Ecosystems: Preservation and Management* (University of Hawai'i Press, 1985); Ronald T. Takaki, *Pau hana: Plantation Life and Labor in Hawaii, 1835–1920* (University of Hawai'i Press, 1983); Alan C. Ziegler, *Hawaiian Natural History, Ecology, and Evolution* (University of Hawai'i Press, 2002).

URBANO FRA PALEO
UNIVERSITY OF EXTREMADURA

United States, Middle Atlantic

MIDDLE ATLANTIC STATES of the United States border the Atlantic Ocean or have port cities that are accessible to it. These states include Connecticut, Delaware, Maryland, New Jersey, New York Pennsylvania, and Rhode Island. All have temperate climates with four distinct seasons, though winters are colder in the western areas, with summers hotter in the coastal areas.

CONNECTICUT

Connecticut is 110 miles long and 70 miles wide. With only 5,544 square miles, it is one of the smallest of the American states. The highest point in Connecticut is Mount Frissell (2,380 feet), located in the extreme northwest corner of the state in the Berkshire Hills. Connecticut is bounded on its western boundary by New York. Rhode Island separates it from the Atlantic. To the north lies Massachusetts, with the Long Island Sound on the south.

Connecticut has five different land regions: The Taconic Section, the Western New England Upland region, the Connecticut Valley Lowland, the Eastern New England Upland and the Coastal Lowland. The Coastal Lowlands are a section of the New England Coastal Lowlands that cover the coast of New England. In Connecticut they are a narrow strip of land that is only 6 to 16 miles wide. The area is found along the southern shore beside Long Island Sound. The area has low ridges, beaches, some swampy areas, and harbors and its shoreline is 618 miles. Most of Connecticut's population lives in the Coastal Lowlands. The two major exceptions are the cities of Hartford and Waterbury.

The Eastern New England Upland area covers most of eastern Connecticut. The area is the southern end of the same land formation that extends through New England to Maine. It is a heavily forested area with many rivers, valleys, and low hills. It is a fertile region where farmers grow tobacco, corn, potatoes, oats, blueberries, and wheat. Poultry farming is also an important agricultural activity.

The Connecticut Valley Lowland runs through the center of Connecticut and is on average only 30 miles wide. The area is distinguished by basalt lava ridges and low hills. The Connecticut River is wide as it flows through the area on its way from Massachusetts to the sea. There are many small rivers that are tributaries and the area is fertile. Farmers raise potatoes, vegetables, corn, strawberries, blueberries, and other fruits. Grass is plentiful, contributing to dairy farming.

The Western New England Upland covers the western third of Connecticut and stretches into Massachusetts and Vermont. It rises from 1,000

Jersey in a northeast to southwest direction. While the plateau is only 20 miles wide it is where about three-fourths of the people live, mainly in industrial cities.

South of the Piedmont Plateau is the Atlantic Coastal Plain. It covers most of the southern and eastern parts of the state. Most of the area is characterized by gentle rolling lowlands that are not much above sea level. Many places in the area are farmed to produce truck crops. The Delaware River forms the western border with Pennsylvania and Delaware. The eastern area is a zone in which the sandy soil produces little besides pine trees. Much of the southern area is swampy and marshy close to the coast or to the rivers, forested, and thinly populated. Along the Atlantic coast are excellent beaches such as the beach at Atlantic City. The beaches have become popular enough to stimulate the development of a number of resort cities and towns.

PENNSYLVANIA

Pennsylvania lies west of New Jersey across the Delaware River. It is bordered on the south by Maryland and on the north by New York. Its main cities are Philadelphia and Pittsburgh, which are at its western and eastern ends. It has an area of 46,058 square miles (119,751 square kilometers).

Pennsylvania's Lowlands are in the northwest and the southeast. The lowlands in the northwest are part of the Erie Lowland region that borders Lake Erie. The land has a rich soil that produces potatoes, grapes and other crops. The southeastern Lowlands area is a corner of the Atlantic Coastal Plain. The Delaware River is at sea level where the Schuykill River empties into the Delaware at Philadelphia. Beyond it to the west is the Piedmont Plateau area.

Most of Pennsylvania is covered by the Appalachian Mountains. The Blue Ridge Mountains, the Pocono Mountains, and the Allegheny Mountains are part of the Ridge and Valley Region. Much of the region has produced hard anthracite coal. The Wyoming Valley is an important area in eastern Pennsylvania and was the site of the Pennamite Wars.

In western Pennsylvania the Allegheny and Monogahela Rivers flow together at Pittsburgh to form the Ohio River. The western region is a major coal producing area. It was also in northwestern Pennsylvania that the first oil well was drilled in 1859.

DELAWARE AND MARYLAND

Delaware is bordered by the Delaware Bay and River to the east and by the Atlantic Ocean. It is bordered by Maryland to its west and by a small area touching Pennsylvania in the north that is a part of the Piedmont Plateau. Some dairy farming and large estates are located in this hilly area. The only part of Delaware that is not part of the Atlantic Coastal Plain is its northern tip. Most of the state is barely above sea level. The state produces soy beans, corn, and other crops in its sandy loam soil. Seafood is also an important product.

Maryland is a relatively small state with a varied terrain. It has five land areas: The Atlantic Coastal Plain, the Piedmont, the Blue Ridge, the Appalachian Ridge, and the Appalachian Plateau. These areas are physically similar to those of states to the north, but there are some variations in the fauna and flora due to the milder temperatures of its more southern climate.

SEE ALSO: Appalachian Mountains; Cities; Coal; Petroleum; Potatoes; United States, Northeast; Wheat.

BIBLIOGRAPHY. James Kavanagh, *New York Birds* (Waterford Press, Ltd., 2001); Rene Laubach and Charles W.G. Smith, *Nature Walks in Connecticut: AMC Guide to the Hills, Woodlands, and Coast of Connecticut* (Appalachian Mountain Club Books, 2004); Raymond Leung, *Connecticut Birds* (Waterford Press, 2001); John H. Long, ed., *Connecticut, Maine, Massachusetts, Rhode Island: Atlas of Historical County Boundaries* (Scribner, 1995); Sylvia McNair, *Rhode Island* (Scholastic Library Publishing, 2000); Glenn Scherer and Don Hopey, *Exploring the Appalachian Trail: Hikes in the Mid-Atlantic States—Maryland, Pennsylvania, New Jersey, New York* (Stackpole Books, 1998); Alex Wilson and John Hayes, *Quiet Water Massachusetts, Connecticut, and Rhode Island: Canoe and Kayak Guide* (Appalachian Mountain Club Books, 2004).

Andrew J. Waskey
Dalton State College

United States, Midwest

THE MIDWEST REGION of the United States generally refers to lands that were part of the Northwest Territories during the time of the Articles of Confederation. Midwestern states include Illinois, Indiana, Iowa, Michigan, Minnesota, Ohio, West Virginia, and Wisconsin. Much of the region is flat prairie lands, but other land types surround it. The great Mississippi and Ohio Rivers flow through it.

Ohio is bounded by Pennsylvania and West Virginia in the east and southeast, Lake Erie in the north, the Ohio River in the south, and Indiana in the west. If a line were drawn across Ohio from the northeast to the southwest, the area north of the line is in the Central Plains, the area to the south is part of the Appalachian Plateau. The northern half of Ohio is part of the Till Plains and the Great Lakes Plains. The Great Lakes Plains are known for Lake Erie fruits, which include wine grapes and vegetable crops. The Till Plains are part of a great corn-growing belt. The glaciers in the ice ages flattened the land and left it a level plain with rich soils. South-central Ohio has an extension of the Blue Grass Region of Kentucky that extends across the Ohio River. Southern and eastern Ohio, as part of the Appalachian Plateau region, is hilly and rough and remains natural and wild, like West Virginia across the Ohio River.

West Virginia is a mountainous state that is sparsely settled. Its Eastern Panhandle is near Baltimore and the Shenandoah and Potomac Rivers. The Northern Panhandle is a crossroads of rivers and highways and rolling countryside along the Ohio River. Down river near Parkersburg is a region in which oil and gas were extracted in the early days of petroleum pumping. Northern West Virginia is a part of the Appalachian Ridge and Valley Region with valleys between long running ridges. It is forested with many caves and underground streams and covers the eastern sixth of the state. The Allegheny Front is a rugged divide in the Appalachian Plateau Region, which makes up most of the territory of West Virginia. The area is rugged with flat-topped plateaus. Coal is mined across the state.

The state of Indiana has an area of 36,420 square miles (94,328 square kilometers). It has a temperate climate and is flat like most of the Midwest region. Central Indiana is part of a major corn-growing belt. Its flat terrain was formed in the ice age by glaciers; the soil they left behind is the Tipton Till, which is made up of finely ground sand and gravel. While generally flat, Indiana's highest point is in the Tipton Till at 1,257 feet (383 meters) above sea level. In southern Indiana, the Southern Hills and Lowlands extend from the Ohio River to the beginning of the Tipton Till region. It is hilly because it escaped the glaciers. It is an area of caves and bedrock outcroppings of limestone, which is mined in quarries. The confluence of the Wabash and Ohio Rivers is the state's lowest point. It is also an area of coal and petroleum. Northern Indiana is part of the Great Lakes Plains. The regions were scraped flat by the glaciers, but some places were missed and remain high and hilly. The North Lake and Moraine Region in the northeast has beautiful scenery; the moraines form high ridges. The land ends in northern Indiana at Lake Michigan. The area just south of the lake has enormous sand dunes that were deposited as its waters retreated and winds from the lake have continued to build the dunes. Those dunes further south are the oldest and are wooded; those dunes next to the Lake form the Indiana Dunes National Lakeshore whose bare piles of sand are like those seen on ocean beaches—sandy, with little vegetation.

North of Indiana and Ohio is Michigan, which has a watery boundary. The Lower Peninsula is shaped like a kitchen mitten with the thumb on the eastern side. It is bordered by Lake Huron on the east and by Lake Michigan on the west. Michigan is separated from Canada by the Windsor River near Detroit. Michigan's Upper Peninsula extends from Wisconsin like an eastward pointing icicle. Lake Superior is on the northern side and Lakes Michigan and Huron are on the state's southern side. In the southeast are 35 miles of sand dunes at Sleeping Bear Dunes National Lakeshore. Cliffs are found along the 3,288-mile (5,292-kilometer) shoreline of the state in many places. The Upper Peninsula is rocky, mountainous, and heavily forested. The trees are basswood, birch, beech, butternut, elm, hickory, maple, oak, poplar, witch-hazel, and others. Moose, wolves, wolverines, porcupines, and other animals are found in the region. It was the center of copper and iron ore mining in earlier times. The Lower Peninsula is generally flat because of glacial action in the ice ages. Because of

the surrounding lakes, the climate is milder in winter. However, in western Michigan, and often elsewhere, lake effect snows create large snowfalls in some areas. The two major landforms in Michigan are the Superior Upland and the Great Lakes Plains. The Superior Upland covers the western half of the Upper Peninsula. The eastern part of the Upper Peninsula and all of the Lower Peninsula are part of the Great Lakes Plains. These are a part of a larger area called the Interior Plains, which cover much of the Midwest. The Lower Peninsula, especially in the southern half, has excellent farmland where blueberries and other crops grow well.

Illinois is located in the central United States with its major city, Chicago, bordering Lake Michigan. It is bound by Indiana in the east, Wisconsin in the north, Kentucky and the Ohio River in the south, and Iowa and Missouri in the west. The state covers a land area of 56,345 square miles (145,934 square kilometers) and has five land areas. In the extreme southern tip, along the Ohio and Mississippi Rivers, is a strip that is part of the Gulf Coastal Plain. Because it is like the delta of the Nile River, early pioneers called it Egypt. The southern part of the state is flat, and rises into hills in the north.

In the north is another strip of land that extends 70 miles from where the Wabash River enters the Ohio River to the Mississippi River—a hilly area called the Shawnee Hills. The area is also occasionally called the Illinois Ozarks. This Shawnee Hills region has river bluffs, hills, valleys, and heavily wooded areas; it is also a region in which many orchards have been planted. The Shawnee Hills vary from south to north from five to 40 miles (eight to 64 kilometers). The heights of the hills vary from 300 to over 1,000 feet (91 to over 325 meters).

North of the Shawnee Hills are the Central Plains or Till Plains. It is a vast fertile region that is filled with corn, soybean, and wheat stretching to the horizon during the growing season; prairie grass covers the unfarmed areas. The plains were created by the glaciers of the ice age. The Central Plains covers 90 percent of Illinois and through it runs the Illinois River to the Mississippi. The area has been called the Great Lakes Plains because Lake Michigan once covered the area. North of Chicago is an area of small hills, marshes, and lakes. In the far northwest corner of Illinois is an area that was not covered by the glaciers, which is called the Driftless area. The Illinois section covers only a small area, but it is filled with tall hills and steep valleys.

North of Illinois is Wisconsin. It is bordered by Minnesota in the west, Lake Michigan and Lake Superior in the east and north, and by Iowa in the south. It has an area of 56,153 square miles (145, 436 square kilometers) and has moderate temperatures in the southern areas in summer but cold winters. The landforms in Wisconsin are numerous. In the southeast are the Eastern Ridges and Lowlands, which extend from Illinois north to Green Bay. These plains were formed by glaciers and glacial till covers limestone ridges. The area is fertile and extensively employed in farming—the Door Peninsula is the center of potato growing.

West of the Eastern Ridges and Lowlands in the south is the Western Uplands region, which includes the previously discussed Driftless area. It is a beautiful area of wooded hills, lakes and rivers. Dramatic sandstone and limestone bluffs line the Mississippi River. Lead was mined in the area in earlier decades. From the Eastern Ridges and Lowlands west to the St. Croix River is the Central Plains area and on its southern section is the Wisconsin Dells region, which is a scenic gorge on the Wisconsin River. Covering most of northern Wisconsin is the Northern Highlands region, which is heavily forested and dotted with numerous lakes. It slopes to a steep cliff beyond which is the Lake Superior Lowland, a flat plain along Lake Superior.

West of the state of Wisconsin and northern Illinois is Minnesota, which has a land area of 84,402 square miles (218,601 square kilometers). It borders North and South Dakota in the west, Iowa in the south, and Wisconsin in the east; Canada (Manitoba and Ontario) and Lake Superior are to the north. Along the Mississippi River, extending from Wisconsin and Illinois is the Driftless area. To the west is the Young Drift Plains, which is a region of gently rolling farmland created by glaciers. Most of the glacial deposits, called drifts, left rich farmland. However, in some areas, moraines (or their remains) are rocky, sandy, and not suited to farming. They extend across the state to the Red River and north to Manitoba. All across the state are numerous lakes left by glacial action, which act as an enormous breeding ground for birds. In the far southwest corner of Minnesota

is a portion of the Dissected Till Plains. The northern Superior Upland is part of the Canadian Shield. It was scoured by glaciers but they were unable to destroy the hard, ancient rock. The northeastern Arrow Head Region pointed toward Lake Superior is the location of the Iron Ranges.

Iowa is to the west of Illinois. In the northeastern part of the state is the western part of the Driftless area. The Dissected Till Plains cover the southern part of the state to Nebraska. The whole region is filled with glacial till that has been eroded to give the land some rolling qualities. Iowa also has the Loess Hills along the Missouri River. The Young Drift Plains cover the flat central and northern sections of Iowa. This state has some of the best farmland in America.

SEE ALSO: Agriculture; Dunes; Glaciers; Great Lakes; Mining; Mississippi River; Prairie.

BIBLIOGRAPHY. Jim Dufresne, *50 Hikes in Michigan: The Best Walks, Hikes, and Backpacks in the Lower Peninsula* (The Countryman Press, 2003); James Halfpenny, *Scats and Tracks of the Midwest: A Field Guide to the Signs of Seventy Wildlife Species* (Globe Pequot Press, 2005); James Kavanagh, *Wisconsin Birds* (Waterford Press, 2001); Doug Ladd, *Tallgrass Prairie Wildflowers: A Field Guide to the Common Wildflowers and Plants of the Prairie Midwest* (Globe Pequot Press, 2005); John Madson, *The Elemental Prairie: Sixty Tallgrass Plants* (University of Iowa Press, 2005); Nature Conservancy, *The Nature Conservancy's Guide to Indiana Preserves* (Indiana University Press, 2006); Shawn Perich, *Wild Minnesota: A Celebration of Our State's Natural Beauty* (MBI Publishing Company, 2005); Stan Tekiela, *Birds of Illinois Field Guide* (Adventure Publications, 2002); Stan Tekiela, *Trees of Wisconsin Field Guide* (Adventure Publications, 2002).

Andrew J. Waskey
Dalton State College

United States, Mountain West

THE MOUNTAIN WEST region encompasses four states: Colorado, Wyoming, Montana, and Idaho. These states represent the northern, middle, and southern Rocky Mountain physiographic regions in the United States. This area is one of the most diverse regions in the United States because it includes eight different physiographic provinces: The Great Plains, the northern Rockies, the middle Rockies, the southern Rockies, the Wyoming Basin, the Columbia Plateau, the Basin and Range, and the Colorado Plateau. This diversity in geology and geomorphology also translates into differences in climate, vegetation, and wildlife habitat. In addition, the natural resources in this area attracted the early pioneers and the different amenities are attracting a new and expanding western population.

This portion of the United States occupies over 432,538 square miles (1.12 million square kilometers) and stretches approximately 860 miles (1,400 kilometers) along the crests of the Rocky Mountains. The northern border of Idaho and Montana follows latitude 49 degrees north separating the United States from Canada, while the southern border is Colorado at latitude 37 degrees north. The eastern limit of this region is longitude 102 degrees west—the eastern border of Colorado—and the western edge is longitude 117 degrees west—the western border of Idaho.

The Mountain West is a region dominated by the Rocky Mountains, a complex of mountains, valleys, and basins formed during the Cretaceous Period (140–65 million years ago). However, portions of the southern Rockies were uplifted more than 3.9 billion years ago during the Precambrian Period. The backbone of the Rockies was created by a combination of igneous and metamorphic rock and the edges are tilted sedimentary layers forming long ridges or hogbacks. The movements of the earth's crust have created a series of folded and faulted mountains. The Rockies have also experienced several volcanic episodes and numerous intrusions, lava flows, and other magmatic features are evident throughout the four states. The various uplifts, folds, and faults have occurred over more than three billion years and the mountains have experienced a multitude of erosional periods creating several flat basins filled with sedimentary material, the most notable of which is the Wyoming high basin. Finally, the three ice and intermittent ice advances have scarred the mountains with spectacular glacial features and remnants still present today in Glacier, Rocky Mountain, Grand Teton, and Yellowstone National Parks.

Adjacent to the Rocky Mountains to the east is the Great Plains, an area of low relief except for outlier uplifted mountains. The majority of the geologic formations were created during the Mesozoic and Cenozoic periods (225–70 million years ago). The surface material is mainly the erosional deposits from the Rocky Mountains. In addition, the northern portions display the large impacts of the continental glaciers, and in western Montana, both continental glacial and the long reach of alpine glacial features.

The western side of the Rocky Mountains border several physiographic provinces. The northern Rockies are adjacent to the Columbia Plateau, an area dominated by volcanic materials of the Miocene-Pliocene periods (25–2 million years ago). Some of the lava flows extend over 100 miles (160 kilometers) and experienced folding and faulting creating ridges and steep hillsides. To the west of the middle Rocky Mountains is the great expanse of the Basin and Range Province, a large area occupying over 297,800 square miles (771,300 square kilometers) of recently faulted mountains and valleys. This province consists of a series of parallel mountain ranges with wide valleys of low relief. The southern Rocky Mountains in Colorado blend into the Colorado Plateau—a long, high area above 4,920 feet (1,500 meters) of mainly horizontal sedimentary rock eroded into steep-walled valleys, exposing folded and faulted rock formations. Scattered throughout the area are igneous structures, including large shield and conic volcano mountains, lava-capped mesas and tables, and lava flows.

The mountains, plateaus, valleys, and plains in this region display the radical topographic variations of the area. Colorado has the highest average elevation among the coterminous states of over 6,800 feet (2,070 meters). Colorado also has 59 peaks that soar over 14,000 feet (4,268 meters), also known as the "14'ers." Wyoming also has a very high average elevation of 6,700 feet (2,040 meters), yet does not have the multitude of high peaks similar to Colorado. Both Montana and Idaho have peaks that are slightly over 12,500 feet (3,811 meters), yet they also have low lands that are below 1,800 feet (549 meters). Overall, there are more than 20 major mountain ranges in the Mountain West, along with many isolated mountains. To the east and west of the Rocky Mountains are the flat plains and plateaus that generally slope away from the high peaks.

The elevation changes, differences between windward and leeward sides of mountains, the rain shadow effect, and the large latitudinal differences between the northern and southern portions of the four states provide diversity in climatic conditions. The northern portions of Idaho and Montana experience the most diversity in temperature and precipitation. Parts of Idaho receive over 60 inches (1,500 millimeters) of precipitation while adjacent valleys may receive less than five inches (125 millimeters) with temperatures ranging from 48 degrees C to negative 51 degrees C. All of the states experience high temperature differences between mountain passes and lowlands along with large ranges in precipitation, both rainfall and snowfall. The eastern portions of Colorado, Wyoming, and Montana display the westerly longitudinal progression of drier conditions in the Great Plains. The orographic effect of the Rocky Mountains produces higher precipitation on the western slopes—the windward side—and drier conditions on the eastern slopes—the leeward side.

The precipitation regime of the Rocky Mountains produces the headwaters of six major drainage systems: the Missouri, Arkansas, Rio Grande, Snake, Bear, and Colorado Rivers. The Missouri and Arkansas Rivers are some the largest contributors to the Mississippi River. From the southern portion of the Rocky Mountains, the Rio Grande drains southward into the Gulf of Mexico. The flow of the Rio Grande is interrupted by a number of reservoirs and over-allocation to municipalities, agriculture, and industries. From the western slope of the Rockies, the Colorado, Bear, and Snake Rivers drain across the Colorado Plateau, the Basin and Range, and the Columbia Plateau provinces. The Colorado River has its beginnings in northern Colorado and western Wyoming, traversing more than 1,429 miles (2,300 kilometers) to the Gulf of California in Mexico. The Colorado River, like the Rio Grande, feeds into a number of reservoirs created by large dams, the most notable of which is the

Hoover Dam producing Lake Mead. Similarly, the Snake River has a number of dams and reservoirs along its route to the Columbia River regulating its maximum and minimum flows. Finally, the Bear River is only a small and short river compared to the other five rivers, traveling only 347 miles (560 kilometers). It is also different from the other rivers because of the fact that it drains into a closed drainage basin, the Great Salt Lake, with no outlet to the ocean or sea.

The vegetation patterns correspond to the elevation and precipitation complexes of the Rocky Mountains, the Great Plains, the Basin and Range, and Plateau provinces. An increase in elevation creates cooler temperatures along with slopes away from direct solar radiation, while precipitation will generally be higher on the windward sides of mountains and lower on the leeward side. There are five vegetation zones through this region: Prairies, foothills, montane, sub alpine, and alpine. East of the Rocky Mountains, the Great Plains are dominated by mixed-grass prairies in the northern portions (Montana and Wyoming) and short grass prairies in the south extending into Colorado. The mixed grass prairie includes little bluestem, needle grasses, wheat grasses, sand-reeds, and gramas. The short grass prairie also has buffalo grass, ring grass, needle-and-thread, June grass, and galleta. The foothills zone is a transition from the prairies to the montane and can have a mix of sagebrushes and scattered woodlands of ponderosa pine, limber pine, juniper, pinion, Gambel oaks, and shrubs. The lower and warmer montane zone is dominated by Douglas firs. In the higher and cooler montane zone, lodge pole pines and dispersed aspen stands are found. The sub-alpine zone varies between spruce fir, white spruce, Englemann spruce, and white pine, depending on micro-climatic conditions. The alpine region transforms trees into shrub-like krummholz with wind, growing-day length, and soil conditions. In addition, grasses, sedges, sagebrushes, mosses, and lichens along with hundreds of flowering plants add to the vegetation diversity in this highest elevation zone, in conditions similar to the Arctic.

Adding to the complexity of the vegetation pattern are two major factors: The micro and local variations in topography, climate, soils, and geology and the drainage patterns with their corresponding riparian and river ecosystems. Changes in solar radiation, small patches of soil nutrients, and exposed bedrock generate mixes of vegetation in short distances. Access to moisture in the floodplain allows plants to extend their regime into the drier portions of the plains, the higher elevations, and the southern slopes of mountains. The major drainage systems flowing out of the Rocky Mountains extend the riparian species of broad-leaf cottonwoods, alders, and willows far into the adjacent plains and plateaus for hundreds of miles.

The combination of the physiography, climatic conditions, soils, fauna, and vegetation create broad areas of similar characteristics called ecoregions. The four-state mountain area is part of the Dry Domain and has elements of the Great Plains, southern Rocky Mountain Steppe, middle Rocky Mountain Steppe, northern Rocky Mountain Steppe, and Intermountain Semi-Desert and Desert. These elements are the basic components that compose the major habitats of the area and explain the considerable number of mammals, birds, and amphibians found throughout the area. The ecozones, coupled with the U.S. Fish and Wildlife Service's Gap Analysis program—a series of statewide terrestrial vertebrate inventories using GIS and remote sensing techniques to model species distribution and species richness—provide regional details of fauna through the area. The number of terrestrial vertebrates found range from a low of 375 in Idaho to almost 600 in Colorado. The majority of these are bird species (60–70 percent) with mammals (24–28 percent) the next largest category. Amphibians and reptiles are found throughout the region and account for 10–20 percent of the total species. Fish species, though not part of the Gap Analysis, are critical to the region's aquatic ecosystem. There are approximately 30–50 fish species; however, this number will vary based on the dominance of native and non-native varieties.

POPULATION

The Mountain West is one of the fastest growing areas in the United States. Between 2000 and 2005 the population of the United States grew by 5.3 percent while Colorado and Idaho statewide grew by over 8.5 percent, and portions of Montana and

Wyoming grew at rates approaching this number. Typical of the west, populations are centered in urban areas, with extremely low densities found in between the concentrations.

Colorado has the largest population (4.67 million) in the region with the majority along the eastern slope, the Front Range, in Denver (554,636), Colorado Springs (360,879), and Aurora (276,393). Idaho's population (1.43 million) is more dispersed across the state, but it is still in urban areas with Boise in the west-central portion the largest (185,787) and the other cities located in the central portions of the state: Nampa (51,867), Pocatello (51,466), and Idaho Falls (50,730).

The highest density of large cities in Montana (935,670) are along the eastern slopes of its western mountains; Missoula (57,053), Great Falls (56,690), and Butte–Silver Bow (34,606). Montana's largest city, Billings (89,847), is located in the west-central plains. Wyoming has the smallest population (509,294) with only Cheyenne (53,011) and Casper (49,644) as population centers. However, the next tier of cities is dispersed across the state with no concentrations except along the historic railroad lines. The population migration into the Mountain West is predominately to the urban areas; however, rural second homes are scattered in the more recreation- and amenity-oriented landscapes.

ECONOMY

Historically, mining, agriculture, forestry, and ranching were the focus of economic activities in the Mountain West. However, as the population shifted to urban areas in the mid-20th century, city-centered economic activities began to increase in importance. The U.S. Bureau of Economic Analysis identifies 10 major domestic production activities: Agriculture, forestry, fishing, and hunting; mining; utilities; construction; durable goods manufacturing; nondurable goods; wholesale trade; transportation; information; and fire. Of these activities, three categories can be considered non-urban: Agriculture, forestry, fishing, and hunting; transportation; and mining. As a percent of the total gross domestic production, agriculture, forestry, fishing and hunting are a major production activity in Montana and Idaho (eight to nine percent), while mining is dominant in Wyoming (47 percent) and important in Colorado and Montana (eight to nine percent). Transportation is also significant in Wyoming and Montana (nine percent). In all four states, recreation is a very important component of the economy, though it is not considered a production activity but rather a service industry. These activities are important because of the amount of land they require and their role in the domestic economy. Crop and grazing lands dominate the landscape in all four states, with forested areas the next largest category.

PUBLIC AND PRIVATE LANDS

The Mountain West is part of the transition area between the dominance of public and private land ownership. To the east, the majority of the lands are in private ownership, while to the west and south more of the land is in public ownership. This is an important factor because as more lands become public, it decreases the potential for private enterprise, but increases protection of natural resources. Idaho has the highest percent of public lands (50 percent), mainly in the northern half of the state. Wyoming (42 percent), Colorado (37 percent), and Montana's public lands are located in the western portions of each state. Each of the four states has approximately the same amount of public lands, between 38–43,700 square miles (98–113,000 square kilometers), however, because of their size differences, the percentages change. The public lands are managed by several federal agencies, mainly the U.S. Forest Service, the U.S. Bureau of Land Management, the U.S. National Park Service, and the U.S. Fish and Wildlife Service.

There are 50 national forests, 34 national parks, national historic sites, national trails and national monuments, and five national recreation areas distributed between the four states. Lands managed by the Bureau of Land Management (BLM) are generally distributed in the plains and plateaus between the mountains, sometimes referred to as the "checker-board" area, for the pattern of every other land section (one square mile).

The diversity of the landscape, variety of fauna and flora, and the importance of water resources make the Mountain West one of the most vital ar-

eas to the United States. The physical amenities and natural resources of the Mountain West will continue to attract people for residency, employment, or recreation. However, the intensity of activities has the potential to have large impacts on the environment. The conservation management of federal lands, the responsibility of private enterprise to sustain the environment, and the participation of local citizens to monitor their impacts can mitigate these impacts. The types of interactions between the environment and the population are critical to sustain this area for future generations.

SEE ALSO: Mountains; National Parks; Prairie; Rivers; Rocky Mountains; Rural Gentrification; Tourism.

BIBLIOGRAPHY. R. Bailey, *Ecoregions of the United States* (U.S. Department of Agriculture, U.S. Forest Service, 1980); J.S. Baron, ed., *Rocky Mountain Futures: An Ecological Perspective* (Island Press, 2002); D. Blood, *Rocky Mountain Wildlife* (Hancock House Publishers, 1976); C. Hunt, *Physiography of the United States* (W.H. Freeman and Co., 1967); L. Kershaw, A. MacKinnon, and J. Pojar, *Plants of the Rocky Mountains* (Lone Pine Publishing, 1998); D.H. Knight, *Mountains and Plains: The Ecology of Wyoming Landscapes* (Yale University Press, 1994); M.J. Mac, P.A. Opler, C.E. Puckett Haecker, and P. D. Doran, *Status and Trends of the Nation's Biological Resources* (U.S. Department of the Interior, U.S. Geological Survey, 1998); W.D. Thornbury, *Regional Geomorphology of the United States* (John Wiley & Sons, Inc., 1965).

Jeffrey Durrant and Mathew Shumway
Independent Scholars

United States, Northeast

THE NORTHEASTERN UNITED States is often referred to as New England and includes the states of Maine, Massachusetts, New Hampshire, and Vermont. Most of the area is part of the northern end of the Appalachian Mountains where winters are cold and snowy. The states that occupy the region were settled by the Puritans and their descendants. Other immigrant groups have also come in the 19th and 20th centuries from Ireland, Portugal, Italy, and French-speaking Canada

Maine is in the northeast corner of the United States and is the largest of the New England states. Its city of Eastport is the most easterly of all American cities. Along with New Hampshire and Massachusetts, Maine is bound by the Atlantic Ocean on its eastern side. On the west, it is bounded by New Hampshire and by the Canadian province of Quebec. The Canadian province of New Brunswick bounds it on the northeastern side.

Maine's coast is in the first of three land regions, which are the Coastal Lowlands, the Eastern New England Upland, and the White Mountain Region. The Coastal Lowlands extend from the Atlantic coast between 10 and 40 miles inland and form a strip along Maine's Atlantic shoreline. It is part of a larger region of land that stretches along the entire New England coast.

The Coastal Lowlands area does not rise much above sea level. It has broad sandy beaches that give way to salt marshes. During earlier ice ages, it was depressed from much greater heights by the great weight of glaciers. The southern part of Maine's coast has numerous sandy beaches. Old Orchard Beach is a hard packed sandy beach that is 11 miles (18 kilometers) long. It is one of the longest of all sandy beaches on the Atlantic coast of the United States. The coast is a lowland area in which farming of cranberries, blueberries, beef cattle, and poultry is done. The northern part of the Coastal Lowlands is rocky and dominated by high cliffs. Deposits of sand, gravel, granite, and limestone are mined for construction and other uses.

Maine is best known for its rocky northern coast. Along the Maine coast are over 400 islands that are from two to 25 square miles (five to 65 square kilometers) in size. There are also thousands of smaller islands. The largest is Mount Desert, which is about 100 square miles (260 square kilometers). There are numerous ports with deep water where ships harbor along Maine's 3,000-mile long coastline. Fishing for lobster and other deep sea fishing are major industries.

The Eastern New England Upland area occupies the middle part of Maine. It is a belt between 20 and 50 miles (32 to 80 kilometers) wide and is part of an uplifted shelf that extends from Connecticut

to Canada. The northern area covers Maine's entire border with New Brunswick except for the thin Coastal Lowlands strip. Some areas of this region are several thousand feet above sea level. Dairy and beef cattle are raised in the area. The Aroostook Plateau in the northeast has deep, fertile soil; excellent for farming, it has made Maine famous as a potato producing state. Forestry is important in the region as it is in other parts of Maine. Its mountains look green all year long because of the vast number of trees covering nearly 90 percent of the state. The Eastern New England Upland area has a great number of lakes south of the Aroostook Plateau. In the center of the Upland area, mountains cut across, as do swift-moving steams fed by melting snows.

The White Mountain Region occupies western Maine. It extends into New Hampshire and Vermont. In the north, the region is only about five miles (eight kilometers) wide. In the south, it is 30 miles (48 kilometers) wide. Logging is a major occupation in the area. Maine's tallest mountains and hundreds of lakes occupy the area's valleys. Mount Katahdin, rising to 5,268 feet (1,606 meters), is the highest mountain in Maine. In addition, 97 other mountains exceed 3,000 feet (910 meters). Among the region's wildlife are moose, beavers, foxes, lynxes, martens, minks, raccoons, black bears, deer, and over 300 kinds of birds. The climate in the region of the White Mountains is cool to cold because of Arctic or coastal sea winds. Snow covers the region in winter making skiing a popular sport in Maine, Vermont, and New Hampshire.

Vermont is about 160 miles long and 80 miles wide. Vermont covers 9,615 square miles (24,900 square kilometers). Its land regions are mostly mountainous. Northeast Highlands are part of an area shared with Maine and New Hampshire. The area is covered by mountains composed of granite, which rise from 2,700 to 3,300 feet (823 to 1,010 meters) above the sea level of the Atlantic coastline. Gore Mountain, at 3,330 feet (1,015 meters), is the tallest. The Northeast Highlands area is cut with many fast-flowing mountain streams.

The Western New England Upland region covers most of the eastern part of Vermont. The area extends into Western Massachusetts and is also called the Vermont Piedmont. Part of the area is composed of broad farmlands with streams that flow into the Connecticut River Valley. Farming in the valley produces apples, strawberries, and other crops. The lowlands of the Western New England Upland contain numerous lakes in the northern area. The region rises to the west in the western hills, which connect with granite outcroppings. These turn into granite hills along the boundary with the Green Mountains.

The Green Mountains region is a spine covering central Vermont. The highest peak, Mount Mansfield, is 4,393 feet high (1,339 meters) and a number of other peaks are nearly as tall. Because of the beauty and the winter snows, the region is the center of Vermont's tourist industry. It is also an area that is logged for timber and mined for a variety of minerals. In the north, the Green Mountains decline to become lower mountain ranges. In southern Vermont between the Green Mountains and the Taconic Mountains is the Vermont Valley. The valley is narrow with small rivers and river valleys that stretch for miles north from Massachusetts into central Vermont. The Baton Kill and Waloomsac rivers flow through the Vermont Valley.

The Taconic Mountains cover a narrow strip in southwestern Vermont. Equinox Mountain (3,816 feet), Dorset Peak (3,770 feet), Little Equinox Mountain (3,320 feet), Mother Myrick Mountain (3,290 feet), and Bear Mountain (3,260 feet) are found in the Taconic Mountains of Vermont from which flow swift streams into areas with beautiful lakes. The Champlain Valley borders Lake Champlain in northwestern Vermont. The area is sometimes called the Vermont Lowland—a fertile area that is farmed extensively with dairy operations. It also has numerous apple orchards surrounded by corn, hay, oats, and wheat fields. Burlington, Vermont's largest city, is located in the region.

New Hampshire is about 190 miles long and 70 miles wide. Canada forms its northern border and Massachusetts is on its southern border. In the east, New Hampshire has a short coastline on the Atlantic Ocean. Maine lies to the east and the Vermont state line marks its western boundary. New Hampshire covers 9,351 square miles (24,100 square kilometers). There are three land areas in New Hampshire.

The coastal lowlands are in the extreme southeastern corner of the state and are part of the larger

New England Coastal Lowlands. The area extends inland from about 15 to 20 miles (24 to 32 kilometers). The coastal area has a number sandy beaches that are used for recreation. Great Bay on the coast is a stopover for migratory birds, especially geese and ducks. The rivers of the area have been a source of water power in the past.

The two remaining regions are the Eastern New England Upland and the White Mountains, which are divided by the Merrimack River that rises in the White Mountains and flows across the upland area and empties into the sea near Boston, Massachusetts. The Upland region covers most of the southern, western, and eastern parts of the state. The Connecticut River Valley forms its western boundary. The Hills and Lakes area surrounds the Merrimack River on the west, east, and northern sides. The White Mountain Region is in the northern part of the state. It is a rugged mountain range, heavily forested with spruce and fir and yellow birch. Mount Washington (6,288 feet) is the tallest in New England and is famous for extreme weather. In 1934, the Mount Washington Observatory recorded winds at 231 miles per hour.

Massachusetts is 190 miles long and 50 miles wide at its most distant points and covers 8,284 square miles (21,456 square kilometers), excluding nearly 1,000 square miles of coastal water areas. New Hampshire and Vermont border Massachusetts in the north. It is bordered in the south by Connecticut and Rhode Island. The Atlantic Ocean forms the east coast and New York forms the western border. The coastline is 192 miles (309 kilometers) long.

Massachusetts has six land areas. In the east is the Coastal Lowlands region. It makes up the eastern third of the state. This area includes the long sandy peninsula of Cape Cod, Nantucket Island, Martha's Vineyard, the Elizabeth Islands, and other smaller islands. The region is a popular summer resort and has several excellent harbors including Boston. Swampy areas are found along the coast. At the end of the last ice age, glacial deposits were left to dot the area. The middle of Massachusetts is covered by the Eastern New England Upland extending from Maine to New Jersey. The Massachusetts portion is about 50 miles wide. Moving westward, the area rises to 1,000 feet in height (300 meters) and then gradually slopes westward to the Connecticut River Valley. The Connecticut Valley, drained by the Connecticut River, is long and narrow. It is about 20 miles (32 kilometers) wide and has rich soil and mild temperatures that make it productive farmland. The Western New England Upland is the major area west of the Connecticut River Valley. It is about 20 to 30 miles (32 to 48 kilometers) across from the Connecticut Valley to the Berkshire Valley. The Berkshire Hills area is an extension of the Green Mountains of Vermont. The Berkshire Hills are over 2,000 feet (610 meters) in height. The northern part of the Upland region near Vermont has the highest mountains in Massachusetts, such as Mount Greylock, at 3,487 feet (1,064 meters) above sea level. The narrow Berkshire Valley lies between the Berkshire Hills and the Taconic Mountains; its green pastures make it a major dairy area. The Taconic Mountains of Massachusetts are the sixth landform. They were formed during the late Ordovician Period (440 million years ago).

SEE ALSO: Atlantic Ocean; Beaches; Coastal Zone; Fisheries; Lakes; Mountains; Potatoes; Rivers; Tourism.

BIBLIOGRAPHY. Rennay Craats, *Guide to New Hampshire* (Weigl Publishers, Inc., 2001); Gene Daniell, ed., *White Mountain Guide* (Appalachian Mountain Club Books, 2003); Green Mountain Club, *Long Trail Guide: Hiking Vermont's High Ridge* (Green Mountain Club, Inc., 2003); John O. Hayden, *50 Hikes in Vermont: Walks, Hikes, and Overnights in the Green Mountain State* (The Countryman Press, 2003); Christopher J. Lenney, *Sightseeking: Clues to the Landscape History of New England* (University Press of New England, 2003); Terrell S. Lester et al., *Maine: The Seasons* (Knopf Publishing Group, 2001); John H. Long, ed., *Connecticut, Maine, Massachusetts, Rhode Island: Atlas of Historical County Boundaries* (Scribner, 1995); Paul M. Searls, *Two Vermonts: Geography and Identity, 1865–1910* (University of New Hampshire Press, 2006); Tom Seymour, *Foraging New England: Finding, Identifying, and Preparing Edible Wild Foods and Medicinal Plants from Maine to Connecticut* (Falcon Press Publishing, 2002); Christina Tree and Diane E. Foulds, *Vermont: An Explorer's Guide* (The Countryman Press, 2003).

Andrew J. Waskey
Dalton State College

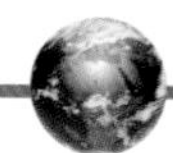

United States, Pacific Northwest

THE PACIFIC NORTHWEST is a region of varied landforms and differing climatic conditions consisting of two states: Oregon and Washington. As of 2005, Oregon is 98,386 square miles and ranks as the 9th largest state; it has an estimated population of 3,641,056 and is the 27th most populated state. Also as of 2005, Washington is 71,303 square miles and ranks as the 18th largest state; it has an estimated population of 6,287,759 and is the 15th most populated state. Both states are located on the west coast of North America within a latitudinal band containing a humid coastal climate. Precipitation amounts are particularly high along the coasts of both states and especially on the windward slopes of the Coastal Range and the Cascade Mountains. High pressure over the eastern Pacific Ocean blocks the moist air from moving farther south. This guarantees maximum precipitation along the coastal areas of Oregon and Washington. On some occasions, the amount of rainfall can be extraordinarily excessive and flooding can be a serious problem. In November 2006, areas of western Washington received over 15 inches of rainfall, an amount that brought severe flooding to homes and businesses. Televised news accounts showed automobiles and even houses caught up in fast moving rivers. Earthquakes are not infrequent, and the massive eruption of Mount St. Helens in the 1980s is a reminder of the dynamic geological nature of the region.

To the east of the prominent north-south trending mountain ranges, precipitation diminishes significantly. The coastal areas of the two states receive between 60 and 80 inches of precipitation annually while areas inland range from 10 to 20 inches annually. In fact, both states contain areas of true desert in their eastern areas. Vegetation in the region includes Douglas fir, redwoods, spruce, red cedar, and hemlock within the alpine coniferous forest. Grasslands are found within the steppe east of the mountains.

The Pacific Northwest was identified by Joel Garreau in his 1981 book *The Nine Nations of North America* as lying within two regions: Ecotopia, a borrowing from a 1975 book of the same name by Ernest Callenbach, and the Empty Quarter, an expanse of the interior stretching from the border with New Mexico north through the arid and mountainous western United States and into Canada and Alaska. The Empty Quarter is characterized by having the lowest population densities of any area of comparable size on the continent. The term *Ecotopia* is derived from Callenbach's combination of the two words *ecological* and *utopia*. The term characterizes well the attitudes of the people in the region and identifies their love of nature and overwhelming support of environmental awareness and protection. The coastal area has also been called "Cascadia" after the Cascade Mountains and because of the multitude of rivers flowing out of the highlands into the Pacific Ocean.

The fast flowing and voluminous rivers in the region are valuable sources of power for the generation of electricity. Within the Columbia Basin, hydroelectric power is used in the processing of aluminum, a metal used extensively in the manufacture of airplanes. With several major operations of the Boeing aircraft company being located in the Puget Sound region, it comes as no surprise that this world class manufacturer of airplanes is the most important customer for aluminum produced locally. Accessibility to substantial amounts of aluminum within the Pacific Northwest is one reason why Boeing remains within this region despite its significant number of rainy days each year, a climatic condition not ideal for operating aircraft.

The Pacific Northwest was visited by Western explorers and settlers much later than other regions in North America. In 1778, Captain James Cook explored along the Pacific Coast in search of the long-sought Northwest Passage. The historic Lewis and Clark expedition of 1806 was the first to enter the region by land. A number of American Indian groups were encountered by early explorers in the region. Among them were the Klamath, Nez Perce, Bannock, and Chinook. By the 1840s, settlers were arriving by way of the Oregon Trail, an activity that waited until a boundary dispute with the United Kingdom could be resolved. Railroad construction in the 1880s brought a much-needed surface connection to points east for the shipment of the region's first important products: Lumber and wheat.

Industrial production in the Pacific Northwest received a powerful impetus with the opening of the

Bonneville Dam in 1943. Power generation from this structure allowed for the expansion of existing industries and development of new ones. Water control also assisted in the expansion of the agricultural sector in the region. The variety of agricultural products grown in the region rivals the variability of the landforms. Washington alone ranks highly in an abundance of food products. Prominent in its product inventory are hops, spearmint oil, lentils, peas, cherries, plums (and prunes), onions, cranberries, strawberries, grapes, carrots, potatoes, apricots, asparagus, and apples.

Agriculture in Oregon's fertile Willamette Valley is equally productive. Cattle raising, dairy operations, apples, peppermint, and potatoes are just a few of the important agricultural products in the state. Both states are important in lumber production but controversies over the years about harvesting, clear cutting, and alleged mismanagement of the resource have clouded the industry. Salmon fishing has long been an economic mainstay in the Pacific Northwest. The country's third major wheat growing area is on the Columbia Plateau. The area is referred to as the "Palouse" and both spring and winter wheat is grown there.

Lumber operations have been a constant economic activity in the region. Emerging from this industry in recent years is a burgeoning furniture manufacturing complex. The wood products industry employs over 50,000 workers, many of them in the booming furniture-manufacturing sector. Perhaps the greatest single economic endeavor has been the emergence of Microsoft under the leadership of Bill Gates, an unparalleled software producing operation that has made Gates the richest man in the world. But Gates is not alone in acquiring a massive fortune. The Puget Sound area in Washington is reportedly the home of more billionaires than any other place in the world. The accumulation of wealth from fast growing computer enterprises has created something of a rich man, poor man society in which the gap between the haves and have-nots is increasing.

A prominent urban corridor, or conurbation, has grown over the years in the intermountain valleys from Eugene, Oregon north to Seattle and beyond to Vancouver, British Columbia. This corridor is home to nearly eight million people in Oregon and Washington. Its annual economic output exceeds $200 million and represents a wide range of industrial and service activities. The corridor has become an important center for electronics and computer related activities, which is a reflection of the emerging information age seen in many regions of the United States. The region is also a favored retirement location for people seeking an association with both true wilderness and the dynamic of a stimulating and ecologically aware urban populace.

Seattle is the region's largest city, a claim it has held since the late 19th century when the region became an important supplier of lumber and wheat to the rest of the country. Initially, Seattle was a logging center but it quickly gained in prominence as a regional growth center once it was included in the national railroad system. The rise of Boeing has guaranteed Seattle a measure of urban prominence since the 1920s. Portland, Oregon has access to the interior by way of the Columbia River Valley. As such, it has long been an important shipping point for food products to the interior and to overseas locations. Both Seattle and Portland have populations in excess of one million. In many ways the Pacific Northwest is more closely tied to other Pacific Rim countries than to points east in the United States and Canada, an attribute shared with California and British Columbia. Trade with Japan, China, and countries in southeast Asia continues to increase every year.

SEE ALSO: Desert; Hydropower; Lewis and Clark Expedition; Mountains; Northern Spotted Owl; Pacific Ocean; Salmon; Timber Industry.

BIBLIOGRAPHY. Ernest Callenbach, *Ecotopia* (Bantam Books, 1975); David Peterson Del Mar, *Oregon's Promise: An Interpretive History* (Oregon State University Press, 2003); Jane Claire Dirks-Edmunds, *Not Just Trees: The Legacy of a Douglas-fir Forest* (Washington State University Press, 1999); John C. Hudson, *Across This Land: A Regional Geography of the United States and Canada* (Johns Hopkins University Press, 2002); Philip L. Jackson and A. Jon Kimerling, eds., *Atlas of the Pacific Northwest*, 9th ed. (Oregon State University Press, 2003); Hill Williams, *The Restless Northwest: A Geological Story* (Washington State University Press, 2002).

Gerald R. Pitzl, Ph.D.
New Mexico Public Education Department

United States, Southeast

THE SOUTHEASTERN STATES of Georgia, South Carolina, North Carolina, and Virginia border the Atlantic Ocean on their eastern shores. Georgia is the southernmost state in the region and borders Florida. Virginia is the northernmost and borders Maryland to its north and east.

The general geography of these states comprises three parts. Each has a coastal plain that begins at the Atlantic Ocean and which extends west to a second area—a plateau region called the Piedmont. The third region is the Appalachian Mountains, which run from Canada to Georgia.

Each of these states is traversed by rivers that flow from the mountains to their Atlantic coast. The rivers were the highways used by the pioneers after the beginning of settlements as they moved westward to eventually across the mountains. In Virginia, the rivers all flow into the Chesapeake Bay. From north to south, the rivers are the Potomac, the Rappahannock, the York, and the James. These rivers all have important tributaries upstream.

The Chesapeake Bay is a vast estuary surrounded by Virginia and Maryland. It begins with the entrance of the Susquehanna River into its northernmost reaches. The Susquehanna's watershed feeding into the Chesapeake includes parts of New York, Pennsylvania, Maryland, the District of Columbia, and West Virginia. The length of the Chesapeake is 455 miles (304 kilometers) from the Susquehanna River's entrance south to the Atlantic Ocean. Geologists believe that the bay was formed at the end of the Eocene era when it was hit by a major bolide at the lower end of the Susquehanna about 35 million years ago. The Chesapeake is relatively shallow with brackish water. The word *Chesapeake* is an Algonquin word for shellfish, which abound in the bay along with oysters, crabs, and numerous species of fish. Most of the bay is surrounded by Virginia. The James, York, Rappahannock, and Potomac Rivers form peninsulas extending into the bay. The Northern Neck is the peninsula formed by the Potomac and the Rappahannock Rivers. The York and the Rappahannock form the Middle Peninsula. The James and the York Rivers form The Peninsula. The James is the longest river in Virginia with a length of 340 miles. The climate of the Chesapeake Bay is humid and subtropical. It has hot, humid summers and mild, rainy winters. However, severe winters do occasionally occur in which the more brackish parts of the tributary rivers freeze over solid enough for lightweight automobiles to drive across. The Virginia eastern side of the Chesapeake Bay is called the Eastern Shore. It forms a fourth peninsula with its eastern side on the Atlantic Ocean and its western side on the Bay.

Hampton Roads Harbor is located at the southern end of the Chesapeake Bay and is one of the world's best natural harbors. South of it along the Virginia–North Carolina border is the Dismal Swamp. The swamp is a large wetlands area with a rich variety of wildlife such as Virginia deer, rabbits, raccoons, foxes, bobcats, bears, and numerous species of birds, snakes, frogs, and turtles. A portion of the swamp has been preserved in far southeastern Virginia and northeastern North Carolina as the Great Dismal Swamp National Wildlife Refuge. Attempts to drain, log, and farm the Dismal Swamp were made in colonial times, but for the most part it is still in primitive condition. In its center is Lake Drummond. Its soils are so complicated by organic materials that crops from its soils have produced unusual results, such as mottled colors in cotton.

The Atlantic Coastal Plain in all of the southeast states extends inland until it reaches the Fall Line. Ships can navigate the rivers from the ocean to the Fall Line, where rocky rapids make anything but limited journeys by canoe almost impossible. The Atlantic Coastal Plain widens from north to south because the Appalachian Mountains run from northeast to southwest. It is broadest in Georgia where it crosses into Florida, and as it moves westward across south Georgia, it becomes the Gulf Coastal Plain that extends into Texas as far south as the Rio Grande River and westward to Del Rio. In the southeast, the Atlantic Coastal Plain was at the time of the first English settlements covered with forests that were a mixture of pines and a variety of deciduous trees. These were, and still are, harvested for use in paper pulp production and furniture making. The soil is usually sandy loam because the sand was deposited when the area was under the Atlantic Ocean. In South Carolina and in some places in Georgia, there are sand hills that are the ancient sand dunes of the ocean. The Atlantic Coastal Plain

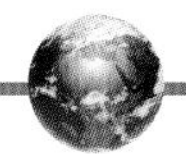

is farmed extensively making use of modern chemical fertilizers. Bright-leaf tobacco, peanuts, cotton, corn, sweet potatoes, and, in recent decades, soybeans have become common crops. Large numbers of hogs are also raised.

In North Carolina, the eastern part of the state juts into the Atlantic Ocean. A spear point of barrier islands called the Outer Banks protects the coastline. They have three capes: Cape Hatteras, Cape Lookout, and Cape Fear. The beaches on the Outer Banks are popular tourist destinations and are the site of frequent storms that have wrecked many ships. Between the Outer Banks and the mainland are a number of islands and vast lagoons. Albemarle Sound and Pamlico Sound are the largest of the brackish water lagoons. They are fed by the Chowan, Roanoke, Neuse, and Tar Rivers. The mainland coast is a low swampy tidewater area.

The land rises between Wilmington at the head of the Cape Fear estuary and Georgetown, South Carolina. In South Carolina, the long bay from the North Carolina state line to beyond Myrtle Beach forms an extensive beach resort area. Between Georgetown and Charlestown there are barrier islands such as Paley's Island and swampy areas that were, until the advent of mechanized rice farming in Texas and Louisiana, major rice growing areas.

The coast of Georgia is home to a number of barrier islands including Sea Island and Cumberland, which is a natural preserve. Inland, the Atlantic Coastal Plain in Georgia is lightly populated. The region is farmed and forested with loblolly pines, which are logged for pulpwood and, in some areas, worked to gather pine rosin.

Just north of the Florida state line lies the Okefenokee Swamp, which has been a national wildlife refuge since 1936. It covers an area of 438,000 acres, which is about 38 miles long by 25 miles wide. The Okefenokee Swamp occupies a vast peat bog that lies in a saucer-shaped depression. Until 7,000 years ago, the depression was part of the ocean floor. The Suwannee River (280 miles long) rises in the swamp, flows west out of Georgia to cross Florida, and empties into the Gulf of Mexico near Cedar Key. The Okefenokee teems with wildlife and is famous for it numerous alligators. The areas to the west and south are very lightly populated and heavily forested.

The Burning of Atlanta

When the American Civil War broke out, Atlanta was a major railway junction and the 12th largest city in the Confederacy. Its cotton warehouses were quickly put to use by Confederate quartermasters to store war materiel for the Confederate armies, and the Atlanta Rolling Mill was used to make iron-plating, including for the CSS *Virginia*. The Confederates had planned to fortify Atlanta, but it was extremely difficult. Lemuel Grant, a local businessman, did built redoubts on some of the approaches to the city. The interlinked earthworks and trenches could not stop a determined Union attack.

In 1864, the Union forces did attack Atlanta, and on September 1, the Confederate General John Bell Hood, after holding out for four months, was forced to evacuate the city. Before he did so, he gave orders for public buildings to be destroyed. General William Sherman of the Union forces then captured the city. He occupied it for two months, and then, as he left, he ordered out the civilian population and gave his men instructions to burn the city down. As he later described it, "Behind us lay Atlanta, smouldering and in ruins, the black smoke rising high in the air, and hanging like a pall over the ruined city." He spared only the churches and hospitals.

The burning of Atlanta, made famous around the world by the book and film *Gone with the Wind*, was not just a random act of revenge against a city that had provided so much support for the Confederacy. It also came soon after Lincoln had won the 1864 presidential election, with Lincoln facing a Democratic platform that was urging for a truce to end the war.

The Atlantic Coastal Plain changes into the Gulf Coastal Plain as it moves westward. It ends in Georgia at the Chattahoochee River, which is the boundary between Georgia and Alabama for the southern half of the state. The Chattahoochee changes its name to the Apalachicola River when it enters

Santa Fe before turning to the east and then north to Raton before crossing into Colorado.

In the colonial era, Santa Fe was a favored location because of two specific site characteristics: First, the settlement was on the Santa Fe River and close to the Rio Grande, insuring access to reliable water sources; second, with the Sangre de Cristo Mountains to the north and east and the Jemez Mountains 30 miles to the west, it was easily defended from attacks by Native Americans hostile to the Spanish.

Currently, the Native American groups reside primarily on reservations and pueblos within the four-state region. The largest single group is the Navajo, whose homeland is centered in the Four Corners area, the place where four southwestern states meet: New Mexico, Colorado, Utah, and Arizona. Within the Navajo area in northern Arizona is an enclave of Hopi. In addition, there are a number of Apache tribes in New Mexico and Arizona. Utes in Colorado and Papago in southern Arizona are also Native Americans, adding to the cultural diversity within the region. Utah is the homeland of the Mormon religion—the Church of Jesus Christ of Latter-day Saints—and Salt Lake City is the primary metropolitan center in the state.

Economic activity within the region has similarities and some decided differences. In Nevada, where the federal government still owns nearly 90 percent of the land, the forms of agriculture found in the Midwest and other regions to the east did not develop. The Homestead Act in the 19th century, which provided 160-acre parcels of land to prospective farmers, did not fit the vast expanses of arid land within Nevada. Nonetheless, land adjacent to adequate water sources was selected for agricultural pursuits. Agricultural activity within the Phoenix area grew slowly until the advent of the refrigerated railroad car, which allowed for the shipment of fresh and frozen food products nationwide. Concurrent with this important technological innovation was the realization that the Phoenix area was an ideal retirement area and a haven for those suffering from respiratory ailments. Within a few short decades, Phoenix grew dramatically in population. This metropolitan area is also the center for a thriving electronics industry and other high value added manufacturing activities.

Economic activity in Arizona was based on the so-called Five C's: Cotton, copper, cattle, citrus, and climate, with the latter referring to the reputedly ideal weather situation encouraging tourism. Cotton production gradually migrated west out of the old Cotton Belt in the south following the onslaught of the boll weevil early in the 20th century. The warm and sunny environs, especially in Texas and Arizona, were ideal for cotton growing as long as adequate water was available. With irrigation water diverted to this activity and other agricultural products, cotton growing flourished. Copper mining is an important activity as well. In addition to mining the copper ore, the concentrating and smelting of the raw product also takes place close to the mine sites because only a small percentage of the ore has copper within it and it cannot be economically shipped great distances. Cattle is important to the Arizona economy, especially in recent decades during which the building of feed lots near large urban markets occurred concurrently with a significant shift away from open grazing on federal lands. Citrus growing, especially orange groves, grew in prominence in Arizona as the activity gave way to urban development in southern California. Tourism continues to be an important industry in Arizona and throughout the four-state region.

The natural wonders of the Colorado Plateau, a vast area of geologic uplift that saw the rise of Pike's Peak and associated mountains in the state, continue to draw tourists from around the world. The Grand Canyon, Zion, Bryce, Capital Reef, and the Canyonlands region near Moab, Utah, exhibit dramatic eroded sandstone and shale landforms found nowhere else in the world. Also unique are the flourishing clusters of hotels and gambling establishments in Las Vegas, Lake Tahoe, and Reno, which attract hundreds of thousands of tourists annually. Early in its history Nevada became important in mining. Near the end of the 19th century, the famous Comstock Lode played out and the Nevada economy was decimated. By 1910, however, other mining operations began and the future of the state's economy was no longer in danger.

Nevada currently is the center of a controversy focused on the proposed permanent storage of spent radioactive fuels from government projects and over 120 nuclear power plants around the country. Yucca

Mountain, a remote site 100 miles northwest of Las Vegas, was identified in the 1970s by the Environmental Protection Agency as the favored place for the underground storage of the spent nuclear fuel. The proposal has met with considerable objections by environmental groups and the State of Nevada from the onset. Representatives of the state objected strenuously to a Nevada site for disposal because the state has no nuclear plants within its borders. In addition, experts are now suggesting that the whole question of burying nuclear wastes be reexamined to determine if more efficient ways can be found to dispose of this dangerous nuclear by-product. Nevada is also the home of the famous Hoover Dam, a gigantic structure on the lower Colorado River that provides both river control and hydroelectric power within the region and beyond. Hoover Dam is one of a number of major water control structures installed on rivers in the southwest. In recent years, proposals have been made by environmentalists and other scientists that the era of great dams was no longer valid and that plans be made to eliminate the majority of them in the western United States.

The border with Mexico in Arizona and New Mexico has become the center of an international controversy surrounding the illegal immigration of primarily Latinos into the United States. Estimates suggest that as many as one million illegal aliens annually have entered the United States over the last three decades. There is a genuine sociological push-pull process in operation in this movement: The pull is, of course, the perceived economic opportunities within the highly productive and wealthy United States, and the push is the high unemployment and lack of sufficient numbers of jobs within Mexico. The issue has defied mutually agreeable resolution through the years and remains a sore point in U.S. relations with the affected Latin American countries, as well as in domestic politics. There is a tragic aspect to the situation along the border: over the years, perhaps thousands of people attempting to enter the United States have died due to exposure in the inhospitable desert environments. In addition, attempts to smuggle people across the border in packed semi-trailers have resulted in many deaths. The existence of an illegal cross border drug trade only exacerbates an already serious set of problems plaguing the United States and Mexico.

A number of Mexican migrants have moved to the expanding dairy production region of eastern New Mexico. The region has grown dramatically with the purchase of extensive and relatively inexpensive areas of land in the region by Californians interested in establishing dairy operations. In 2005 New Mexico ranked as the fourth largest dairy state in the country. The optimism of the state's dairy industry is best exemplified by its stated goal to surpass Wisconsin and become the leading dairy state in the United States. New Mexico also has extensive grazing opportunities for cattle and sheep. There are even a few llama raising operations in the state.

A variety of agricultural products are grown in sunny New Mexico. Hay predominates and is an essential ingredient in the successful dairy and cattle raising industries. Pecans are grown in the southern part of the state and chiles, both red and green, are found in abundance in New Mexico. In fact, the official state question is "Red or green?" referring, of course, to which chili pepper a customer prefers when ordering a restaurant dish.

Fresh water availability is a constant concern within the region and in contiguous western and Midwestern states. From the mid-1990s to the present, a serious drought has gripped the region. One of the immediate consequences of the drought was the weakening of the piñon and other coniferous trees followed by an invasion of the pine beetle. The insect is able to penetrate the bark of a weakened tree and set up a colony. The response of the tree is literally to shut down its activity. Within four weeks of a pine beetle invasion, the needles on a mature piñon will be brown and the tree dead. It is estimated that virtually all the piñon in Arizona were lost and a good share of the New Mexico stand was similarly affected. Another outcome of the drought impacted the annual and legally required delivery of water from New Mexico to Texas. In order to ensure that the correct amount of water was available, the New Mexico state government purchased agricultural land along the Pecos and took it out of production, allowing for the required delivery of water to Texas.

SEE ALSO: Cattle; Cotton; Desert; Mexico; Mountains; Native Americans; Steppe.

BIBLIOGRAPHY. Clay Anderson, *Arizona Grand Canyon* (Hunter Publishers, Inc., 2001); John C. Hudson, *Across this Land: A Regional Geography of the United States and Canada* (Johns Hopkins University Press, 2002); Donald W. Meinig, *Southwest: Three Peoples in Geographical Change, 1600–1970* (University of New Mexico Press, 1988); John A. Murray, *Cactus Country: An Illustrated Guide* (National Book Network, 1997); National Park Service, *El Camino Real de Tierra Adentro: National Historic Trail* (U.S. Department of the Interior, 2002).

GERALD R. PITZL, PH.D.
NEW MEXICO PUBLIC EDUCATION DEPARTMENT

United States, Texas

TEXAS IS THE largest of the continental 48 American states. It lies at the western end of the Gulf of Mexico and the Southern Gulf Coastal Plain. It is also at the southern end of the North American Great Plains. The state is 266,807 square miles (691,030 square kilometers) and it extends over 800 miles from the Texas state line with Louisiana near Beaumont to El Paso in the west, which is close to New Mexico. It extends 737 miles (1,186 kilometers) from its northern extreme near Dalhart in the panhandle to its southernmost point near Brownsville on the Rio Grande River, which separates it from Mexico.

The topography of Texas can be compared to four steps. Beginning at sea level at the Gulf of Mexico the land slowly rises until it meets the second step at the Edwards plateau, which holds the Texas Hill Country. The third step is the Stockton plateau and the High Plains. The fourth step is at El Capitan (8,751 feet or 2,667 meters), which is part of the Guadalupe Mountains and the Rocky Mountains in El Paso.

There are five land regions in Texas. The Gulf Coastal Plain includes the coast, which stretches 367 miles (591 kilometers) from Louisiana to Mexico. It is a region of marshes, coastal wetlands, bays, tidal flats, marshes, dunes, beaches, coastal prairies, and barrier islands. Texas's barrier islands protect the mainland from tidal surges during the hurricanes that come every summer or so. These barrier islands, lagoons, and bay shores were the home of the Karankawa Indians at the time of the arrival of the Spanish and later European explorers. There are 17 barrier islands, five of which are major islands. These islands include Galveston, Padre, Mustang, and Matagorda. Much of Padre and Matagorda are still pristine preserves. The Aransas National Wildlife Refuge, which includes part of Matagorda, is the winter home of the endangered Whooping Crane. The largest barrier island is Padre and the largest bay is Galveston.

The Gulf Coastal Plain extends northward up the Sabine River to near Shreveport. In its lower reaches the Sabine River is a petroleum-rich area of swamps and bayous separating Texas and Louisiana. The Big Thicket, an area rich in biodiversity, extends from Beaumont (site of the 1902 Spindletop oil gusher) northward. The northernmost part of the Gulf Coastal Plain is forested with pines, oaks, and other trees common to its more eastern parts. At Kilgore the great East Texas Oil Field began pumping in the 1930s. The southern part of the plain extends south to Mexico and west to about Del Rio. The southern part is less forested and more open. In South Texas the Gulf Coastal Plains extend from Corpus Christi to Brownsville and then west to the Edwards Plateau, the southern end of which is at Del Rio. The area is one of huge ranches including the famous King Ranch. Javelina (*Peccary angulatus*) or collared peccary are often hunted in the region. Along the Rio Grande from Laredo to Brownsville orchards of oranges and grapefruits are an important part of the economy.

The Gulf Coastal Plain joins the Prairie Plains that extend from Red River in the north to San Antonio. The two plains merge in the Cross Timbers, which is a long narrow strip of forest that extends from Oklahoma deep into central Texas. The trees are smaller post oaks, blackjacks, and other less valuable trees. This area is also rich in petroleum resources in the Austin Chalk Formation, around Luling, and elsewhere. The Prairie Plains end in central Texas at the Balcones Escarpment centered at Austin. The Balcones Escarpment is the eastern edge of the Edwards Plateau, a desiccated region covered by cedars and shrubby plants. Like much of western Texas it appears desolate; however, the high mineral content of the thin, rocky soils and the food value of shrubs and grasses support cat-

Big Bend National Park in Texas is occasionally visited by jaguars at the northern end of their range.

tle, mule deer, and other forms of life such as black ground squirrels and road runners on ranches and in wild areas. Rainfall amounts decline to desert levels in the Trans-Pecos area into which the Chihuahuan Desert extends. There, the thorny plants increase and trees are replaced by shrubs. Ranchers in the area turn to the prickly pear cactus to support their cattle in times of drought. The spines are burned off with flame throwers, allowing cattle to eat the succulents for food and moisture.

The Prairie Plains turn into the Rolling Plains west of Fort Worth. The area is hilly with increasing shrubby plants and numerous cattle ranches. The Great Plains join the Rolling Plains in the Panhandle and they extend to the Rio Grande River. The southern end covers the Stockton Plateau, which is west of the Pecos River and extends to the east to about Del Rio. The Panhandle and the Rolling Plains are often referred to as the cap rock region because the rock is impermeable. The central area of the Great Plains includes the Monahans sand hills, a large dune area west of Odessa. It was described by Spanish explorers who also explored the Llano Estacado, a dry treeless area of the High Plains extending about 250 miles by 150 miles.

The Permian Basin is a large area of oil-rich sedimentary rock. For most of its geologic history Texas was under the ocean, and in the Permian geologic era the Permian Basin was surrounded by a vast coral reef. It became a trap for enormous quantities of oil and gas. The western part of the reef is now the Guadalupe Mountains, most of which lie in New Mexico and include the Carlsbad Caverns. However, the southern part of the Guadalupe Mountains extends into Texas and rises to 8,749 feet (2,667 meters) on El Capitan at the southern end. The land drops dramatically from El Capitan to salt flats at its base. The peak was used as a landmark by stage coaches of the Butterfield Overland Mail in pioneer days.

The region of far southwestern Texas is an extension of the Basin and Range Region of the Rocky Mountains. It includes the Fort Davis Mountains where in the upper elevations antelope mingle with cattle on the ranch lands. Other mountains are the Glass Mountains, the Santiago Mountains, and the Tierra Vieja Mountains. The southern portion of the area is rugged, isolated, and virtually uninhabited.

Along the Rio Grande River the area where the river turns north is called the Big Bend, much of which is now a National Park. The park is usually filled with visitors during the mild days of Christmas and New Years. The Big Bend region, high in elevation, is occasionally visited by jaguars at the northern end of their range and by mountain lions. The course of the Rio Grande River often passes through canyons of spectacular beauty.

El Paso is the largest city in southwest Texas. The area is arid and abounds with barrel cacti and other cacti. To the north of the city are the Hueco Tanks. The tanks are large natural rock formations that act as basins for trapping rain water. They supplied water to Native Americans and to travelers on stagecoaches, and were the site of one of the last Native American battles.

SEE ALSO: Cattle; Drilling (Oil and Gas); Livestock; Rio Grande River; Rocky Mountains; United States, Southwest (Arizona, Nevada, New Mexico, Utah).

BIBLIOGRAPHY. Wayne H. McAlister, *Matagorda Island: A Naturalist's Guide* (University of Texas Press, 1993); Laurence Parent, *Big Bend National Park* (University of Texas Press, 2006); Darwin R. Spearing, *Roadside Geology of Texas* (Mountain Press Publishing Company, 1991); D. Gentry Steele, *Land of the Desert Sun: Texas' Big Bend County* (Texas A&M Press, 1999); Joe C. Truett and Daniel W. Lay, *Land of Bears and Honey: A Natural History of East Texas* (University of Texas Press, 1994).

Andrew J. Waskey
Dalton State College

Universal Soil Loss Equation (USLE)

THE UNIVERSAL SOIL Loss Equation (USLE) is one of the most widely used empirical models for estimating long-term average soil loss. Wischmeier and Smith introduced this equation in 1978; the U.S. Soil Conservation Service first utilized USLE on a large scale. Soil erosion has for a long time been a major concern, particularly because of its negative impacts on agriculture and sedimentation of reservoirs. A range of natural and anthropogenic factors influence soil erosion processes, which occur in all landscape types and under a range of land use systems. USLE was intended for cropping systems, but is also applicable to nonagricultural conditions (such as construction sites). USLE was developed to predict annual soil loss rates from a particular field or single slope, where it is used to compare soil losses from a specific cropping and management system to "tolerable soil loss" rates. Moreover, USLE is used to evaluate the effectiveness of alternative cropping systems and management practices in reducing soil loss, and determining optimal levels of cropping and maximal tolerable slope.

The USLE consists of six factors reflecting effects of precipitation patterns, topography, vegetation, land management, and soil characteristics. Specifically, USLE computes expected surface erosion on a particular slope as:

$$A = R \times K \times LS \times C \times P$$

A represents the potential long-term average annual soil loss in tons per acre per year; *R* is the rainfall and runoff factor (erosivity); *K* is the soil erodibility factor; *LS* is the slope length-gradient factor; *C* is the crop/vegetation and management factor; and *P* is the support practice factor.

USLE is used worldwide and has come under increasing criticism. However, every model makes assumptions that limit its usefulness to certain conditions. First, USLE is intended to estimate sheet and rill erosion only. It is not calibrated to account for soil loss from gully and channel erosion, or wind erosion. It does not identify areas susceptible to landslides. Second, USLE was designed to estimate average annual soil loss from a particular field. Its predictive capability is best at evaluating soil loss at that scale and less applicable to others. Particularly in areas of high spatial and temporal variability of input parameters, uncertainty associated with results may be high. Third, USLE does not identify areas of sediment deposition. Fourth, USLE requires judgment in applying values for the variables. Differences in judgment will account for differences in the assessment of field conditions, making the comparison of estimates difficult. Fifth, USLE was designed for use in the Midwestern United States, but has been widely used and misused throughout the world.

Revisions of USLE, which more accurately predict soil loss, have been implemented since the 1990s. The revised USLE (RUSLE) was released in the early 1990s and afterwards underwent several revisions. These revised models take advantage of new research about soil erosion processes and relationships between the variables, and of the capability of computer technology and geographic information systems. In spite of its limitations, which are well known, USLE remains a widely accepted model. USLE is easy to apply for assessing relative erosion potentials under different site conditions and land management practices. However, it is a best available estimate of annual average soil loss, rather than an absolute value.

SEE ALSO: Soil Erosion; Soil Science; Soils.

BIBLIOGRAPHY. K.G. Renard et al., "RUSLE: Revised Universal Soil Loss Equation," *Journal of Soil and Water Conservation* (v.46, 1991); K.G. Renard et al., "Predicting Soil Erosion by Water. A Guide to Conservation Planning with the Revised Universal Soil Loss Equation (RUSLE)," in *Agriculture Handbook No. 703* (U.S. Department of Agriculture, Agricultural Research Service, 1997); T.J. Toy, G.R. Foster, and K.G. Renard, *Soil Erosion: Processes, Prediction, Measurement and Control* (Wiley & Sons, 2003); United Nations Environment Programme (UNEP), "Guidelines for Sediment Control Practices in the Insular Caribbean," *CEP Technical Report No. 32* (UNEP, 1994); W.H. Wischmeier and D.D. Smith, "Predicting Rainfall Erosion Losses: A Guide to Conservation Planning," in *Agriculture Handbook 537* (U.S. Department of Agriculture, Agricultural Research Service, 1978).

Wiebke Foerch
University of Arizona
Adane Abebe
University of Siegen, Germany

Ural Mountains

THE URAL MOUNTAINS extend for approximately 1,000 miles in a generally north-south orientation in Russia from the Kirgiz Steppe in Kazakhstan to the Kara Sea on the Arctic Ocean. Novaya Zemlya, an island in the Arctic Ocean, is an extension of the Urals. This mountain range is geologically quite old and dates from the Carboniferous period. The Urals formed when the Siberian plate impacted on the more massive Laurasian Plate. Because of their age the mountains are worn down and reduced through erosion. The average height of the range is 3,000–4,000 feet and the highest peak is Mount Narodnaya at 6,214 feet. The Urals mark the unofficial but traditional boundary between Europe and Siberia.

The Middle Urals are densely forested and rich in mineral wealth. The famous Urals Industrial Region was developed in the Middle Urals region and is based on mining and industry. The region is comprised of the Chelyabinsk, Sverdlovsk, Kurgan, Orenburg and Perm oblasts. The Middle Urals region has extensive deposits of iron, copper, chromium, bauxite, lead, zinc, gold, platinum, potassium, magnesium, asbestos, and other important minerals. There are major oil fields in the region but no coalfields. Industries within the region reflect its mineral wealth: Metallurgical, chemical, and heavy industries are dominant. The Urals Industrial Region developed rapidly during World War II with the movement of industrial capability from the Russian Plain in the west to prevent it from attack by German forces. Industry and mining give way to pasture lands in the Southern Urals.

Ekaterinaberg, an industrialized city of over 1.3 million people, is the self-proclaimed capital of the Urals. The city has experienced a variety of environmental contamination problems resulting from years of accumulated pollution from the multitude of industries both within the city and the surrounding region. There is a high degree of water pollution from heavy metals and deposits of tailings from the many mines. The problem of thermal pollution is also found in streams and lakes near the Beloyarskaya nuclear power plant, which is 15 miles from Ekaterinaberg.

Scientists in the region are well aware of the dangers of environmental degradation in the Urals Industrial Region. The Ural Environmental Union, a composite of government officials, environmental activists, and scientists has taken the lead in gathering information on pollution problems and in developing programs to rectify the situation. Another organization, the Institute of Industrial Ecology, is leading a program to set environmental priorities for the Urals. The program is called "The Assessment of Priorities for Middle Urals' Environmental Pollution Protection." Its initial focus area is the Chelyabinsk and Sverdlovsk oblasts. The plan calls for a series of environmental studies at local levels with the results subsequently extrapolated to the region. The project includes development of methods for establishing priorities for future environmentally related projects and to ensure sound economic growth as well.

International collaboration is an important part of the effort and experts from the United States will play a role in implementing pollution protection safeguards learned in addressing comparable problems. Eradicating environmental degradation in the

Urals and establishing effective programs to avoid serious levels of pollution in the future will not be an easy task. Decades of environmental disruption and the accumulation of a wide variety of pollutants presents an enormous challenge. Also, there is the need to maintain high levels of industrial activity in the region while environmental remediation is underway.

SEE ALSO: Environmentalism; Industry; Mining; Mountains; Nuclear Power; Pollution, Water; Russia (and Soviet Union).

BIBLIOGRAPHY. Edward A. Keller, *Introduction to Environmental Geology*, 3rd ed. (Prentice Hall, 2004); John J.W. Rogers and P. Geoffrey Feiss, *People and the Earth* (Cambridge University Press, 1998); Les Rountree, Martin Lewis, Marie Price, and William Wycoff, *Diversity and Globalization: World Regions, Environment, Development*, 3rd ed. (Prentice Hall, 2005); Denis D.J. Shaw, *Russia in the Modern World* (Blackwell, 1999).

Gerald R. Pitzl, Ph.D.
New Mexico Public Education Department

Uranium

URANIUM IS A metallic-white element, number 92 in the periodic table, with the symbol U. It is found in the earth in the form of minerals such as pitchblende, carnotite, and urainite. Attention is focused on uranium because of its central role in the nuclear energy industry. Martin Heinrich Klaproth discovered uranium in 1789, naming it after the recently identified planet Uranus. However, it was not isolated until 1841. It is the heaviest naturally occurring element and it exists in 16 different isotopes, which have different configurations of nuclear particles. Naturally occurring uranium consists of a mixture of three of these isotopes, with more than 99 percent being uranium-238. Uranium is radioactive, which means that it emits particles and ultimately deteriorates into lead over thousands of years. The time taken for half of the atoms of uranium in a sample to become lead is called the half-life.

Henri Becquerel discovered the property of radioactivity in 1896. In 1938, Otto Hahn and Fritz Strassmann determined that the slow bombardment of uranium by neutrons could lead to the breakdown of atoms and the emission of additional neutrons, which leads to the chain reaction known as nuclear fission. This atomic technology was rapidly adapted for military purposes and in 1945 the United States exploded two nuclear devices on cities in Japan. As well as furnishing a variety of missiles and weapons, nuclear power was also used to generate electricity and to transport submarines.

Uranium represents approximately two parts in a million of the earth and the uranium-235 isotope is needed for the nuclear fission process. Consequently, although uranium represents a much more productive source of energy than fossil fuels, securing a steady source of the material is expensive and problematic. Uranium is used as the basis for the production of the heavier transuranium products that are used for fission activities. The radioactivity represents a severe health problem for people in the proximity of the metal and the possibility that it can be used as a weapon of unparalleled power means that considerable care must be taken in its sale and transportation.

Speculation surrounds the fate of the uranium used in the former Soviet Union, much of which is believed to be in an insecure situation. In the former Soviet Union, at Chernobyl in the Ukraine, the world's worst nuclear plant accident occurred and generations of Ukrainians are being poisoned by the still virulent radiation. The danger of further incidents, with the problem of disposing of waste products that remain toxic for thousands of years, continues to bedevil the use of uranium in electricity generation. However, the present rate of global environmental degradation and climate change means that it is likely to remain under consideration for the foreseeable future. Currently, it is estimated that the demand for low-cost uranium supplies for energy production will exceed supply over the next few decades unless significant new deposits are found. This includes secondary sources of uranium, which use recycled uranium in different forms.

SEE ALSO: Nuclear Power; Nuclear Regulatory Commission (NRC) (U.S.); Nuclear Weapons.

BIBLIOGRAPHY. Greenpeace, www.greenpeace.org (cited July 2006); International Atomic Energy Agency, www.iaea.org (cited July 2006); Broder J. Merkel and Andrea Hasche-Berger, eds., *Uranium in the Environment: Mining Impact and Consequences* (Springer, 2005).

JOHN WALSH
SHINAWATRA UNIVERSITY

Urban Ecology

URBAN ECOLOGY IS the study of the urban ecosystem as an ecological unit that is part of the larger global ecosystem; it is also known as the ecology of cities and towns. Urban ecology examines the relationships between the urban and natural systems and interactions among the biotic components—including humans. It also involves the study of the urban ecosystem's impact on other ecosystems, seeking to understand relationships with the rural system, particularly transfers of matter and energy and the complementary functions the surrounding space provides. The relationship with the global system derives from the contribution of the city to global environmental change and the use of renewable and nonrenewable natural resources.

As an interdisciplinary field of study, there is a debate among fields regarding an identifiable specific approach. Ecology understands urban ecology as a subfield of the discipline that studies the ecological relations in the unique human-modified environment, adopting a traditional approach and applying conventional theories, or with the perspective of integrating the human system and humans as another species. From this point of view, we can examine the budget of matter and energy flowing through the urban ecosystem and observe a unit that almost wholly depends on external sources of system inputs—food, fuel, water, and building materials—a heterotrophic ecosystem that does have parallel on earth. The outputs—solid waste, wastewater, combustion gases, and heat—are the results of industrial respiration, which is a metabolic process of what resembles a large living organism. The system inputs have varied distant origins while the outputs have varied destinations, thus the city depends on areas much larger than its own surface. This is the ecological footprint, the equivalent area of land required to support the provision of inputs and process the outputs. It measures the dependence of the city on other ecosystems and its sustainability; the more area required the more unsustainable the city.

Sociology and anthropology understand urban ecology as a division of human ecology, the study of how humans relate to their environment. The concept has a sociological origin in the Chicago School of Sociology of the 1920s. Robert E. Park and Ernest W. Burguess conceived the human ecology approach to explain the urban development and spatial segregation within the city of Chicago as the result of the intervention of social and economic forces. Cities were regarded in their theory as environments governed by competition and accommodation forces, inspired by ecology and the ecological factors intervening in a natural ecosystem. From this point of view, groups compete for a scarce resource—the land—and this struggle leads to the division of the urban space into areas with homogeneous social and economic characteristics and to the appropriation of the most valuable areas by the higher rent groups. Naturalist Edward O. Wilson developed the notion of biophilia to define the nongenetic emotional affiliation of human beings to nature and other life forms for having lived within a biological world. This helps to understand the preference for living close to nature, moving to the suburbs, valuing natural landscapes, and ultimately, supporting conservation of ecosystems and species and desiring to manage them efficiently. Biophilia can be promoted by education and experience with wildlife or discouraged by living within a completely industrial environment.

Political studies approach urban ecology by highlighting the role of institutions and identifying economic and social processes as forces implicated in environmental changes at various spatial scales, focusing on man as an agent of change. The field develops into political ecology. Urban ecology becomes a policy to increase the sustainability of the urban system by minimizing the impact on natural systems, promoting restoration plans of degraded habitats and conservation of those areas with a natural habitat. Urban ecology is seen as the conceptual basis of the process of sustainable urban

development, a policy to build green cities or ecocities. The goal is to achieve healthier and more liveable cities, maintaining biodiversity and more efficiently managing space and resources, so that the community has not only architectural, social, and economic assets but also environmental ones.

THE URBAN ECOSYSTEM

The urban ecosystem is the result of human alteration of the natural system and is an environment intensely modified by social and economic development. The components of the urban habitat—climate, soils, hydrology, and biodiversity—are different from the surrounding nonurban areas. However, differences in the urban ecosystem are observed in line with the model of urbanization. Sprawl—the spatially-extended urban expansion over rural land—produces a city model based on high rates of energy consumption and associated disadvantages: Increased travel times and transport costs, pollution, traffic congestion, and broad transformation of the countryside.

The model has a major advantage over the dense city model: The conservation of large undeveloped land areas interspersed with the urban fabric, which locally retain former environmental attributes. These spaces are extremely varied. Green spaces such as natural parks, urban river corridors, formal gardens, recreational areas, sports parks, and street trees are important environmental assets for urban communities as they provide both recreational opportunities and ecosystem services, supporting local species and maintaining air quality. Some are remnants of past landscapes while others have been planted.

The course of urbanization produces a fragmentation of natural habitats, that is, wetland, grassland, woodland, and agricultural land turning into interstitial open spaces of semi-natural vegetation or farm land surrounded by urban, industrial, commercial, or residential land uses. Other areas with significant green space but dedicated to other uses are airports, cemeteries, golf courses, scientific and technological parks, and university campuses. Water courses, water bodies, lakes, and reservoirs are aquatic ecosystems that may or may not be attached to vegetated areas. Other areas offer a much lower environmental quality with low to medium concentrations of waste or pollution and highly variable physical, chemical, and biological soil characteristics: Derelict land, vacant lots, and brownfields—former industrial or commercial lands. Still other areas are highly degraded by dumps or densely crossed by infrastructures, yet they attract and keep some wildlife species.

THE CITYSCAPE

The mixture of built forms with the residual rural landscape yields a differentiated landscape known as a cityscape. The variety originates from combining land use categories, densities of population, and types of vegetated areas: Paved industrial, residential, commercial, grassy residential, dispersed residential, forested residential, agricultural, green areas, and vacant sites. The extensive network of roads, railroads, distribution pipelines, power lines, and dikes act as barriers that create habitat fragmentation by obstructing normal animal movement, increasing mortality, and isolating habitats and populations, threatening species with extinction. Essentially, this struggle for survival takes place in suburban areas, for city centers have already gone through this phenomenon.

There are a number of alterations and pressures acting upon species and humans in the built environment. The urban climate—the result of the modifications produced by the built structures and the combustion of fossil fuels—is defined by the heat island effect, which is an increase in temperature of up to nine degrees F (five degrees C), an increase in precipitation between five and 10 percent, a reduced total solar radiation, a higher precipitation runoff, a lower precipitation infiltration rate caused by the broad extent of pavement and other impervious surfaces, and a varied regime of wind speeds according to street orientation with respect to predominant wind directions.

AIR, WATER, AND LIGHT POLLUTION

Air pollution is caused by emissions derived from the burning of fossil fuels and industrial processes used by heavy industry, urban traffic, and household heating systems. Concentrations of complex atmo-

spheric pollutants, gases, and suspended particulate matter create a toxic atmosphere. The deposition of those airborne pollutants, by either wet or dry processes, produces an acidification and nitrification of soils and water bodies. The application of fertilizers to home lawns and urban gardens at similar rates to agriculture adds large quantities of nitrogen and phosphorous to the water bodies. The leading causes of reduced water quality are nonpoint source stormwater, or polluted runoff; sewer overflows; and nontreated wastewater point discharges flowing into water courses. They increase the loadings of nitrogen and phosphorous nutrients, raise temperatures, reduce visibility, and decrease dissolved oxygen levels, producing a decline in wildlife, hypoxia, and, eventually, harmful algal blooms. Artificial night lighting, a form of pollution produced by the illumination of buildings, commercial signage, and streetlights, reaches miles around cities, and has the biological effect of disturbing nocturnal species. Light pollution goes in all directions and interferes with feeding, predation, reproduction, and other activities of populations.

URBAN WILDLIFE

Disturbance is responded to with the settlement of tolerant and adapted species that represent earlier stages of ecological succession in the urban environment. The high frequency of disturbance often causes retrogression, thus these species are tolerant to stress. Species richness—the number, evenness, and relative abundance of species—sharply declines in the urban environment from the human disruption and habitat destruction, while total populations increase, creating a decline in ecological diversity. Specialist species are replaced by fewer synanthropic generalist species, some of them invasive, and benefiting from the more favorable climate, food abundance, and habitat. It has been observed how some wild avian, mammalian, or amphibian species adjust to the new urban environment by occupying emerging and growing ecological niches, a response termed synurbanization. These species are adapted to reduced territories, nonmigratory patterns, frequent changes in habitat and diet, and daily contact with people, for they possess an ecological, demographic, and behavioral plasticity differentiated from the rural populations. This response is progressively increasing diversity in urban wildlife.

MANAGEMENT

A range of measures and approaches have been conceived to implement an urban ecology policy. Habitat management includes landscape and watershed restoration of native forests, riparian woodland, and urban intertidal wetlands. Methods applied are revegetation with native species, stabilization of banks and slopes by reducing runoff, cleanup, and remediation. These measures bring back the functions of earlier habitats in densely settled regions, including water filtration and improvement of water quality by removing pollutants, flood control, and providing green space for recreation. Other measures are linking fragmented urban habitats isolated from each other with a corridor network to strengthen their ecological value, expanding forested areas, facilitating safe wildlife crossing and free movement, and preventing road kill by means of underpasses and overpasses. The creation of new habitat with bioengineering techniques to mitigate the destruction of another has not been demonstrated to be completely successful since it can fail to compensate for the loss because some properties have vanished due to the rarity or complexity of the initial ecosystem.

Urban management measures include regulation of development intensity and patterns using improved land use control and zoning, sewage and stormwater runoff treatment, landfill restoration, and land remediation by removal of contaminants in soil, groundwater, and sediments as well as implementation of transportation policies such as slower speed limits, efficient public transportation, car sharing, park and ride programs, and encouragement of change in fuel consumption.

Economic inequity and segregation across urban neighborhoods has environmental and social implications. Land values are higher next to designed or preserved green spaces, while affordable housing is limited to areas close to infrastructures, industrial areas, commercial areas, and impoverished areas with poorer recreational assets. The socioeconomic status of a family is a limiting factor to accessing housing in middle-class neighborhoods. The exposure

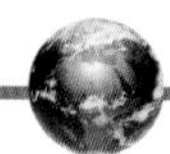

to poor ecological conditions in urban middle and low income neighborhoods may decrease the ability to assess the significance of biodiversity, and, thus, decrease support for diversity.

SEE ALSO: Anthropology; Cities; Ecology; Ecosystem; Eutrophication; Human Ecology; Political Ecology; Pollution, Air; Pollution, Water; Sociology; Urban Gardening and Agriculture; Urban Growth Control; Urban Parks Movement; Urban Sprawl; Urbanization.

BIBLIOGRAPHY. Marina Alberti, John M. Marzluff, Eric Shulenberger, Gordon Bradley, Clare Ryan, and Craig Zumbrunnen, "Integrating Humans into Ecology: Opportunities and Challenges for Studying Urban Ecosystems," *BioScience* (v.53/12, 2003); Jürgen Breuste, Hildegard Feldmann, and Ogarit Uhlmann, eds., *Urban Ecology* (Springer-Verlag, 1998); Oliver L. Gilbert, *The Ecology of Urban Habitats* (Chapman and Hall, 1989); Gordon McGranahan and David Satterthwaite, "The Environmental Dimensions of Sustainable Development for Cities," *Geography* (v.87/3, 2002); Rutherford H. Platt, Rowan A. Rowntree, and Pamela C. Muick, eds., *The Ecological City: Preserving and Restoring Urban Biodiversity* (University of Massachusetts Press, 1994); Herbert Sukopp, Makoto Numata, and A. Huber, eds., *Urban Ecology as the Basis of Urban Planning* (SPB Academic Publishing, 1995).

Damian White
Independent Scholar

Urban Gardening and Agriculture

WITH HALF THE world's population now living in urban areas and most of the world's best cropland already under cultivation, urban gardening and agriculture is rapidly becoming an important source of food for city dwellers. Researcher Margarida Correia notes, "urban agriculture marks a return to early cities, where food production was part and parcel of daily life."

Urban agriculture has the following general characteristics: It is located within or on the fringe of a town, city, or metropolis; grows, raises, processes, and distributes a diversity of food and non-food products; uses and reuses human and natural resources, products, and services largely found in and around that urban area; and in turn, supplies human and material resources, products, and services largely to that urban area. However, urban agriculture is more than just a way to produce food in cities. In terms of the three elements of sustainable development: Environment, economy, and society, urban gardens and agriculture provide a number of significant functions.

From an environmental perspective, urban gardens can make an integral contribution to the amount of green space in cities. Increased green space can assist in reducing airborne pollutants, thereby improving air quality. Urban gardens also create suitable habitat for a number of bird species. Common urban gardening techniques, such as regular crop rotation and crop-mix, discourage pest problems and reduce the need for pesticide use, leading to both health and environmental benefits. Related to this, locally-grown produce does not have to be transported long distances, therefore requiring fewer preservatives. Moreover, gardens absorb rainwater and stormwater, which reduces urban sewer loads. Communities participating in urban gardening initiatives also harvest rainwater and recycle grey-water from their homes for use in gardens. Urban organic waste can also be composted and used in gardens. Rooftop gardens in particular have been noted for their insulating effect, keeping buildings cooler in summer and warmer in winter.

Economically, urban agriculture has proven to be financially beneficial. Community-sourced food products lower family food expenses, especially in cities where most fresh produce is imported, and therefore costly. In addition, some community gardens sell a portion of their produce at local farmers' markets with the proceeds providing funds for related projects.

In many cases, urban gardens are community-driven initiatives. Neighborhoods or other social units collaborate in the maintenance, growing, and harvesting of these gardens, all of which provide excellent vehicles for community integration and pride. Other community benefits also emerge from these initiatives, such as youth programs, local

school activities, and job training for local residents and youth. Some gardens donate produce to local food banks, addressing access to fresh food in less advantaged communities. As experience with urban agriculture expands and diversifies, communities are integrating other social programs, such as prisoner rehabilitation.

Interrelated to all of these factors is the issue of food security. Cuba is frequently cited as an excellent example of how urban gardening and agriculture can be harnessed to provide food for urban and suburban populations. The collapse of the Soviet Union in the late 1980s drastically reduced agricultural inputs sent to Cuba, and in the immediately following years the national caloric intake of Cubans declined by one-third. In response, the number of urban gardens in Cuba has surged to about 8,000 nationwide and statistics indicate that urban agriculture produced 58 percent of the country's vegetables—over 1.7 million tons—in 2000.

Urban gardens have been planted in a diverse range of urban spaces: Patios and balconies, abandoned plots of land, rooftops, schoolyards, and more. Urban gardening initiatives have occasionally been fraught with land-use planning or zoning issues, such as in New York City in the 1990s, when over 100 vacated lots containing community gardens were due to be auctioned off by the city council. Over 60 of these were bought by a local organization and are now protected. In developed countries, the leaders in urban gardening initiatives are Germany and Switzerland. Esslingen in Germany has a bylaw requiring flat and sloping roofs—up to 15 degrees—to be vegetated. In Switzerland, a new law stipulates that all new buildings must relocate the green space taken up by the building's footprint to their rooftops; even existing buildings—some centuries old—are required to vegetate 20 percent of their roof surfaces.

SEE ALSO: Community Gardens; Cuba; Farmers' Markets; Gardens; Germany; Landscape Architecture; Switzerland; Urban Ecology; Vertical Ecology.

BIBLIOGRAPHY. Miguel A. Altieri et al., "The Greening of the 'Barrios': Urban Agriculture for Food Security in Cuba," *Agriculture and Human Values* (v.16/2, 1999); Margarida Correia, "Harvest in the City," *Earth Island Journal* (v.20/3, 2005); International Development Research Centre, www.idrc.ca (cited September 2006); Sonja Killoran-McKibbon, "Cuba's Urban Agriculture: Food Security and Urban Sustainability," *Women and Environments International Magazine* (no.70/71, 2006); L. Mougeot, "The Role of Urban and Periurban Agriculture in Urban Food Security and Poverty Alleviation," presentation given at United Nations Development Programme and Food and Agriculture Organization's "Food for the Cities: Parallel Event at Istanbul+5" (June 2001); Goya Ngan, *Green Roof Policies: Tools for Encouraging Sustainable Design* (Canadian Society of Landscape Architects, 2004); Oxfam International, "Havana's Green Revolution," www.oxfam.org (cited September 2006); Raquel Pinderhughes, "From the Ground Up: The Role of Urban Gardens and Farms in Low-Income Communities," *Environmental Assets and the Poor* (Russell Sage Foundation, 2000); World Resources Institute, "Inexhaustible Appetites: Testing the Limits of Agroecosystems," (July 2001), earthtrends.wri.org (cited September 2006).

Damian White
Independent Scholar

Urban Growth Control

THE TERM *urban growth control* is used to describe a broad set of growth management strategies intended to combat urban sprawl and its social and environmental consequences. For growth control advocates, it is generally the spatial expansion or "growth" of cities that necessitates "control," not the size of its population. Urban growth control may therefore be characterized as a set of land use planning policies meant to limit the suburbanization of metropolitan areas.

The urban growth control movement in the United States is a key element of "smart growth" policies and the Livability Agenda promoted by both Bill Clinton and Al Gore. These agendas are generally interested in producing cities that are compact and dense while at the same time livable, convenient, and pedestrian-friendly. As of 2005, 28 states had institutionalized growth management programs. However, in many cases the most rigorous growth

control legislation is administered through county and municipal governments.

The rapid horizontal expansion of cities to suburban and exurban areas is a major cause of concern for the urban growth control movement. Urban sprawl is characterized by a variety of distinct land use patterns. First, large-scale municipal and commercial developments such as wide streets, broad parking lots, large retail stores, and expansive office parks all consume large land areas. Second, low-density housing developments add urbanized land disproportionately to increased population. Third, homogenous housing and commercial development often result in low structural diversity and uniform building designs. And fourth, segregated, single-use zoning results in commercial, residential, and business zones that are separated by large distances.

There are many effects of urban sprawl. Urban growth control advocates argue that suburban sprawl diverts financial resources away from valuable urban infrastructure by funding new and upgraded road and highway projects; consumes open space with ecological value such as forests and wetlands; subdivides and impinges upon productive agricultural regions; produces a culture of single-occupant vehicle use as residents of suburban and exurban areas become increasingly dependent on automobiles to move between single-use areas and to and from the urban core; creates longer commutes, which in turn raises air pollution levels as well as driver fatality rates; and increasingly segregates the citizenry of metropolitan areas along class, cultural, and racial lines, most notably captured in the movement of white middle-class populations to suburban areas, a phenomenon called White Flight.

Opponents contend that urban growth leads to less traffic because the driving population is spread over a larger area, which in turn leads to lower pollution levels. They argue that urban growth control will lead to higher real estate prices inside the growth boundary, placing a burden on low- and middle-income households, and a loss of freedom by citizens to choose where they live and work.

A variety of growth management policy tool options are available to urban planners attempting to control sprawl. Traditional policy tools include zoning ordinances and land use regulations. Another common policy option necessitates the establishment of certain public facilities such as water, sewage, and electricity as a precondition to suburban development. Still another option consists of infill and redevelopment strategies in the urban core—especially high density housing options, mixed use development, and viable downtown transportation alternatives.

Urban sprawl's large parking lots and stores lead to dependency on cars to move between single-use areas.

Urban growth boundaries (UGBs), also referred to as urban and rural "limit lines," present another zoning policy option for controlling sprawl. The establishment of urban growth boundaries is an intentional effort to control urban sprawl by assigning the area inside the boundary for high-density settlement and the area outside for low-density development. Areas outside of the UGB are often referred to as greenbelts. Low-density development outside the UGB can be a misleading term. While the overall density is low, many growth management plans mandate smaller, dense settlement clusters surrounded by agricultural and/or open space outside the UGB.

Urban growth boundaries are not without controversy. In limiting growth outside of the growth boundary in favor of open space, critics argue that politicians are effectively telling landowners in these areas what they can and cannot do with their property. Legislation promoting UGBs is oftentimes contested by suburban and rural citizens

who argue that their private property loses value under restrictive land use policies. Citing the Fifth Amendment of the U.S. Constitution, these groups argue they should be duly compensated for land "takings." Others, including many urban residents, counter by citing public trust doctrines that stipulate that the ecosystem services provided by preserved spaces are in fact common property to be valued and used by everyone.

SEE ALSO: Land Use Policy and Planning; Public Trust Doctrine; Suburbs; Takings; Urban Sprawl; Urbanization.

BIBLIOGRAPHY. American Planning Association, www.planning.org (cited April 2006); Andres Duany, Elizabeth Plater-Zyberk, Jeff Speck, *Suburban Nation: The Rise of Sprawl and the Decline of the American Dream* (North Point Press, 2001); Dolores Hayden, *Building Suburbia: Green Fields and Urban Growth, 1820–2000* (Vintage, 2004); Jane Jacobs, *The Death and Life of Great American Cities* (Vintage, 1994); Sustainable Communities Network, www.smartgrowth.org (cited April 2006).

Gregory Simon
University of Washington

Urbanization

URBANIZATION, OR THE process by which cities grow, is one of the most important geographic phenomena in the world today. This is because the proportion of the world's population living in urban settlements is growing at a rapid rate, and many of the most significant economic, social, cultural, political, and environmental processes are increasingly occurring within and between the growing numbers of cities in the world today.

According to the United Nations (UN), by 2003, there were already 372 cities of a million or more people, 39 cities with over five million residents, and 16 cities over 10 million. Projections for the future suggest that, in 30 years, about 60 percent of the world's population will live in cities, though what defines a "city" and "urban" is subject to debate, because different countries use very different definitions. Still, even the most conservative estimates suggest that the world's urban population will grow from 2.86 billion in 2000 to 4.98 billion by 2030. Future projections suggest that, by 2030, there will be 500 cities with more than 1 million residents, 50 cities with over five million residents, and 20 cities over 10 million. Thus, we are living in an increasingly urbanized world with larger and larger urban populations.

GLOBAL TRENDS

While all parts of the world will become increasingly urbanized, there is a striking difference in the trends and projections on a regional basis. In certain parts of the world, like in Africa, Asia, and the Middle East, urban growth is taking place at a dramatic pace. In these regions, many of the largest cities are growing at annual rates between 4 and 7 percent, which means that they will double in size in only 10 to 18 years. For the most part, urban growth in these regions is a consequence of internal migration by massive numbers of rural residents seeking a better life in urban areas, as agricultural development problems persist, and cities become the engines of economic growth.

In other parts of the world, particularly in the largest cities of North America, Europe, and Latin America, there are strong indications that metropolitan growth rates have been slowing down. Of course, many of the countries in these regions already have high levels of urbanization; according to their own national definitions, between 75 and 90 percent of the population live in urban areas, so rates of urban growth are likely to be slower.

Even so, there is evidence to suggest that the largest cities in these regions are beginning to grow less rapidly, and in some cases, to lose their populations to mid- and small-sized cities. One case in point: Almost all the cities in Latin America with a million or more inhabitants, despite continued increases in the absolute number of residents, have had much slower population growth rates in the past two decades, including Mexico City, Buenos Aires, São Paulo and Río de Janeiro. So despite the fact that cities like São Paulo and Mexico City are still among the largest cities in the world, mid- and small-sized cities have begun to grow very rapidly, gaining population both from large urban centers and from rural areas. The same process is being repeated in North America and Europe as well.

Four main factors have prompted the slowing down of growth in many of the world's largest cities. First of all, governments have begun to employ deliberate decentralization policies because the unequal allocation of people and resources in large cities produces serious problems and regional unbalance. Second, many industrial plants have moved out of larger cities into smaller cities either by policy or to take advantage of cheaper land and labor. Third, improvements in telecommunication and transportation technologies have further increased the dispersion of manufacturing factories and residences away from principal cities. Finally, there is an overall trend toward suburbanization and population deconcentration, where low-density, ex-urban settlements, with their own shopping malls, factories, office parks, and entertainment facilities, predominate. Thus, the vast majority of new urban growth in North America, Europe, and Latin America has begun to, and will continue to occur in mid- and small-sized cities.

While Mexico City is still among the largest cities in the world, its population growth rate has begun to slow.

SOCIAL POLARIZATION

According to the UN, no matter where urban growth occurs, the economic contributions of cities are, and will continue to be, critical. Urban-based economic activities account for more than 50 percent of Gross Domestic Product (GDP) in all countries and up to 80 percent in more urbanized countries in Latin America, or more in Europe. Thus, cities and towns are not only the loci of production, but they are also the loci of the most important impacts of globalization and, hence, will be the places of change and expectations for the future.

Yet, ironically, today's cities are marked by social polarization, which may unwittingly place the economic and social futures of cities at risk. Social inequality is an integral and inevitable part of everyday life. In fact, in most cities today, this social inequality is actually on the rise. Many scholars argue that these trends are not merely incidental, but are inscribed as part of the global economy where the prosperity of the elite rests on the exploitation of the poor.

One manifestation of this social polarization is the fact that today's cities are characterized by great inequalities in income distribution, with the richest 10 percent of the population often earning 30–40 percent of the total income, and the poorest 50 percent earning less than 25 percent of the total income. With few exceptions, between 1960 and 2000, the majority of countries experienced a continued concentration of income within their populations. While absolute incomes for all groups have grown, according to the UN, the population in the top quintile has grown much faster than those in the bottom two quintiles.

This rising inequality in income has in turn created markedly different housing situations for the

rich and the poor in cities. While many of the richest residents separate themselves in gated communities, the poorest residents live in homogeneously poor public housing or informal housing communities. A long-term housing crisis has emerged in many cities of the world, which means many of the poorest families settle land illegally and build their own homes informally, especially on the fringes of major urban centers. A vast majority of these families hold no legal title, and many live in housing that is considered substandard or even unfit for settlement.

While land title regularization programs and access to more efficient formal markets have improved conditions for some, for many, the only option for affordable housing remains the informal sector. Thus, access to low-cost, quality land and housing is a major concern for most cities across the globe, where population growth and rapid urbanization predominate.

ENVIRONMENTAL DISTRESS

Most cities are also undergoing severe environmental distress. This distress is caused by a lack of resources, inadequate attention, and inefficient management practices. As a result, a variety of environmental problems exist in cities today. These include problems involving:

- Public land management: The constant reduction of green areas, which causes an excessive impermeability of soil and an increase in critical areas of flooding. The illegal occupation of watersheds is also causing groundwater contamination.
- Public transportation planning: Inadequate public transportation alternatives, including the expansion of subway networks and bus lines, means that the percentage of car users continues to rise, which leads to air, noise, and water pollution.
- Air quality: The lack of strict practical and short-term measures and policies, along with enforcement of environmental standards, causes overall high levels of industrial air pollution. A highly motorized and congested transport system also results in high levels of particulate air pollution.
- Sewerage: The delay in the completion of sewage master plans means that a significant percentage of residents are not connected to the sewer system. Thus, only a small percentage of sewage receives some sort of treatment in wastewater treatment plants.
- Water: Even though most residents have access to piped water, most water sources and waterways within cities are contaminated. Because of pollution problems coupled with growing demand, maintaining reliable supply is a problem, especially for those populations that live in areas prone to flooding.
- Solid waste: The thousands of tons of solid waste created every day by millions of residents means that conventional and formal methods for disposal are cost-prohibitive. This has led to open burning of undisposed waste, as well as soil, groundwater, and surface water contamination through run-off and leaking.

In all cases, environmental degradation is associated with ill-health effects, for both adults and children. In many cities today, residents suffer from respiratory ailments, vehicular deaths (from poor transport planning), and industrial accidents (from inadequate occupational safety standards). In the most impoverished pockets, where poor water quality, overcrowding, substandard housing, and underventilation prevail, adults and children also suffer and die from infectious diseases including diarrhea, tuberculosis, cerebrospinal meningitis, schistosomiasis, and skin infections. Because so many residents in today's cities are poor, the result is a highly "unlivable" environment.

TOWARD "LIVABLE" CITIES

Given that urbanization is inevitable, there is a dire need to come up with ways to make cities more "livable." Of course, "livable" means different things to different people, but most would agree that "livability" can be defined as an equitable urban environment that assures jobs close enough to decent housing with wages commensurate with rents. "Livability" also means having access to the services that make for healthful surroundings and personal satisfaction and happiness. Importantly,

"livability" depends on people having an effective say in the control and management of their urban environment. Thus governance, or the ways in which governmental and nongovernmental organizations (NGOs) work together, as well as the ways in which political power is equitably distributed in cities, becomes essential to creating "livable" cities.

In the end, it is an open question whether the economic advantages that can be found in the city result in improved quality of living for most residents, who are not a part of the elite population. Even though many cities have become centers for economic growth and wealth generation, most evidence suggests that a handful of people at the top of the social hierarchy reap most of the benefits. Few cities in the world today provide decent livelihoods and healthy habitats for a majority of ordinary people. On the contrary, income inequality is rapidly growing and environmental resources are being degraded on a great scale. Consequently, the changing form, economic base, environmental condition, and social structure of cities will continue to be of immense importance.

SEE ALSO: Automobiles; Cities; Heat Island Effect; New Urbanism; Pollution, Air; Pollution, Water; Poverty; Public Land Management; Sewage and Sewer Systems; Suburbs; Sustainable Cities; Transportation; Waste, Solid; Wastewater; Water Demand.

BIBLIOGRAPHY. Gary Bridge and Sophie Watson, eds., *The Blackwell City Reader* (Blackwell Publishers, 2002); Peter Evans, ed., *Livable Cities? Urban Struggles for Livelihood and Sustainability* (University of California Press, 2002); Peter Marcuse and Ronald van Kempen, eds., *Globalizing Cities: A New Spatial Order?* (Blackwell Publishers, 2000); Saskia Sassen, *Cities in a World Economy,* 2nd ed. (Pine Forge/Sage, 2000); Allen Scott, ed., *Global City-Regions* (Oxford University Press, 2002); John Short and Kim Yong Hyon, *Globalization and the City* (Addison Wesley Longman, 1999); United Nations Human Settlements Programme (UN-Habitat), *The State of the World's Cities, 2004/5* (UN-Habitat, 2005).

Emily Skop
University of Texas, Austin

Urban Parks Movement

PUBLIC URBAN PARKS are a product of a reform effort that emerged in the mid-19th century to ameliorate the living conditions of working people. In the United States, the best known park advocate was Frederick Law Olmsted, who, with his partner Calvert Vaux, conceived of and promoted the construction of Central Park in New York City (1858) and the Emerald Necklace in Boston (1878–80), as well as some of the most notable parks in other large cities in the United States. Never easy to fund, the case for parks was always pitted against the potential for profit from the undeveloped real estate, and the possibility that parks would attract lower classes into more affluent areas. Related conflicts continue to this day.

Over the course of the 20th century, the case for constructing urban parks has ebbed and flowed, evolving with changes in the affluence and demographic composition of neighborhoods. At the beginning of the century, interest in providing access to nature in densely urbanized areas began to shift toward an outdoor recreation model, wherein recreation facilities were developed to encourage fitness, team sports, and activities aimed at the acculturation of immigrant communities. Natural or naturalistic spaces were then less important, and were encroached upon by tennis courts, baseball fields, recreation halls, and other facilities. At the same time, there was an increasing interest in the preservation of nature and wilderness far from the urban centers. This movement is well known as the conservation movement, largely formulated under Theodore Roosevelt's administration (1901–09).

THE LATE 20TH CENTURY

As the country became more affluent after World War II and the federal government underwrote suburbanization, the groundwork was laid for the emergence of the environmental movement, including greater concern about ecological processes and the need to preserve wilderness and open spaces—including at the suburban fringe. Conservation approaches of the Roosevelt Progressive Era that espoused the use and long-term sustainable management of natural resources were replaced by a

Urban parks have come full circle from relieving crowded living conditions to remedying environmental problems.

politics of preservation for ecological values and for leisure. It is also during the late 1960s and into the 1970s that new models were experimented with, such as conservation easements, greenways, community gardens, and land trusts in and near urban areas. Large-scale national recreation areas adjacent to cities were also created from the newly established Land and Conservation Fund (1964) to offer natural settings for outdoor recreation for urban dwellers. There was a general shift in appreciation toward more natural settings that offered contact with local indigenous ecosystems such as the Santa Monica National Recreation Area, dominated by coastal chaparral and oak woodlands.

What constituted a park became harder to define, and the umbrella under which such as concept could be categorized broadened. Meanwhile, older cities were being depopulated by the middle class, and investments were being made in parks in suburbs, combining large open spaces with recreation facilities; in urban fringe open spaces; and in the preservation of remote "wild" lands. Urban parks, including such well-known ones as New York City's Central Park, suffered from lack of funds as many large cities went through fiscal crises in the 1970s.

With the rise of the environmental movement, there was an increasing recognition that natural processes, especially at the urban fringe, needed to be protected. Ian McHarg's *Design With Nature* (1970) was one such important intellectual milestone. McHarg pointed out that development could be designed to minimize environmental impacts if natural environmental processes were understood and considered in siting subdivisions. He pioneered the use of overlay maps showing streams and sensitive riparian corridors, for example, and where development could take place that would have the least ecological impact. While McHarg's analysis was influential intellectually, it remained at the margin of planning practice. Yet, it was important because it supported challenges to sprawl, and contributed to emerging efforts to preserve ecologically important (and other) open spaces in suburbanizing environments.

Large-scale subdivisions at the urban fringe, especially those catering to the middle class and upper middle class, felt obliged to provide open spaces and parks as part of the amenity package. They followed a formulaic offering of lawns, playing fields, meandering bicycle paths, and recreation facilities. In deteriorating urban cores, another phenomenon was developing: the rediscovery of urban gardens for food self-sufficiency on vacant and abandoned parcels. Neither of these different trajectories corresponded—understandably—to the early mission of urban parks to provide relief from insalubrious and crowded living conditions and an aesthetic respite from the industrial city. With urban diversity came an increasingly disparate set of approaches to public open spaces addressing the multiplicity of urban settings—the older, poorer urban core, the more affluent residential neighborhoods, older established suburbs, and expansion on the urban fringe.

THE 21ST CENTURY

By the turn of the 21st century, the variety of approaches to public open spaces has grown considerably, and encompasses diverse ideologies about nature, the role of public spaces, and the place (and type) of recreation in an urbanized context. The rise of new urbanism and return to the urban core of large cities have revived interest in urban parks as places for relief from city pressures, a place for nature, and other functions such as stormwater mitigation. The new sciences of conservation and restoration biology have also been a factor in reassessing the function and location of urban parks, leading

urban processes. Jacobs observed that high-density, irregular urban neighborhoods with a high level of pedestrian activity and interaction were diverse and innovative places. She critiqued the sterility of large public works projects, which encouraged urban exodus into sprawling suburbs and fragmented exurbs. Jacobs was on the winning side of preventing highway construction in Manhattan in the early 1960s. Some of Jacobs's ideas have found their way into new urbanism, with its elements of mixed land use and pedestrian-friendly neighborhoods.

Current and future urban planning may be focused on three areas: Core central cities through selected redevelopment or gentrification of decayed neighborhoods, sprawling suburbs through the rezoning of developments to have mixed uses with closer proximity of work and residence, and fragmented exurbs through the clustering of lots and the retention of common area open space. Examples of all three development types can be found in numerous locations around the earth, and viewed from space using satellite photography.

SEE ALSO: Garden Cities; Landscape Architecture; Land Use Policy and Planning; New Urbanism; Sustainable Cities; Urban Sprawl; Wright, Frank Lloyd.

BIBLIOGRAPHY. Peter Calthorpe and William Fulton, *The Regional City: Planning for the End of Sprawl* (Island Press, 2001); Jane Jacobs, *The Death and Life of Great American Cities* (Random House, 1961); David Rusk, *Inside Game, Outside Game: Winning Strategies for Saving Urban America* (Brookings Institution Press, 1999); Stephen V. Ward, ed., *The Garden City: Past, Present and Future* (E & FN Spon, 1992); Frank Lloyd Wright, "Broadacre City: A New Community Plan," in R. Legates, ed., *The City Reader* (Routledge, 1996).

Ron McChesney
Ohio Wesleyan University

Urban Sprawl

THE TERM *urban sprawl* is is used widely and inconsistently and is usually associated with negative connotations of urban expansion. The definition means, variously, unplanned loss of agricultural and other land to sub(urbanization) where there is minimal or no coordination of service and infrastructure provision, through to planned urban expansion that provides appropriate services and infrastructure but converts land to urban uses. The term could be more accurately expressed as *suburban sprawl.* As Richard Peiser notes, the term *sprawl* is used to mean different things, including "the gluttonous use of land, uninterrupted monotonous development, leapfrog discontinuous development and inefficient use of land." In terms of sustainability, each of these problems labeled *sprawl* invokes different solutions in order to make cities more sustainable.

HISTORICAL CONTEXT

The use of the term *urban sprawl* has increased as people have become more concerned about the environmental and social impacts of urban expansion and are advocating that cities become more sustainable. This is not to say that urban expansion is new. For example, relative to English cities, some American and Australian cities were spread over large areas in the 19th century. Whereas today the expansion of cities is often seen by governments, planners, and many ordinary people as a problem, the spread of the city was understood as being beneficial for health, sunlight, and to reduce the risk of disease. Spreading the city out was one way of overcoming the damp, unhealthy, overcrowded conditions of older European cities. It was also considered part of the moral health of citizens to garden and demonstrate pride in maintaining their dwelling and yard. This way of thinking was also important in England, where one of the common elements of many planned towns in the 19th and early 20th centuries (including Saltaire, Bourneville, Port Sunlight, Letchworth, and Welwyn Garden City) was the provision of space, gardens, and access to sunlight. Importantly, the reduction in urban densities was often accompanied by clear urban boundaries to prevent the city "spilling over" onto other land uses.

TRANSPORTATION

The spread of cities is closely related to the means of transport, and influenced by factors such as to-

pography, population growth, and industrial development. When walking was the only available and affordable means of transport, the urban density was very high. Improvements in transport enabled people to commute over longer distances. The expansion of cities such as Melbourne in the 19th century and Los Angeles in the early 20th century was due largely to the provision and affordability of train and/or tram/streetcar transport. The transport infrastructure in many cities was developed as a way of selling land for residential use. The later arrival of the automobile accelerated this process because it enabled infilling between rail lines and the outward growth of the city beyond the rail lines.

Los Angeles, which was once promoted as the vision of a healthy, wealthy, and uncrowded lifestyle, became associated with terms such as *automobile dependence.* Los Angeles has become a metaphor for sprawl—the specter of freeway cloverleaf interchanges is raised in many cities as a warning of what could occur if a city was permitted to "sprawl." Los Angeles is certainly spread out. It extends 131 miles (212 kilometers) along its east-west axis and covers 2,814 square miles (7,287 square kilometers) of land. Whether this spreading of an urban area is seen as positive or, as is implied through the use of the term *sprawl,* negative, depends on how an individual assesses the economic, sociocultural, and environmental costs and benefits of this form of urban development.

The spread of suburbia brings economic costs and benefits. In some cases, landholders on the fringe of the city sell their land after years of farming because this is their only means of supporting themselves in retirement. Various U.S. states now have programs to retain agricultural land and prevent its conversion to urban uses. The economic benefits also accrue to automobile manufacturers and associated industries and to construction and white goods industries. The economic costs of urban expansion include the loss of agricultural productivity, the costs to provide infrastructure and services for residents in outer suburban locations, the potential for transport congestion as commuting is predominantly automobile-based, and the potential loss of productivity if commuting times are longer.

The sociocultural benefits of urban expansion include the possibility of larger houses and more entertaining space, safe space for children to play, and the provision of space for other activities such as gardening. It is also less likely that residents from different socioeconomic classes would mix, which is seen as desirable by many people when selecting accommodation. On the negative side are aspects including the loss of community, the experience of social isolation (particularly for women), and the loss of identity as previously separate towns are "gobbled up" by urban sprawl. The process of urban expansion is also seen as self-perpetuating, in that low-density development means many modes of public transport are not viable and therefore people without access to private transport become further trapped and isolated in these dispersed locations where there are often insufficient or inappropriate amenities.

ENVIRONMENTAL IMPACTS

The environmental aspects of urban expansion are now perceived as mostly negative. This is because environmental regulation and technology, in the form of pollution abatement devices, noise insulation, and so on, have reduced the negative impacts of many urban and industrial activities. Previously, the spatial separation of perceived incompatible activities was considered crucial. Now, spatial separation is often perceived as "sprawl," which is a significant factor to consider in relation to sustainable cities because it has an impact on issues such as water, transport, and biodiversity and because it is one of the most visible aspects of the relationship between cities and other physical environments.

There has been significant debate in recent years about the desirability of various urban forms. At its simplest, the debate has been characterized as a compact-versus-dispersed-city debate. The dispersed city has higher environmental costs in biodiversity loss, its contribution to climate change, the provision of more infrastructure, and so on but is generally better for local air quality because pollution can often be dispersed by winds. The claim many advocates of centralized or compact urban forms make about the more efficient use of infrastructure is often disputed, because the infrastructure in many existing urban areas is dated or in poor condition.

The outward expansion of cities readily enables the provision of new infrastructure.

Recent attempts to reduce urban sprawl include Smart Growth and some new urbanist developments. These initiatives do not entirely reduce sprawl and have been criticized as adding to the problem or creating other problems. It is important to ask: What is the problem to be solved? It is apparent that over the past 50 years there has been a significant move from perceiving crowding and high urban densities as a problem to the understanding that the outward growth of cities is a problem. The use of the term *sprawl* is often very loose, but it draws attention to important issues that will occupy urban planners and policy makers for many years as the world's population grows and the need for sustainability is increasingly recognized.

SEE ALSO: Cities; New Urbanism; Suburbs; Sustainability; Sustainable Cities; Transportation; Urban Growth Control; Urbanization; Urban Planning.

BIBLIOGRAPHY. Lionel Frost, *The New Urban Frontier: Urbanisation and City Building in Australasia and the American West* (UNSW Press, 1991); George Gonzalez, "Urban Sprawl, Global Warming and the Limits of Ecological Modernisation," *Environmental Politics* (v.14/3, 2005); Phil McManus, *Vortex Cities to Sustainable Cities: Australia's Urban Challenge* (UNSW Press, 2005); Peter Newman and Jeffrey Kenworthy, *Sustainability and Cities: Overcoming Automobile Dependence* (Island Press, 1999); Richard Peiser, "Decomposing Urban Sprawl," *Town Planning Review* (v.72/3, 2005); Bernard Salt, "LA Likeness Lingers in La-La Land," *Australian* (July 28, 2005); Michael Southworth, "New Urbanism and the American Metropolis," *Built Environment* (v.29/3, 2003); Karen Till, "New Urbanism and Nature: Green Marketing and the Neotraditional Community," *Urban Geography* (v.22/3, 2001); Harriet Tregoning, Julian Agyeman, and Christine Shenot, "Sprawl, Smart Growth and Sustainability," *Local Environment* (v.7/4, 2002).

PHIL MCMANUS
UNIVERSITY OF SYDNEY, AUSTRALIA

Uruguay

LOCATED IN THE southern part of the continent of South America, Uruguay has a land area

The Paper Mill Protests

In 2003, ENCE, a Spanish company, received permission from the government of Uruguay to build a cellulose processing plant on the Uruguay River at Fray Bentos. Two years later, a Finnish multinational company, Metsa Botnia, was also given approval to build a similar plant. As the Uruguay River is shared by both Argentina and Uruguay, and Argentines were worried about possible pollution of the river, protests began. Argentines argued that the use of the river was governed by a bilateral treaty, and the Uruguayan government would have to seek permission from the Argentine authorities, which they had not done. The Uruguayans denied that there would be any environmental damage to the river.

On April 30, 2005, some 40,000 Argentine activists from the town of Gualeguaychu, north of Buenos Aires, blocked the Libertador General San Martín Bridge connecting the town to Fray Bentos as a protest. Soon Argentine politicians became involved and on December 23, the bridge was again blocked, this time along with another nearby bridge. The Uruguayan chancellor Reinaldo Gargano criticized Argentina for violating the regulations of Mercosur—the regional trade grouping—which allowed for free circulation of goods.

In early 2006, the situation escalated with the involvement of international environmental groups such as Greenpeace. The border was again blockaded by the Argentines, in particular against the Finnish mill. Protests increased during the year and finally ENCE announced that it would relocate its works elsewhere. With Uruguay trying to protect one of the largest foreign investments in the country, the dispute continues.

of 68,039 square miles (176,220 square kilometers) and an estimated population in 2006 of 3.4 million, most of them concentrated in Montevideo, the capital. Uruguay is a flat and fertile plain interrupted only by small elevations in the south and east that do not exceed 1,640 feet (500 meters). The coast is low and sandy, and the climate mild with occasional strong winds. The dense fluvial network, dominated by the Uruguay River, and adequate precipitation explain why around 70 percent of the country is constituted by natural pastures grazed by the livestock introduced by the Spaniards in the 17th century,

Many of the environmental problems of Uruguay, a predominantly agrarian country and increasingly also a tourist destination, have been traditionally related to transboundary pollution from neighboring Brazil, especially the acid rain produced by the coal power plant of Candiota, which affects approximately one-fifth of the country. Soil erosion by wind has also been a traditional concern for ranchers (especially in the Department of Canelones), but recently it has begun to affect the expanding agricultural areas planted with soya beans on the highly erosion-prone soils of the eastern side. In 2003, only 0.4 percent of the total land area of Uruguay enjoyed some degree of environmental protection.

Water pollution by food processing industries (chiefly meat and meat products) is also significant. Critical areas in this respect are the Santa Lucia Basin (providing around 60 percent of the urban water supply of the country) and the urban basins near Montevideo. The capital suffers from air pollution originated by the oil refinery and the thermal power plants located in the vicinity. However, Uruguay obtains most of its energy from hydropower produced in the big dams of Rio Negro and Salto Grande (shared with Argentina) on the Uruguay River. In periods of drought, coal and oil power plants supply the energy needed.

During the 1990s, the expansion of forest land, advised by the World Bank and encouraged by the state with the objective of attracting foreign companies and developing the pulp and paper sector, was one of the key environmental issues in Uruguay. Rapid growth species such as eucalyptus have been introduced and benefits provided to foreign investors in the form of economic subsidies for plantations on specially designated areas. In turn, this has created wood and cellulose surpluses for export and possible carbon sinks for climate change policies.

However, the momentum gained by forestry faces opposition as well, especially from Argentina. In 2006 the proposal to build two paper and cellulose factories on the Uruguay River raised strong opposition by the neighboring country (a large proportion of the border between Argentina and Uruguay is formed by the Uruguay River). These factories represent the highest investment in Uruguay's history (about $1.8 billion) and are planned to produce more than 1.5 million tons per year of paper and cellulose. On the other side of the river, Argentine citizens argue that the pollution caused by these factories would ruin their agricultural and tourist activities. Argentina threatened to take this case to the international court of The Hague.

SEE ALSO: Acid Rain; Argentina; Brazil; Hydropower; Plantation Forestry; Pollution, Air; Pollution, Water; Pulp and Paper Industry; Soil Erosion; Tourism.

BIBLIOGRAPHY. Central Intelligence Agency, "Uruguay," *World Factbook*, www.cia.gov (cited April 2006); Timothy Doyle, *Environmental Movements in Minority and Majority Worlds: A Global Perspective* (Rutgers University Press, 2005); United Nations Development Programme, "Human Development Report: Uruguay," hdr.undp.org (cited April 2006); World Bank, "Uruguay," *Little Green Data Book*, lnweb18.worldbank.org (cited April 2006).

David Sauri
Universitat Autònoma de Barcelona

Use Value versus Exchange Value

USE AND EXCHANGE value are two different measures of the value of resources, goods, or services for humans. *Use value* refers to the actual use of something; for example, fruits and vegetables have a use value in providing nutrition for people. *Exchange value* refers to the price on the marketplace; for example, a commercial farmer grows crops for

their exchange value. Anything that does not exist in commodity form has no exchange value, even if it has use value. This may be because property rights and/or markets do not exist, either because they are difficult or impossible to establish (e.g., clean air to breathe, many kinds of knowledge), or because there has been little or no interest in a given society to develop them (e.g., many societies did not provide for market exchange of land until they were colonized). Also, people lacking means of exchange (i.e., the poor) may place a high use value on basic necessities but be unable to pay for them. Therefore, the highest exchange value can be realized from selling a commodity to rich people, even if they have little need for it. Thus, exchange value is not an accurate measure of use value.

Mainstream economics usually assumes that exchange value *is* the best available measure of use value, however. This has both environmental and social implications. Environmentally, economic "efficiency" leads to treating noncommodified components of nature as if they had zero value, including clean air and water, scenic beauty, biodiversity, the "existence value" of species, and human health. The destruction of such values is ignored by indicators of economic welfare and economic growth (e.g., Gross National Product). Economic growth policies that ignore these adverse effects promote environmental destruction. Cost-benefit analysis attempts to overcome this problem, but only by arbitrarily defining exchange values of things for which no markets exist.

In social terms, the assumption that exchange values are a good measure of use values helps to justify an allocation of resources that favors the wealthy. The concept of Pareto efficiency claims that the most efficient allocation of resources exists when any change in allocation would lead to the loss of economic welfare of at least some individuals. This ignores that, if some amount of money were transferred from rich to poor people, the latter would surely gain more use value than the former would lose. Hence, arguments for economic efficiency tend to justify an unequal distribution of resources, such as highly unequal land ownership favoring large export-oriented plantations that typically apply large amounts of agrochemicals, at the expense of small peasants trying to eke out a living on the marginal land not occupied by the plantations. A large portion of peasant production is subsistence-oriented, meaning that it is devoted to use and not exchange value, and hence tends to be regarded as "unproductive."

The distinction between use and exchange value also has a gender dimension, in that women working in the household, or in subsistence-oriented agriculture, are often classified as unproductive because they do not generate exchange value. The greater difficulties they may face as a result of environmental degradation (e.g., long treks to collect fuelwood) are therefore also underestimated in economic calculations.

In fact, even in the wealthiest countries, a very large portion of production occurs within households (e.g., cooking) or within contexts of mutual aid and gift exchange (e.g., the free exchange of knowledge among scholars). Hence, as particularly pointed out by J.K. Gibson-Graham, a focus on the money economy (and thus exchange value) alone ignores a huge portion of the real economy, severely restricting any attempts to effect positive change.

In principle, the most severe discrepancies between use and exchange values could be overcome if: (1) most things of use value were made into commodities, and (2) income distribution were made more equal. However, it is hardly desirable to make everything into a commodity, and a larger degree of equity in income is hard to achieve. Furthermore, even use value is a utilitarian concept; people may wish to preserve spiritual or other values in nature that cannot be reduced to use value (because nature is not there just to serve humans). Hence, nonmarket mechanisms, including actions by government as well as civil society, are needed to preserve many environmental resources that have little or no exchange value.

SEE ALSO: Cost-Benefit Analysis; Economics; Historical Materialism; Markets; Marx, Karl; Resources; Subsistence.

BIBLIOGRAPHY. Herman E. Daly and Joshua Farley, *Ecological Economics: Principles and Applications* (Island Press, 2004); Paul Ekins and Manfred Max-Neef, eds., *Real-Life Economics: Understanding Wealth Creation* (Routledge, 1992); J.K. Gibson-Graham, *The End*

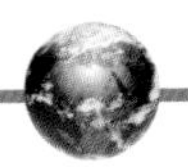

of Capitalism (as We Knew It): A Feminist Critique of Political Economy (Blackwell, 1996); E.F. Schumacher, *Small Is Beautiful: Economics as If People Mattered* (Blond and Briggs, 1973).

Wolfgang Hoeschele
Truman State University

U.S. Geological Survey

THE U.S. GEOLOGICAL Survey was established on March 3, 1879, during the last minutes before the close of the final session of the 45th Congress, when President Rutherford B. Hayes signed the bill appropriating money for the Survey. The inclusive bill included a brief section establishing the new agency, the U.S. Geological Survey. Under the Department of the Interior, it was created to oversee the "classification of the public lands, and examination of the geological structure, mineral resources, and products of the national domain." The legislation to create the Survey developed from an 1878 report of the National Academy of Sciences, which had been requested by Congress to provide a plan for surveying the American Western Territories.

By 1867, America's emerging industries were making huge demands on its natural resources, so the Commissioner of the General Land Office, J. Wilson, declared that the development of geological characteristics and mineral wealth was of the highest concern to the American people, and Congress authorized western explorations in which geology would be the principal objective. These General Land Office surveys were to include a study of the geology and natural resources along the 40th parallel route of the Transcontinental Railroad under the auspices of the Corps of Engineers and a geological survey of the natural resources of the new state of Nebraska. Clarence King, the first director of the U.S. Geological Survey, would later say:

> 1867 marks, in the history of national geological work, a turning point, when the science ceased to be dragged in the dust of rapid exploration and took a commanding position in the professional work of the country.

The Geological Survey was fashioned to unify and centralize the work undertaken by these important field surveys across the American West. From 1868 to 1870, the King and Hayden Surveys received funding for exploration in Wyoming and Colorado, and in 1869 the bureau was placed directly under the Secretary of the Interior. In 1870, Hayden presented to Congress a plan for the geological and geographical exploration of the Territories of the United States. With Congressional authorization, the Hayden Survey became the Geological and Geographical Survey of the Territories.

By 1870, two more surveys had taken place—Professor of Geology John Wesley Powell with a party of nine men left Green River, Wyoming, in three small boats to explore the unknown canyons of the American southwest under private sponsorship. Between 1867 and 1868, he had explored the Rocky Mountains in Colorado and eastern Utah and decided to explore these unknown canyon lands in boats. In a legendary and troubled trip down the Green and Colorado Rivers, Powell and five remaining members completed the journey through the Grand Canyon on August 13, 1869. In 1870, Professor Powell received an appropriation of $10,000 from Congress to make a second trip down the Colorado, being required only to report his results to the Smithsonian Institution. On June 10, 1872, Congress appropriated another $20,000 for completion of the survey.

In 1869 and 1871, expeditions were led by Lieutenant George Wheeler, an Army Engineer who explored California, Nevada, and Arizona. He surveyed the American West from south and east of White Pine, Nevada, to the Colorado River to create maps of wagon roads and military sites. Two years later, he was sent to explore the land south of the Central Pacific Railroad in eastern Nevada and Arizona. The extensive maps and information from these four surveys represent the foundation for the establishment of the Geological Survey.

The responsibilities of the bureau include exploratory surveys of geologic structure, the preparation of geological and topographical maps; the examination, classification and evaluation of natural resources; water studies to provide irrigation and water power; the organization of public lands; and the investigation of natural hazards, all related to

the publication of papers, bulletins, and maps based upon these surveys. In 1962 the Survey was authorized to conduct surveys on private lands. The Survey also serves the United States by:

> providing reliable scientific information to describe and understand the Earth; minimize loss of life and property from natural disasters; manage water, biological, energy, and mineral resources; and enhance and protect our quality of life.

SEE ALSO: Army Corps of Engineers; Geology; Hydropower; Irrigation; Minerals; Mining.

BIBLIOGRAPHY. Mary C. Rabbitt, "The United States Geological Survey: 1879–1989," *U.S. Geological Survey Circular 1050*, www.usgs.gov (cited January 2007); U.S. Department of The Interior, *A Brief History of the U.S. Geological Survey* (U.S. Department of The Interior, 1974); U.S. Geological Survey Committee of Review, David Jack Cowen, and National Research Council, *Weaving a National Map: A Review of the U.S. Geological Survey Concept of the National Map* (National Academy Press, April 2003).

Thomas Paradise
University of Arkansas

Usufruct Rights

THE LEGAL TERM *usufruct right* is derived from ancient Roman law and means the temporary right to use to the property of another as long as the nature of the property is unchanged. In terms of natural resources, when a landholder gives or sells the usufruct rights to another person, that person may use the resources on the land in various ways, including planting and harvesting crops, grazing livestock, and collecting forest products. *Usufruct rights* refer to levels of control a person has over property, while not owning the property. In addition to usufruct, or use rights, other levels of control a person might have over property include the right of exclusion (the right to prevent other people from using the resource) and the right of alienation (the right to sell and give away the resources). A person with usufruct rights to natural resources may also have the right to exclude other people from using the resources, and may (but not always) have the right to sell the use of the resource.

Property rights define relationships between people and mediate their use of property in terms of "who can (and cannot)" "do what" and under "what circumstance" to the specific property. As such, property rights are usually considered social relationships between people with respect to a thing, but not between people and a thing. Consequently multiple aspects of social and cultural institutions influence the configuration of property rights, including usufruct rights.

The usufruct rights to lobster fisheries in northeastern Nova Scotia provide an excellent example illustrating the social nature of property rights. Describing the system of property rights that characterize the valuable lobster fisheries in St. George's Bay, Nova Scotia, John Wagner and Anthony David found the term *kindness* was used to refer to usufruct rights. This term could be traced back to 18th century when a large number of Scottish people immigrated to Nova Scotia. A *kindness* was considered as the right of occupation and use of the land, but not the right of ownership. Transferred to the fishing industry, the right to harvest lobsters in certain parts of the bay was also based on *kindness*. In today's generation these rights are referred to as "gentlemen's agreements." As a form of usufruct right to resources, *kindness* is best understood as a cultural system in which resources users are motivated to perpetuate and sustain the system of access to resources (effectively their economic livelihoods) through social and cultural institutions that sustain family and community values. This property system provides a well-defined set of rules that ultimately constitute an essential component of an effective management system.

Usufruct rights are often associated with customary law and common pool resources, which were common in preindustrialized America and Europe and are still active forms of law and resource control in many developing nations. Oftentimes, usufruct rights to hunt, forage, and plant crops on commonly held property are seen as an insecure and an economically inefficient form of tenure. Market-driven arguments posit that when resources are managed under a usufruct system, it obstructs investment aimed

at improving the long-term quality or value of the land. As a result nation states around the world have sought to privatize land ownership. The most classic case of privatizing land while removing all use rights took place between the 12th and 19th century with the enclosure of the English commons. The motivating force was one of commoditization of land and its resources. As the English landowners brought together their scattered parcels of land and privatized the ownership of the land, former ways of subsistence living that used the commons for grazing, hunting, and collection of fuel were abolished. From a state perspective, privatizing land is a critical part of establishing state legitimacy in that it creates a census of people and their holdings, generates a class of taxpayers, and is believed to introduce land security. However, this process of land settlement also erases all usufruct rights, and as a result a class of landless people has emerged around the world.

In many parts of the world farmers still rely on usufruct rights and other customary forms of resource tenure for their subsistence livelihood. Contemporary empirical studies increasingly demonstrate that, in both industrialized and developing countries, common pool resources and usufruct rights associated with them have been and continue to be a successful way to manage natural resources.

SEE ALSO: Enclosure; Land Tenure; Private Property; Property Rights; Resources; Subsistence.

BIBLIOGRAPHY. Amity Doolittle, *Property and Politics in Sabah, Malaysia: Native Struggles over Land Rights* (University of Washington, 2005); Bonnie J. McCay and J.M. Acheson, eds., *The Question of the Commons: The Culture and Ecology of Communal Resources* (University of Arizona Press, 1990); Robert McNetting, "Unequal Commoners and Uncommon Equity: Property and Community among Smallholder Farmers," *Ecologist* (v.27/1, 1997); John Wagner and Anthony Davis, "Property as a Social Relations: Rights of 'Kindness' and the Social Organization of Lobster Fishing among Northeastern Nova Scotian Scottish Gaels," *Human Organization* (v.63/3, 2004).

Amity A. Doolittle
Yale School of Forestry and
Environmental Studies

Uzbekistan

ECOLOGY AND UZBEKI national identity are intimately tied together. The forced cultivation of cotton—a cash crop that requires large amounts of water—and the diversion of the Aral Sea's two main feeder rivers, the Amu Darya and the Syr Darya, by Soviet authorities made the resulting Aral Sea environmental disaster a major, if not predominant, national concern. The Aral Sea, which is divided in half by Kazakhstan in the north and Uzbekistan in the south, was at one time the fourth-largest body of landlocked water in the world. In addition to providing a steady supply of water in a relatively arid region, the Aral supported a productive and important fishing industry. Perhaps even more important for Uzbeki national identity, the Aral Sea has had an important historical significance for the Uzbekis.

Now, however, cities like Moynaq that were on the banks of the Aral in the 1960s are some 100 miles away from the shrinking sea. Since the 1960s and large-scale, seemingly deliberate Soviet diversion of water away from Central Asia, the Aral Sea has lost around 60 percent of its volume. With accelerating demands from agriculture and industry the water level drops some 11 inches a year. If current trends continue, there will be nothing left of the mighty Aral but a vast salt desert.

Aggravating the Uzbeki sense of national ecological betrayal even further was the final shelving by Soviet authorities in the 1985 of the Sibaral Project, a project that promised to divert Siberian rivers back into the Aral.

Some of the first independent Uzbeki intellectuals and dissidents against Soviet rule used the Aral Sea as a symbol of Soviet exploitation. Sagdulla Karamatov wrote the novel *The Last Sand Dune* in 1983 as a veiled protest against Soviet policies. The writer Mamadil Makhmudov in *Today and Tomorrow* called the Aral environmental crisis the result of "limitless demands, unjustice and unfairness" by central Soviet authorities and modern Russia. The poet Zulfia Mominiva in *I'm Grateful for Your Lessons* used the name of the Aral—*ar* means dignity, *al* means to take away—to protest the destruction of Central Asian identity and dignity for the sake of northern, centralized power. The drying of the

Aral Sea was like the drying of the Uzbeki spirit under Soviet oppression. The demands for quotas of cotton and the exploitation of Uzbeki labor along with Uzbeki water caused further popular divisions with central rule and continued Russian attempts to dominate the Central Asian region.

Despite popular dissidence and national anguish, the Aral Sea continues to evaporate. The current climate of largely totalitarian rule by President Karimov has not led to any significant changes in the Aral Sea crisis. Nor does it seem likely that a weakened Russia would be willing to divert its water supplies to the independent state of Uzbekistan. In addition, Uzbekistan faces the potential for new environmental crises with the expansion of the oil and gas sector and the prospect of unsustainable industrial and urban development.

SEE ALSO: Aral Sea; Cotton; Kazakstan; Lakes; Ob-Irtysh River; Russian (and Soviet Union); Water Demand.

BIBLIOGRAPHY. Tom Bissell, *Chasing the Sea* (Pantheon Books, 2003); Peter Craumer, *Rural and Agricultural Development in Uzbekistan* (The Royal Institute of International Affairs, 1995); Arun Elhance, "Conflict and Cooperation Over Water in the Aral Sea Basin," *Studies in Conflict and Terrorism* (v.20, 1997); Rusi Nasar, "Reflections on the Aral Sea Tragedy in the National Literature of Turkestan," *Central Asian Survey* (v.8/1, 1989); Anita Sengupta, *The Formation of the Uzbek Nation-State* (Lexington Books, 2003).

Allen J. Fromherz, Ph.D.
University of St. Andrews

Vaccination

VACCINATIONS ARE USED to prevent infectious diseases. A weakened form of the infectious pathogen is used to stimulate the body's immune system to manufacture antibodies enabling the body to defend itself against the infectious pathogen. Vaccinations work because the immune system in the bodies of animals, birds, or humans is able to utilize the immunogen that is introduced to make antibodies. The antibodies destroy or neutralize the infectious agent whether it is a virus, bacteria, fungi, or some other pathogen. There are four types of vaccines. Those that contain bacteria or viruses that have been altered are called live attenuated vaccines. Vaccines that contain only parts of the infectious bacteria or virus are component vaccines. Killed vaccines use bacteria or viruses that have been killed. Toxid vaccines use toxins that the pathogen makes to neutralize it.

Vaccinations may be given orally or with a hypodermic needle. Oral vaccines may be used because they are cheaper, and do not have the risk of an injury or infection from injection. The polio, rotavirus, brucellosis, and cholera vaccines have been successfully administered orally. Many vaccines given to children, especially those under two years of age, have been given orally.

Vaccination with hypodermic needles is also widespread. Alexander Wood and Charles Gabriel Pravaz invented this form in 1853; prior to that time a cut was made in the skin for inoculation. The vaccine shots may be administered in the muscles that surround the stomach. Rabies vaccination is administered in this manner. Other vaccinations are administered in the shoulder muscles of the arms, or in the hips. Whether a vaccination is administered orally or with a hypodermic needle generally depends upon where the vaccination can be most productive in triggering the immune system to work. Many vaccines are not absorbed well if given in the stomach. Others are more effective if given orally rather than hypodermically.

The development of vaccines began in the late 1700s after it was noticed that milkmaids who developed cowpox were immune to smallpox. The practice of inoculating people with infectious material from a mild, but active, case of smallpox was developed before the advent of vaccinations. Edward Jenner, who coined the term *vaccine* from the Latin word for cow—*vacca*—introduced the safer method of inoculation with cowpox, which

eventually led to banning smallpox vaccinations by the middle of the 19th century.

Opposition to vaccinations has occurred in many times and places. The mandating of compulsory vaccinations by governments has added to the controversies. During the Colonial era, the Boston printer James Franklin used his newspaper to attack inoculations by distorting the number of deaths in the Inoculation Controversy. More recent controversies have arisen in response to a number of vaccines. The manufacture of vaccines seems to inevitably involve the use of materials to which a few individuals are extremely sensitive. Deaths or medical injuries have occurred, but while tragic for those individuals who die or are injured, the vast majority of people benefit because they do not die from the disease or suffer harm from its effects. In recent decades, the use of mercury in vaccinations may have contributed to the rise of autism. Other materials used to manufacture vaccines have had negative effects.

Vaccinations may be given to immunize against a disease, but others are given to minimize a disease already contracted. Louis Pasteur's first vaccination was administered to a child who had been injured by a rabid dog. The weakened form of the rabies virus he administered triggered the immune response to work more rapidly than the original infection. Therapeutic vaccines against the HIV/AIDS virus, non-Hodgkin's Lymphoma, and other diseases have had some success.

SEE ALSO: Antibiotics; Disease; Pasteur, Louis; Smallpox.

BIBLIOGRAPHY. James Keith and Keith Colgrove, *State of Immunity: The Politics of Vaccination in Twentieth-Century America* (University of California Press, 2006); David Kirby, *Evidence of Harm: Mercury in Vaccines and the Autism Epidemic—A Medical Controversy* (St. Martin's Press, 2006); Richard Neustaedter, *The Vaccine Guide: Risks and Benefits for Children and Adults* (North Atlantic Books, 2002); Peter Parham, *The Immune System* (Taylor & Francis, Inc., 2004); Stanley A. Plotkin and Walter A. Orenstein, *Vaccines* (Elsevier Health Sciences, 2003).

Andrew J. Waskey
Dalton State College

Values, Environmental

VALUES REPRESENT AN individual's judgment about what is valuable or important based on his or her principles or standards. Environmental values, ethics, and worldviews are human social-psychological constructs informed by people's inner experiences and their personal reasoning about nature. Environmental ethics are the moral judgments and attitudes that guide people in the way they behave toward nature. By comparison, value systems and worldviews are the reference frameworks through which people interpret their experiences and make them meaningful. The terms *environmental ethics* and *environmental values* are often used interchangeably. While environmental values are said to be formed early on in life, a person's worldviews are based on his or her broader social and political experiences and are, therefore, formed later in life. Worldviews relate to people's beliefs about the reality of the world, how the world behaves, their notions of justice, and what they think is right and wrong. Collectively, the values, ethics, and worldviews held by groups of individuals shape social identity and culture.

The origin of the concept of values expressed as an environmental ethic can be traced to Aldo Leopold's book *A Sand County Almanac and Sketches Here and There*. Although published shortly after his death in 1949, the book received greater popularity when re-released in 1970. The 1970 edition also included several influential essays by Leopold such as "The Land Ethic" in which he proposed that the causes of the ecological crisis were philosophical. Leopold also discussed an evolution in ethics from a focus on relationships between individuals, to relationships between individuals and society, to relationships between individuals, society, and the environment. Leopold expressed this environmental ethic as: "A thing is right when it tends to preserve the integrity, stability, and beauty of the biotic community. It is wrong when it tends otherwise."

Two other landmark papers published in the journal *Science* in the late 1960s also fostered the debate on environmental ethics: Lynn White's "The Historical Roots of our Ecologic Crisis" (1967), and Garrett Hardin's "The Tragedy of the Commons" (1968).

Environmental worldview theory emerged in the late 1970s and was underpinned by social scientists Riley Dunlap and Kent Van Liere's New Environmental Paradigm (NEP) scale, published in 1978. The NEP was developed in response to the anti-environmental worldview referred to by Dennis Pirages and Paul Ehrlich in 1974 as society's Dominant Social Paradigm (DSP). The NEP continues to be used extensively in research on environmental worldviews and is considered to be one of the more popular scales for measuring environmental beliefs and charting public attitudes toward the natural environment. In the early 1990s, social scientists Paul Stern and Thomas Dietz were among the first to propose a value-basis theory for environmental concern. These conceptual links have since been expanded to incorporate considerations of attitudes and beliefs.

Social scientists have demonstrated that environmental norms, values, and attitudes are correlated with environmental behaviors. These research findings support the premise that developing an understanding of people's values can help to understand the strength of their commitment to environmental issues and also predict when their environmental attitudes will be translated into environmentally relevant behaviors. Environmental value scales represent the most common form of measurement used to predict people's activities, consumer behavior, and/or economic sacrifice made to protect the environment. Environmental value scales can be broadly categorized into the following three forms: (1) values based on self-interest; (2) values based on concern for others; and (3) values based on concern for ecosystems.

Self-interest values are represented by egocentric and egoistic values. These values support the extraction and use of nature by individuals to enhance their own lives and the lives of their families. While they promote the protection of environmental aspects that provide personal benefits, they oppose the protection of environmental aspects that result in high personal costs. Garrett Hardin's theory of the "Tragedy of the Commons," where farmers' actions are governed by their self-interest, is underpinned by an egocentric ethic.

Values based on one's concern for others are represented by homocentric, anthropocentric, and social-altruistic values. These values support nature for its role in maintaining or enhancing the quality of life for humans including, for example, its role in providing clean air, water, and fossil fuels. As human-centered value orientations, these values promote social justice and maximizing the social good for all people. They support the view that people's attitudes toward nature or environmental policies should be judged on the basis of how human beings are affected by them. As such, they consider the well-being of other living creatures to be of lesser, if any, importance.

Values based on one's concern for ecosystems or the biosphere include biocentric, biospheric, and ecocentric values. These values assign intrinsic worth to all aspects of the environment (inanimate and animate) and they consider the survival of all living and nonliving things as components of healthy ecosystems to be of primary importance. Ecocentric value orientations value nature (in the form of ecological wholes such as species, ecosystems, and the biosphere) for its existence, aesthetics, and spiritual value, regardless of its ability to satisfy human needs. Ecocentrics identify with a connectedness between themselves and nature, as exemplified by Aldo Leopold's land ethic in his book *A Sand County Almanac*. Similar to environmental value scales, environmental worldviews range from those that are more anthropocentric (humans ruling over and manipulating nature) to those that are more ecocentric (humans connecting with and coexisting equally with nature).

Anthropocentric worldviews include the technocentric, mechanistic, cornucopian, and the accommodationist or managerialist worldviews. The technocentric worldview is considered the dominant worldview in Western organizations and is the assumed underlying position for conventional scientific method. It considers the world to be objectively knowable through the study and measurement of its parts, and that technological advances can overcome environmental problems. Similarly, the cornucopian worldview sees humans, through their ingenuity and technology, using nature to provide indefinitely for their needs and wants. The accommodationist or managerialist worldview sees using improvements in environmental legislation and environmental or ecological management practices as the way to accommodate or manage human impacts on nature.

Ecocentric worldview orientations include the ecocentric, communalist or ecosocialist, and gaianist

or utopian worldviews. People who contend that the natural world consists of ecosystems that should be managed as such typically hold the ecocentric worldview. It sees the world and organizations and communities in it as being organized into interdependent systems. The communalist or ecosocialist worldview shows concern for ecologically sustainable development and distributive social justice. It also sees small-scale technologies directing environmental management and production, and providing the economic resources for all people to sustain an equitable standard of living. The gaianist or utopian worldview is an extreme ecocentric worldview that considers the land and all living things to be equal and which promotes the rights of nature. This is reflected in the deep ecology movement initiated by Norwegian philosopher Arne Naess in 1973, and later expanded by other writers including George Sessions, Bill Devall, and Warwick Fox.

Just as people's environmental ethics, values, and worldviews influence their environmental behavior; they also inform their reasoning on what environmental sustainability means and how sustainable development can be achieved. Two of the first documents to popularize the notion of sustainable development, the World Commission on Environment and Development's (1987) report "Our Common Future" and the United Nations' (1992) "Agenda 21" promote an anthropocentric ethic, in the form of intergenerational anthropocentrism, or ecologically sustainable economic development, as the preferred global ethic for achieving sustainable development.

Education for sustainable development is a key process in fostering the environmental ethic promoted by such international conferences and charters. Education for sustainable development builds upon environmental education, which first gained international recognition at the UN Conference on the Human Environment in Stockholm in 1972. It also draws from other associated disciplines including values education to move beyond a focus on environmental concerns to utilize an interdisciplinary approach encompassing human, social, and economic factors. While such education is traditionally concentrated within schools and associated institutions aimed at children and young people, educationalist Darlene Clover states that it is also important to provide such learning opportunities for all ages, including adults. Clover highlights the value of providing community-based education that is experiential and develops people's relationships with each other, their community, and the natural environment. John Fien expanded the notion of experiential learning through education for the environment, whereby education becomes a social change process based on translating knowledge into action.

Educating people so that they may hold ethics and values consistent with ecologically sustainable development can involve a variety of techniques including formal and informal approaches and public awareness raising and advocacy. By using a combination of teaching approaches, it is anticipated that participants will be better equipped to develop a life-long approach to learning and the ability to adjust their ethics and values so that they may remain appropriate in changing contexts.

SEE ALSO: Environmentalism; Ethics; Land Ethic; Religion; Social Ecology.

BIBLIOGRAPHY. Darlene Clover, *The Greening of Education* (UNESCO, 1997); John Fien, *Education for the Environment: Critical Curriculum Theorising and Environmental Education* (Deakin University, 1993); Garrett Hardin, "The Tragedy of the Commons," *Science* (v.162, 1968); Linda Kalof and Terre Satterfield, eds., *Environmental Values* (Earthscan, 2005); Aldo Leopold, *A Sand County Almanac and Sketches Here and There* (Oxford University Press, 1949); Carolyn Merchant, *Radical Ecology: The Search for a Livable World* (Routledge, 1992); Timothy O'Riordan, *Environmentalism* (Pion, 1981); Dennis Pirages and Paul Ehrlich, *Ark II: Social Responses to Environmental Imperatives* (Freeman, 1974); Mikael Stenmark, *Environmental Ethics and Policy-Making* (Ashgate Publishing, 2002); UNESCO-UNEP, *The Tbilisi Declaration* (UNESCO-UNEP, 1978); Lynn White, "The Historical Roots of Our Ecologic Crisis," *Science* (v.155, 1967).

Timothy F. Smith
Commonwealth Scientific and
Industrial Research Organization
Lionel V. Pero
University of Queensland
Dana C. Smith
Th!nk

Variability (Natural, Patterns of, Climatological)

WEATHER IS CONSTANTLY changing. Climate, in contrast, is a record of the variations in weather over long periods of time. For example, the climate of the Ice Age, which, in terms of geologic ages was a mere few thousand years ago, was extremely cold; since then the global climate has warmed considerably. However, the long-term trends in the climate, whether of localities, continents, or even of the globe, vary in discernable patterns. One factor that has affected the variability of the climate has been continental drift. Over millions of years, continents moved from warmer regions to colder regions and vice versa. However, of interest to climatologists are variable patterns that are more intermediate.

From the 1940s until the early 1970s, the climate in North America was cooling; since then the climate has been warming. Today, many scientists fear that global warming is occurring because of human activities such as the burning of fossil fuels, especially coal, natural gas, and petroleum. The problem with this assessment is that over the last 10,000 years there have been periods of warming and cooling in North America. In addition, studies of the growth rings of trees have shown variations in the patterns of growth that indicated periods that were wetter and periods that were dryer. The general conclusion climatologists and other scientists have made is that there are variations in the patterns of rainfall, temperature, winds, and other meteorological phenomena. These variations have a number of causes.

The oceans covering over 70 percent of Earth's surface are a major influence on the earth's climate; they affect the weather constantly. As the sun strikes the surface of the oceans and sea—especially in the equatorial regions—it warms them, causing evaporation. The resulting cloud covers are driven by the winds onto land masses where they interact with the colder air masses of the polar regions, causing rain and snow. The oceans also act as heat traps by absorbing vast amounts of heat, which is then slowly released and, in the case of the warm currents such as the Gulf Stream, transported to colder regions.

Many climatologists, meteorologists, and other scientists believe that changes in the oceans can create long-term patterns in the weather. Changes in evaporation rates can affect the salinity of the oceans. Even small changes can produce significant variations in the patterns of the climate.

Oceans affect the weather constantly; clouds driven onto colder land masses can cause rain and snow.

It is essentially the way in which heat is distributed and redistributed in changing patterns that causes variability in the climate. Studies of the way in which the oceans absorb and then spread the heat from the sun via ocean currents, hurricanes, typhoons, and by moisture patterns are pointing to a better way of understanding climatic variations.

Because there are naturally occurring global patterns of warming and cooling, many scientists are of the opinion that the term *global warming* is misleading. The better term to use to reflect variation in global climate is *global climate change*. This term has been suggested as a means for including the naturally occurring changes that are due to more than

just the release of vast amounts of carbon dioxide into the atmosphere since the beginning of the Industrial Revolution in the late 1700s.

Of interest to scientists are variations in the rainfall patterns occurring in the African Sahel where the rainfall varies widely. For some periods it is too wet and at other times extended droughts affect the lives of millions living on the margins of the Sahara Desert. The ultimate goal is to not only aid the people of the area, but to understand the types of patterns that can destroy civilizations.

SEE ALSO: Climate; El Niño–Southern Oscillation (ENSO); Global Warming; North Atlantic Oscillation (NAO); Weather.

BIBLIOGRAPHY. John D. Cox, *Climate Crash: Abrupt Climate Change and What It Means for Our Future* (National Academies Press, 2005); Brian M. Fagan, *Little Ice Age: How Climate Made History (1300–1850)* (Basic Books, 2001); Brian M. Fagan, *Long Summer: How Climate Changed Civilization* (Basic Books, 2005); Tim Flannery, *The Weather Makers: How Man Is Changing the Climate and What It Means for Life on Earth* (Grove/Atlantic, Inc, 2006); Eugene Linden, *The Winds of Change: Climate, Weather, and the Destruction of Civilizations* (Simon & Schuster Trade, 2006).

Andrew J. Waskey
Dalton State College

Vegetarianism

VEGETARIANISM, IN ITS broadest definition, is a dietary pattern where meat, fish, and poultry are excluded. Frequently, vegetarians will exclude dietary products that include animal by-products and derivatives; however, *vegetarian* is a broad term that can have varied definitions and meanings. Some self-defined vegetarians may include animal products ranging from eggs and dairy, to seafood, and even occasional meat. Vegetarian dietary patterns are found throughout the world today and are influenced by a myriad of factors ranging from religious beliefs, economic influence, meat availability, environmental beliefs, and ideological beliefs.

In less-developed nations, large populations have primarily vegetarian diets due to economic factors, not choice.

Broadly speaking, vegetarians can be broken into four primary groups: lacto ova vegetarians, lacto vegetarians, vegans, and fruitarians. Lacto ova vegetarians do not consume meat, fish, and poultry, but do eat eggs and dairy products. Lacto vegetarians do not consume meat, fish, poultry, and eggs, but do eat dairy products. Vegans do not consume any meat, fish, poultry, dairy, eggs, or other animal-made food products, such as honey. Vegans, as well as many vegetarians, are also likely to avoid animal products in their clothing (wool, leather, and silk, for example), grooming, and cosmetic products, and other products. Fruitarian diets are vegan, but specific in that only fruits and vegetables that are defined as fruits are consumed.

In many parts of the world, particularly in less-developed nations and areas that do not lend themselves to meat production, large populations may consume a diet that is primarily vegetarian in nature not by "choice" per se, but because of circumstance. In agriculturally-based environments where horticultural production is limited these resources are more beneficial for human consumption than animal consumption, thusly limiting the capacity to develop strong animal-based agriculture and limiting meat available

for human consumption. In such populations, whose diets are primarily vegetarian with occasional meat eating, meat consumption is frequently aligned with holidays and special events.

In the industrialized developed world, meat consumption is quite high. The United States leads the world in meat consumption (red meat, poultry, and fish) with an average annual consumption of 195 pounds per person. Within the United States research suggests that 2.5–5 percent of the population identify as vegetarian and that the numbers of vegetarians are on the rise. While in many parts of the world vegetarian dietary habits may be shaped by environmental and economic factors limiting the availability of meat, vegetarians in the United States often make an active choice in their dietary patterns. While various factors influence this choice—the healthful benefits of a diet low in animal products, ethical beliefs about animals, and religious beliefs—the most influential belief affiliated with American vegetarians is the belief that vegetarianism is beneficial for the environment.

As the developed world continues to increase meat consumption, there has been a parallel growth in the production of meat. Today, significant portions of the world have been transformed to enable cattle raising. In Central America, over approximately the last 50 years, a quarter of the rainforest loss has been to beef production. In addition to the rainforest loss, this beef production also impacts the environment further, as it is shipped to its primary consumer markets in the United States and Europe via the consumption of fossil fuels in transport and the output of toxic exhausts. Cattle farming has, as a practice, immediate environmental impacts.

In the United States, cattle farms consume approximately one-half of the annual water used, while simultaneously being a major source of water pollution via the tons of organic farm waste. Additionally, the intensive farming of beef production in the United States also consumes high levels of fossil fuels (with coinciding emissions) via the transportation of grains and food for the animals, the removal of animal waste, the transportation of animals, their slaughtering, and the transportation of meat.

Worldwide, the nation with the highest percentage of intentional vegetarians is likely India, with only 30 percent of the population consuming meat regularly, 20 percent being strict vegetarians, and the remaining 50 percent being occasional meat eaters. Many Indians who adhere to a strict vegetarian diet do so in part because of religious beliefs. In addition to religious beliefs, one's economic situation may prohibit the purchase and consumption of meat for many. Socio-environmentally, for those with little economic means, cattle may be more useful as a source of labor, dairy, and dung (that may be used as a fire source) than as meat.

SEE ALSO: Cattle; Food; Livestock; Meat; Religion.

BIBLIOGRAPHY. Alan Beardsworth and Teresa Keil, "The Vegetarian Option: Varieties, Conversions, Motives, and Careers," *Sociological Review* (v.40, 1992); Linda Kalof, Thomas Dietz, Paul C. Stern, and Gergory A. Guagnano, "Social Psychological and Structural Influences on Vegetarian Beliefs," *Rural Sociology* (v.64, 1999); Michael Pollan, *The Omnivore's Dilemma: A Natural History of Four Meals* (Penguin Press, 2006); F.J. Simoons, *Eat Not This Flesh: Food Avoidances from Prehistory to the Present* (University of Wisconsin Press, 1994); United States Department of Agriculture (USDA), "Agriculture Fact Book 2001–2002," www.usda.gov (cited June 2006); USDA, "Country Profile: Passage to India," www.fas.usda.gov (cited June 2006).

Daniel Farr
College of St. Rose

Venezuela

LOCATED IN THE northeastern part of the South American continent, Venezuela has a land area of 353,839 square miles (916,445 square kilometers) and an estimated population of 25.7 million people in 2006. The country can be divided into four main physiographical units: the old Guyana Massif in the southeast with maximum elevations of 9,843 feet (3,000 meters); the Andes to the west (maximum elevations around 16,404 feet [5,000 meters]); the coastal ranges in the north formed by two mountain chains separated by a tectonic plain, and the large flat plains of the Llanos drained by the Orinoco

river and its tributaries in the south. The Orinoco is one of the largest rivers of South America, running 1,333 miles (2,150 kilometers) and forming a delta of more than 9,653 square miles (25,000 square kilometers). The variety of natural conditions produces a high biodiversity with ecosystems ranging from the coastal mangroves of the Atlantic to the rich savanna formations of the Llanos to the xerophitic communities of the arid north. Venezuela is one of the top 20 countries of the world in plant and animal diversity.

Venezuela remains an oil-dependent nation, and many of the environmental problems of the country derive from this condition. From 1929 to 1970, Venezuela was the largest world exporter of oil and still holds enormous reserves of oil and natural gas in the Orinoco Delta. Oil spills have contaminated large parts of the Maracaibo Lake, killing fish and forcing the closure of some coastal resorts, and Lake Valencia is seriously affected by the discharge of untreated wastes. Air pollution is common in cities such as Caracas, Maracaibo, and Valencia. In these and other urban areas, 30 percent of the population lacked wastewater facilities in 2000. In rural areas, population without water sanitation exceeds 50 percent, although only one-tenth of the total population of Venezuela is rural.

Venezuela has the highest percentage of protected land of any Latin American nation. In 2003 it was estimated that more than 70 percent of the country enjoyed some environmental protection. Eleven natural sites (including Ramsar Sites and a Reserve of the Biosphere) cover more than 2.5 million acres (one million hectares). A particularly emblematic protected area is the Imataca Forest Reserve (bordering Guyana) for its natural and cultural diversity (it is the home of at least five indigenous groups) Despite this, the country is losing its rain forests at a fast pace (more than 2.5 million hectares disappeared between 1990 and 1995, or twice the average rate for tropical South America). Moreover, there is increasing evidence of soil degradation in the pastures of the Llanos due to overgrazing by cattle.

In December 1999, Venezuela suffered the worst environmental catastrophe of its history and one of the worst episodes of this kind of Latin America. Heavy rains coupled with landslides on the hills in the state of Vargas (near the Caribbean) killed approximately 30,000 people and left more than 500,000 homeless. The city of La Guaira, where informal settlement on steep slopes was and remains widespread, took the hardest toll (25,000 dead or missing). Uncontrolled urbanization may also be behind the closing in February 2006 of the main highway (and economic backbone of Venezuela with a circulation of more than 50,000 vehicles a day) between Caracas and La Guaira. The highway was closed because of the high risk of failure of several bridges whose pillars have been undermined by wastewater coming from the numerous slums surrounding this communication network.

SEE ALSO: Biodiversity; Deforestation; Ecosystems; Hazards; Natural Gas; Oil Spills; Overgrazing; Petroleum; Pollution, Air; Pollution, Water; Rain Forests; Sewage and Sewer Systems; Urbanization; Wastewater.

BIBLIOGRAPHY. Central Intelligence Agency, "Venezuela," *World Factbook,* www.cia.gov (cited April 2006); Steven L. Hilty, *Birds of Venezuela,* 2nd ed. (Princeton University Press, 2003); United Nations Development Programme, "Human Development Report: Venezuela," hdr.undp.org (cited April 2006); World Bank, "Venezuela," lnweb18.worldbank.org (cited April 2006); Yale University, "Pilot 2006 Environmental Performance Index," www.yale.edu/epi (cited April 2006).

David Sauri
Universitat Autònoma de Barcelona

Vernacular Housing

THE TERM *VERNACULAR* simply means native to a place. Anthropologists describe vernacular architecture as "ordinary building," which is distinguished by its "communally sanctioned" qualities and its "intensity of social representation," as opposed to individual expression. Vernacular architecture reflects the everyday lives of ordinary people and their relationships with the built environment. It embodies the social and cultural values, customs, and practices of a particular place, often providing insights into history. Due to its cost-effective use

of local building materials and techniques, and its climatic and environmental sensitivity, vernacular architecture is considered an important component of sustainable development. Vernacular housing is a subset of vernacular architecture. It refers to individual dwellings built using traditional building styles, as well as the aggregation of such dwellings into larger settlements.

Various building principles, practices, and elements together characterize vernacular housing. Ancient builders are known to have used solar principles and other local climatic characteristics not only for individual dwellings but also for groups of dwellings. For instance, in ancient Greek towns, most buildings had stuccoed walls with few openings. Shadows kept them cool despite the bright sun. In residential buildings, windows were restricted to upper levels to ensure safety. Similarly, traditional dwellings in cold climates were usually sited just below the brow of the hill on a southward slope. The north face of the buildings had few openings whereas the southern façade had the main openings to maximize the benefits of limited sunshine. Such common-sense approaches, which are the foundation of vernacular building traditions, have inspired what is called "green" building today. In a search for innovations to promote sustainability, green builders have begun to adapt vernacular techniques and materials to achieve energy and cost efficiency.

Global housing demand for the projected population of nine billion people by 2050 is expected to have severe social and environmental implications. Promoting vernacular housing, particularly in developing countries, is seen as a long-term sustainable solution to the housing problem given its environmental and cultural sensitivity. Vernacular architecture is estimated to make up almost 90 percent of the world's housing stock. Although not much of the housing seen today in the United States and Europe is vernacular, in parts of Asia, Africa, and Latin America, vernacular architecture still accounts for a majority of the buildings. Since vernacular building traditions are still prevalent in parts of the world with rising populations, they are expected to be the dominant housing pattern in this century.

While vernacular design is mostly guided by unwritten rules, there are also the more formalized, almost normative, vernacular building principles such as those represented by Feng Shui manuals in China or the Vaastu Shastra principles in India, which are immensely popular today. Also, the work of some architects, such as Hassan Fathy in Egypt and Laurie Baker in India, exemplify the skillful interpretation and expression of vernacular traditions in contemporary architecture. For instance, Fathy is credited with reviving the Nubian vault—an ancient building technique dating back to the Pharaonic times—which uses architectonic elements, and the ancient craft of *claustra* or lattice designs in mudwork. His design for Gourna village in Egypt is an example of contemporary vernacular housing that meets the needs of Egyptian Islamic society through a clear demarcation of the private, economic, and religious lives of the community. Similarly, Laurie Baker's use of brick or mud walls, lime mortar made from seashells, recycled materials, and woven bamboo floors showcases the cost-effectiveness, sociocultural sensitivity, and environmental appropriateness of vernacular architecture today.

Technological advances, urbanization, and increased consumption, which accompany processes of globalization, have resulted in cultural changes. They have also created a range of environmental problems—from the depletion of natural resources to the excessive generation of pollution and waste. Consequently, the built environment, both as a cultural category and as a consumer of energy and resources, has a major role to play in addressing some of these issues. Architects, engineers, and planners have been promoting green building technologies in response to the growing environmental crisis.

However, critics point out that efforts to use such technologies in the past, especially for low-income housing in the developing world, have often failed because they tend to impose building types and standards without considering cultural values, local needs, and expectations. They argue that the success of green building technologies in the future will require them to be adaptable to cultural values and local customs. The current anthropological focus in architecture may be a step in that direction as it seeks to study the dynamic processes of living and transcend the simplistic reading of vernacular as just an "organic" physical form.

SEE ALSO: Sustainable Development; Urbanization.

BIBLIOGRAPHY. Lindsay Asquith and Marcel Vellinga, eds., *Vernacular Architecture in the Twenty-first Century: Theory, Education and Practice* (Taylor and Francis, 2006); R.L. Kumar, "What Is Architecture?" in Ashis Nandy and Vinay Lal, eds., *The Future of Knowledge and Culture: A Dictionary for the 21st Century* (Penguin Viking, 2005); Paul Oliver, *Encyclopedia of Vernacular Architecture of the World* (Cambridge University Press, 1997); Paul Oliver, *Dwellings: The Vernacular House Worldwide* (Phaidon, 2003); Amos Rapoport, *House, Form and Culture* (Prentice Hall, 1969).

Priyam Das
University of California, Los Angeles

Vertical Ecology

CLIMATE IS VERTICAL. The fauna and flora of an area varies not only with the latitude of a region, but also with its elevation. For example, along the Mississippi River walnut trees and red squirrels flourish on the river's floodplains; however, gray squirrels and hickory nut trees flourish higher up on ridges above the river. While the distance between the floodplain and the heights of the river's banks is relatively small, the same vertical ecology occurs around the world. Vertical ecology has significant implications for the niches that are occupied by animals that are specialist feeders like the panda bear, rather than generalist feeders like the raccoon.

Mountain elevations allow different fauna and flora to flourish. In the southern Appalachian Mountains of Georgia, North Carolina, Tennessee, and Virginia the fauna and flora differ from that of the coastal plains and the Piedmont Region. The high elevations of the Appalachians (over the 4,500-foot, or 1,370-meter, level) have a climate and plants that are more like southern Canada. The higher mountains in New Mexico (Cloudcroft) surrounded by the Chihuahuan Desert and Arizona (Chiricahua Mountains) by the Sonoran Desert ecosystem are sometimes called "islands in the sky." In the summer, extreme temperatures and lack of water are fatal to all but desert plants. However, on the tops of the mountains that reach 7,000–9,000 feet, there are trees, springs, animals, and other plants.

The *tepui* (mountain plateaus) in Venezuela are another example of unique ecologies. Extremely isolated, each *tepui* has its own unique set of plants that flourish in its moist environment. Another form of vertical ecology is found in tropical rain forests. The plants on the ground are not the same as those in the canopy. Vertical ecology also occurs in marine ecology. Sea plants, fish, shellfish, and other creatures vary widely with the depths of the water; those in the relative shallows are different from those in ocean depths.

Human beings have long adapted themselves to vertical ecologies of their respective domains. The Apaches in Arizona and New Mexico would spend the colder months of the year in the warmer desert areas and the hot summer in the mountain elevations. The change in location would also allow them greater opportunities for farming, hunting, and fishing. Indigenous people in the Andes Mountains have long practiced agriculture that uses the vertical climate of the region.

Vertical ecological systems are threatened by global warming. As the temperature increases, the warmer ecology advances up the mountainsides; eventually the tops of the mountains may be overrun and alpine fauna and flora may struggle for survival.

SEE ALSO: Cloud Forests; Ecology; Ecosystem.

BIBLIOGRAPHY. K. Derek Denniston, *High Priorities: Conserving Mountain Ecosystems and Cultures* (Worldwatch Institute, 1995); K. Dettner et al., eds., *Vertical Food Web Interactions: Evolutionary Patterns and Driving Forces* (Springer-Verlag, 1997); Colin S. Reynolds and Yasunori Watanabe, eds., *Vertical Structure in Aquatic Environments and Its Impact on Trophic Linkages and Nutrient Fluxes* (E. Schweizerbart'sche Verlagsbuchhandlung, 1992).

Andrew J. Waskey
Dalton State College

Victoria, Lake

LAKE VICTORIA (or Victoria Nyanza) is the largest freshwater lake in Africa, and the second largest

freshwater lake in the world. The lake's total surface area amounts to nearly 27,000 square miles, about the same size as Ireland. Located in Tanzania and Uganda, it feeds the Nile River its greatest supply of water. The lakeshores are highly irregular and numerous reefs and islets are scattered across the surface. More than 200 species of fish are represented in the lake and many have economic value, especially the tilapia. The water surface is almost 4,000 feet above sea level and the lake reaches a depth of nearly 300 feet. The lake was first formed some 400,000 years ago and is vulnerable to rapid climate change. It has dried out more than once, most recently 17,000 years ago.

Large amounts of water, frequently in excess of the existing legal frameworks, are being extracted from the lake, as is also the case for the other great lakes of the continent. The water is used to help in irrigation and in the production of electricity. Birds and fish have been negatively affected by the loss of habitat and this has had an impact on the wider biosphere. It has become more difficult to navigate boats across the lake and this has led to a reduction in the level of trade and commerce. Some parts of the lake's shores are among the most densely settled in all of Africa; many inhabitants have become dependent on the lake for income and food. This process of environmental degradation seems most likely to have begun with the European colonization of the area, which was marked by the large-scale cutting down of trees to create plantations. The lake was named after the British Queen Victoria by the explorer John Hanning Speke, who, like many colleagues, was searching for the source of the Nile.

Degradation has noticeably intensified over the last three decades. The breakdown of previous legislative frameworks with the collapse of the East Africa Community has contributed to the lack of regulations governing fair use of the lake. It is hoped that the creation of the Lake Victoria Fishing Organization, in conjunction with other international agreements, will add a measure of control. The success of many types of fish in the lake, including alien species introduced illegally, has directly led to the extinction or near-extinction of many other species. Remaining species display declining diversity and health and many once-economically important species have disappeared from markets completely.

SEE ALSO: Colonialism; Lakes; Nile River; Tanzania; Uganda.

BIBLIOGRAPHY. Neil Ford, "Africa's Lakes Are Drying Up," *African Business* (v.320, 2006); F. Witte et al., "Species Extinction and Concomitant Ecological Changes in Lake Victoria," *Netherlands Journal of Zoology* (v.42/2–3, 1992).

John Walsh
Shinawatra University

Vietnam

SINCE THE *Doi Moi* (renovation) free market reforms of the late 1980s, the Socialist Republic of Vietnam has experienced remarkable economic growth evidenced by an 8 percent Gross Domestic Product (GDP) and 14 percent industrial growth rates per annum; rapid urbanization (4.5 percent per annum); and dramatic increases in the use of motorized vehicles and the manufacture and use of chemicals. Population has tripled over the past 50 years, and now stands at 83,689,518 (July 2005 estimate). Together, these factors have led to significant environmental problems, especially on densely populated coastal plains.

Coastal waters are polluted with suspended solids, nitrite, nitrate, heavy metals, grease, and oil, in some cases at levels four times greater than the

Vietnam is in the top 10 countries for biodiversity, but agricultural expansion and other activities threaten this.

Vietnamese standard. Freshwater is in increasingly short supply as a result of rapid industrial and urban growth and is often polluted by untreated industrial and municipal wastewater discharged directly into rivers and lakes. There exist serious water pollution problems in Ha Noi, Ho Chi Minh City, Hai Phong, and Hue. Freshwater resources, however, are now regulated by the Water Resources Law (effective 2000). Wastewater treatment facilities have been built in the four largest cities and others will be developed in cooperation with international agencies.

Most urban areas experience serious air pollution problems. For example, Ho Chi Minh City's estimated 28,000 factories generate airborne dust content that exceeds Vietnamese standards by 2.1–6.0 times and lead content that is 1.4–3.4 times World Health Organization standards. Greater use of motor vehicles and inadequate fuel and emissions standards have exposed millions to airborne lead, carbon monoxide, and sulphur dioxide. A range of remedial strategies, however, are being pursued, including tightening fuel quality specifications and setting maximum emission limits for motor vehicles.

Solid waste disposal brings unique problems in Vietnam. For instance, as one of the world's largest manufacturers of athletic shoes, Vietnam generates tons of shoe leather waste each day. Solid waste collection efficiency is very low. Only about half of the generated waste is collected, the rest being scattered into waterways and unsafe dumping grounds. Annually, Vietnam produces more than 15 million tons of waste and 80 percent of that is municipal waste. Open dumping is the most popular disposal method but of the country's 91 disposal sites only 17 are sanitary landfills, and 49 have been identified on a national list as hotspots with high environmental and human health risks.

Land degradation is a major issue, particularly in upland areas. Causes include poor logging practices, insecure land tenure, salinization, acidification, pollution, and organic reduction. Agricultural yields now depend on fertilizers and pesticides—use of which increased 200 percent from 1992 to 2002—to the extent that management of agrochemicals is an environmental concern of high priority.

Most of Vietnam's virgin forest and forest with rich standing volume has now been degraded. The government aims to protect 9.6 million hectares of existing natural forests, however, and to recover five million hectares of open lands in the next 20–30 years. Moreover, 150,000–200,000 hectares of new forest, of improving quality, are planted each year.

Vietnam is one of the world's 10 most biologically diverse countries, but that astonishing biodiversity is under threat from forest clearance, illegal wildlife trade, agricultural expansion, and dam and road construction. In response, a protected areas system comprising national parks, nature reserves, and protected landscape areas has been established.

The Vietnam government faces considerable challenges balancing continuing rapid development with effective environmental management. That it is working toward this end is signified by the country's Socio-economic Development Strategy 2001–2010, which gives shared emphasis to economic growth, social equality, and environmental protection, and by the commitment of at least one percent of the state budget to environmental activities from 2006. However, although Vietnam is making considerable progress establishing environmental regulatory systems, a host of legal, institutional, and funding limitations continue to make enforcement a problem.

SEE ALSO: Agent Orange; Land Degradation; Pollution, Air; Vietnam War.

BIBLIOGRAPHY. National Environment Agency, "State of the Environment in Vietnam 2005," www.nea.gov.vn (cited April 2006); Dara O'Rourke, *Community-Driven Regulation: Balancing Development and the Environment in Vietnam* (MIT Press, 2003); United Nations Environment Programme, "State of the Environment in Vietnam 2001," www.rrcap.unep.org (cited April 2006); United States–Asia Environmental Partnership, "Vietnam Accomplishments," www.usaep.org (cited April 2006).

Nguyen Van Loi
Iain Hay
Flinders University of South Australia

Vietnam War

THE VIETNAM WAR resulted from the liberation attempt by the Vietnamese people from French co-

lonial rule. Several factions among the Vietnamese cooperated in the struggle including the Viet Minh, which was especially popular in the north of the country and which came under the control of Ho Chi Minh. The eviction of the French in the mid-1950s led the Viet Minh to declare that it was a Communist movement and intended to unite the whole country under Communist rule. A brief period of fragile peace ended with civil war, which was accompanied by the widespread movement of refugees. United States military forces intervened on a massive scale on the side of the West-leaning South Vietnam.

Over the next two decades of intense fighting, approximately one million Vietnamese civilians, 900,000 North Vietnamese soldiers, and 200,000 South Vietnamese combatants were killed, with many wounded and dispossessed. American losses were about 47,000, with other allied forces losing smaller contingents.

The war broadened to neighboring Cambodia and Laos, both of which had hoped to remain neutral, but ended up with Communist governments. The people of Laos, across whose unpoliced borders the so-called Ho Chi Minh Trail passed, became, per capita, the most heavily bombed people in the world. Eventually, U.S.–South Vietnamese troops were overwhelmed by massive popular support for the Viet Minh and their international backers. American interests abandoned South Vietnam and paved the way for the creation of a unified, Communist Vietnam in 1976. Countless thousands of South Vietnamese attempted to escape overland and by boat. Other refugees, such as the Hmong of Laos who fought on the side of the CIA, await relocation to a friendly environment 30 years after the war ended.

The impact of the war on the environment was enormous and remains a significant hindrance to development today. It includes the unexploded ordnance that litters much of the land and claims a steady stream of victims, and landmines that are particularly effective at blowing off limbs, especially legs. The legacy of Agent Orange has been hugely problematic. It was a mixture of herbicides sprayed by U.S. airborne forces from low altitude in great volumes and was aimed at destroying foliage and crops that could be used by the Viet Minh troops and their supporters. Agent Orange is linked with a long series of congenital deformities, miscarriages, cancers, and other deadly illnesses. U.S. troops and their representatives brought a successful lawsuit against manufacturers of Agent Orange chemicals and received a substantial out-of-court settlement. However, this was only one of a range of chemicals used against the people of Vietnam and its neighbors. Estimates of the numbers of deaths caused by the remnants of these chemicals in the years since the war ended exceed 50,000.

SEE ALSO: Agent Orange; Cambodia; Laos; Vietnam.

BIBLIOGRAPHY. Nigel Cawthorne, *Vietnam: A War Lost and Won* (Arcturus Publishing, 2006); Philip Jones Griffiths, *Agent Orange: Collateral Damage in Vietnam* (Trolley, 2004); The Nordic News Network, "Environmental Conference on Cambodia, Laos, and Vietnam," www.nnn.se (cited July 2006).

John Walsh
Shinawatra University

Virgin Islands

THE VIRGIN ISLANDS are a group of around 90 small islands and islets that form part of the West Indies and are within 50 miles of Puerto Rico. The islands are divided into two groups, one of which is administered by the United Kingdom (UK) as former colonies; the second group, which had been the Danish West Indies, was purchased by the American government in 1917 and is administered by the United States. The Virgin Islands are often considered to be an extension of the Lesser Antilles Islands. The islands are actually the peaks of mountains that are mostly underwater; the total surface area of the islands is around 190 square miles and the population is a little over 100,000. Annually, as many as two million tourists visit the islands.

The islands were originally settled with slave labor to produce sugar cane in plantations; this industry is no longer competitive. Few alternatives exist for the islanders apart from the tourism industry, which has been successful enough to attract

migrant workers from other parts of the Caribbean, leading to some ethnic conflict in recent years.

Tourism has had in some cases a significantly negative impact on the physical environment, as motorboats, divers, and related activities have damaged marine life. The many coral reefs represent a particular attraction. These problems have been exacerbated by a succession of hurricanes, which have devastating effects on many parts of the West Indies. The UK and U.S. governments provide direct assistance to the island groups for which they are responsible, but have been able to do little to address the problems of scarce clean water and fundamentally weak economies.

A large petroleum refining plant is located on one of the U.S. Virgin Islands and attempts are being made to diversify the economy in terms of manufacturing and international finance. This latter issue is controversial because of the islands' reputation, perhaps in some cases unfairly earned, for being linked with tax avoidance and money laundering. The Virgin Islands also suffer from problems such as HIV/AIDS, crime and drug smuggling, and the many nonpoliced beaches and coves make illicit activities comparatively easy to hide.

Climate change leading to intensification and prevalence of hurricanes and related phenomena represent significant threats to the security of the islands. Rapid development of roads, second homes, and related infrastructure on tourist destination islands has contributed to sedimentation and other forms of environmental degradation.

SEE ALSO: Caribbean Sea; Hurricanes; Puerto Rico; Tourism.

BIBLIOGRAPHY. Isaac Dookhan, *History of the Virgin Islands* (University Press of the West Indies, 2000); Lee H. Macdonald, Donald M. Anderson, and William E. Dietrich, "Paradise Threatened: Land Use and Erosion on St. John, U.S. Virgin Islands," *Environmental Management* (v.21/6, 1997); Caroline S. Rogers and Jim Beets, "Degradation of Marine Ecosystems and Decline of Fishery Resources in Marine Protected Areas in the U.S. Virgin Islands," *Environmental Conservation* (v.28/4, 2001).

John Walsh
Shinawatra University

Viruses

VIRUSES ARE PRIMITIVE biological infectious agents that live only in the cells of bacteria, plants, and animals. A specific virus invades and reproduces inside of a specific living cell until the cell explodes, spewing hundreds of copies; however, a virus cannot live outside of a cell. Viruses may be spherical, rod-shaped, or in the case of those that attack bacteria, like a screwdriver with clasps. Viruses are so tiny that they can only be seen by means of an electron microscope. Viruses are pseudo-life forms and do not match the commonly used definition of life. They do not have a cell structure, and must reproduce inside of another living cell. When expelled in search of a new host, they are inert until they can connect with a new host. They have characteristics of life forms while in the cells they infect, but not during times outside of an infected cell.

Every plant and animal is susceptible to viruses. Tens of thousands of viruses have been identified using electron microscopes. However, efforts to create a viral taxonomy have not succeeded for several reasons. Their origins are still obscure, and there is little in the way of a fossil record, so they are hard to place in the established domains of biological classification. Several domain names have been suggested such as *Acytota*. Organizations like the International Committee on Taxonomy of Viruses (ICTV) are working to find an organizing scheme. Viruses occur in plants and animals. Tobacco mosaic, a very thoroughly studied virus, is a viral disease in tobacco plants. The virus mottles the leaves and causes them to lose value. The recognition of viruses in plants began in the 1600s in the Netherlands. Tulip break virus was one of the most well-known plant viruses, which causes the petals to become ornamentally variegated.

The study of viruses took a major step when Louis Pasteur was able to use attenuated rabies viruses to make a vaccine against the disease in 1884. In 1892, Dmitri Ivanovski was able to isolate tobacco mosaic viruses, but he was not able to identify them specifically because of the limitations of microscopes. However, his work demonstrated the existence of a disease agent that was not bacterial. Marinus Beijerinck, a Dutch botanist, contributed the name *virus*, using a Latin word for poison. Vi-

rology, or the study of viruses, developed rapidly in the early 1900s from work done on viruses by Frederick William Twort and Felix d'Herelle. From their studies, many scientists studied bacterial viruses (phages). In 1935, Wendell M. Stanley identified protein as a part of the chemical makeup of some viruses, which enabled him to crystallize them. Since then, scientists have found that some viruses have a deoxyribonucleic acid (DNA) genome, and others have a ribonucleic acid (RNA) genome around which is a protein coating (capsid). Other viruses have lipids or proteins in their structure.

A single virus (viron) will have a DNA or RNA core surrounded by a protein coating. In some cases, there is additional protein of lipid material present. For example, all of the viral hemorrhagic fevers (arenaviruses, filoviruses, bunyaviruses, and flaviviruses) are RNA viruses covered with a fatty lipid coating. In 1911, it was discovered that viruses could cause tumors in chickens. Since then, other tumor-causing viruses have been isolated and described. In the 1980s, researchers linked some human cancers with viruses. Since then, Pap smears have become routine tests for early detection of the papilloma virus that causes cervical cancer in women. Most cancers in humans do not have a viral origin, but a DNA-type virus causes some cancers. The list of diseases caused by viruses is long. Smallpox (variola), yellow fever, mumps, measles, chicken pox, rabies, influenza, herpes, polio, and hepatitis have long plagued humans.

Antibiotics do not work with viruses once infection has been established. The ability to crystallize viruses enabled vaccines to be developed against polio (poliomyelitis) in the 1950s. Vaccinations do provide protection, but with serious limits. For some viruses no vaccine yet exists. Vaccinations against viruses have to deal with the problem that viruses mutate frequently. Influenza viruses are a global source of infection that can easily reach epidemic proportions. The Spanish Influenza Pandemic at the end of World War I killed millions of people. A source of influenza infections lies in the exchange of viruses that occurs between birds and animals, especially swine in southeast Asia. The exchanges provide opportunities for the viruses to mutate. This in turn means that new vaccines have to be developed to provide protection against the changed viral agent. The common cold is caused by a viral infection, as are a number of other viral infections. The cold virus is highly infectious, but rarely deadly. However, some viruses that cause Ebola, Marburg, and Lassa fevers have extremely high mortality rates that can only be overcome by isolation.

Global air travel since the 1960s is making it possible for new kinds of viruses to emerge from remote places. Influenza such as SARS has been spread in this fashion. Other emerging viruses include the HIV/AIDS virus, which is now killing millions of people globally. The global spread of viruses includes those that infect animals and humans. After the West Nile virus appeared in the United States, it killed millions of birds and a few humans. Viruses spread through casual contact, the ingesting of food from infected sources (hepatitis from oysters), insect bites, or even animal bites. Airborne currents spread respiratory viruses, such as the hanta virus from rodent droppings. Others spread by body fluids. Careful sanitation helps to reduce the rate of infection.

The human immune system fights viral infections in several ways. In infections like measles with a high fever, lymphocytes use antibodies to cover the virus's capsid, or by destroying cells infected with the virus. Mucus is used to capture and expel large amounts of respiratory viruses, while interferon, a protein-like substance made by the body, fights other viruses. Some viruses suppress the immune system and spread in the body rapidly. A few viruses move slowly. Others, like herpes, may be dormant for a long time and then have sporadic outbreaks. Viruses cost billions of dollars annually because of the damage they cause to crops and animals. However, viruses have been used to control insects and invasive species. Rabbits, a serious invasive species in Australia, have been controlled with the myxoma virus.

SEE ALSO: Acquired Immune Deficiency Syndrome (AIDS); Disease; Influenza; Parasites; Pasteur, Louis; Vaccination.

BIBLIOGRAPHY. S.J. Flint et al., *Principles of Virology: Molecular Biology, Pathogenisis, and Control of Animal Viruses* (ASM Press, 2003); Laurie Garrett, *Coming Plague: Newly Emerging Diseases in a World Out of Balance* (Penguin Group, 1994); J.B. McCormick, Susan Fisher-Hoch, and L.A. Horvitz, *Level 4: Virus Hunters*

of the CDC (Barnes & Noble Books, 1999); J.H. and E.G. Strauss, *Viruses and Human Disease* (Elsevier Science & Technology Books, 2002); E.K. Wagner and M.J. Hewlett, *Basic Virology* (Blackwell Publishers, 2003).

ANDREW J. WASKEY
DALTON STATE COLLEGE

Volatile Organic Compounds

VOLATILE ORGANIC COMPOUNDS (VOCs) are a set of chemical substances that tend to enter gaseous state during normal ground level atmospheric conditions. Owing to various chemical properties, VOCs have numerous uses in industry and in preparing consumer goods of different types, but since they also easily enter the atmosphere, their impact upon health must also be carefully investigated and, where necessary, regulated by government.

VOCs may be produced naturally, through the waste products of animals, or through by-products of hydrocarbons such as petroleum and its derivatives. VOCs may enter land and water sources as contaminants or become present in indoor air, increasing pollution of the air and possibly causing negative health impacts. Somewhat unfortunately, the tendency to increase energy efficiency in housing has led to a greater proportion of air retained inside accommodation, which has led to more indoor air pollution. This can cause minor symptoms, such as eye watering, headaches, and nausea, and more serious effects such as organ damage and cancer.

The main categories of VOCs include mostly carbon-based molecules such as hydrocarbons and aldehydes. A significant outdoor naturally occurring pollutant is methane, which, escaping into the atmosphere, is an important contributor to greenhouse gas global warming. In the United States, the Environmental Protection Agency (EPA) has led research to determine the presence of VOCs both indoors and outdoors in a range of different locations. The presence of VOCs indoors has been found at up to five times the level of outdoor pollutants.

The possibility of VOCs in commercially available products causing negative health impacts has led to a burgeoning industry in lawsuits relating to possible negligence. Consequently, much effort is being spent on defining what are and what are not VOCs and to what extent separate sub-categories of the chemicals should be permitted in domestic use. This is likely to increase in the near future as the long-term health impacts of exposure become clearer and the introduction of new chemical substances and their interaction with existing products is studied more intensely. This will in turn stimulate the creation of new technologies to deal with problems caused by VOCs and the meaning of regulations necessary to supervise their production. Both domestic occurrence and workplace hazards will need to be included in these evaluations as the range of products emitting VOCs is increasing.

SEE ALSO: Environmental Protection Agency (EPA); Global Warming; Methane; Pollution, Air.

BIBLIOGRAPHY. Alpha Barry and Diane Corneau, "Effectiveness of Barriers to Minimize VOC Emissions Including Formaldehyde," *Forest Products Journal* (v.56/9, 2006); Environmental Protection Agency, www.epa.gov (cited January 2007); R.E. Hester and R.M. Harrison, eds., *Volatile Organic Compounds in the Atmosphere* (Royal Society of Chemistry, 1995).

JOHN WALSH
SHINAWATRA UNIVERSITY

Volga River

THE VOLGA RIVER, long characterized as "Mother Volga" and renowned as the cultural heart of Russia, rises in the Valdai Hills northeast of Moscow and runs 2,300 miles in a sweeping arc to the south before reaching its complex delta on the Caspian Sea. The Volga is fed by more than 200 tributaries and drains a watershed comprising 40 percent of European Russia, that portion of the country reaching from its western boundaries to the Ural Mountains. The vast Volga watershed embraces 40 percent of the Russian population, 45 percent of the country's industry, and half of Russia's major agricultural sector. An important transportation route in Russia for centuries, the Volga

has been compared to the Great Lakes in North America for its key role in economic development. Currently, the Volga carries nearly 70 percent of all cargo on Russia's inland waterways. There are hundreds of ports and industrial docks along its course, and eight major complexes with dams, hydroelectric generating plans, and reservoirs line its bustling banks. The dams and reservoirs have transformed the Volga into a series of expansive lakes.

Boats on the Volga can reach the Black Sea through the Volga-Don Canal, and access to St. Petersburg and the Baltic Sea is possible through the Volga-Baltic waterway, which links the river with Lakes Ladoga and Onega in the north. The Volga reaches Moscow via the Moscow Canal and the Moscow River. Because of Russia's increased connection with the European Union, negotiations have been underway to allow access to the Volga and other Russian inland waterways by other European countries.

The Volga has been subjected to a great degree of pollution from a variety of sources. Industrial wastes, runoff of agricultural chemicals, and infusions of silt from deforested lands have seriously endangered the river. The construction of dams along the river has made it difficult for fish to reach spawning grounds and chemical changes in the water from pollutants have damaged the fishing industry. Pollutants from the Volga entering the Caspian Sea have damaged the immune system of thousands of seals and greatly reduced the fish catch. Especially vulnerable has been the sturgeon, the source of the Russian delicacy caviar. Although the delta of the Volga has thousands of individual streams serving as filters to cleanse the waters, this natural process does not trap all the pollutants carried in the waters.

The environmental degradation of the Volga has attracted international attention. In 2003 President Vladimir Putin stated his intention to double Russia's gross domestic product by 2010. Economists and environmentalists both expressed concern about the difficult task of balancing economic growth and protection of the environment in the Volga basin. Monitoring this situation is now the task of CABRI/Volga, a multinational organization dedicated to addressing risk management in the Volga basin. Institutional members represent the Russian Federation, Germany, The Netherlands, Greece, Italy, France, Hungary, and Malta. The acronym CABRI stands for Cooperation Along a Big River, and the organization has advocated for strong water management and coordination among groups administering environmental protection programs in the Volga basin.

SEE ALSO: Caspian Sea; European Union; Russia (and Soviet Union); Ural Mountains.

BIBLIOGRAPHY. Micha Bradshaw, *A New Economic Geography of Russia* (Routledge, 2006); Tim McNeese, *The Volga River* (Chelsea House Publications, 2005); Lawrence Oliphant, *The Russian Shores of the Black Sea in the Autumn of 1852: With a Voyage Down the Volga and a Tour through the Country of the Don Cossacks* (Adamant Media Corporation, 2000).

Gerald R. Pitzl, Ph.D.
New Mexico Public Education Department

War on Drugs

THE WAR ON DRUGS is the title of the policy of aggressively pursuing the production, distribution and use of illegal drugs that are abused for pleasurable effects. Since 1971, when President Richard M. Nixon launched the War on Drugs, thousands of people who were in some way involved with illegal drugs have been killed, and millions more have been arrested and imprisoned.

Natural drugs from herbs and plants number in the thousands. Most of these are taken as tonics, stimulants, or medicinally; while others are used for religious purposes. With the advances of chemistry in the 20th century many new drugs were developed. At first medicinal drugs were controlled by governments in order to protect people from unscrupulous purveyors of quack remedies. Then, as new synthetic drugs were sold for their pleasurable effects, many governments began to control the manufacture and distribution of synthetic drugs as well as natural drugs in order to protect the public from drug abuse.

All drugs have some "dramatic" impact on the body after being ingested, inhaled, or injected and reaching the target receptors in the body. Drugs may be classified as depressants, stimulants, steroids, hallucinogens, or opiates. Some forms of these, such as alcohol, nicotine, and caffeine, are legal, and others have legitimate medical uses.

Enormous social problems have arisen from the abuse of drugs. People who are addicted often squander their resources, lose moral restraint, and soon engage in a variety of criminal activities. If the drug use reaches a serious level it impedes occupational performances so that productivity is lost, or, in the transportation industry, lives and cargo may be endangered. Drug users involved in work that requires a security clearance may become vulnerable to blackmail.

The value of the illegally sold drugs around the world is estimated to be in the hundreds of billions of dollars. Criminal groups such as the Mafia in Italy have smuggled heroin from Afghanistan or Burma; while drug cartels have smuggled cocaine from South America. The War on Drugs has hindered the traffic in illegal drugs but it has not eradicated it.

ENVIRONMENTAL EFFECTS

The War on Drugs has used a number of environmentally damaging tactics to stop drug trafficking. These have included burning poppy fields, coca tree plantations, and marijuana patches. More damaging

has been the use of herbicides to destroy crop areas held by armed local farmers, gangs, and, in some areas, ideologically-driven guerilla bands. In Columbia the aerial fumigation program of the U.S. government has delivered enough herbicides to growing areas to damage fragile ecosystems in some areas of the Amazon Basin and to negatively affect the health of people in the area.

In some areas of Columbia, Mexico, the United States, and other countries, deforestation and destruction of hundreds of thousands of acres of crops have resulted. In many cases the deforestation is in delicate rainforests and cloud forests. Some critics believe that the negative environmental impact of the War on Drugs may soon exceed the costs of drug addiction.

SEE ALSO: Afghanistan; Cocaine; Columbia; Deforestation; Drugs; Herbicides; Opium (and Heroin); Poverty.

BIBLIOGRAPHY. Antonio Escohotado, *Brief History of Drugs: From the Stone Age to the Stoned Age* (Inner Traditions, 1999); Mike Gray, *Drug Crazy: How We Got into This Mess and How We Can Get Out* (Taylor and Francis, 2000); Michael Kerrigan, *War Against Drugs* (Mason Crest Publishers, 2002); Tamara L. Roleff, ed., *War on Drugs* (Thompson Gale, 2004).

Andrew J. Waskey
Dalton State College

Wars

MANY PEOPLE THINK of peace as simply the absence of war. Peace scholars, however, recognize peace entails far more. "Positive peace" refers to the absence of war as well as conditions of social justice, including full human rights for all. The ability to live in a sustainable environment would clearly be a part of this positive peace.

Every war has impacted the physical, chemical, biological, and human or social environment in a variety of ways, both directly and indirectly. In recent years, environmental damages wrought by warfare have worsened significantly. This is due to the increased intensity of modern warfare as well as the use of new, more destructive, technologies. In addition, scarcity and unequal distribution of needed resources, such as clean water and arable land, have contributed to conflicts both between countries as well as within specific nations. Experts have warned that rapid depletion of these resources will only exaggerate the likelihood that environment will be a precursor for conflict. While all forms of warfare are environmentally damaging, civil war has been found to be more harmful than wars between nations. This is likely due to a number of factors, including the extended length of civil wars.

WAR AND NATIONAL INFRASTRUCTURES

One way war wreaks havoc on the environment is by degrading a country's infrastructure. Water supply systems and sanitation services, for instance, are often contaminated or rendered completely unusable by bombs or bullet damage to pipes. This then leads to contamination of drinking water, associated with a number of diseases, some of them fatal, to humans and animals. In the current war in Iraq, unreliable electricity due to warfare has led to sewage backups, and waste is being dumped into the Tigris River, Baghdad's only source of water.

Countries experiencing depleted infrastructures from warfare must prioritize their reconstruction efforts, and environmental damages often end up near the bottom of the list. Many countries ravaged by war have limited, if any, hazardous waste treatment facilities or other means to take care of environmental problems. In Kuwait, Iraqi forces destroyed sewage treatment plants during the 1991 Gulf War, resulting in over 50,000 cubic meters of raw sewage discharged daily into Kuwait Bay.

Sometimes the destruction of parts of a country's infrastructure is by design. In World War II, destruction of dams and dikes was common. In Sarajevo, soldiers cut off electricity and water pumps. Destruction of facilities designed for war production can lead to a host of other problems.

IMPACT ON PLANTS AND ANIMALS

War also threatens biodiversity. Historically, examples of deliberate destruction of crops and forests can be found in the conflicts between Israelites and Phi-

listines in the 12th century B.C.E. Genghis Khan also authorized the destruction of crops and forests in his conquering of China. Military machinery and explosives damage forests and habitats, which in turn disrupts the ecosystem, leading to erosion as well as concerns about safe water and food. For instance, approximately 35 percent of Cambodia's forests were destroyed by warfare over two decades.

The destruction of oil wells, generally a feature of conflicts in the Middle East, has brought a number of forms of environmental damage. In the 1991 Gulf War, Iraq destroyed more than 700 oil wells, releasing approximately 10 million barrels of oil into gulf waters. The desert of Kuwait, said to be a healthy area prior to the war in 1991, is coated with oil residues that affect water permeability, seed germination, and microbial life. In addition, it took months to cap the oil wells, so crude oil released into the sea killed marine birds and mammals, while the oil itself formed petrochemical lakes. Toxic smoke and fumes from oil spills killed migratory birds. Veterinarians claimed to have seen birds literally dropping from the sky.

As already noted, civil war is perhaps even more devastating to plants and animals than is war between nations. In Angola, decades of civil war have left national parks and wildlife reserves with only 10 percent of their 1975 wildlife levels, a dramatic reduction in the region's biodiversity.

CHEMICAL AND BIOLOGICAL WARFARE

Chemical and biological warfare is especially damaging to the environment. The United States used the pesticide DDT in World War II, primarily in the Pacific. One naval officer reported that the first use of DDT in the Pacific completely destructed the animal and plant life there.

The U.S. military's use of toxic defoliant Agent Orange in Vietnam between 1962 and 1971 destroyed approximately 14 percent of the forests in South Vietnam, including up to 50 percent of the mangrove forests. Agent Orange also resulted in the loss of freshwater fish in Vietnam, as well as half of the commercial hardwood trees and many other rubber trees. Agent Orange contained dioxin, one of the worst carcinogens. Thus in addition to environmental destruction, use of Agent Orange has been linked with birth defects, spontaneous abortions, chloracne, skin and lung cancers, lower IQ, and emotional problems in children.

Biological warfare also poses tremendous environmental risks. Biological weapons were prohibited by the 1925 Geneva Protocol, as well as by the Biological Weapons Convention (BWC) of 1972, which had been signed by 134 nations by the mid 1990s. Biological weapons are still a concern, however, as some nations have continued to develop and use them regardless of international law. Historically, aggressors have spread the bubonic plague, anthrax, typhoid, cholera, dysentery, and a host of others. In 1346, rats and fleas were released during war in what is now the Ukraine. Between 1754 and 1767, the U.S. military infected Native Americans with smallpox, both unintentionally through contact as well as intentionally through the distribution of infected blankets. During the 1937 Sino-Japanese war as well as during World War II, the Japanese experimented with a number of types of biological warfare. Most notably, the conducted experiments on the Chinese, giving them plagued food items as well as intentionally contaminated water sources. More recently, concerns that Iraq had developed, stockpiled, and even used biological weapons was a major impetus for the U.S. waging war.

NUCLEAR WARFARE

Nuclear weapons and facilities are also devastating to the environment. The most notable example is the U.S. bombing of Hiroshima and Nagasaki in 1945, which destroyed over ten square miles of land. In 2003, an estimated 200 plastic barrels containing uranium were stolen from the Tuwaitha nuclear plant in Iraq. Poverty-stricken residents dumped the contents into rivers, then used the barrels to store their water, cooking oil, and other basic amenities. These substances not only harm those who immediately ingest them, but seep into the ground, air, and water and food supplies. It is projected that thousands of hectares of Iraqi land is contaminated from depleted uranium used in the first Gulf War. Lake Karachai in the South Urals is considered the most contaminated body of water on earth due to nuclear testing and production.

Most of what is known about releases of radiation involves the United States. Several major production

sites have been found associated with severe environmental contamination, including the Hanford Nuclear Reservation in Washington, the Oak Ridge Reservation in Tennessee, the Rocky Flats Plant in Colorado, and the Savannah River Plant in Georgia. All of these sites have been involved in accidental releases and continued emissions as part of their daily production.

WAR PREPARATION

Weapons production, testing, and maintenance are also destructive to the environment. Fuels, paints, solvents, heavy metals, pesticides, and PCBs, cyanides, phenols, acids, alkalies, and propellants are the waste products of the production, maintenance, and storage of conventional, chemical, and nuclear weapons and of military machinery. Producing semiconductors and other electronic components of weapons and equipment involves many highly toxic chemicals. Likewise, readying troops takes a tremendous toll on large pieces of land. It is estimated that NATO maneuvers in West Germany cost $100 million in damages to crops, forests, and private property per year in the 1980s.

The U.S. military is said to be the largest producer of hazardous materials in the United States and possibly even in the world. More than 7,000 former military properties in the U.S. are being investigated for toxic contamination, and almost 100 bases are already on the Superfund National Priorities List. Military testing in Alaska's Eagle River Flats, near Anchorage, has released high levels of toxic chemicals and contaminants into the soil, air, and water.

WAR REMAINS

The remains of the technology of war are also destructive to the environment. Land mines remain in many countries such as Vietnam and Cambodia, and in addition to the threat they pose to humans, they make agricultural production on the land impossible. It is estimated that some 70 to 100 million antipersonnel land mines are still active world-wide, and another 100 million exist in stockpiles.

Prior to the mid 1980s, there was little public attention to the potential for environmental damage of stockpiled chemical weapons in the United States. Until the late 1960s, surplus weapons were routinely dumped in oceans, burned in the open air, or buried. In 1986, Congress mandated the destruction of the U.S. stockpile of chemical weapons. While the Act stipulated that destruction of these weapons needed to involve environmental protections, it is unclear precisely how well this has been done.

INDIRECT EFFECTS

There are also many indirect environmental effects of war. Since land, roads, and bridges are often destructed, many crops are spoiled. Since people are no longer able to safely live in some areas of war-ravaged countries, over-use of other land contributes to soil degradation, deforestation, desertification, and many other environmental problems.

In many nations, refugee camps are created after a war. It is estimated that there are some 17 million refugees and 25 million internally displaced persons in the world today. These camps are likely another source of environmental damage, causing deforestation, loss of endangered species, water pollution, air pollution, and depleting sanitation systems. Deforestation occurs when land is cut for campsites, housing, and for cooking and heating. Overgrazing can accelerate soil erosion and the silting of rivers and streams. Disposal of solid wastes is difficult, so refugee camps often become breeding grounds for flies, rodents and other pests.

NEW WARS AND THE ENVIRONMENT

Terrorism as a form of violent conflict also poses great threats to the environment. In 1995, the Japanese cult Aum Shinrikyo planted the nerve agent sarin on subways in Tokyo. Raids of Aum's labs showed they were developing the botulin toxin, anthrax, cholera, and Q fever as well. The attacks on the World Trade Center in New York City on September 11, 2001, released asbestos and other hazardous chemicals into the air and land that the EPA is still working on cleaning up.

Yet another form of warfare with effects on the environment is the war on drugs in Latin American. In Columbia, the U.S. began aerial spraying of herbicides in an effort to destroy coca and poppy crops in 2000. While it is still to be determined if this is an effective measure in the effort to reduce drug use

and trafficking, no doubt the widespread use of herbicides will damage the soil, plants, and animals beyond the targeted crops.

COSTS OF ENVIRONMENTAL CLEANUP

The cost of repairing the environmental damage of warfare is tremendous. The United States estimates that the cost of nuclear waste management and decontamination from the cold war alone are between $200 an $350 billion, while estimates of the cleanup costs for toxic wastes at U.S. military bases range between $20 and $40 billion. Cleanup costs from the 1991 Gulf War are tremendous. Just to decontaminate 200 hectares of land (of the thousands) will cost four to five billion dollars. Cleanup of the oil released into the Gulf is projected to cost more than $700 million.

Unfortunately, the environmental damage wrought by warfare generally goes unpunished. Although the UNEP labeled Iraq's lighting and dumping of oil in Kuwait in 1991, "one of the worst engineered disasters of humanity," no one was ever tried or punished. The Bern Protocols I and II of the 1977 additions to the Geneva Conventions of 1949 could potentially be used to hold countries' responsible for environmental damages wrought by warfare, but they only apply to half of the nations in the world. The Declaration of 1972 on the Human Environment established that nations have a responsibility to ensure their actions do not cause damage to the environment. It prohibits the targeting of dams, dikes, and nuclear power plants if doing so would release dangerous materials that would endanger civilians. It also prohibits the complete destruction of items required for human survival, including food, agricultural areas, livestock, and drinking water. Some have recommended a fifth Geneva Convention that would specifically address environmental damage in the course of war.

The situation may be changing, however. In 1992, the Rio declaration denounced environmental destruction during war and demanded states respect international law regarding the environment. 1996 marked the First International Conference on Addressing the Environmental Consequences of War. The Chemical Weapons Convention and the Treaty to Ban Landmines, both in 1997, are also tremendous steps toward greater consideration of war's impact on the environment. Unfortunately, several major countries have refused to sign the landmine treaty, which required the destruction of stockpiled landmines within four years of signing and the complete cleanup of all landmines within ten years. Most notably, the world's largest producer of landmines, the U.S., has refused to sign, although the U.S. has stopped production of new landmines.

THE ENVIRONMENT AS A CAUSE OF WAR

Environmental problems can also be part of the cause of war. Thomas Homer-Dixon identifies six types of environmental change that impact violent conflict: water and land degradation; deforestation; decline in fisheries; global warming; and ozone depletion. Norman Myers maintains there are five types of environmental problems that determine or exacerbate conflict: access and availability of water; deforestation; desertification; species extinction and gene depletion; and greenhouse gases.

Some claim the 1967 Arab-Israeli war was, in large part, due to water scarcity. Some groups have greater access to needed and desired resources because of the way geographical boundaries were created by colonial powers, as is the case in much of Africa. Socioeconomic scarcity involves unequal distribution or purchasing power and property rights. Environmental scarcity, in contrast, refers to resources that are becoming scarce because of humans' failure to use sustainable methods. Clearly, underdeveloped countries are at greater risk for both environmental problems and violent warfare. Scarcity undermines the states' capacity to provide for its citizens, and it also leads to economic and political demands.

In the Nigerian Delta, pollution from oil production has caused environmental damage that has disproportionately impacted the native Ogoni peoples. Gas-flaring, pipe leakage, dumping, and spills have all impacted the Ogoni, harming the soil, water, vegetation, and wildlife. The primary oil company in the region, Shell, operates in more than one hundred countries, but forty percent of all their recorded oil spills are in Nigeria. Between 1982 and 1992, 1,626,000 gallons of oil were spilled in 27 different instances. While oil executives live in lavish surroundings from their profits, the Ogoni live in abject poverty. Some Ogonis have organized to protest the degradation of their lands for the profit

of oil executives. Their efforts have met with much resistance, most notably when military dictator Sani Abacha had activist Ken Saro-Wiwa and eight others hung in 1995. In 1999 alone, more than 200 people were killed in oil-related riots.

Not only does environmental scarcity lead to violent conflict, but the reverse is true as well. In Liberia and Sierra Leone, fighting destroyed forests. Rebel groups then exploited the scarce timber resources to further finance their warfare. The Food and Agricultural Organization (FAO) and the World Food Programme (WFP) say that civil strife is a tremendous threat to food security in Africa and other developing areas. Although war generally does terrific damage to the environment, it is possible that some positive can come from it as well. Some post-war or former war manufacturing areas are now beautiful nature preserves. Rocky Mountain Arsenal is now one of the nation's premier wildlife refuges, home to some 300 species of wildlife and visited by some 50,000 people each year. Similarly, the U.S. military has turned over approximately 100,000 acres of land by 2000 in Illinois, Maine, California, and northern Virginia to various federal agencies. The demilitarized zone in Korea is home to endangered species and migratory birds and is considered one of the most plentiful in Asia.

SEE ALSO: Agent Orange; Nuclear Weapons; Vietnam War.

BIBLIOGRAPHY: Ken Conca and Geoffrey Dabelko, eds., *Environmental Peacemaking* (Woodrow Wilson Center Press, 2002); E. Conteh-Morgan, *Collective Political Violence* (Routledge, 2004); Paul Diehl and Nils Petter Gleditsch, eds., *Environmental Conflict* (Westview Press, 2001); Barry Levy and Victor Sidel, eds., *War and Public Health* (American Public Health Association, 2000); Max Manwaring, ed., *Environmental Security and Global Stability* (Lexington Books, 2002); E. Russell, *War and Nature* (Cambridge University Press, 2001); Judith Shapiro, *Mao's War Against Nature* (Cambridge University Press, 2001); United States Department of State, "Hidden Killers: The Global Problem With Uncleared Landmines" (Department of State, 1993).

Laura L. Finley, Ph.D.
Florida Atlantic University

Waste, Human

HUMAN WASTE (SOMETIMES referred to as *raw sewage*) is becoming an increasing problem for people, animals, and the environment. Untreated human waste, as it runs off into streams, rivers, lakes, and oceans, is causing significant problems for humans and marine life. People are negatively affected by human waste because it contributes to illness and disease. Raw sewage is a significant factor in the sickening of over one million people annually. The bacteria, viruses, and parasites common in human waste can lead to diseases such as hepatitis (a liver disease), meningitis (an inflammation of the membranes that cover the brain and spinal cord), and cholera (an acute intestinal infection). It is estimated that each year approximately 900 people die from contamination of this kind. Exposure to human waste can be caused by drinking contaminated water; swimming in oceans, lakes, and streams that have raw sewage in them; and eating food that has been in contact with human waste.

Not only is the toll on humans significant, but human waste contamination in oceans is also causing problems for marine life. When untreated human waste makes its way into the oceans, it can poison shellfish. It is also a contributor to "dead zones" in coastal waters where oxygen levels are too low to sustain life. The nitrogen found in human waste (which is also found in fertilizers and emissions from vehicles and factories) contributes to this problem. Nitrogen, emitted into the ocean, fertilizes microscopic plant life and causes it to flourish. When the plankton die, they fall to the ocean floor and are digested by microorganisms. This process removes oxygen from the water and creates the dead zones.

Human waste seeps into the water primarily through septic tanks and antiquated sewage systems in municipalities. Many homes in the United States have septic systems in which wastewater is piped into the septic tank from the home and the excess flows out into the ground, where it is absorbed. It is estimated that one person puts out seven pounds of nitrogen a year into a septic tank, and about half of this nitrogen reaches the water table. Antiquated sewage systems are very costly to modernize and expand. Many cities have sewage pipes that were laid in the 1800s, and most lack the funding neces-

sary for improvements. It is estimated that human waste seeps into streams and lakes 40,000 times every year as a result of this.

There are ways to minimize the impact of human waste on the environment. When developing new areas, cities must continue to build sewage systems that can accommodate a surge in population. Better maintenance of current sewage systems (such as having crews check lines to keep tree roots and grease clogs out of the system) helps. In addition, new regulations provide incentives to control human waste contamination in water.

SEE ALSO: Disease; Groundwater; Marine Pollution; Pollution, Water; Septic Systems; Sewage and Sewer Systems; Wastewater.

BIBLIOGRAPHY. Tom Vanden Brook, "Sewage Pouring into Lakes, Streams," *USA Today* (August 20, 2002); Centers for Disease Control, www.cdc.gov (cited April 2006); Cheryl Lyn Dybas, "Dead Zones Spreading in World Oceans," *Bio Science* (v.55/7, 2005); Elizabeth Weise, "As Suburbs Grow, So Do Environmental Fears," *USA Today* (December 28, 2005).

Margaret H. Williamson
Gainesville State College

Waste, Nuclear

THE PROCESSING OF nuclear material such as plutonium and enriched uranium for energy production results in spent material. A nuclear power station with a 1,000 megawatt capacity will typically produce in excess of 20 tons of spent fuel annually. This material must be disposed of as safely as possible because it is highly radioactive and dangerous to human health, causing cancer and other illnesses. However, the material may sometimes be reprocessed. The fission materials produced by the original processing may be removed and the spent material can either be recycled or collected into weapons-grade plutonium or uranium.

Approximately 270,000 tons of nuclear waste has already been produced globally, and current projections are for an additional 12,000 tons to be produced annually for the next 25 years. In addition to nuclear power plants, nuclear waste is produced by nuclear-powered surface ships and submarines, and by some private research institutions. In all of these situations, regulations exist to ensure safe handling. In the case of the accidental meltdown at the Chernobyl nuclear power plant in the Ukraine, it is not clear whether regulations were properly policed.

Spent material is processed in a separate facility from where it was used. It is placed in steel canisters with additional overcoats; the canister is then welded shut. This method is considered the safest means of dealing with the spent fuel. However, depending on the type of spent fuel involved, and the methods by which it has been treated, alternative means of disposal are also possible. For example, French nuclear power station technicians have devised a special method for nuclear waste disposal. Solid residue is first melted and then formed into borosilicate blocks that solidify within steel canisters approximately one meter tall and up to almost a half meter in diameter. The steel canisters are then deposited into a safe repository.

Despite accidents such as those at Chernobyl and Three Mile Island in the United States, the demands for nuclear energy are great and will only intensify in the future as a result of problems with fossil fuels such as emission of greenhouse gases and the secure sourcing of oil. Demand for nuclear energy will presumably be greatest in those countries with the highest requirements for energy and those that do not have access to alternative energy sources. Many countries are unable to store nuclear waste within their own borders because of geological, geographical and political reasons. As a result, it is necessary to consider the creation of a cross-border trade in the disposal of nuclear waste.

Understanding of safety issues surrounding nuclear waste has improved significantly since the end of the 20th century. Since nuclear waste is radioactive and slow to decay, it can cause harm for thousands of years. Considerable political and technical controversy surrounds issues relating to the safe storage of the waste. Safety issues must be considered in transporting radioactive material to a desired location. The risks involved include accidental leakage of the radioactive material while en route and attempts to seize the material for purposes of

terrorism or extortion. There is also the threat of geological rupture. For example, if an earthquake occurred, it could break open the containment materials holding the nuclear waste. The method widely considered viable for nuclear waste disposal is to contain the radioactive material within a non-reactive barrier and bury it deep within the earth's crust in a region not known for geological disruption.

The Yucca Mountain project in the state of Nevada in the United States has been designated as one such possible place for nuclear waste disposal. This has, understandably, led to great concern among local residents, as well as those who live near transportation routes. Located some 70 miles from Las Vegas, the site is the only one under active consideration by the U.S. government for the development of high-level nuclear waste disposal. It has been estimated that the use of this facility would require movement of radioactive material through 43 U.S. states and more than 100 cities, and passage through the Great Lakes. Shipments would pass within a mile of millions of American citizens. Throughout the 1990s, the rate of rail accidents involving hazardous material averaged 33 per year. Approximately 10,000 people are also evacuated from their homes each year as a result of the nuclear waste transportation. Accident rates for trucks and barges, which would also be required, are comparable. More than 108,000 shipments of nuclear waste are projected by the U.S. Department of Energy in the next 38 years based on existing trends.

People opposed to the use of nuclear power point to these alarming statistics to argue that it is impossible to guarantee safety. People in favor of nuclear power say that there is little choice but to make sure that the methods of disposal are as safe as possible though some level of risk does exist. Public skepticism about the ability of science to deal with nuclear waste may be a result of the poor image of the nuclear power industry. To create a more positive perception of this industry, there are a variety of ongoing public relations efforts in many nations.

Many scientists have concluded that the practical difficulties involved with deep disposal are surmountable. However, the political controversy surrounding this method of disposal, not to mention the costs involved, means that no such facilities yet exist. Many believe that Australia would be an appropriate repository for the long-term disposal of internationally-produced nuclear waste. They note that Australia is a perfect candidate because of its large central desert area that is lightly populated. Its stable geology and the fact that it is a modern democracy are also pluses. Technical requirements for a nuclear waste repository require a 200-meter-thick barrier between ground level and the disposal area. In addition, there should be predictable and low flows of groundwater and dense sedimentary ground formation. There should also be an absence of any resources, including freshwater. All of these factors are important in choosing a location as they would reduce the likelihood that people would ever live nearby.

Nuclear power undoubtedly leads to some serious risks, including the possible consequences resulting from an accident or malicious intervention. However, it seems likely that there is a need for the use of nuclear energy as an alternative energy source in place of oil and hydrocarbons. It is necessary for science to minimize the risks involved in nuclear waste disposal and for society to determine whether those risks are indeed acceptable.

SEE ALSO: Chernobyl Accident; Greenhouse Gases; Nuclear Power; Nuclear Regulatory Commission (NRC) (U.S.); Nuclear Weapons; Three Mile Island Accident; Uranium; Yucca Mountain.

BIBLIOGRAPHY. Rodney C. Ewing and Alison M. MacFarlane, *Uncertainty Underground: Yucca Mountain and the Nation's High-Level Nuclear Waste* (MIT Press, 2006); Tom Quirk, "The Safe Disposal of Nuclear Waste," *Institute of Public Affairs Review* (v.57/2, 2005); Peter D. Riley, *Nuclear Waste: Law, Policy, and Pragmatism* (Ashgate Publishing, 2004); Sierra Club Nuclear Waste Task Force, "Deadly Nuclear Waste Transport," www.sierraclub.org (cited November 2006).

John Walsh
Shinawatra University

Waste, Solid

SOLID WASTE, ALSO called *trash* or *garbage*, commonly refers to domestic waste. Waste generated

Modernization and higher standards of living are often followed by an increase in paper and plastic waste.

from households includes food scraps, paper, newspaper, clothes, packaging, cans, bottles, grass clippings, furniture, paints, batteries, and more. In developing countries, it is often contaminated by hospital waste, industrial waste, and other hazardous waste. The World Health Organization (WHO) defines waste as "something which the owner no longer wants at a given place and time, and which has no current perceived value." According to P.R. White, waste often contains the same materials found in useful products; it only differs in its lack of value.

Solid waste is also the term used internationally to describe nonliquid waste materials arising from domestic, trade, commercial, industrial, agricultural and mining activities, and from public services. Solid waste comprises countless different materials: Dust, food, packaging, clothing and furnishings, garden waste, agricultural waste, industrial waste, and hazardous and radioactive waste, to name a few. Municipal solid waste includes wastes that result from municipal functions and services such as street waste, dead animals, and abandoned vehicles. In waste management practice, however, the term is applied in a wider sense to incorporate domestic wastes, institutional wastes, and commercial wastes that arise in an urban area.

The quantity of waste generated depends on the socioeconomic conditions, cultural habits of the people, urban structure, density of population, extent of commercial activity, and degree of salvaging at source. Some of the factors that contribute to an increase in solid waste generation are growth in Gross Domestic Product (GDP), rise in disposable incomes, and a structural change in the pattern of production. The kind of waste generated and, hence, the way it should be handled changes with modernization and urbanization. People in rural areas tend to generate different kinds of waste; there is always an increase in paper and plastic and a decrease in ash and earth content in waste as a society urbanizes. Another difference between developed societies and developing countries is the difference in the amount of organic waste generated. Developed countries relying more on packaged and canned food have shifted organic waste production from domestic to industrial sources (where the food is packaged or canned).

Poor solid waste management, especially uncontrolled dumping, can cause health problems and environmental problems such as pollution of surface and groundwater from leachate production. If waste is not managed properly, unhygienic conditions put people at risk of acquiring infections of the skin and of the gastrointestinal and respiratory tracts. Poor waste management or accumulated garbage can trigger epidemics of foodborne infections. Uncovered and mismanaged waste attracts flies, mosquitoes, and rodents, leading to the spread of vectorborne diseases. Health hazards can be caused by the presence of human excreta, hospital and clinical waste (including medicines, syringes and infected human parts), and hazardous waste from small-scale industries. Therefore, appropriate solutions—minimization at source and appropriate disposal, whether through recycling, reuse, composting, incineration, or disposal at landfill—are necessary.

Integrated waste management is one of the recommended ways to handle waste effectively. It is a complex, multi-stage process that covers generation, collection, storage, transportation, and disposal of waste from beginning to end. Effective waste management involves all stakeholders, including the communities that generate the waste. Changes in policies or methods usually require changes in people's behavior; therefore, municipalities are finding ways to involve communities in coming up with

innovative ways of dealing with the challenges of increased waste generation.

A solid waste management system should not only ensure human health and safety, but it should also be both environmentally and economically suitable. To be environmentally sustainable it must reduce the environmental impacts of waste, including energy consumption; pollution of land, air, and water; and loss of amenities as much as possible. To be economically sustainable, waste management options should be such that the cost is acceptable to the community, including private citizens, businesses, and government.

Environmental and economic objectives cannot always be achieved at the same time, and a balance, called the Best Practicable Environmental Option, often needs to be struck to minimize the overall environmental impacts of the waste within an acceptable level of cost.

SEE ALSO: Disease; Garbage; Landfills; Pollution, Water; Recycling; Waste, Human; Waste Incineration.

BIBLIOGRAPHY. V.I. Grover, S.G. McRae, W. Hogland, and B.K. Guha, *Solid Waste Management* (Oxford and IBH Ltd., Delhi, 2000); J.R. Holmes, ed., *Managing Solid Wastes in Developing Countries* (John Wiley & Sons, 1984); M. Suess and J.W. Huismans, "Management of Hazardous Waste," *European Series* (v.14, 1983); G. Tchobalogious, H. Theisen, and S.A. Vigil, *Integrated Solid Waste Management* (McGraw-Hill, 1993); P.R. White, M. Franke, and P. Hindle, *Integrated Solid Waste Management: Lifecycle Inventory* (Blackie Academic & Professional, Glasgow, Chapman and Hall, 1995).

Velma I. Grover
Indpendent Scholar

Waste Incineration

HISTORICALLY, AND EVEN today in some developing countries, the easiest way of disposing of waste has been to burn it in open air. Incineration of raw waste has been practiced throughout the history of humanity in the most crude form of incineration: Indiscriminate burning. The indiscriminate open burning of waste on a large scale, however, causes air, water, and soil pollution. Current incineration technology has come a long way from open burning to sophisticated incinerators. The purpose of burning has also changed from simply getting rid of waste to reducing waste volume and recovering energy. Nevertheless, adverse environmental impacts like air pollution and water pollution have made this process unpopular.

As solid waste constitutes a low-grade fuel, it has become a tradition in recent years, wherever possible, to recover part of the energy content of waste. The energy recovered can be used for energy requirements of the waste facility itself or for heating or residential, industrial, or commercial power generation. Incinerators and waste-to-energy facilities are more common in urban industrialized areas, largely because of the nature of the urban waste stream; rural waste does not normally have sufficient calorific value to make energy recovery efficient.

Incinerators can be classified by the type and form and waste input, by throughput capacity, by rate of production of heat, by the state in which residue emerges from the combustion chamber, and by the shape and number of furnaces. The key systems involved in incineration are the tipping area, storage pit, equipment for charging the incinerator, combustion chamber, bottom ash removal system, and gas cleaning equipment and boiler, if energy has to be recovered.

The process that takes place inside an incinerator is called pyrolysis, the thermal decomposition of waste at high temperatures in the absence or near absence of oxygen. The products of incineration are in all three forms—solid, liquid, and gaseous. It is important to regulate the temperatures of the furnace depending on the quality of waste so that more hazardous products are not produced upon decomposition and released into the environment.

The impact of emissions from waste incinerators on human health is of great public concern, especially the release of toxins like dioxin. Research has identified numerous toxic compounds emitted in gases and in ashes (e.g., organic pollutants such as chlorinated and brominated dioxins, PCBs and PCNs, heavy metals, sulphur dioxide, and nitrogen dioxide), as well as many unidentified substances of unknown toxicity. This leads to contamination of

the environment and to potential exposure of humans to hazardous pollutants that may cause health problems such as cancers.

Besides polluting air, incinerators emit wastes to water from cleaning equipment. While published data on air pollution through emissions from burning waste, fly ash, and bottom ash is available, emissions to water from incineration remain largely understudied. Wastewater from wet exhaust gas cleaning, however, is known to contain heavy metals, the most significant being lead, cadmium, copper, mercury, zinc, and antimony. Wastewater from wet slag removal equipment contains high levels of neutral salts and also contains unburned organic material from the residue. Based on some of this evidence, it can be argued that use of incinerators ignores the adoption of the precautionary principle.

The advantages of incineration include:

1. Incinerators can be built close to the source of waste, reducing transport costs.
2. Incineration is suitable for many flammable, volatile, toxic, and infectious wastes that should not be land-filled.
3. It produces no methane, unlike landfill sites.
4. It reduces the amount of waste requiring landfill disposal.

On the other hand, the disadvantages of incineration are:

1. High capital and operating costs make it a relatively expensive method of waste disposal.
2. Reliance on incineration could restrict choices of future disposal options, including a proper consideration of waste minimization or recycling.
3. There is significant danger of atmospheric pollution (though some modern incinerators do meet the strict emission criteria); some incinerators generate toxic liquid effluent.
4. The volume of residue is still 40–50 percent of equivalent waste in compacted landfills.
5. It concentrates toxic materials in the residue and most residues still have to be land-filled.

The competitive demand for declining fossil fuel supplies has led to the search for energy from renewable sources, including waste. Recovering energy from waste does hold promise for the future, especially with new technologies for emission control that might meet more strict emissions criteria. Also, it is the basic principle of an ecologically-oriented waste management policy or waste hierarchy that waste should be reused and recycled. When waste cannot be avoided, recycled, or reused, moreover, it has to be subjected to special treatment before its ultimate disposal to prevent any further environmental burden or impact. One of the most flexible ways to do this is the thermal treatment of unavoidable and unrecoverable residues in modern waste incineration plants equipped with special flue gas cleaning processes. In these plants, the pollutants of the flue gases are eliminated to such an extent that they do not cause any environmental burden.

An example of an incinerator with strict emission control is in Vienna, where an eminent painter was even invited to paint a mural on the incinerator to promote it and increase public awareness of the better emission control system in the incinerator. However, for developing countries, incineration should be used with care.

Although incineration generally reduces the quantity of waste going to landfill sites, it is mainly recommended for medical or hazardous waste in such countries, where resources remain scarce and modern equipment is too expensive to procure for standard waste streams.

SEE ALSO: Garbage; Landfills; Pollution, Air; Pollution, Water; Recycling; Waste, Solid.

BIBLIOGRAPHY. M. Allsopp, Pat Costner, Paul Johnston, and David Santillo, "Incineration and Human Health: Characterization and Monitoring of Incinerator Releases and their Impact," in V.I. Grover, V.K. Grover, and W. Hogland, *Recovering Energy From Waste* (India and Science Publishers, Inc., 2002); E. Ardevol et al., "Environmental Tobacco Smoke Interference in the Assessment of the Health Impact of a Municipal Waste Incinerator on Children Through Urinary Thioether Assay," *Public Health* (v.113, 1999); Helmut Loffler, "Thermal Treatment of Waste in Vienna: An Ecological Solution?" in V.I. Grover, V.K. Grover, and W. Hogland, *Recovering Energy From Waste* (India and Science Publishers, Inc., 2002).

Velma I. Grover
Independent Scholar

Wastewater

WASTEWATER IS NOT just sewage. Defined as domestic, industrial, agricultural, and storm water flows that drain into sewage collection systems, wastewater reflects the geographic character of communities and environments. Sewage, or refuse liquid and waste matter produced by residences and commerce, is often labeled "wastewater;" yet sewage is technically limited to discharge channeled by sewer pipes. Wastewater, however, pulls from a broader array of social and environmental sources: Storm drains, overflowing creeks, septic tank leaks, and runoff from parking lots and pavements, the crop field, and the industrial dump site. Wastewater quality and quantity are thus related to the patterns and politics of water availability, governance, and waste-making practices.

Wastewater composition is approximately 99 percent water by weight, but it contains numerous biological, chemical, and material compounds ranging from pathogenic bacteria to pharmaceutical compounds and trash. In large quantities, these compounds produce adverse effects on human and ecological systems. For example, in municipalities with combined storm drains and sewer infrastructure, storm water mixes with wastewater after severe rainfall events, often resulting in combined sewer overflows. These overflows, in tandem with renegade wastewater flows and increased urban runoff, frequently result in poor water quality. For this reason, many laws and regulations (such as the U.S. Clean Water Act) mandate wastewater treatment to decrease environmental contamination and improve water quality.

Wastewater treatment plants intervene at critical points in the water cycle. Although septic tanks are still common in rural areas, the majority of municipal wastewater is treated in large-scale plants. There are no holidays for wastewater treatment: Most plants operate 24 hours per day, seven days per week. Treatment plants are designed to reduce harmful substances and pollutants in wastewater before flows are returned to rivers, oceans, or the broader environment. In general, there are three stages of wastewater treatment: (1) primary treatment (physical removal of floatable and settleable solids; (2) secondary treatment (biological removal of dissolved solids); and (3) tertiary or advanced treatment (removal of nutrients and chemicals).

Primary treatment extracts solid particulates and oils from wastewater. First, influent is screened to remove large objects, such as rocks, corpses, or condoms, which could plug sewer lines or block tank inlets. Next, flows enter a grit chamber and decrease in velocity, allowing sand and grit to fall out. Macerators (revolving cylinders with rotating knife edges) are sometimes used in place of screens to cut solids into smaller, collectable particles. Finally, wastewater is slowly moved through sedimentation tanks (also called clarifiers or settling tanks). Fecal solids settle out in the tanks and are pumped away, while oils, grease, and plastics float to the surface and are skimmed off.

Secondary treatment typically utilizes aerobic biological processes to further degrade the supernatant (remaining flows after primary treatment) and convert nonsettleables to settleable solids. This level of treatment removes approximately 85 percent of the total suspended solids (TSS) in wastewater and is the minimum level of treatment required by the U.S. Clean Water Act. Secondary treatment is a balance of engineering, siting politics, budgets, and local environmental conditions. Secondary systems are classified either as suspended growth or fixed film, although systems may use elements of both. The most common suspended growth option, activated sludge, uses microorganisms to break down organic material via aeration, agitation, and settling. The sludge, which contains fungi, protozoa, and aerobic bacteria, is continually recirculated through the aeration basins to speed the process of organic decomposition. In general, suspended growth systems require less space, but may not be able to handle shocks in biological loading.

In many older plants, fixed film processes are used. For example, wastewater is sprayed into the air (a process called aeration) and allowed to trickle down through coarse media, such as beds of stones or plastic. Microorganisms, attached to and growing on the media, break down organic material as wastewater seeps past. These secondary systems provide higher removal rates for BOD (biological oxygen demand: an indicator of pollutant quality) and are better able to cope with quantity variability, but require large tracts of land and are often

rejected by nearby communities for aesthetic and political reasons.

Tertiary treatment is the polishing stage of wastewater treatment. In response to successful litigation by environmental groups and higher regulatory standards, many wastewater facilities increasingly employ advanced tertiary methods to improve effluent quality. Tertiary treatment includes a broad range of methods, such as physical, biological, or chemical processes to remove nitrogen and phosphorus, carbon adsorption to remove chemicals, and disinfection using chlorine, ozone, or ultraviolet light.

Tertiary treatment is needed to produce reclaimed water: Highly treated and recycled wastewater commonly used for nonpotable and nonagricultural uses (such as the irrigation of parks and public spaces). However, due to increasing population, water consumption patterns, and demand for new supplies, many areas are now considering broader uses for reclaimed water. For instance, water providers in Orange County, California, use reclaimed water for indirect potable recharge: The method of blending reclaimed water with other drinking sources through groundwater recharge or reservoir augmentation. For better or for worse, the debates over reclaimed water use in the municipal sector have focused attention on wastewater treatment plants as key links between water quality and quantity.

Despite advances in engineering and treatment, problems associated with wastewater pollution, management, and disposal continue to plague communities and environments worldwide. Approximately 2.6 billion people lack access to improved sanitation: A broad category that includes ventilated pit latrines, composting toilets, and toilets connected to septic tanks or piped sewers. Although global sanitation coverage rose from 49 percent in 1990 to 58 percent in 2002, access to potable water supply still outstrips sanitation access. The United Nations (UN) Millennium Development Goals aim to halve the proportion of people without access to basic sanitation by 2015; yet the UN estimates that if the 1990–2002 sanitation trend continues, roughly 2.4 billion people will be without improved sanitation in 2015, almost as many as are without today.

Developed nations have not escaped the problems of wastewater either. Many European and North American countries feature excellent rates of sanitation access, well-established institutions, and strong regulatory mechanisms; yet, non-point-source pollution, high rates of water consumption, and excessive waste-generating practices have contributed to wastewater problems in many major cities. For example, on a daily basis, California sends billions of gallons of partially treated sewage into the Pacific Ocean. The sewage usually meets state and federal effluent standards, but increased nutrient loading, urban runoff, and wastewater discharges have caused massive algae blooms, turning coastal waters into toxic soup for marine mammals, fisheries, and recreational users.

SEE ALSO: Recycling; Septic Systems; Sewage and Sewer Systems; Sewer Socialism; Water; Water Demand; Water Quality.

BIBLIOGRAPHY. Robert L. Droste, *Theory and Practice of Water and Wastewater Treatment* (J. Wiley, 1997); George Tchobanoglous, Franklin L. Burton, and David H. Stensel, *Wastewater Engineering: Treatment and Reuse* (McGraw-Hill, 2003); United Nations (UN) World Wide Water Assessment Program (WWAP), *Water, A Shared Responsibility: The United Nations World Water Development Report 2* (UN Educational, Scientific, and Cultural Organization and Berghahn Books, 2006); Kenneth R. Weiss, "Sentinels Under Attack," *Los Angeles Times* (July 31, 2006).

Katharine Meehan
University of Arizona

Water

WITHOUT WATER, LIFE would not be possible on earth: Water is an essential part of any ecosystem and is indispensable for human development, health, and well-being. In December 2003, the United Nations (UN) General Assembly proclaimed the years 2005–15 as the International Decade for Action, Water for Life. Water's molecular arrangement is very simple: Two hydrogen atoms attached to an oxygen atom. The elements are the two most common in the universe. This special substance has many properties, including the ability to change state. Water can be found in nature in three different forms: Solid, liquid, and gas. Solid water is found in glaciers and snow, vapor in the atmosphere, and liquid water in the oceans and seas, rivers and lakes, and underground water. Water on the surface of earth is constantly changing between these three states. These continuous changes create a cycle of repeating events called the water cycle.

This cycle starts when the sun's heat provides energy to water on the earth's surface, causing evaporation into the atmosphere. By this process, water changes from a liquid to a gas. Plants also contribute water to the air by transpiration. Water vapor condenses in the atmosphere to form clouds, and when the weather conditions are adequate, precipitation occurs in the form of rain or snow and water returns to the land and oceans and the cycle can start all over again. In this process, water vapor turns into liquid water (rain) or solid water (snow). Snow stays on the top of mountains for a long time until it finally melts and runs into the streams and rivers. Sometimes, snow turns into ice, and ice becomes a glacier. Most of the rain falls in the oceans and rivers, but some of the precipitation soaks into the ground, forming aquifers that store much drinking water. The water cycle never ends, and it is the only way that earth can be continually supplied with freshwater.

Although 70 percent of the earth is covered with water, only a small part is freshwater. Unfortunately, 98 percent of surface water is in the oceans—the remaining two percent accounts for the freshwater supplies of the world. Ninety percent of this freshwater supply is either in the poles or remains underground. Therefore, humans actually have access to only 0.000006 percent of the water available on the planet. Only 0.26 percent of freshwater resources are available for human consumption. There are also factors that contribute to the diminution of freshwater resources supply and to the rise of demand, all of them reducing water quantity and water quality.

DEMAND-INCREASING FACTORS

The world demand for water increased six-fold between 1900 and 1995. Apart from the water needed for maintaining natural ecosystems, both world population growth and urbanization contribute to this increase by pressing for domestic supply and economic use. Industrial and agricultural use is concurrently increased, which leads to a greater intersectorial competition for this resource.

One of the most worrying factors of this century is world population growth. The population living on the planet increased from 1.5 billion in 1900 to 6.0 billion at present. As a consequence, there has been a worldwide increase in water demand. Within the next 25 years the world's population could face severe water shortages. Greater demand for water for domestic purposes creates conditions in which some sectors do not receive water at all, and some receive it in lower quality and at a high cost. Moreover, this growth produces an increase in water use for industrial operations and for food production worldwide, but especially in poor countries that record the highest agricultural population growth rates. At the same time, excessive water use leads to surface and groundwater resource exhaustion, which produces a chronic shortage.

Water is used for household consumption, agriculture, industry, communication, as an energy source, and for recreation. It also contributes to environmental sustenance. Water used for household purposes represents 10 percent, just a small part of its total consumption. Approximately 90 percent returns to rivers and aquifers as wastewater. Most water is used in agricultural and industrial operations; industry uses between 20 and 25 percent of world freshwater reserves. Increases are expected to occur mostly in countries facing fast industrial growth. Industry only consumes approximately five percent of the total water extracted—the rest becomes wastewater that may contribute to the pollution of water reserves.

For the most part, freshwater is used for agricultural purposes. During the last half of the 20th century, food production increased 25 percent, slowly increasing world nutrition. On the other hand, agricultural irrigation uses at present between 60 and 75 percent of the water consumed on the planet. Irrigation systems are crucial but inefficient as they use and waste large amounts of water. Moreover, most irrigation is done by pumping subsurface water at a rhythm that makes it difficult for aquifers

to recharge and depletes the amount of water available. According to the World Bank, the world's population will need an increase of 55 percent in food production by 2030 in order to survive. Most of this increase will derive from irrigation, and three-fourths of irrigated surfaces will be in developing countries.

SUPPLY-DECREASING FACTORS

Human beings modify the environment and misuse water resources that are also constantly threatened by pollution. At the same time, one-third of world population lacks adequate water supply. All these factors affect the quality and quantity of world freshwater supply. According to the Water Pollution Control Federation, more than 90 percent of drinkable water in the world is groundwater. Human beings extract water to develop domestic, industrial, and agricultural activities. But most of the water extracted from surface or underground sources is wasted or used inefficiently. The Second UN Report on the Development of World Water Resources states that 25 to 40 percent of potable water consumed in the world comes from under the ground. Groundwater is important where surface water is scarce. However, the rhythm of water extraction from aquifers is so fast that it prevents them from being recharged, which leads consequently to their depletion.

Most of the world's water available for human consumption is polluted, mainly by human activity. Industrial and agricultural development affect water quantity and quality as they return bad quality water to the hydrographic system. On one hand, fertilizers and pesticides often pollute the water returned to surface water and groundwater through irrigation. Industry and urban areas also return polluted water to surface water and groundwater. Thousands of effluents are emitted into lakes and rivers without previous treatment, and others form leachate, which is mixed with groundwater. Industries, including mining operations that emit toxic, sulfide, and metallic elements, and the food sector, which uses organic raw material that produces organic pollutants, contribute to water pollution with millions of tons of discharge per year. Moreover, household wastewater, which carries organic matter, pollutants, and bacteria of fecal origin, is discharged without previous treatment.

Although at present, people live more healthily than the previous generation, more than 20 percent of world population do not have access to good quality water. There are more than one billion people who lack access to potable water, and more than two billion do not have access to sanitary sewer systems. Most of these people live in developing countries and have medium to low income—the poorest and most vulnerable population. Developing countries have also historically lacked territorial planning that helps them distribute their populations in a more balanced way. This concentration of millions of persons in large metropolises causes great pressure on the environment with a lessening of water quality. Especially in Latin America, most people live in huge cities where the water is usually contaminated, which becomes a permanent threat to their health.

At the same time, these countries use fewer resources, while developed countries consume resources in excess. For instance, in Canada the average use of water for a typical family is 91 gallons a day and in Europe the average is around 42.9 gallons a day, while in Africa it is 5.2 gallons a day. The UN estimates that around 50 percent of the freshwater supply systems in developing countries are being lost due to inadequate maintenance and the lack of investments. Water, sanitation, and hygiene have important impacts on health and disease. More than five million people per year are estimated to die from diseases caused by bad quality water use. Diarrhea, schistosomiasis, filariasis, trachoma, malaria, cholera, typhoid fever, and other water-transmitted diseases cause deaths that could be prevented if the population were provided with potable water and adequate sanitary facilities.

One of the consequences of population growth is the simultaneous increase in the demand for forest goods and services. These include the forestry industry, which also produces fuel, wood, and paper, as well as farming operations. When trees are bulldozed or burned, apart from the huge biodiversity loss implied, the lack of vegetal cover exposes the ground to weather erosion, making the ground less fertile. Another consequence of deforestation is the loss of water sources. The forest cover acts as a sponge, retaining

water from precipitation in order to gradually release it later, thus minimizing downstream floods and drought conditions. If the vegetal cover is removed, the leaking is superficial since water never reaches the deep tree roots. It precipitates rapidly and can cause floods. It also causes water quantity and water quality reduction, especially in urban areas.

Floods and droughts are two sides of the same coin affecting the poorest the most. On one hand, bulldozers, road building, and intensive soil use for agricultural purposes often increase erosion and sedimentation. This may cause floods in the intermediate zones of river valleys and a reduction of downstream flow. On the other hand, in arid and semiarid areas, the transformation of habitats for human use, mostly agricultural, and the increase of over-exploitation including overfarming, have led to the degradation of more than 20 percent of the ecosystems with severe results: Desertification, drought, and biodiversity loss. It is necessary to highlight that one-sixth of world population lives in arid and semiarid basins, constituting about three-fourths of the poorest population.

Water resources are inextricably linked with climate, so the prospect of global climate change has serious implications for water resources and regional development. The major cause of global warming is the excessive emission of greenhouse gases in industrialized countries caused mostly by fossil fuel burning. The annual emission is estimated at about six billion tons of carbon, mostly in the form of carbon dioxide. An additional two billion tons, or about 25 percent of total emissions of carbon dioxide, are thought to be a consequence of deforestation and forest fires. These excessive emissions accumulate more heat near the planet, leading to a more unpredictable climate coupled with sea level rise and over warming. The negative consequences of global warming are catastrophic: Desertification and drought increase, bad harvests, melting of polar ice cover, coastal floods, and replacement of the principal vegetation regimes.

UNEVEN DISTRIBUTION

The small fraction of freshwater accessible to humans is extremely unevenly distributed. Some researchers refer to regions with water "scarcity" and water "stress," usually defined as regions with less than 1,000 and 1,667 cubic meters per person per year, respectively. Quantities depend on precipitation that is scarce and light in arid regions. In many countries like Australia the availability of water is seasonal. In absolute values, the largest volumes of water resources are those of Asia and South America. They do not fully reflect water availability within the continents, as they differ so much in area and population number. Also, groundwater resources may play an important role in contributing to the total volume of renewable water sources.

However, these resources are unevenly distributed, both among countries and within them. On the continental scale, Europe appears to have abundant water resources. But there are many water resources in the Nordic countries and central and eastern Europe, while in western/central Europe they are scarce. In Africa, there are deserts in the northern and southern sub-regions of the continent where almost no rain falls. But in tropical humid areas in the eastern, western, and central sub-regions there is too much water. The UN declared in its World Water Development Report that by the middle of this century, at worst, seven billion people in 60 countries will be water-scarce, at best, two billion people in 48 countries.

SCARCITY AND CONFLICT

At the national or local level, water conflicts are related to access to and use of water among different users and sectors. On one hand, the way water supply is distributed among industry, agriculture, and urban activities has an impact on those sectors' development. On the other hand, the way sanitation facilities and protected water supplies are allocated and distributed affects people's health and livelihoods. At the international level, conflicts are related to countries sharing water resources. Almost 40 percent of the world population lives in countries that share river basins or aquifers. According to Aaron Wolf, water resources have played a role in shaping political forces and national boundaries. Westing has suggested that "competition for limited...freshwater...leads to severe political tensions and even to war," and Michel Klare and other authors give many examples of water conflicts.

The most important cases are related to the Nile, Jordan, Euphrates, Tigris, and Indus Rivers. At 4,184 miles long, the Nile River is the longest river in the world. It has three mayor tributaries and flows from east Africa to the Mediterranean through nine countries: Tanzania, Burundi, Rwanda, Congo, Kenya, Uganda, Ethiopia, the Sudan, and Egypt. The Nile serves as a constant source of water for these countries. It has a vital role in agriculture and it also plays a major role in transportation. Access to and control of the Nile's waters has already been defined as a vital national priority by these countries, so conflicts among neighbors cause tension and instability in the region. Though the conflict still remains among the main actors: Sudan, Egypt, and Ethiopia, it is probable that all the countries in the Nile basin will be affected while the population continues growing and water needs increase. There were some agreements and treaties to share the use of river water, but these were only bilateral, without including all of the countries of the Nile Basin.

The Jordan River flows through Syria, Lebanon, Israel, Palestine, and Jordan and ends in the Dead Sea. In the Ghawr Valley, it defines the border between Jordan and Israel, and Jordan and Palestine, a region with plenty of ideological, religious, and geopolitical differences and historical rivalries. Israel was created in 1948 although surrounding Arab states did not recognize its claim to the land. The region is extremely arid. The rainy season is short, and rains are insufficient for basic agriculture, so these countries rely on the Jordan River—especially Israel, because it is the only natural and clean river it has access to. According to the Arab States, Israel is using the river illegally.

Turkey, Syria, and Iraq share the Euphrates-Tigris water basin. The Euphrates and Tigris Rivers begin in Turkey, but then the Euphrates River goes southwest to Syria, and the Tigris River enters Iraq. Both rivers flow through the Mesopotamian region and join up again near Qurna, in Iraq, to form the Shatt al-Arab and end in the Persian Gulf. These nations suffer from serious water scarcity; therefore the Euphrates-Tigris Basin plays an important role in economic development and freshwater supply. The three countries have built dams in order to produce electricity, stop water from flowing, and for irrigation purposes. But the completion of these dams caused serious tensions among Turkey and the downstream countries, Iraq and Syria.

The Indus River is the longest in South Asia. Originating in Tibet, it runs through Kashmir in India, and then flows through Pakistan. The Indus is a strategically vital resource for agricultural production, industries, and water supply in Pakistan and India. The river is also sacred for Hindus who live in both countries. In 1947 India was divided into two separate states, India and Pakistan, and the basin was also divided, creating a conflict over the use of the river's resources. The Treaty of Indus Water signed by both countries is considered an example of water conflict resolution, although territorial disputes for Kashmir remain.

In all these cases, river systems are situated in arid regions with severe water scarcity, and the only source of water supply for economic and social development has to be shared, so access to the river's waters has already been defined as a vital national priority. Also, these basin systems are situated in the middle of historically tension-filled regions. Water's uneven distribution and scarcity exacerbates the existing crisis over transboundary water resources. There are presently 261 international river basins, and 145 nations have territory in shared basins. While conflict can be dangerous, it also carries the possibility of producing creative cooperation. The need for integrated and cooperative solutions with special emphasis on the economic, environmental, and security dimensions of integrated water management is necessary to find a nonconfrontational way to resolve water disputes.

SUSTAINABLE WATER USE

Sustainable water resource management has involved several groups. Industrial and agricultural sectors have to improve water use strategies, use clean technologies, re-use water, reduce contamination in their operations, and take care of groundwater storage. Reform in the economic sector alone is not sufficient—water institutions and policies also need reform. The water crisis is related to a crisis of governance because weaknesses in governance systems have greatly impeded progress toward sustainable development and the balancing of socioeconomic needs with ecological sustainability.

The International Conference on Water and the Environment in Dublin in 1992 declared that popular participation should be a cornerstone of government policies in order to achieve sustainable water planning. Citizens have to reclaim the chance to participate in decision-making; ensuring sustainable use of water will require changes in both the economic-political approach and in societal attitudes.

SEE ALSO: Aral Sea; Nile River (and White Nile); Population; Tigris-Euphrates River; Wastewater; Water Conservation; Water Demand; Water Harvesting; Water Law; Waterlogging; Water Markets; Water Quality.

BIBLIOGRAPHY. M. Barlow and T. Clarke, *Blue Gold* (Stoddart Publishing, 2002); R.O. Collins, *The Waters of the Nile* (Clarendon Press, 1990); P.M. Kennedy, *Preparing for the Twenty-First Century* (Random House, 1993); M.T. Klare, *Resource Wars: The New Landscape of Global Conflict* (Henry Holt and Company, 2001); United Nations (UN) World Water Assessment Program, *First United Nations World Water Development Report: Water for People, Water for Life* (UNESCO and Berghahn Books, 2003); UN World Water Assessment Program), *Second United Nations World Water Development Report: Water, a Shared Responsibility* (UNESCO, 2006); A.H. Westing, *Global Resources and International Conflict: Environmental Factors in Strategic Policy and Action* (Oxford University Press, 1986); A. Wolf, "Conflict and Cooperation along International Waterways," *Water Policy* (v.1/2, 1998).

Veronica M. Ziliotto
Universidad de Buenos Aires

Water Conservation

WATER CONSERVATION AIMS at influencing water utilization in order to achieve more efficient, equitable, and sustainable water consumption levels. It is mainly a tool for on-site and watershed approaches, since it originally focused on conserving water as stored soil moisture and groundwater. With increasing population and water demand, conservation of this limited resource becomes ever more important. The notion that freshwater is a finite resource arises as the global hydrological cycle on average yields a fixed quantity of water per time period. Humans cannot yet alter this overall quantity significantly, although it frequently is depleted by human-made pollution. Drought, water stress, and scarcity are often the strongest incentives for end-users to implement water conservation strategies. As water demand continues to grow, a choice must be made: Either to augment water supplies or to limit demand—the latter is more effective.

USER BEHAVIOR

Agriculture, especially irrigation, is globally the largest user of freshwater, consuming about 70 percent of all water withdrawn from rivers, lakes, and groundwater. Efficient water management practices and improved irrigation technologies are crucial in reducing water use in agriculture. Irrigation scheduling involves managing the soil water reservoir in such a way that water is available when crops need it. It is necessary to determine all variables of the water cycle. The easiest method of soil moisture monitoring is to observe the soil appearance at various soil depths within the root zone. Other methods to measure soil moisture content, like tension-meters, determine the suction head a plant needs to abstract water from the soil. Monitoring air temperature, precipitation, air humidity, and evapotranspiration is important to determine how much water is available for the crop and helps to estimate when and how much water should be used during irrigation.

In developed countries, sophisticated digital recording systems are used to control and monitor water availability, as well as water applications. Many types of irrigation systems are available, such as sprinkler, center-pivot, furrow, and flood irrigation systems. However, drip irrigation is the most efficient method as far as water and nutrient applications are concerned, and it can be effectively applied on uneven terrain and in high-value greenhouse crops. With all irrigation systems, drainage of irrigation water is a crucial component of avoiding water pollution. Inefficient irrigation and drainage management, together with inappropriate irrigation, lessens water quality and causes severe soil salinity problems.

In areas where sprinkler systems are used, care should be taken to ensure that the system is not over-designed; furthermore, sprinkler use in hot climates is a cause of increasing soil salinity, since precipitating water droplets evaporate easily, leaving salt crystals behind. Automatic systems should incorporate an override to prevent the sprinklers from operating during wet periods. Plants with similar water needs should be grouped together so that they can be watered for the same length of time and in the same amounts. In cases where expensive irrigation systems are used, they should be equipped with a soil moisture controller that will restrict irrigation to when it is needed.

In small-scale schemes, irrigation should be restricted to the early mornings when evaporation is lowest and crop water demand is highest. Many varieties of grasses used in the lawns of housing areas in developed countries are not drought-resistant and require regular irrigation, though drought-resistant species are available. Small trenches can harvest natural runoff and irrigation water to the areas where needed. Creating micro-basins around specific plants will enable them to be watered individually. Spreading mulch reduces the water lost to evaporation by up to 70 percent, as well as preventing excessive runoff, inhibiting weed growth, and supplying nutrients to the soil. Drip irrigation uses significantly less water than normal irrigation systems and is equally effective. With increasing prices for piped water, the investment for the drip systems may pay off quickly. While small-scale farmers in developing countries often lack resources to invest in pumps or sufficient energy for technical irrigation, low-cost irrigation technologies (such as treadle pumps and drip irrigation) are becoming more widespread.

An increasing number of industrial enterprises in developed countries implement efficient water consumption programs. Industrial ground and surface water utilization and the use of water taken from public supply networks have been reduced considerably, mainly in industrialized countries. More and more industrial enterprises are transferring to a rational use of water through the multiple use of water or closed production cycles. Water-saving reduces overall costs and saves energy at the same time (for example, energy saved from a reduced use of water pumps and a reduced amount of water that needs to be cleaned afterwards). Water management plans should consider if groundwater can be substituted by surface water, and if the use of drinking water quality is necessary for different production processes.

Domestic water consumption can be greatly reduced through individual technical measures, such as flow-limiting taps or water-stop flush buttons for toilets. Installation of individual meters (for a household) to monitor usage and the costs related to it is essential. Modern washing and dishwashing machines are significantly more water efficient than earlier models (down from 150 to less than 60 liters in 25 years, reaching less than 30 liters per washing cycle in Europe). Moreover, changing personal behavior could halve total water consumption; for instance, through the replacement of a seal from a dripping tap, and stopping the tap while brushing teeth. It is important to encourage end-users to install water-saving devices at the time of their investment, for example, when they build a house or factory. End-users will only do so if they are aware of the water-saving options and the benefits in terms of cost savings. End-users would need to be encouraged to invest in water-efficient fittings through the amendment of by-laws and codes that regulate building practices. However, the best tool in developed countries for reducing water consumption and waste is to increase the price of water for a high standard product.

MEASUREMENT OF LOSSES

In developing countries, most water supplies are unmetered. In many instances, water standpipes or blocks of houses have never been fitted with meters, or they have broken. In these cases, neither water departments nor individual end-users know how much water is being used. Effective billing cannot take place, and water demand management plans cannot be implemented effectively. The calibration, repair, and replacement of meters are important components of a water conservation strategy. A periodic calibration of system supply and customer meters provides a more accurate measurement of the water supply and use. Furthermore, unlicensed use of water, water losses through broken pipes, and water wastage can only be determined if appropriate metering takes place.

COST RECOVERY

In many countries, fees or taxes do not cover the costs of providing water. This may result in low service levels, water coverage insufficiencies, and under-funded operating costs. As a consequence, infrastructure deteriorates and service quality declines. Inadequate cost-recovery will result in an inability to operate and maintain existing supplies with consequent increases of leakages, supply interruptions, and likely deterioration in the quality and quantity of the water supplied. This leads to increased public health risks, with possible increases in disease, morbidity, and mortality rates.

The commercialization of water provision aims at introducing appropriate water tariffs that consider the full costs for water provision. At the same time, it is known that users are only willing to pay a price for a quality product that is available when needed. It is known from rural and peri-urban areas in Africa, with water provision from vendors and public stand posts, that prices for potable water are often more than five times higher than in high-income areas with house connections. Willingness to pay often exceeds the ability to pay. Therefore, block tariffs and cross subsidies are essential for sustainable and cost-covering water provision at sufficiently high technical and service standards. An important factor in cost-recovery is the setting of adequate standards of service. It has been shown that consumers are willing to pay for good quality services and are prepared to pay increased costs for improved services in terms of water quality and supply continuity.

Commercialization should not be misunderstood as the privatization of public water providers. For investments in low-income areas, government subsidies remain important. Many water suppliers argue that in order for them to raise the capital required to improve service quality, tariffs which reflect the cost of doing this need to be charged immediately; which may lead to unaffordable prices for many. However, from a public health point of view, it is vital that service quality improvements in poor areas should be implemented immediately. There is a significant risk that users will disconnect from an expensive, but poor quality service. This will inevitably lead to greater health risks as unprotected water sources are used for water supplies.

CLOSED WATER LOOP CONCEPT

The closed water loop concept is a management tool within water demand management. At the scale of the household, neighborhood, community, industry, or institution, water can be managed as a closed loop. Water inputs of various qualities can be brought into the closed water loop for the various water applications where the water quality is matched with the intended application requirements. Every drop of water can be used at least twice before it is sent out of the loop. After water is used, the generated wastewater is segregated according to the level and type of contamination it contains. The wastewater streams are treated and the recycled water is kept in the loop and used in the appropriate applications.

At the scale of the household and residential buildings, the highest quality water is reserved for drinking, food preparation, and hygiene requirements. Water of lower quality can be used for landscaping or toilet flushing. Grey-water is separated, treated, and kept in the household water loop for landscaping or toilet flushing. Wastewater from the toilets and kitchen can be treated in a septic tank followed by a sub-surface wetland. The sub-surface wetland can aid the treatment process and be built within the household landscape to grow ornamental plants. The treated effluent can be applied though sub-surface irrigation networks to irrigate trees and to create habitat. This concept offers the potential of increased water conservation worldwide.

While water conservation (and conservation in general) is largely regarded as a positive environmental activity, there are important caveats. In many cases, conservation of water in water-scarce environments by individuals or communities may leave more water available in the system to allow growth or expansion of new settlement or economic activity, with concomitant water use problems as well as other undesired environmental outcomes. In other words, individual efforts at conservation may simply lead to increased consumption elsewhere. Sometimes called "Jevon's Paradox," this critical approach to conservation does not deny or refute the importance of reduced consumption for overall sustainability of ecosystems, however. Water con-

servation is therefore a complex problem not only in technical terms (i.e., how) but also in political and economic ones (i.e., why or why not).

SEE ALSO: Irrigation; Septic Systems; Wastewater; Water; Water Demand; Water Harvesting; Water Law; Water Markets; Water Quality; Watershed Management.

BIBLIOGRAPHY. R.G. Allen et al., "Crop Evapotranspiration: Guidelines for Computing Crop Water Requirements," *Irrigation Drainage Paper 56* (Food and Agriculture Organization, 1998); M.A. Blackmore et al., "A Comparison of the Efficiency of Manual and Automatic Dishwashing for the Removal of Bacteria from Domestic Crockery," *Journal of Consumer Studies and Home Economics* (v.7/1, 1983); J. Boberg, *Liquid Assets, How Demographic Changes and Water Management Policies Affect Freshwater Resources* (RAND Corporation, 2005); W.J. Cosgrove and F.R. Rijsberman, *Water Vision: Making Water Everybody's Business* (World Water Council, 2001); B. Gumbo et al., "Training Needs for Water Demand Management," 4th WaterNet/WARFSA Symposium: Water, Science, Technology and Policy: Convergence and Action by All (Gabarone, 2003); *Toward Water Security: A Framework for Action* (Global Water Partnership, 2000).

Wiebke Foerch
University of Arizona
Ingrid Althoff and Gerd Foerch
University of Siegen, Germany

Water Demand

WATER DEMAND IS defined as the volume of water requested by natural and human users to satisfy their needs. Water sustains human health, well-being, food production, and economic development. But less than three percent of earth's water is freshwater, out of which nearly 70 percent is stored in glaciers and icebergs, and is not available for direct human use. The freshwater that is available comes from precipitation, surface, or groundwater sources.

Due to rapid population growth, potential water availability worldwide decreased from 12,900 cubic meters per capita per year in 1970 to less than 7,000 cubic meters in 2000. In densely populated parts of Asia, Africa, and central and southern Europe, current per capita water availability is between 1,200 and 5,000 cubic meters per year. The global availability of freshwater is projected to drop to 5,100 cubic meters per capita per year by 2025. It is estimated that three billion people will be in the water scarcity category of 1,700 cubic meters per capita per year by 2025.

Approximately two-thirds of total water consumption is used by the agricultural sector, 20 percent is consumed by industry, and 10 percent by private households. However, regional differences exist. In Europe, more than half of the water consumption is used by industry. In Asia and Africa, the agricultural sector consumes more than 85 percent of the available water. It has been estimated that the global water demand will rise by 20 percent for agriculture, about 50 percent for industry, and 80 percent for private households until 2025.

AGRICULTURAL WATER DEMAND

Plants require adequate water at the right time for establishment and growth. Crops have specific water requirements that vary depending on local conditions. Sources of water for crop production are rainfall, shallow groundwater, and irrigation water, which is water diverted from surface flows or groundwater. For instance, water withdrawals for agriculture account for about 91 percent of all water withdrawals in the Middle East and North Africa and 95 percent in Central Asia. Drylands, where irrigation plays an important role, have the highest level of water withdrawal for agriculture.

In most developing regions, hundreds of millions of people are served by hand-pumps fed by boreholes, hand-dug wells, or communal stand-posts or yard-taps fed by elevated tanks and distribution networks. The daily task of fetching and carrying water remains a major burden for women and girls throughout the developing world; valuable time is lost for schooling and education. The world's irrigation areas totaled approximately 253 million hectares in 1995. By 2025, they are expected to reach about 330 million hectares. Therefore, irrigation systems urgently have to be modernized to reduce water consumption and wastage.

Thingyan

In tropical countries where water is not in short supply, there are a number of festivals celebrating its abundance. The annual Burmese New Year Water Festival, Thingyan, takes place in the Burmese month of Tagu (roughly mid-April), and covers the four or five days before the New Year. During Thingyan, people douse each other with water. Thingyan has its origins in Hindu mythology in which Arsi, the King of the Brahmas, was beheaded and the head of an elephant placed on his body—making him Ganesha. The story was that the original head was so powerful that if it was thrown into the sea, the water would evaporate immediately. As a result, it is the duty of the people to douse everything and everybody with water.

On the first day of Thingyan, people start preparing for the event with everybody from older people to very small children, but especially the youth, getting buckets, pots, water pistols, and hoses ready. On the second day, *a kya nei*, the water festival begins in earnest. People go up to others and pour water over them, usually in a good-humored manner. Most local people make sure not to damage tourists' cameras and the like, and wait for the moment when a tourist puts away his or her camera. Only monks and pregnant women are spared. When Thingyan coincides with the Christian Easter, the Christian churches change their services to midnight to allow worshipers to come and go from the churches without being soaked to the skin.

From wooden balconies built on all the major roads around Yangon and other cities, large numbers of people aim their hoses at people and passing cars. Some young people remove the doors of their cars, to allow the driver and passengers to be doused, and to let the water flow out afterwards. This continues for several days, sometimes with fire trucks using fire hoses. For the diplomatic corps, they receive a symbolic splash of water from the foreign minister.

INDUSTRIAL WATER DEMAND

In the industrial sector, the biggest share of freshwater is stored in reservoirs and dams for electrical power generation and irrigation. Industrial uses account for about 20 percent of global freshwater withdrawals. Of this, 57–69 percent is used for hydropower and cooling in nuclear power generation, 30–40 percent for industrial processes, and 0.5–3 percent for thermal power generation. Not all withdrawals are consumptive, as water is recharged into the water cycle after use. The volumes of industrial water demand are quite different within individual branches of industry and from country to country, and depending on the technology of the manufacturing process.

Increasingly, industrial enterprises convert to more rational water uses, for example through water reuse and closed water cycles. Water saving reduces overall costs and saves energy. Environmental pressures and water pricing have stimulated an increasing amount of recycling and reuse by industries in the developed world, but so far, there has been less progress made in developing countries.

DOMESTIC WATER DEMAND

Around one-tenth of global water consumption directly meets the needs of private households. This includes, besides drinking water, the use of water for cooking and hygienic purposes. Another important domestic use of water is for productive purposes around the household, including activities such as growing vegetables and fruit trees or giving water to small stock. Enormous regional differences exist: In rural areas of dry land regions in Africa, water consumption amounts to less than 20 liters per day per capita, while the United States on average it reaches 295 liters per day per capita. According to the World Health Organization (WHO), a minimum of 25 liters per capita a day is needed to meet basic needs (for drinking water, cooking, hygiene).

The classic domestic water cycle of better-off urban residents involves house connections to deliver enough high-quality water for all lifestyle needs, and sewer connections to take away the wastewater for centralized treatment and return to watercourses. In many crowded peri-urban settlements, construc-

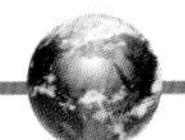

tion of water mains and sewers remains unattractive for government and private investors, and therefore impractical. There, residents depend on communal water points or water vendors, and on a range of often unhygienic ways of disposing of solid and liquid waste. This situation may be changing: One United Nations (UN) Millennium Development Goal is the aim to halve the proportion of people who lack access to hygienic means of sanitation by 2015.

WATER POLLUTION PROBLEMS

Water pollution is the contamination of streams, lakes, groundwater, bays, or oceans by substances harmful to nature and humans. The quality of water sources is often negatively affected by pollution from agriculture or industry. Excess fertilizer applications, improper disposal of hazardous materials from industry, municipal dumping, poorly constructed septic tanks, or poorly managed transport systems are major pollutants. Plants and animals require water that is moderately pure, and they cannot survive if their water is loaded with toxic chemicals or harmful microorganisms. If severe, water pollution can kill large numbers of fish, birds, and other animals, in some cases killing an entire species in an affected area. People who ingest polluted water and consume polluted aquatic resources become ill, and, with prolonged exposure, may develop cancers or bear children with birth defects.

Around 80 percent of all diseases in developing countries are caused by the lack of access to clean drinking water and sufficient sanitation. Two of five persons in the world do not have access to an adequate sewage system. Approximately 1.1 billion people do not have access to clean drinking water. About 90 percent of the wastewater in developing countries infiltrates freely into the ground, or runs into rivers, and returns into the water cycle.

SEE ALSO: Pollution, Water; Wastewater; Water; Water Conservation; Water Harvesting; Water Quality.

BIBLIOGRAPHY. H. Bouwer, "Integrated Water Management for the 21st Century: Problems and Solutions," *Journal of Irrigation and Drainage Engineering* (v.128/4, 2002); F. Fischer and A. Hossein, "Optimal Water Management in the Middle East and Other Regions," *Finance and Development* (v.38/3, 2001); R.B. Jackson et al., "Water in a Changing World," *Ecological Applications* (v.11/4, 2002); J. Lundquist et al., *New Dimensions in Water Security: Society and Ecosystem Services in the 21st Century. Land and Water Development Division* (Food and Agriculture Organization, 2000); I.A. Shiklomanov and J.C. Rodda, *World Water Resources at the Beginning of the Twenty-First Century* (Cambridge University Press, 2003).

Wiebke Foerch
University of Arizona
Ingrid Althoff and Gerd Foerch
University of Siegen, Germany

Water Harvesting

WATER IS ESSENTIAL to life, therefore, it is important that adequate supplies of water are available. However, water supplies should be developed in such a way that ecosystem functioning and the hydrological cycle are not negatively affected. Within this context, water harvesting has received increasing attention worldwide. The harvesting of rainwater refers to the collection of water from surfaces on which rain falls, and subsequently storing this water for later use. For as long as humans have occupied and cultivated dry lands, water harvesting has been practiced. It has been the basis of living and has allowed the establishment of civilizations in dry lands. Water harvesting has provided for drinking, domestic needs, livestock, crops, pastures, and trees, and a way to replenish groundwater levels. There is a range of rainwater harvesting systems, operating at large and small scales. The choice of system depends on physical and human considerations, and matching the needs of the farmers with environmental, economic, and political conditions.

Water harvesting can occur in multiple contexts and at a variety of scales. The collection of rainwater from the rooftops of buildings is suitable for urban areas and requires little investment or technology, as only roof gutters and storage tanks are needed to capture rainwater. Measures may be needed to keep insects away from the stored water, to avoid increases in waterborne diseases such

qualities. It is believed to have been introduced to the United States in 1884 at an exhibition in Louisiana. It arrived in Africa in 1879, in Asia in 1888, and in Australia in 1890.

Water hyacinth populations have been managed with herbicides, hand pulling, mechanical harvesting and biological control. Hand pulling works for small infestations, but is too labor-intensive to control larger populations. Herbicides, including copper sulfate, 2,4-D, and glyphosate, reduce populations of water hyacinth but damage other organisms in the ecosystem. Several of water hyacinth's natural enemies, including insects and fungi, have been used to control populations with varying degrees of success. Five of these biocontrol agents are used in the United States: Two weevils, a moth and two types of fungi. Attempts to use explosives and fire to keep populations in check have not been successful.

Some populations of water hyacinth have been successfully controlled. In the 1950s, water hyacinth occupied 126,000 acres of Florida's waterways. A combination of herbicides, harvesting, and biocontrol methods reduced Florida's water hyacinth population to 2,000 acres.

Although the harm caused by the species generally outweighs its benefits, a few uses have been found for water hyacinth. It has been fed to pigs, used to remove toxins from sewage, and made into paper. One study in Bangladesh found arsenic in water can be removed by water hyacinth, producing safer drinking water.

SEE ALSO: Herbicides; Invasions, Biological; Plants; Native Species; 2,4-D; Weeds.

BIBLIOGRAPHY. Michael S. Batcher, *Element Stewardship Abstract for* Eichhornia crassipes *(Martius) Solms* (The Nature Conservancy, 2000); Columbia University, "Introduced Species Summary Project: Water Hyacinth (*Eichhorinia crassipes*)," www.columbia.edu/itc/cerc/danoff-burg/invasion_bio/invbio_home.html (cited December 2006); Florida Department of Environmental Protection, "Weed Alert: Water Hyacinth (*Eichhornia crassipes*)," www.dep.state.fl.us/lands/invaspec (cited December 2006); Mir Misbahuddin and Atm Fariduddin, "Water Hyacinth Removes Arsenic from Arsenic-contaminated Drinking Water," *Archives of Environmental Health* (November–December 2002); World Conservation Union, "*Eichhornia crassipes* (Aquatic Plant)," www.issg.org (cited December 2006).

Denise Quick
Community College of Vermont

Water Law

GIVEN THE PRIME importance of water in all human activities, available water resources need to be protected, conserved, and managed in terms of both quantity and quality, for which water legislation becomes critical. Two important issues water law deals with are the ownership of water resources and the nature and distribution of water rights (which are usually usufructuary rights).

Early codifications related to water are found in the Pharaonic Water Regulations (of ancient Egypt), in the Laws of Manu (or Manava-Dharma-Shastra) in India, in the Hammurabi Code (of Babylon), in Chinese water regulations, and in Roman and Moslem law. The philosophy of the early water regulations depended (as it still does today) on geo-climatologic and physical factors, as well as on the social, technical, economic, and political situation of the countries or areas concerned. Thus, in regions where water was abundant, water regulations were largely directed toward defense against the harmful effects of water (e.g., flood control); in areas where water was scarce, regulations were concerned with the need to conserve available water supplies and with efficiency in allocation.

The old water codifications are not just of historical interest, but have also had considerable influence on current legal regimes in water. For instance, the principles of early Chinese water law (which are based on a belief of a close inter-connection between the human order and the natural cosmic order) have influenced water regulations in China, Japan, Korea, and Vietnam, at least until recently. But it is ancient Roman law that has exerted the greatest influence on the legislation of practically all modern nations. Hence, it is useful to briefly consider the form that it took.

Early Roman law recognized three classes of water rights—private, common and public. Under private rights, the owner of the land owned everything located above and below the land. This was the precursor to the riparian doctrine now followed in many countries, according to which use of such waters was private, unlimited and unrestricted, subject to sale, acquisition or transfer of the land over or under which the waters are located. That is, the riparian doctrine links control over water to control over land. Common rights permitted the use of water that was not yet occupied or without any owner to everyone without any limit or permission. In the case of public water (i.e., water owned by the state), use was subject to the state's control.

The doctrine of public trust, which is found today in some countries, is derived from this. The idea here is that the particular characteristics of water resources (e.g., its unbounded nature) and its importance in different facets of life mean that it is not justified to make it an object of private ownership. Instead, water should be included under the public domain, which implies that the state should protect the resource for enjoyment by the general public, rather than permit its use for private ownership or commercial purposes.

These early Roman principles took three major directions. The first is found in a number of European countries such as Spain, France, and Italy as well as Cambodia, Laos, and Indonesia. Water law in these countries derives from the Napoleon Code (a code of law adopted in France in 1804). Water could be public (subject to government control) or private (freely utilizable on the basis of the riparian doctrine).

The second variant is the water law of the "Common Law" countries, which is derived from the English application of the original Roman law. This Common Law of England is found in the United Kingdom (at least until recently), the eastern United States, and many former British colonies. Here, the original principles of Roman law are basically followed, although the use of water could be limited via court decisions, administrative ordinances, or regulation.

The third direction that Roman law took was the so-called new American doctrine of prior appropriation, which is found in some of the western states of the United States. According to this, water rights are vested with the first claimant and user. Judicial decisions in the United States have now limited this doctrine by the provisions of "correlative rights" and "beneficial uses of water." Some of the tenets included under this are: Water is not to be obtained for speculation or let run to waste, that the end use must be generally recognized and socially acceptable, that water is not to be misused, and that the current use must be reasonable as compared to other uses. Since the relevance of water rights is related to the availability of the resource, many water laws (other than those in the western United States) now have provisions that require the effective use of water. For instance, the notion that water rights risk forfeiture if not used according to the terms of a license or permit and is found in the laws of a number of countries such as Germany, Spain, and Mexico, although the terminology varies.

These different legal regimes have varying implications for equity. For instance, riparian rights—as traditionally constituted—usually have a negative impact on downstream users, in spite of the requirement that upstream riparians should not reasonably interfere with their rights. Similarly, the doctrine of prior appropriation is unfair to latecomers.

Apart from changes in the basic regimes of water rights over time, the content and scope of water legislation has also undergone change over time. Older water laws tended to promote water utilization, and were more concerned with punishing those who would harm existing uses or structures. For instance, an ancient article of water law traceable to the Code of Hammurabi of the Babylonian era (dated around 1700 B.C.E.) reads: "If anyone opens his irrigation canals to let in water, but is careless and the water floods the field of his neighbor, he shall measure out grain to the latter in the proportion to the yield of the neighboring field." However, as a result of population growth and technological progress, water laws have begun to deal with newer questions such as priorities across different uses, setting of quality standards, conservation of water, and prevention and regulation of pollution. For instance, the policies of the 1992 Mexican Water Law include the preservation of water quality and the promotion of sustainable development. In the United States, the Safe Drinking

Water Act of 1974 directs the Human Health Sub-Committee of the U.S. Environmental Protection Agency to ensure that both public and noncommunity water systems meet minimum standards for protecting public health.

Note that water law usually differs in the case of surface water and groundwater. In general, legislation pertaining to surface water has been most clearly and explicitly articulated. In the case of groundwater, the traditional Roman rule or English rule (that the owner of surface land was also the owner of the water under the ground) is usually followed. But with the advent of modern technology and the consequent overexploitation of groundwater, many countries have begun to regulate and control groundwater too as public property or by invoking the police power of governments.

Water law, whether dealing with surface water or groundwater, with questions of ownership and allocation, or regulation of use, varies not only across different nations, but also often within nations. This is especially true in those cases where provincial or local governments have jurisdiction over water. Furthermore, different kinds of water issues are also often divided under different ministries such as environment, agriculture, industry, health, and so on. As a result, many countries now have a plurality of laws relating to water, often conflicting with each other and known only to the administration of a particular agency of water resources development. In spite of this plurality of laws, provision to deal with various aspects—such as ownership, pollution, and coordination between different uses—is still inadequate, and/or the institutional structure necessary to ensure effective implementation of the laws is missing.

Apart from water legislation at the national level and within nation-states, water laws are also formulated at the international level. International water law derives from a number of sources: Conventions (bilateral or multilateral treaties over sharing of water resources), international customs (general principles of international behavior recognized by most nations), and judicial decisions. Two important principles found in many water treaties at the international level are: (1) the principle of equitable utilization, which states that the uses and benefits of a shared watercourse should be divided in an equitable manner, and (2) the requirement that a state—through its actions affecting an international watercourse—may not significantly harm other states. Both principles are found in the 1966 Helsinki rules governing the uses of waters of international rivers, and the Convention on the Law of Non-Navigational Uses of International Watercourses (adopted by the United Nations General Assembly in 1997). There is also a settled requirement under international law (which would also include the law on international watercourses) that states cooperate, consult, and negotiate in cases where the proposed use of a shared resource may have a negative impact on their rights and interests.

Apart from statutory law (the body of law laid down in acts of legislature and in subordinate legislation), water law (like any other law) also derives from other sources such as local uses and customs. Such laws and rules which are based on long-standing practice, and are not codified in written form, are called customary law. It is important to take these into account while preparing formal legislation, as otherwise customary users could be marginalized.

At present, water legislation is in a state of flux in many countries, partly as a response to the perception of a water crisis, but more often as part of a larger reform agenda (such as structural adjustment programs at the macro level). One important change is that state ownership as well as state involvement in development and distribution of the resource is slowly giving way to private decision-making. This in turn is leading to new kinds of rights such as the nonconditioned water rights in Chile, which do not have the requirement of effective and beneficial use. Another recent and important landmark in water legislation is found in South Africa, where an explicit right to water in the constitution is matched with an explicit right in implementing legislation (the Water Services Act of 1997 and the National Water Act of 1998).

SEE ALSO: Prior Appropriation; Riparian Rights; Water Conservation; Water Demand; Water Harvesting; Water Markets.

BIBLIOGRAPHY. Dante Caponera, "Water Legislation in Asia and the Far East," *Water Resources Series* (v.35

1968); K.K. Framji, B.C. Garg and S.D.L. Luthra, *Irrigation and Drainage in the World: A Global Review* (International Commission on Irrigation & Drainage, 1981); Stephen C. McCaffrey, " Water, Politics and International Law," in Peter Gleick, ed., *Water in Crisis: A Guide to the World's Fresh Water Resources* (Oxford University Press, 1993); Miguel Solanes, "International and Legal Issues Relevant to the Implementation of Water Markets," in *Water Policy International*, www.thewaterpage.com (cited April 2006); Sanjay Upadhyay and Videh Upadhyay, *Handbook on Environment Law: Water Laws, Air Laws and the Environment* (LexisNexis, 2002).

Priya Sangameswaran
Centre for Interdisciplinary Studies in Environment and Development, India

Waterlogging

WATERLOGGING IS A term used to describe the saturation of ground or wood. The saturation renders the ground, wood, or other object unfit for use. When an area of ground is waterlogged, water saturation is so complete that the land cannot be used for a variety of activities, such as sports or recreation. Heavy rains or floods can waterlog ground for a period of time. Ground that is waterlogged has a water table that is virtually the same as the surface of the ground. In some cases, this can have positive consequences. For example, rice shoots are planted in ground that is flooded, and as the water recedes the rice continues to grow, but it would be damaged by waterlogging near harvest time.

Most agricultural production is harmed by waterlogging, with salinization a consequence. Many crops need oxygen in their growing process. Soil that is loose and allows the presence of air aids the growth of some plants, and waterlogging blocks the oxygen from the plant. If the roots of plants are blue-black, they are exhibiting a typical sign of waterlogging. Other signs are a smell of sour rotting or leaves that turn yellow and wither or whose midribs turn dark. Evergreens are very prone to leaf coloration changes that turn leaves brown. In addition because the plant cannot take in water properly, shoots at the extremities may die and bark may easily peel off of the shoots. Plant growth is also stunted.

Herbaceous plants are intolerant of waterlogging and may fail to sprout. Or they may sprout in the spring, open leaves, and then die. If the ground in an area or region has an impermeable layer, such as a clay layer under a thin topsoil, waterlogging can occur. Areas with a water table that is perched on top of an impermeable clay lay will drain only slowly in the direction of the lowest level. Agriculture in such an area can suffer from heavy rains. When fields are irrigated excessive waterlogging can occur. Fields that are poorly drained or that have soils that absorb and retain water are prone to this condition. If the irrigation comes from canals that seep into the water table, water is raised to the surface and thereby harms the crops in the affected fields.

Researchers have estimated that approximately 10 percent of the arable land in the world is waterlogged. This has caused a crop loss of approximately 20 percent in the affected areas. In contrast to farmland that suffers from waterlogging, wild lands, swamps, and other forms of wetlands benefit from it. The great basin in south Florida in which the Everglades and Lake Okeechobee are situated are areas in which waterlogging is extremely beneficial and necessary. The wetlands of the area constitute a form of natural wealth because they serve as natural pollution filters, water sources for the recharging of aquifers, places that support fisheries, and protection from storms and storm surges.

SEE ALSO: Floodplains; Floods and Flood Control; Swamp Land Acts; Water; Wetlands.

BIBLIOGRAPHY. S.K. Gupta ed., *Crop Production in Waterlogged Saline Soils* (Scientific Publishers, 1997); Edward Malthy, *Waterlogged Wealth* (Earthscan, 2006).

Andrew J. Waskey
Dalton State College

Water Markets

WITH DEMAND FOR water fast outstripping its supply, experts are turning to water markets as an

innovative strategy to manage water. Like all markets, water markets allow for the trade—buying and selling—of water for commercial or noncommercial uses. The price of water in a water market is determined by the exchange of water rights. Such rights are either customary or established through laws and regulations and they define entitlements to and ownership of water.

Meeting demand by expanding existing water systems is becoming increasingly difficult due to prohibitive environmental and economic costs. In countries where water use exceeds its natural recharge, groundwater levels are dropping—reducing available water and raising the cost of pumping, or causing salt water intrusion and contamination. Moreover, water is highly underpriced, leading to wasteful practices and lack of funds to maintain water infrastructure. Experts, therefore, argue that pricing water is a good way to check demand and reallocate water from current uses for a more efficient management of the resource.

In developed countries, the most active forms of water markets are those where water for irrigation is sold to water districts under long-term contracts specifying the quantity to be delivered and its cost per acre-foot. Such markets are best established in the western United States (states such as Colorado, California, Utah, and Nevada). Many states in the U.S. Pacific northwest have also adopted laws facilitating the trading of water to increase in-stream flows. For instance, to protect endangered salmon, an environmental organization in Oregon pays farmers to use less water for irrigation so that more water is available in the rivers. However, water markets are still in their infancy and there are various factors that limit their growth. For example, buyers and sellers in water markets often cannot find trading partners or lack adequate market information on pricing and terms of trade. Moreover, to protect water rights, government agencies require all transfers to undergo an approval process, which can often be lengthy and expensive.

In developing countries, water markets have primarily emerged in response to the scarcity of water due to different reasons—harsh climate; drought; pollution; lack of access due to social, economic or political reasons; or the failure of public water providers. They operate at various scales ranging from private vendors who buy water from farmers and landowners and sell it at costs determined by distance and demand to multinational corporations that sell bottled water. In most informal settlements where there is little or no access to formal water supply systems, people depend on water vendors and often pay as much as 10 to 20 times more per liter. The regulatory framework within which water markets function in developing countries is weak and can often lead to groundwater overdraft where farmers and landowners are eager to make quick profits by selling large quantities of water. Since there is no mechanism to monitor the quantity of water being withdrawn, it is difficult to minimize wastage. Moreover, where property rights regarding water are less defined or enforced, it is very difficult to ensure responsible individual behavior.

More problematically, the record of implementing water markets in the developing world suggests that the outcomes for the poor can be disastrous. Privatization of water in Chile has led to some abridgment of access rights of poorer citizens. More dramatically, when the water system in Cochabamba, Bolivia, was privatized and put under the control of a consortium led by a global corporation, International Water Limited (IWL) (itself in-part owned by the U.S. company Bechtel Enterprise Holdings), prices quickly rose to a point that made basic access to water a near impossibility for some of the poorest citizens, and led to protests that paralyzed the city and government until a reversal of policy occurred. Water markets can therefore be quite controversial.

SEE ALSO: Groundwater; Irrigation; Markets; Underdeveloped (Third) World; Water Demand; Water Law.

BIBLIOGRAPHY. K. William Easter, Mark W. Rosegrant, and Ariel Dinar, eds., *Markets for Water: Potential and Performance* (Kluwer Academic Publishers, 1998); V. Galaz, "Stealing from the Poor?: Game Theory and The Politics of Water Markets in Chile," *Environmental Politics* (v.13/2, 2004); Ganesh Pangare, Vasudha Pangare, and Binayak Das, *Springs of Life* (Academic Foundation, 2006); Sandra Postel, *Last Oasis: Facing Water Scarcity* (W.W. Norton, 1992); Property and Environment Research Center, www.perc.org (cited April 2006).

Priyam Das
University of California, Los Angeles

Water Quality

FRESHWATER IS A scarce but essential resource, and its quality is of utmost importance as increasing numbers of people and living organisms depend on it for survival. While water scarcity from lack of quantity often receives more attention, water quality becomes more critical than quantity when available water is degraded or polluted. Water quality is important in the ways that it affects human health, livelihoods, agriculture, industry, recreation, and ecosystem services.

Lack of water quality can thus jeopardize socioeconomic development and environmental sustainability, and the availability of clean and good quality water is increasingly recognized as a key factor for sustainable development. Water quality issues are a serious problem in much of the developing world, where lack of access to clean and safe water leads to high rates of morbidity and mortality (e.g., two million children die each year due to inadequate sanitation and clean water). Globally, 1.1 billion people do not have access to safe drinking water, making water quality a serious global concern.

Water can be polluted from a variety of sources, both human-made and natural. Important sources of water pollution can be microbial (viruses, bacteria), chemical (metals, salts, pesticides/herbicides, solid waste), and radiological. Water quality can be measured using a number of parameters: pH, salinity, oxygen content, turbidity, color, odor/taste, dissolved chemicals, total suspended solids, biochemical oxygen demand, and dissolved oxygen. Common water quality treatments include aeration, chemical treatment, filtration, and ultraviolet light treatment. Water quality can be degraded via point source pollution (e.g., oil spills) or diffuse pollution (e.g., agricultural wastewater seepage).

Due to the connectivity of groundwater and surface water sources, the pollution of one may threaten the water quality of another connected source. As such, water pollution containment and monitoring is challenged by the flow and connective nature of water, as well as by increasing numbers of sources and types of pollution. How water quality is managed thus reflects society's priorities in water use and management, and the value placed on water quality. Water that is safe for organisms (plants and animals) to survive in, as well as for human use, is at the center of much of the environment-development debates; poor quality water affects different groups of organisms and human society differently across temporal and spatial scales. Given the dialectical nature of human-environment relationships, poor water quality that affects ecosystems also affects society, and vice-versa.

What is deemed to be acceptable levels of pollution of a water source depends on its use, linkages to other water sources, and costs of alternative water usage as well as cleanup or reduction of polluting sources. For instance, agricultural wastewater and industrial effluents can pollute a variety of water sources, making them unsuitable for domestic water purposes as well as aquatic species survival. Pathogens and microbial quality issues are important to humans in drinking water and the spread of waterborne diseases that can affect human society; similarly, overloading of organic matter and chemicals can reduce the ability for aquatic species to survive (e.g., by increasing the biochemical oxygen demand [BOD] to break down pollution). Water quality is generally monitored and regulated through systems of permits and fines that can act as deterrents to pollution or degradation of water sources. Water quality issues become a problem when different uses of a water source are directly threatened. Drinking water quality usually receives the most attention in water quality discussions. When a water source that provides drinking water is contaminated or polluted, it generally becomes important to address that more quickly than nonconsumptive water.

Irrigation water's quality, however, also needs to be monitored and ensured to prevent crop and soil damage and contamination. Dependency of livelihoods directly on water quality is also an important factor in how people value and organize around water quality issues. For instance, farmers who need good quality water for agricultural production are more likely to be concerned than those whose livelihoods are not directly dependent on irrigation water quality. Similarly, recreationalists may place greater importance on clearer water in lakes or rivers, while governments may deem that it is economically not viable to maintain such quality levels.

Societal power relations are reflected in the ways water quality is assessed, monitored, and judged.

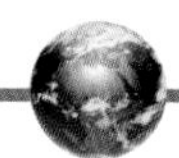

Different societies will place different priorities and valuations on the water quality desired, and thus in the different levels of allowable quantities of pollutants in the water. As such, there aren't universal water quality indicators that are enforced, but there are international guidelines on safe levels and degradation indicators. These guidelines generally are followed by national governments and water authorities in areas of drinking water, wastewater treatment, recreational water facilities, and agricultural production. For instance, the World Health Organization (WHO) provides details of safe and allowable limits of the many pollutants (biotic and abiotic) for drinking water quality in order to maintain human health.

There can be different interpretation of the same data and quality issues, however, depending on the position of the viewer as well as broader societal understandings of what is deemed safe or unsafe. Notions of acceptable risk come to the fore, as different societies and people will perceive risk or threat from water quality differently. As such, different countries may follow slightly different sets of guidelines in monitoring and evaluating water quality for the different uses. In the United States, the Environment Protection Agency (EPA) is largely responsible for monitoring and evaluating water quality and setting guidelines. The Clean Water Act of 1977 is an example of one of the important regulatory mechanisms by which the EPA monitors and control wastewater pollution from industrial sources.

Conflicts over water uses can stem from the different valuations of water. For instance, the same water source may provide drinking water supplies as well as receive industrial and agricultural wastewater, thus necessitating management of the water source so that its quality is maintained for multiple uses. Economics, as well as value systems, also influence water quality issues due to the costs involved in maintaining or returning to a certain level of quality. Similarly, the attention to scale is important, as the scale of a water quality problem will influence the scale of treatment or management needed, and the number of actors involves as well as ecosystems influenced. As such, water quality management necessitates sufficient flexibility and responsiveness in surveillance, quality control, and management mechanisms in order to address different societal and environmental needs.

Given the growing scarcity of good quality water, increasing focus is being given to reusing water and increasing productivity from limited quantities of water. While such technological solutions provide important ways to use scarce water more efficiently and productively, questions remain about social access to safe water and the role of water quality in the broader political economy of development. As such, water quality is as much an environmental and technological question as a political and developmental one.

SEE ALSO: Clean Water Act (U.S. 1972); Safe Drinking Water Act (U.S. 1974); Sewage and Sewer Systems; Wastewater; Water; Water Conservation; Water Demand; Water Law.

BIBLIOGRAPHY. J. Bartram and R. Balance, eds., *Water Quality Monitoring—A Practical Guide to the Design and Implementation of Freshwater Quality Studies and Monitoring Programmes,* www.who.int (cited April 2006); International Water Management Institute, www.iwmi.cgiar.org (cited April 2006); J.W. Kijne, R. Barker, and D. Molden, eds., *Water Productivity in Agriculture: Limits and Opportunities for Improvement,* CAB International, www.iwmi.cgiar.org (cited April 2006); J. Perry and E. Vanderklein, *Water Quality: Management of a Natural Resource* (Blackwell Publishers, 1996); J. Trottier and P. Slack, eds., *Managing Water Resources: Past and Present* (Oxford University Press, 2004); United States Environmental Protection Agency/United States Agency for International Development (USAID), *Guideline for Water Reuse* (USAID, 2004); World Health Organization, *Water, Sanitation, and Health,* www.who.int (cited April 2006).

Farhana Sultana
University of Minnesota

Watershed Management

A WATERSHED IS the land from which water drains into a stream, river, lake, or other body of water. All land, and the humans and wildlife found on that land, are part of a watershed. The term *watershed* is commonly used in North America, and

is equivalent to *drainage basin*, a term used in Europe. Watershed management is a process of managing an area in order to protect and rehabilitate land and water and associated aquatic and terrestrial resources through human activities (intervention). This is done while recognizing the benefits of orderly growth and development with the aim of contributing to the environmental, social, and economic health of the area for sustainable development. To manage and protect the watershed better, plans should be made based on the areas delineated by watersheds and not the political boundaries; or in other words, many water quality and ecosystem problems are best prioritized, addressed, and solved at the watershed level rather than at the municipal level or the level of a single body of water or individual discharger.

The watershed management approach requires crossing traditional boundaries and considering various uses of water when making a policy, and considering all the point as well as nonpoint sources of pollution. This is important to solve local watershed pollution problems, since all of them are interrelated and can be best dealt with using an integrated approach. This means that the watershed approach essentially coordinates a framework for environmental management that focuses efforts to address problems of ground and surface water flow within hydrologically defined geographic areas. According to the Ohio Watershed Academy, watershed management is a process for managing water resources that involves integrating sound science and social values, incorporating stakeholder involvement, and making management decisions that are appropriate for local conditions.

The watershed approach goes beyond just hydrological- and science-based decisions to good governance involving all stakeholders in managing local watersheds. The basic concept has been extended to incorporate participatory decision making, which brings additional benefits, as informed users apply local self-regulation in relation to issues such as water conservation and watershed protection far more effectively than regulations and surveillance can achieve. The stakeholders involve public and private sectors, including all the users of water such as industry, farmers, fishermen, institutions, shopping malls, and the community. The watershed approach also requires a gender-balanced approach in planning, which means involving both women and men in decision making. Since all watersheds are unique in the sense of living and nonliving organisms present in the area, and how they interact with each other, understanding of local conditions and use of local knowledge are both important in the watershed approach. The emphasis is also on broadening decision making to take into account overall social and economic goals, including the achievement of sustainable development.

As discussed by the U.S. Environmental Protection Agency (EPA), there are six phases of management to achieve watershed management goals: 1) identification of issues and data gathering; this phase should have a multiyear strategy to portray existing information on physical, chemical, biological, and habitat conditions and comprehensively monitor waters, 2) as an outcome of watershed planning processes, new or revised water quality standards for the waters within a watershed can be formulated to reflect agreements made by the stakeholders to meet the watershed goals like adopting precisely defined uses given the chemical, physical, and biological characteristics of the water body, 3) planning or/and prioritizing; the watershed approach should take into consideration findings and priorities established under pre-existing initiatives, 4) each watershed partnership should develop management options and set forth a watershed or basin management plan that should set objectives, identify indicators, and set forth milestones, 5) due to the participatory nature of watershed approaches, responsibility for implementation of watershed plans will fall to various parties relative to their particular interests, expertise, and authorities, and 6) monitoring and evaluation; the watershed management cycle should include monitoring to ascertain the environmental and socioeconomic impacts of watershed plans. Progress should be reported and results of monitoring should help guide decisions about continued implementation.

Watersheds usually cover vast amounts of land, both public and privately owned; it can be difficult to bring all the stakeholders to the table to come up with a plan to manage these huge areas. To understand the interactions among different living and nonliving organisms in a watershed can also be

challenging; this is further complicated by the study of how human activities affect watershed functions. As there are so many different users in the watershed, there is always a conflict between different stakeholders and between stakeholders and management goals; resolving these conflicts can be tough, but is necessary to accomplish to achieve watershed management. Also, watershed management requires time and resources to generate interest and to build relationships between stakeholders. Funding agencies and stakeholders may grow impatient with the lack of observable outcomes.

Watershed management has evolved and become an integrated and comprehensive approach to addressing a broad range of water protection issues. This approach allows the evaluation of the important links between land and water, between surface and groundwater, between water quality and water quantity, and between watershed management and municipal planning.

SEE ALSO: Lakes; Pollution, Water; Riparian Areas; Rivers; Water; Water Conservation; Water Law.

BIBLIOGRAPHY. A.O. Akan, *Urban Hydrology, Hydraulics, and Stormwater Quality: Engineering Applications and Computer Modeling* (John Wiley & Sons, 2003); K.N. Brooks et al., *Hydrology and the Management of Watersheds* (Iowa State Press, 2002); I.W. Heathcote, *Integrated Watershed Management: Principles and Practice* (John Wiley & Sons, 1998).

Velma I. Grover
Independent Scholard

Watt, James G. (1938–)

JAMES G. WATT served as the U.S. Secretary of the Interior from January 23, 1981, to November 8, 1983, during President Ronald Reagan's first term. Watt is a Wyoming native and a staunch Republican. Prior to being appointed to the Interior position, Watt drew the attention of President Reagan for his work as founder and leader of the Mountain States Legal Foundation (MSLF), an organization that supported and encouraged the expanded extraction of oil, timber, and mineral resources. The MSLF was identified as a strong antienvironmental organization. Earlier in his governmental career Watt worked as Secretary to the Natural Resources Committee and Environmental Pollution Advisory Panel of the U.S. Chamber of Commerce. In 1969, Watt received appointment as Deputy Assistant Secretary of Water and Power Development in the Department of the Interior. He also worked briefly for the Federal Power Commission before founding the MSLF.

During his tenure as Secretary of the Interior, Watt was widely criticized for his positions on environmental programs. In April 1981, just three months after his appointment, the Sierra Club and other environmental organizations were calling for his dismissal. Over a million signatures demanding his dismissal were presented to Congress as evidence of general displeasure with his stance on the environment. Watt cut funding for the Endangered Species Act and strongly suggested expanding the offering of oil and gas leases in wilderness areas and in offshore regions.

Watt is remembered for his statements linking stewardship of the land and religion. In February 1981 in testimony before the House Interior Committee, Watt said, "I do not know how many future generations we can count on before the Lord returns; whatever it is we have to manage with a skill so as to leave the resources needed for future generations." There is a hint of the conservationist in this statement, but further statements suggested otherwise. Watt stated on one occasion that "We will mine more, drill more, cut more timber." It is this philosophy that quickly caught the attention of environmentalists. Watt's political stance was never in doubt; in 1982 he stated, "I never use the words Democrat and Republican. It's liberals and Americans."

Watt was politically linked with Anne M. Gorsuch, administrator of the Environmental Protective Agency, who reduced the effectiveness of the organization and sought to ease environmental regulations on industry. Environmental organizations such as the Sierra Club and the National Audubon Society were able to significantly increase their membership ranks in response to the actions of the two perceived antienvironmentalists. Watt was forced to leave his post following comments he made about his staff to members of the U.S. Com-

mission on Fair Market Value Policy for Federal Coal Leasing to the U.S. Chamber of Commerce on September 21, 1983: "We have here every mixture you can have. I have a black, a woman, two Jews and a cripple. And we have talent." Within weeks of this statement, Watt resigned.

When President George W. Bush appointed Gale Norton as Secretary of the Interior, environmentalists were again concerned. Norton had served as a staff member under Watt in the MSLF and it was feared that she would replicate the environmental stance taken by Watt. Under President Bush, environmentalists saw a resurrection of many of the goals Watt furthered in the early 1980s. The notion of placing production ahead of conservation, presented in a speech by Vice President Richard Cheney in 2001, is a clear reflection of the Watt philosophy. Watt himself commented on the Bush position in an interview for the *Denver Post* in 2001, "Everything Cheney's saying, everything the president's saying—they're saying exactly what we were saying 20 years ago, precisely."

SEE ALSO: Bush (George W.) Administration; Department of the Interior (U.S.); Endangered Species Act (U.S.); Reagan, Ronald Administration; Sagebrush Rebellion; Sierra Club.

BIBLIOGRAPHY. S.P. Bratton, "The Ecotheology of James Watt," *Environmental Ethics* (v.5/fall, 1983); G.C. Coggins and D.K. Nagel, "Nothing Beside Remains: The Legal Legacy of James G. Watt's Tenure as Secretary of the Interior on Federal Land Law and Policy," *Boston College Environmental Affairs Law Review* (v.17/spring, 1990); Carolyn Merchant, *The Columbia Guide to American Environmental History* (Columbia University Press, 2002).

Gerald R. Pitzl, Ph.D.
New Mexico Public Education Department

Weapons of Mass Destruction (WMD)

THE MEANING AND definition of *weapons of mass destruction* (WMD) continue to evolve over time and with technology. Though the phrase was first used in a *New York Times* article in 1937, referring to a saturation bombing during the Spanish Civil War, the first administrative use of the term came when the United Nations established the Atomic Energy Agency in 1946. Originally referring only to atomic weapons, through treaties and international conventions, WMD has come to include all types of nuclear, biological, chemical, and toxic weapons. Today an exact definition of WMD is nonexistent, varying by place and policy. However, in general WMD are broken down into the following four categories of weaponry: Nuclear, biological, chemical, and radiological.

Due to the longevity and range of destruction that they are capable of unleashing, nuclear weapons indisputably pose the gravest risk to the living environment. Though only used twice in warfare, in Japan in 1945, nuclear weapons have been detonated thousands of times around the world by countries testing their nuclear weapon technology—China, France, India, Pakistan, Russia, the United Kingdom, and the United States. Other states have been pursuing, or have declared that they possess, nuclear weapons (i.e., Iran, Israel, and North Korea.) Hypothetically, if enough nuclear weapons are detonated at approximately the same time, a "nuclear winter" would be the result. This would entail a drastic cooling of global temperatures due to particles in the atmosphere blocking the sun's radiation from reaching the surface. There is a good chance that no life forms would survive such an event on the planet Earth. The state possessing the most nuclear missiles in the world, and thus from an environmental standpoint the most dangerous state to ecological longevity, is the United States.

Biological weapons are the oldest of the contemporary WMD. They include the use of any poisonous or toxic pathogens for military advantage. However, the military usefulness of biological weapons is dubious. Though potentially resulting in the deaths of thousands of people, animals, and natural fauna, there would be little possibility of preventing biological WMD from afflicting one's own forces or population. Moreover, biological weapons take longer to infuse themselves than many other types of WMD, making them largely inefficient for conventional military campaigns. Nonetheless, over the

Hans Blix

Hans Blix was born in 1928 in Uppsala, Sweden, to a family from the Norwegian nobility. He studied at Uppsala University, Columbia University, and the University of Cambridge in Britain, from where he graduated with a doctorate in law. He was an associate professor in international law until 1961, when he was appointed to the Swedish delegation to the United Nations (UN), a post he held until 1981. He also served as a member of the Swedish delegation to the Disarmament Conference in Geneva. He was the foreign minister of Sweden from 1978 until 1979.

In 1981, Hans Blix was appointed head of the International Atomic Energy Agency, spending much of his time traveling around the world inspecting nuclear facilities. This involved visits to many well-known plants, and also to Iraq and North Korea. He went to Osiraq near Baghdad several times before the Israeli aerial attack in 1981 that destroyed much of the plant. The Israelis had claimed that nuclear weapons were in the early stages of development there, although Blix never found any evidence of this.

In January 2000 Hans Blix was appointed the first executive chairman of the UN Monitoring, Verification, and Inspection Commission, a post he held until June 2003. During this time, he was involved in the search for Iraqi weapons of mass destruction (WMD). Although no stockpiles were found, he did locate some missiles that the Iraqis had been making illegally—and which they dismantled. Blix was always extremely critical of Saddam Hussein, the Iraqi leader, who tried to hide his weapons after the Gulf War of 1990–91, but also felt too much was made of WMD as the excuse for the invasion of Iraq in 2003.

last quarter century biological agents have become the most readily available WMD for use in bioterrorism (e.g., anthrax attacks in the United States during 2002).

Chemical weapons are far more efficient WMD than biological ones, but often far more deadly. Their effects are often immediate and severe, their effects taking hold through breathing, ingestion, or skin contact. Chemical weapons are unique in the fact that it is rare that the weapon system delivering them is the cause of carnage (unlike in nuclear or conventional weapon attacks). Instead, toxic agents are dispersed by the weapon delivery system. Unlike nuclear weapons, chemical weapons are relatively cheap and easy to produce. Over 70 different chemical agents are known to have been created. It is presumed that numerous countries maintain stockpiles of chemical agents. Several states are known to have used chemical agents in battle over the past 50 years. International treaties have largely been ineffective in controlling the development of chemical weapons, partially due to the fact that treaties are based on chemical structures and countries can create new chemical weapons that are undetectable.

Radiological weapons are a relatively new addition to WMD and may better be classified as weapons of mass hysteria. None are known to have ever been used in warfare or terrorist attack. Often referred to as "dirty bombs," models illustrate that radiological weapons would likely do more psychological harm to a community than ecological devastation. The ingredients for such a bomb would likely come from nuclear power waste, and regardless of what type of radioactive material was used, the radiation would either dissipate too quickly to cause widespread damage or it would take a long time to exterminate local living organisms.

The potential environmental impact of WMD is enormous. Obviously with a nuclear Armageddon, humankind's longevity on the planet would be placed in jeopardy. Ecological destruction would be complete and limit the chances of continued life on earth. Even the detonation of a single nuclear weapon has been shown to have a devastating impact on the environment—most of these weapons make areas uninhabitable for decades, if not hundreds of years. Radiation exposure lies behind many diseases, many longitudinal, as were witnessed in Japan after World War II. Though biological and chemi-

cal weapons are both cheaper and easier to create than nuclear weapons, their impact on local environments may be just as devastating. Capable of killing all life forms and, in the case of biological weapons, diffusing via communicable means, these WMD may pose the greatest risk for humans in the future. Finally, radiological weapons do not pose as much of a risk to the environment as their sources do—that is, nuclear waste from power plants. However, if placed in an urban area and or set off without detection, the impact on human life could be devastating.

Though numerous treaties have been passed on almost all WMD, the fact is that there is no commission to enforce compliance. Moreover, now that the United States has determined that it is justifiable to preemptively strike states that may have WMD, it appears that diplomacy may no longer be a viable option for supervising and controlling the diffusion of WMD around the world. Nuclear nonproliferation had largely been effective throughout the Cold War era, but since the decline of the Soviet Union has become a major concern for Western states around the world, as many "rogue states" choose to pull out of the treaty. WMD are also an enticing weapon for terror organizations due to their ubiquity and potentially devastating effects.

SEE ALSO: Radioactivity; Nuclear Weapons; Wars.

BIBLIOGRAPHY. Anthony H. Cordesman, *Terrorism, Asymmetric Warfare, and Weapons of Mass Destruction: Defending the US Homeland* (Praeger Publishers, 2001); Global Security, "Weapons of Mass Destruction," www.globalsecurity.org/wmd (cited March 2006).

Ian Alexander Muehlenhaus
University of Minnesota

Weather

WEATHER IS THE condition of the atmosphere in a local environment or region over a short period of time, ranging from an exact instant to a few days. This is distinct from climate, which refers to persistent atmospheric systems over larger areas and greater time periods. The atmospheric conditions may be hot or cold, dry or humid, rainy or dry, windy or calm, cloudy or clear in combination.

The atmosphere is the envelope of air that surrounds the surface of the earth. It has four layers: They are the troposphere, the stratosphere, the mesosphere, and the thermosphere. The term atmosphere also means the weight of the air pressing against the earth at any given point on earth. At sea level the weight of the air is 14.7 pounds per square inch (1.03 kilograms per square centimeter) of surface. At places below sea level such as the surface of the Dead Sea the atmospheric pressure is greater than one atmosphere. On mountaintops the atmospheric pressure is less. This natural feature of the weight of the atmosphere at various places on the earth's surface is an important feature in the weather and in the climate.

Climate differs from weather. Weather is the immediate atmospheric conditions. Climate is the average of the weather over an area for a long period of time. Almost all weather takes place in the troposphere that extends from the surface to six to ten miles above the surface of the earth. Most of the atmosphere, water vapor and heat are in this layer.

Weather conditions involve temperature, air pressure, wind, and moisture. A weather report will combine all of these to show the weather as it is currently, or more importantly as it is expected to be in the hours or days ahead in order that people may respond appropriately.

Temperature readings measure the amount of heat in the atmosphere. The heat comes from sunlight shining on the earth. However, sunshine does not strike the earth everywhere with the same effect. At the equator the sun shines directly on the earth making the weather at the equator warm. The further from the equator and the closer to the North Pole or the South Pole the more the sun strikes the earth's surface at an angle. In addition, as the earth rotates around the sun in its yearly circumnavigation it presents either the Northern or the Southern Hemisphere to the sun more directly, which produces summer or winter.

As the sun strikes the earth's surface it may strike water or land. Since the oceans and the continents present different surfaces to sunshine the effect generates different amounts of heat. In desert regions

the heat is reflected away from the surface of the earth. At night the radiant heat quickly turns the desert into a chilly place. However, in land areas of moisture and extensive plant growth heat is absorbed by the plants and retained in the locality. In the oceans the heat is absorbed and distributed by currents, evaporation, and by reflection in a different pattern.

Another factor affecting the temperature of the atmosphere is the greenhouse effect. This occurs because the carbon dioxide given off by humans, animals, in natural springs, or by industry blocks some of the escape of radiant heat creating conditions similar to a green house or to a thermal cover on a swimming pool.

The varying temperatures on the surface of the earth create atmospheric pressure differences. As air is heated it expands and rises. The heating of the earth at the equator, and especially in the equatorial ocean waters, sends warm air upward. In contrast, the air at the poles is colder and more condensed than warm air. The effect is that warm air creates areas of low atmospheric pressure (lows) because the weight of the atmosphere at that locality is less than in areas where low temperatures condense air and create places of high atmospheric pressure (highs). The condensing of air at the poles and the rising of air in the equator creates pressure differences between the two regions. The rising air of the equator moves toward the poles and the sinking air of the poles moves toward the equator creating wind.

Winds, which are named for the direction from which they flow, blow from highs to lows. Conversely, as cooler air moves toward a low it forces the rising warmer air to move upward more rapidly. The higher the elevation the warmer air reaches, the more it cools and contracts. The moisture in the warmer air then condenses into clouds and precipitation.

Most of the rain that falls on earth is from water vapor that evaporated from the oceans. Humidity is the amount of water vapor in the atmosphere. When a volume of air has absorbed as much moisture as it temperature and pressure will allow, then it is saturated. The dew point is the temperature at which a volume of air is saturated with moisture. If a given volume of air continues to cool beyond its dew point then moisture is converted into dew, or if cold enough, into frost. Moisture-saturated air that is cooled to the dew point that is near the ground or the ocean may also create fog. If the cooling continues, then the moisture in the air condenses into either liquid precipitation (rain) or frozen precipitation (hail, sleet, or snow).

Most places on the surface of the earth have changing atmospheric patterns that vary with the seasons. At the poles the seasons are long periods of daylight in the summer or total darkness in winter. The freezing temperatures make the poles both cold deserts where most of the moisture has been squeezed out by the lower temperatures. The weather in tropical zones is marked by periods in which it is hot and dry followed by rainy seasons.

Weather reports are of great importance to farmers, sailors, and others operating in the open such as military field commanders. Weather predictions can also be of great value to people in areas that are vulnerable to extreme weather such as tornadoes, hurricanes, blizzards or heat extremes. The weather affects humans directly and constantly. In cold weather heating for homes or businesses is needed; in hot weather protection from extreme temperatures or from dehydration is necessary. In rainy or stormy weather shelter and protective clothing are necessary to prevent hypothermia or injury or death from lightning or strong winds.

Weather affects agriculture, industry, transportation, and communications. Storms such as violent thunderstorms, tornadoes (cyclones), and hurricanes (typhoons), or even dust and sand storms can create disasters for civilization in many areas of the world.

SEE ALSO: Atmosphere; Climate; Drought; Greenhouse Effect; Hurricanes; Precipitation; Thunderstorms; Tornadoes.

BIBLIOGRAPHY. Edward Aquado and James Burt, *Understanding Weather and Climate* (Pearson Education, 2006); Marq De Villiers, *Windswept: The Story of Wind and Weather* (Walker and Company, 2006); Storm Dunlop, *Weather* (Smithsonian Institution Press, 2006); Terry J. Jennings, *Weather Patterns.* (Smart Apple Media, 2005); David M. Ludlum, *National Audubon Society Field Guide to North American Weather* (Alfred A. Knopf, 1998); Vincent J. Schaefer and John A. Day, *A*

Field Guide to the Atmosphere (Houghton Mifflin Co., 1981); Jack Williams, *The Weather Book* (Knopf Publishing Group, 1997).

Andrew J. Waskey
Dalton State College

Weather Modification

WEATHER MODIFICATION IS the process by which, deliberately or otherwise, human intervention changes or attempts to change weather conditions. Deliberate interventions include attempts to weaken hurricanes or to change patterns of precipitation for agricultural purposes. Inadvertent interventions include the widespread burning of carbon fuels that leads to global warming and has various impacts upon weather systems. Speculation exists as to whether the use of weather modification in warfare is a viable concept. The limited success with which weather modification has been completed for peaceful purposes would suggest that it is not possible.

Perhaps the most common attempt to modify the weather comes from the use of substances such as dry ice or silver iodide to seed clouds with supercool particles to alter rainfall. Experiments in this area began on a systematic basis with the work of Vincent J. Schaefer and Irving Langmuir, recipient of the 1932 Nobel Prize for chemistry. They determined that super-cooled clouds could be made to dissipate in certain circumstances or else to begin precipitation because ice crystals are formed that are too large to remain part of clouds. The use of this technology can help to organize precipitation at a time deemed suitable for agricultural or social purposes (to prevent rain during parades, for example, which Soviet scientists attempted).

In the United States, this has caused a number of state-level bodies to pursue legislation to determine ownership of clouds because of concerns over access to needed water resources. However, the degree to which the precipitation can be affected remains largely unpredictable, owing in part to the complexity of atmospheric conditions. Nevertheless, this has not deterred governments of countries around the world from trying to modify cloud behavior by spreading particles from airplanes, distributing them by rockets or artillery, or otherwise causing them to be lifted on air currents. The ability of scientists to modify weather conditions systematically and consistently remains low, although it is possible that Soviet scientists were able to influence the fall of hail on a predictable basis. Consequently, the hope that it will be a useful weapon in the struggle against intensified and more prevalent dangerous weather conditions such as hurricanes and droughts remains impractical.

SEE ALSO: Atmospheric Science; Climate Modeling; Climatology; Weather.

BIBLIOGRAPHY. Stanley A. Changnon, "Inadvertent Weather Modification in Urban Areas: Lessons for Global Climate Change," *Bulletin of the American Meteorological Society* (v.73/5, 1992); Congressional Research Service, *Weather Modification: Programs, Problems, Policy, and Potential* (University Press of the Pacific, 2004); John A. Day and Vincent J. Schaefer, *A Field Guide to the Atmosphere* (Houghton Mifflin, 1998); Weather Modification Association, www.weathermodification.org (cited July 2006).

John Walsh
Shinawatra University

Weeds

JOHN LELAND, a natural historian, defines a weed as a "plant out of place." The Bureau of Land Management (BLM) states that weeds constitute "any plant growing where it is not wanted." Weeds can be native or nonnative, invasive or noninvasive, and noxious or nonnoxious. According to the Animal and Plant Health Inspection Service of the U.S. Department of Agriculture, a weed is "any plant that poses a major threat to agriculture and/or a natural ecosystem." This particular definition is significant because it highlights the economic and ecological impact that weeds can have on human communities, landscapes, and natural ecosystems. Some weeds have been categorized as "noxious" by

federal agencies and state and local governments. A noxious weed is a plant that poses a major threat to agriculture, public health, recreation, wildlife, or property. The BLM defines noxious weeds as those plants that are "competitive, persistent, and pernicious" and out of place.

CHARACTERISTICS AND PROLIFERATION

Weeds can be found worldwide. A few of them, like plantain and henbit, have medicinal qualities. Others, such as dandelion, can be eaten. Most weeds, however, are nuisances and some pose severe problems to human communities and natural ecosystems. Each year, landowners spend billions of dollars on herbicides in a never-ending fight to eliminate weeds in their yards. Yards and lawns are not the only place to find weeds. In the United States, it is estimated that over 5,000 nonnative, exotic, and alien plants (which include many weed species) have escaped into natural ecosystems. Some of these weeds have been naturalized and assimilated into the forest community. Others, though, pose a severe threat to both land and aquatic ecosystems.

One of the most significant factors in the proliferation of weeds is human activity. Weeds came to North America with the early explorers and settlers. American colonists, for example, brought along with them plants for food and medicinal purposes, including dandelions, which, over time, became "naturalized" weeds. Some weeds came to the New World accidentally as stray seeds in livestock and human food supplies. Other American weeds, like crabgrass and Bermuda grass, came from early global networks involving Africa and the slave trade. During the late 19th century, a whole array of weeds, like kudzu, honeysuckle, and wisteria, came to the United States as ornamental plants. Commercial ventures, too, are responsible for the great profusion of weeds. The fish aquarium industry, for example, is directly linked to the accidental introduction of the Asian aquatic plant hydrilla to native ecosystems in the United States. Today, the introduction of weeds to new areas has significantly increased due to globalization and an exploding human population.

Although there exists an immense variety of weeds throughout the world, they all share certain characteristics. For one, weeds are hardy survivalists and are able to thrive in new habits due to their ability to adapt to adverse conditions, such as poor soils and extreme climates. For this reason, weeds have been identified in some of the world's harshest natural environments. Since many invasive and noxious weeds are rapid colonizers of barren soil, they tend to favor habitats that have been disturbed through human activities and natural occurrences.

European purple loosestrife has cost U.S. taxpayers $45 million annually in efforts to control its spread.

In addition to being resilient, weeds are aggressive, fast-growing plants, and they have a tendency to out-compete natural vegetation for essential nutrients, space, and light. Superior competition among invasive weeds may be enhanced by a lack of natural predators or the diseases that, in the weeds' natural environment, would tend to keep them under control. Once established, aggressive weed species encroach on native vegetation and eventually endanger local plant communities. In some areas, weeds have completely replaced native plant species. Dense weed infestations tend to retard natural succession and reforestation and thus decrease biodiversity. According to the Nature Conservancy, 42 percent of the species listed as endangered or threatened under the Endangered Species Act are connected to invasive weeds and exotic plants.

Weeds thrive, also, because of superior reproduction capabilities. One of the most common means of reproduction among weeds is through seeds that have high germination rates, although this is not the only way that weeds propagate. In the southern states, Johnson grass utilizes both seeds and an expansive rhizome root system to form new colonies. In addition to seed and root reproduction, hydrilla can also replicate asexually through its leaves or leaf particles. To enhance their survival, some weeds, over time, hybridize with native plant species, altering their genetic make-up. Hybridization occurs when human activity inadvertently brings together two similar plant species. In many instances, the hybrid offspring is more competitive—and destructive—than either parent.

DANGERS TO THE ENVIRONMENT

Weeds present many problems to both human and natural environments. In the United States alone, "alien" weeds make up nearly 65 percent of the total weed flora and the costs associated with their control are enormous. In 1994, it was estimated that Americans spent $20 billion on weed control. In that same year, agricultural losses due to weeds exceeded $10 billion.

The impact of weeds on ecological systems is tremendous. In many instances, they completely alter natural functions within ecosystems by modifying nutrient and hydrology cycles. In this respect, weeds have the potential to change natural habitats. One example of habitat alteration concerns the hybrid cattail, which has replaced native white top and wild rice in North American wetlands. European purple loosestrife, which came to North America during the 19th century as an ornamental plant, has invaded most of the wetland areas in the United States. Today, the plant has negatively impacted wetlands by reducing native plant species and threatening wildlife habitat. Consequently, native plant destruction caused by the European purple loosestrife has directly contributed to a decline in the box turtle population, due to a reduction in its natural food supply. It has also cost taxpayers $45 million annually in efforts to control its spread. Not only do weeds pose threats to endangered species of plants and animals, they also prevent or retard the natural succession of native vegetation by forming dense mats of infestation. Certain species of weeds, such as leafy spurge, are also toxic to cattle. Additionally, weeds like thistle have reduced native forage plants in pastures, rangelands, and forests and have had a significant impact on the ability of cattle to graze.

SEE ALSO: Biodiversity; Colonialism; Dandelions; Ecosystems; Endangered Species; Herbicides; Invasions, Biological; Invasive Species; Lawns; Pesticides; Water Hyacinth; Wetlands.

BIBLIOGRAPHY. Bureau of Land Management Weeds Website, www.blm.gov/weeds (cited April 2006); Federal Noxious Weed List, www.aphis.usda.gov (cited April 2006); "Invasive and Exotic Species," www.invasive.org (cited April 2006); John Leland, *Aliens in the Backyard: Plant and Animal Imports into America* (University of South Carolina Press, 2005); David Pimentel, Lori Lach, Rodolfo Zuniga, and Doug Morrison, "Environmental and Economic Costs Associated with Non-Indigenous Species in the United States" (1999), www.news.cornell.edu (cited April 2006).

Clay Ouzts
Gainesville State College

Wells

WELLS ARE HOLES dug, punched, or drilled into the ground to access water or some other liquid such as oil or brine, or gas such as methane. The technology for digging wells has developed from shallow wells dug with hand tools to wells that are drilled thousands of feet into the surface of the earth. Water wells are dug because there is no surface or groundwater available, or because the supply of groundwater is insufficient for the needs of people, livestock, or industry. Shallow wells dug by hand were the primary source of water for people for thousands of years. Wells are considered shallow if they descend to less than 50 feet. Shallow wells reach either the water table (water table wells), which is the level at which the soil is saturated with water, or a well may be dug through an impermeable layer of clay or rock to an

aquifer, which is a permeable layer of either soil or rock through which water flows, but which is capped by at least one impermeable layer (aquifer wells).

In many third world countries, shallow wells are still being dug and used by people as a substitute for sharing a waterhole with wild animals. Typically a shallow well is dug using hand tools and local labor. The well is dug during the dry season to the local water table and then a little deeper. Digging during the dry season insures that the lowest point the local water level descends to below ground will be the lowest level for water in the well at any time. The bottom of the well is covered with sand and gravel to inhibit mud incursions. This will keep the water freer from particles of dirt. If suitable rock is unavailable, then mud bricks are made on site. After the mud bricks have been fired and cooled they are solid enough to not dissolve in the water and will last for a long time. The bricks are then used to line the well from bottom to top and are joined together with cement.

Wells such as those built in some African countries are usually covered with a locally made cement top that has a hole in it. PVC pipe is put through the hole in the well top, which may be called a cap. Then a locally made pump, which is like a bicycle pump, is attached to the PVC pipe. When the pump is propelled water flows out of the lip in the top of the well. Wells of this type can cost $300 or less to build. They then provide water for a whole village that may number 100–200 people or more. Wells that are properly dug will be located well away from possible sources of pollution or from runoff that may carry pollution. In addition a good well will have a sanitary seal made of cement grout or bentonite clay and its pump will be designed to protect the wellhead.

Wells can be drilled with machinery as well as dug with hand tools. The drilling of wells began in the mid 1800s. The well diggers of that time adopted a machine that drove or "pounded" a bit into a hole into the ground. The bit was attached to a cable and was dropped repeatedly into the drill hole. The cable tool drilling method was slow and inefficient. The rotary drilling method replaced it in the 1900s. The rotary drill bit is made of tungsten steel or other tough metals. Attached to a drill pipe, it is rotated to grind up rock that it encounters in drilling the well. It operates inside of a larger pipe that is regularly flushed with water to cool the drill bit and to wash out the cuttings, dirt, gravel, and sand. The outer pipe acts as a wall and prevents the drill hole from collapsing, which would block the drill hole. As the drilling progresses the well driller keeps an accurate log of the depths of the drilling and the levels at which water is encountered.

Many drillers use a "down-hole air hammer" when drilling in hard rock. The compressed air blows the crushed rock debris out of the well hole along with any water that has entered the hole. After the well is drilled the hole has to be finished to prevent collapse and contamination by pollutants. Lining it with well casing usually made of steel or plastic seals a well. The casing is smaller than the well hole. The space between the well hole and the casing will be filled with a "grout" to prevent contamination. The "grout" is either cement or volcanic clay called bentonite. The space may also be filled with a special kind of fine rock. Usually, only the top 20 feet are grouted. Well drilling equipment is commercially available in units for individuals who wish to drill wells of depths less than 300 feet that can be used for irrigating lawns, gardens, or other farm or home operations. Proper drilling and maintenance of water wells is needed to ensure that the well's water remains safe. In many locations the accelerating exploitation of well water has had a negative impact on aquifers.

SEE ALSO: Drilling (Oil and Gas); Water; Water Demand; Water Harvesting.

BIBLIOGRAPHY. Michael D. Campbell and Jay H. Lehr, *Water Well Technology: Field Principles of Exploration, Drilling, and Development of Ground Water and Other Selected Minerals* (McGraw-Hill, 1973); Michael Detay, *Water Wells: Implementation, Maintenance and Restoration* (John Wiley & Sons, 1997); Ulric P. Gibson and Rexford D. Singer, *Water Well Manual: A Practical Guide for Locating and Constructing Wells for Individual and Small Community Water Supplies* (Premier Press, 1971); P. Howsam, *Water Wells—Monitoring, Maintenance, Rehabilitation: Proceedings of the International Groundwater Engineering Conference Cranfield Institute of Technology* (Taylor & Francis, 1990).

Andrew J. Waskey
Dalton State College

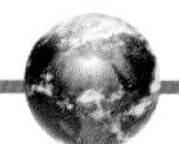

West Nile Virus

WEST NILE VIRUS (WNV) is a vectorborne infectious disease of the family Flaviviridae and is found in both tropical and temperate regions. WNV normally transmits among birds, mosquitoes, and mammals (especially humans and horses). The virus circulates in the blood of birds (reservoir hosts) for a few days after infection. Mosquitoes, particularly *Culex pipiens, C. restuan, C. tarsalis* and *C. quinquefasciatus*, become infected when they feed on infected birds. Infected mosquitoes (vectors) can then transmit WNV to humans and horses through their bites. They also infect other birds when they bite again. The virus is injected from the mosquito's salivary glands into the blood stream where it can multiply and cause illness. It was initially believed that direct human-to-human transmission was impossible and that humans are dead-end "hosts." However, in 2002, the Centers for Disease Control and Prevention (CDC) discovered the transmission of WNV through blood transfusion and organ transplants as well through breast milk, prenatal infection, and occupational exposure.

The typical incubation period for West Nile virus is 2–6 days, although it can be as long as 15 days. Most people (80 percent) infected with West Nile virus will be asymptomatic or experience a flu-like illness. In others, the virus causes West Nile fever. Very few infected people will develop the more severe form of West Nile, i.e., West Nile encephalitis (inflammation of the brain), and meningitis (inflammation of the lining of the brain and spinal cord), both of which can be fatal.

The geographic distribution of this virus has expanded since its discovery in Uganda in 1937 and now includes Africa, Asia, Europe, North America, central and south America, and the Caribbean. In the United States, the virus first appeared in the Bronx borough of New York City in 1999. Since then, it has spread rapidly west and south from its initial focus. By 2003, 45 states and the District of Columbia had reported human cases of WNV. From 1999 through 2001 the CDC confirmed 149 cases of human WNV infection, including 18 deaths. In 2002 the count increased to 4,156 cases and 284 fatalities. WNV outbreaks reached a peak in 2003 when the CDC reported 9,862 case and 264 deaths. At least 30 percent of those cases were considered severe, involving meningitis or encephalitis. However, in 2004, there were only 2,539 reported cases and 100 deaths. In 2005 there was a slight increase in the number of cases, with 2,949 cases and 116 deaths reported. Canada, Israel, and Romania also experienced outbreaks of WNV in recent years.

The distribution of WNV is dependent on the occurrence of susceptible avian reservoir hosts, competent mosquito vectors, and abundance of preferred hosts (humans and horses) for the infected mosquitoes. These factors are in turn affected by potential environmental and social factors. Several studies have highlighted land use and land cover change, elevation, abundance of vegetation, physiographic regions, stagnant water bodies, temperature, precipitation, spatial and temporal differences in periods of drought and rain, close proximity to dead birds and mosquito vectors, and farmland created by irrigation as potential environmental factors. Social factors influencing the distribution and transmission of WNV are human population density, age, income, race, age of housing, mosquito control activities, location of scrap tire stockpiles, and location of organic wastes from chemical industries.

In spite of a plethora of research on WNV, a vaccine for humans is not yet available. However, a vaccine for horses exists. Nevertheless, following precautionary measures as indicated by mosquito control agencies can mitigate the risk of infection. These measures include staying indoors at dawn and dusk when most mosquitoes are active, wearing long-sleeved shirts and long pants during outdoor activities, applying mosquito repellent sparingly on exposed skin, and removing any water holding containers from property.

SEE ALSO: Center for Disease Control; Disease; Malaria; Mosquitoes; Viruses.

BIBLIOGRAPHY. J.S. Brownstein et al., "Spatial Analysis of West Nile Virus: Rapid Risk Assessment of an Introduced Vectorborne Zoonosis," *Vector Borne Zoonotic Disease* (2, 2002); L.M. Bugbee and L.R. Forte, "The Discovery of West Nile Virus in Overwintering *Culex pipiens* (Dipters: Culicidae) Mosquitoes in Leigh County, Pennsylvania," *Journal of the American*

Mosquito Control Association (v.20/3, 2004); Center for Disease Control and Prevention (CDC), "West Nile Virus Activity—United States, 2001," *Morbidity and Mortality Weekly Report* (51, 2002); CDC, "Testing and Treating West Nile Encephalitis in Humans, West Nile Virus" (CDC, 2003); Samantha Gibbs et al., "Factors Affecting the Geographic Distribution of the West Nile Virus in Georgia, USA: 2002–2004," *Vector-Borne and Zoonotic Disease* (v.6/1, 2006); Marilyn Ruiz et al., "Environmental and Social Determinants of Human Risk During a West Nile Virus Outbreak in the Greater Chicago Area, 2002," *International Journal of Health Geographics* (v.3/8 2004).

Debarchana Ghosh
University of Minnesota

Wetland Mitigation

WETLAND MITIGATION REFERS to efforts to reduce the negative impacts of development on wetlands, and has been required by U.S. law under the Clean Water Act (CWA), the National Environmental Policy Act (NEPA), and other elements of environmental law. The CWA §404 permit program, established in 1972, is administered by the U.S. Army Corps of Engineers (Corps) and overseen by the U.S. Environmental Protection Agency (EPA); wetland mitigation is often required in order to obtain a §404 permit to fill or dredge in a wetland. However, in the 1970s mitigation was not consistently required, and Corps leadership did not agree that requiring mitigation was a necessary part of administering the §404 program.

The White House Council on Environmental Quality drafted NEPA regulations on November 29, 1978 defining mitigation as including five components: Avoidance, minimization, rectification, reduction, and compensation. All federal agencies are required to comply with NEPA; thus the EPA adopted a modified version of mitigation on December 24, 1980, known as the "§404(b)(1) Mitigation Guidelines." Because these guidelines are binding on the Corps issuance of wetland fill permits, they formalized the administrative requirement for wetland mitigation. The guidelines require a permit applicant to achieve all practicable avoidance and minimization of impacts to wetlands before a permit can be issued. The Corps can require compensation for any remaining unavoidable impact that constitutes "significant degradation." These three elements constitute mitigation under the CWA, and the Corps must deny a permit where mitigation cannot eliminate significant degradation. The third step, compensatory mitigation, has become an extremely common and important feature of U.S. wetland policy, and has come to symbolize the principle that one environmental feature can replace another, which has become part of the American conception of sustainability (embodied in the policy slogan "no net loss of wetlands" from the late 1980s). This principle is not widely accepted in environmental policy outside the United States, and compensatory mitigation for wetland impacts is rare.

Compensatory wetland mitigation is the restoration, creation, preservation, or enhancement of a wetland using techniques of environmental restoration. As early as 1982, reports began to emerge indicating that compensation sites were of poor quality. These reports became increasingly common, culminating in an infamous report on compensatory mitigation in Florida that found that only 33 percent of required sites had even been constructed. The Corps' limited attention to compliance monitoring continues to be blamed for this. The principal debate over wetland mitigation in the 1980s concerned the Corps' position that extra compensation could reduce the need for avoidance and minimization. The EPA held that all practicable avoidance and minimization must be achieved before compensation can be considered, and this position was formalized into joint EPA and Corps policy in the 1990 Mitigation Memorandum of Agreement (MOA). Since 1990, wetland mitigation has been understood to occur in a sequence: The permittee must first avoid, then minimize, and finally compensate.

The 1990 MOA also established a general preference for compensatory mitigation to occur on the site of the impact, and to be of the same kind of wetland as the impacted wetland; this was known as the "on-site in-kind preference." As various kinds of off-site and third party mitigation methods became more prevalent in the 1990s, such as wetland mitigation banking and in-lieu fee mitigation, the on-site

in-kind preference gradually relaxed. Historically, most compensatory mitigation has been performed by the permittee, and nearly all compensation was on-site in the 1980s. This was in part because the Corps was reluctant to consider the requirement of off-site compensation "practicable" and therefore simply did not require compensation in cases where on-site compensation was not feasible. The practice of consolidating many compensation sites into a single large "bank" site—often constructed in advance of impact—had been used since the U.S. Fish and Wildlife Service issued guidance on the subject in 1981. However, the effort to streamline environmental regulations, led by a report from Vice President Quayle's Council on Competitiveness in 1991, resulted in the turn to market-led approaches to wetland compensation.

The first entrepreneurial wetland bank, selling wetland "credits" that permittees can purchase to satisfy their compensation obligations, was approved in 1992, and commercial mitigation banks quickly came to far outnumber public or single-client banks. As of 2006 permittees could purchase compensation credits in 32 of the 38 Corps districts, rather than construct their own compensation site. In-lieu fee compensation also became a relatively common form of mitigation, in which a permittee pays a set fee per acre of impact into an aggregate account (often controlled by a state agency or a non-profit organization). The fund is then used to fund assorted wetlands-related projects. By 2003, 60 percent of wetland compensation was performed by the permittee, 33 percent was performed at banks, and seven percent was performed through in-lieu fee programs. Furthermore, 67 percent of all wetland compensation was entirely off-site, or had an off-site component (16 percent). The effects of this "mitigation migration" on the hydrology and landscape ecology of wetlands are poorly understood. Recent U.S. Supreme Court cases have restricted the jurisdiction of the CWA, reducing the types of wetland impacts for which the Corps can require mitigation.

SEE ALSO: Army Corps of Engineers (U.S.); Clean Water Act (U.S. 1972); Environmental Protection Agency; Pollution, Water; Restoration Ecology; Water; Water Law; Wetlands.

BIBLIOGRAPHY. William Kruczynski, "Mitigation and the Section 404 Program: A Perspective," in Jon Kusler and Mary Kentula, eds., *Wetland Creation and Restoration: The Status of the Science* (Island Press, 1990); National Research Council, *Compensating for Wetland Losses under the Clean Water Act* (National Academy Press, 2001); Tom Yocom, Robert Leidy, and Clyde Morris, "Wetlands Protection Through Impact Avoidance: A Discussion of the 404(b)(1) Alternatives Analysis," *Wetlands* (v.9/2, 1989).

Morgan Robertson
University of Kentucky

Wetlands

WETLANDS ARE AREAS in which shallow standing water or saturated soil occurs for a period long enough or with enough regularity to influence the development of biotic assemblages and/or soil characteristics. Wetlands can be fresh, brackish or saline, can be inland or coastal, can be connected to or isolated from other aquatic systems, and are generally shallow enough to support rooted vegetation that emerges from the water surface, even if that vegetation is not present at all times. Wetlands are locations of focused ecological flows and changes of state. Many inland wetlands, for example, are areas of groundwater discharge or recharge. In recharge wetlands, nearby contaminants may easily enter the groundwater profile; conversely, in discharge wetlands, rare biotic communities may be supported by the outflow of nutrient-rich groundwater. Coastal wetlands are the site of important ecological interactions between marine and terrestrial environments, where the larval and juvenile stages of many marine animals are supported by nutrient fluxes from the nearby landmass.

Wetlands can be classified by hydrology, which is a function of climate and the wetland's position in a landscape. Wetlands perched high in a watershed will tend to be recharge wetlands, where precipitation and surface water become groundwater. They may be ephemeral, rather than permanent. Because precipitation is nutrient-poor, they may be oligotrophic or have low species diversity. Wetlands in the middle of a watershed will tend to be flow-through systems, experiencing groundwater discharge and recharge, and are more likely to be biotically diverse and permanent. Wetlands low in a watershed are likely to be permanent, experiencing groundwater discharge and/or fringing larger water bodies such as lakes, rivers, and oceans. Fens are wetlands that are largely dependent on groundwater for their hydrology; their ecology is highly dependent on the chemistry of the groundwater, and very sensitive to groundwater fluctuations. Fens in areas of limestone bedrock often receive very calcium-rich water, and are known as calcareous fens or marl flats, hosting a highly specialized flora. Bogs are wetlands that are largely dependent on rainwater for nutrients and hydrology (ombrotrophic), and tend to be low-diversity and nutrient-poor.

Biotic community and habitat can also classify wetlands. The Western folk taxonomy of wetlands relies primarily on vegetative structure, distinguishing "swamps" (forested wetlands) from "marshes" (grassy wetlands). Community composition is often effective shorthand for classifying certain wetland types dominated by characteristic vegetation, as with the cordgrass marshes of the American East Coast, or the mangrove swamps of tropical coasts worldwide. Wetlands serve to store surface water in a landscape, often reducing, delaying, and desynchronizing flood peaks. Coastal wetlands often act to reduce storm surges and coastal erosion; the loss of wetlands in coastal Louisiana was widely cited in 2005 as a culprit in the devastation caused by Hurricane Katrina. Wetlands provide essential habitat for many species, including most game birds and a large number of endangered species. Wetland soils, rich in organic matter, have proven extremely productive under agriculture once the technical challenges to draining were overcome in the early 1900s.

Control over wetlands is often linked to state expansion and the assertion of control over territory.

Wetlands have long been seen as waste areas to be reclaimed or ignored, and the many synonyms have also served as metaphors for undesirability or difficulty, such as: Swamped, quagmire, spewy, or bogged down. Few, even among the European and American Romantic tradition in the 1800s, saw transcendent beauty in wetland areas. They have been associated in many cultures with wilderness or the supernatural and nonhuman, and used as burial places or as refuges from colonial incursion and control. Control over wetlands, using large-scale technologies of drainage, is often linked to state expansion and the assertion of control over state territory (for example, in the Florida Everglades, Italy's Pontine Marshes, and the Tigris/Euphrates marshes in Iraq). Only with the 20th-century environmental movement, and the publication of such popular-science books as Bill Niering's *Life of the Marsh*, have wetlands become the object of societal concern, aesthetic appreciation, and legal protection at the local, national, and international level. They are one of the few ecosystems defined by statute: The U.S. federal government defines them as "those areas that are inundated or saturated by surface or ground water at a frequency and duration sufficient to support, and under normal circumstances do support, a prevalence of vegetation adapted for life in saturated soil conditions" (40 CFR 232.2).

The U.S. Federal Water Pollution Control Act (Clean Water Act) of 1972 established a permit system by which anyone wishing to dredge or fill a wetland must apply for permission to the U.S Army Corps of Engineers. The act also required an accounting of wetland loss in five-year reports, the first of which appeared in 1984 and reported that the continental United States had lost over half of the wetlands present at American independence, a loss rate of 60 acres per hour. Following these reports, wetland protection became a major electoral issue in the 1988 U.S. presidential election, which was marked by George H.W. Bush's extensive use of the campaign slogan "no net loss of wetlands." The U.S. government has assessed the status of wetlands since 1956, and claims that in 2004 there were 107.7 million acres (43.6 million hectares) of wetlands in the conterminous United States, 95 percent of which are freshwater and five percent of which are estuarine or saline.

International wetland conservation efforts are structured around the Ramsar Convention on Wetlands, signed February 2, 1971, in Ramsar, Iran. Signatory countries pledge to designate at least one Wetland of International Importance and to adopt policies and programs that promote wetland ecosystem health and awareness. In late 2006, there were 153 signatory nations and 1,629 wetland sites totaling 145.6 million acres (58.9 million hectares). The North American Waterfowl Management Plan (NAWCA) implements the Tripartite Agreement on wetlands between the United States, Canada, and Mexico, which directs funding and research on wetlands throughout the continent. Both the Ramsar Convention and the NAWCA focused originally on wetlands as bird habitat, but have expanded their scope considerably.

SEE ALSO: Clean Water Act (U.S. 1972); Swamp Land Acts; Water; Watershed Management; Wetland Mitigation.

BIBLIOGRAPHY. William Mitch and James Gosselink, *Wetlands* (Van Nostrand Reinhold, 1993); National Research Council, *Wetlands: Characteristics and Boundaries* (National Academy Press, 1995).

Morgan Robertson
University of Kentucky

Whales and Whaling

WHALES BELONG TO the mammalian order Cetacea (whales, dolphins, and porpoises), which in turn is divided into two extant groups, Mysticeti (baleen whales) and Odontoceti (toothed whales). The former group includes most of the large or "great whale" species, such as the blue whale and humpback whale, all of which are filter feeders. These latter whales use baleen plates (mostly made of the protein keratin) in place of teeth to sieve prey species out of seawater. The mysticete whales breathe through two closable blowholes, as opposed to one blowhole in the toothed whales. The odontocetes include the other "great whale," the sperm whale, and a variety of other families of cetaceans including beaked whales (family Ziphiidae), dolphins (family Delphinidae), and porpoises (family Phocoenidae). The moniker *whale* historically refers to a large cetacean and is not a biological term. Indeed, several "whales" such as killer whales and pilot whales are, in fact, dolphins.

Humans have utilized cetaceans since prehistory. In Europe, the Anglo-Saxon and Nordic peoples conducted hunts for large baleen whales, primarily coastal species such as the now extinct Atlantic gray whale, from at least the 9th century. Indeed, whales were so much a part of Nordic life that several laws were drawn up in the middle ages as to the ownership and disposition of whale carcasses. Whaling also took place in Japanese waters since at least the 3rd century C.E. by "driving"—that is, trapping whales and dolphins in small bays where they were then killed.

Whaling as a commercial activity began with the French and Spanish Basques in the middle of the 11th century, who hunted North Atlantic right whales, bowhead whales, and Atlantic gray whales. Much of the Basque whaling was originally concentrated around the Atlantic coasts of Spain and France; however, as whales became scarce in the Bay of Biscay, whalers expanded their area of activity and were hunting in Canadian waters as early as 1526. Commercial whaling operations by the Dutch and English began in 1610, often using experienced Basque whaling men on their crews. The Danes followed suit shortly after. Colonists of New England began whaling in the early 17th century,

although Native Americans had been practicing whaling since before the arrival of Europeans. Germans began whaling in 1694, but Dutch and British fleets dominated the industry in the 17th century, with the British taking the lion's share in the 18th century.

These whaling activities again focussed on North Atlantic right whales and bowhead whales in arctic and subarctic Atlantic waters. Indeed, the right whale gains its name because it was considered to be the "right" whale to hunt—it was slow, primarily found in coastal waters, it had a thick blubber layer that yielded much oil, and when the animal died it did not sink. Although records and archaeological evidence are scarce, the Atlantic gray whale is believed to have been rendered extinct during this period, with whaling possibly being the final straw for a perhaps already vulnerable species.

Early Basque whalers used every part of the whale, from the consumption of the meat, to the use of the feces as an orange-colored dye for clothing. However, the main product of commercial whaling was whale oil, which was used not only for lighting but also in industrial processes such as soap making. Baleen was also utilized, and in many respects this "whalebone" was the plastic of its time, being strong yet flexible. Many European whalers indeed concentrated on whalebone as a resource, particularly when the ladies' fashions of the day enlarged the market for whalebone-reinforced garments. In other nations, whalebone was used to make household tools, such as brooms in Barbados.

EXPANSION OF COMMERCIAL WHALING

As commercial whaling progressed, the focus shifted to collecting only the most profitable oil and whalebone. The most famous description of commercial whaling is Herman Melville's 1851 novel *Moby-Dick*. The book describes sperm whale hunting, which began in 1712 and had its heyday between 1740 and 1880. The head of the sperm whale contains a fine oil known as spermaceti, which was used to lubricate clockwork and delicate machinery in particular. The stomachs of sperm whales also occasionally contained lumps of a waxy substance called ambergris, also a valuable commodity. It was used as a fixative in the production of perfumes and was literally worth more than its weight in gold. Sperm whale meat was considered inedible and was rarely consumed by whaling crews.

The method used to catch whales in this era involved setting down a number of rowed catching vessels. When close enough to the whale, the harpooner—a practitioner of an extremely skilled, valued, but dangerous profession—would hurl his harpoon into the side of the whale, where its swiveling head would lodge firmly. The wounded whale would then try to escape or dive. However, the harpoon would be attached by rope to the catching boat, which would then act as a buoy, and be dragged along by the fleeing whale in what became known as a "Nantucket sleigh ride."

Eventually the whale would become exhausted or fatigued due to blood loss, and the catcher boat could approach more closely, when it would finish off the whale with a number of strikes from a long, thin whaling lance. This type of catching technology limited the size, species, and locations of whales that could be caught. However, in the mid-19th century, the development of steam-powered whaling ships and catching boats, grenade-tipped harpoons, and cannons to fire them meant that larger and faster whales could now be caught, including species such as blue, fin and sei whales. The methods used to kill whales have remained largely unchanged for the last 150 years. The age of "modern" industrial whaling was born.

MODERN AGE OF WHALING

Typically, many species of hunted whales were brought back to a shore-based whaling station where they were processed and the blubber was rendered. In 1925, another technological innovation, the invention of large, ocean-going factory ships, meant that whales could be processed at sea, and operations were no longer tied to shore bases. This opened up new regions for intensive exploitation, in particular the waters of the Southern Ocean around Antarctica. Other technological innovations such as larger vessels, spotter planes, and the use of sonar to detect whales and drive them to the surface increased the ability of whalers to catch animals.

By 1931, in recognition of the fact that some whale species were in decline and of the potential impacts that this might have on the whaling industry, the main whaling nations negotiated and signed the Convention for the Regulation of Whaling. Due to depletion of the species, bans on whaling were introduced for bowhead whales (1931), southern and northern right whales (1935), and Pacific gray whales (1937).

This convention eventually led to another agreement, the 1946 International Convention for the Regulation of Whaling, which formed the International Whaling Commission (IWC) in that same year. The IWC is now recognized internationally as the competent authority for the management of whale stocks. Under the IWC, more whaling bans were introduced for humpback and blue whales (1966) and sei whales (1979; except in Iceland). Finally, in 1982 the IWC voted to introduce a temporary moratorium on all commercial whaling, which came into effect in 1986. This moratorium was put in place to allow depleted whale stocks to recover and to allow the development of a better and more effective whaling quota system that would result in a sustainable whale catch. It is important to note that the IWC whaling ban only covers commercial hunting of baleen whales (except the pygmy right whale) and sperm whales. It does not stop scientific or subsistence whaling and hunting, commercial or otherwise, of all other cetaceans not controlled by the IWC.

CURRENT WORLD STATUS

Up until the date of the moratorium, over two million whales had been killed through commercial whaling, with many species such as the blue, fin, humpback, and sei whale becoming endangered. Even since the whaling moratorium came into effect, over 25,000 whales have still been killed. When the moratorium was enacted, Norway took a reservation on (opted out of) the ban and is not bound by the moratorium. It resumed commercial whaling in 1993, and in recent years Norwegian whalers have been taking approximately 550–700

The International Whaling Commission

The International Whaling Commission (IWC) was established on December 2, 1946. Since the 1980s it has been the primary mechanism for the ending of commercial whaling around the world. The IWC is a voluntary agreement with organizational headquarters in Cambridge, England, and has 70 member nations that meet annually to discuss limits on whaling. From the 1960s, it imposed quotas on the number and type of whales that could be caught with the aim of allowing the whale stocks to replenish—at the time some species were being hunted almost to extinction. Some countries, notably the Soviet Union, secretly flouted these quotas by massively under-reporting the number of whales killed. Gradually the main whaling nations of Japan, Norway, and Iceland found themselves outnumbered by the anti-whaling countries.

In the 1980s, the IWC voted to end commercial whaling, allowing it to take place on two grounds: Scientific whaling and whaling by aboriginal peoples. Norway started commercial whaling again in 1994, but at massively reduced levels, and Iceland started again in September 2006. The Japanese have never stopped whaling, claiming scientific research purposes. However, critics have seen the research as merely an excuse to continue operating whaling fleets while the meat goes to restaurants and retailers.

In recent years, the Japanese have been persuading Pacific and Caribbean countries to join the IWC and support their attempts to lift bans on commercial whaling. Conservation groups claim that Japanese overseas aid to poor countries in the Caribbean, the Pacific, and Africa has been directly tied to these countries' support in the IWC. There has also been criticism of the eight landlocked countries that are members of the IWC: Mali and Mongolia supporting the resumption of whaling, with Austria, the Czech Republic, Luxembourg, San Marino, Slovakia, and Switzerland opposing whaling. The IWC continues to maintain a moratorium on all commercial whaling.

northern minke whales a year, although a quota of over 1,000 animals was proposed for 2006.

Japan also hunts whales, even though its government agreed to the whaling moratorium. They do this by using a provision in the convention that allows whales to be killed for scientific research. After samples of blubber and stomach contents are taken from killed whales, meat is processed and sold in Japanese markets for human consumption. Japan currently hunts northern minke, Bryde's, sperm, and sei whales (151, 50, 10, and 50, respectively, in 2003 and 160, 51, three, and 100, respectively, in 2004) in the North Pacific and Antarctic minke whales (443 in 2003, 441 in 2004) in the Southern Ocean for scientific purposes. There are currently proposals to double the take of minke whales in the Southern Ocean and also to add fin whales, and eventually humpback whales, to the list of species being hunted in the Antarctic scientific whaling program. In 2003 Iceland also started a scientific whaling program, catching 37 animals in 2003 and 25 in 2004.

The sale of whale meat in Japan and Korea has some controversy attached; for example, genetic analyses have discovered the meat of endangered blue whale and protected J-stock minke whale being sold illegally. In addition, whale meat sales have provoked some environmental health concerns: Recent research has shown that meat being sold for human consumption in Japan had extremely high levels of mercury. Average contamination levels in meat were 22 and 18 times higher than health regulation limits permitted by the Japanese government, with some samples exceeding these limits by up to 200 times.

Another aspect of whaling is so-called aboriginal whaling. This is a type of hunt that is allowed by the IWC for aboriginal, indigenous, or native peoples that have a nutritional and cultural need for whale meat, with all products of the hunt to be consumed locally. Currently there are subsistence quotas allocated for Bering Sea bowhead whales (used by American and Russian natives), eastern Pacific gray whales (also used by American and Russian natives), Atlantic humpback whales (used in St. Vincent and the Grenadines), and north Atlantic fin and minke whales (used by Greenland natives). A hunt by the Makah Tribe of Washington State has not been conducted in recent years due to domestic legal issues. Although less controversial than commercial and scientific whaling, there are also some problematic issues with respect to aboriginal whaling in some areas; for example, the bowhead hunt by Alaskan natives occurs despite the bowhead whale being considered endangered under the U.S. Endangered Species Act.

Since the 1980s nations with a voting history of pro-conservation or antiwhaling tendencies, such as the United States, most European countries, New Zealand, and Australia, have been in the majority. However, in recent years the number of countries with a pro-commercial whaling stance has increased at the IWC, as have calls to lift the commercial whaling moratorium on populations that are considered to be showing signs of recovery. In the meantime, in Western countries there is strong and increasing societal opposition to a resumption of commercial whaling; in 1999, a survey found that less than one-fifth of Americans supported whaling of even abundant whale species, with 70 percent stating that they were opposed to the killing of whales on moral grounds. A similar survey in Scotland reported that 96 percent of the members of the public interviewed were opposed to whaling.

SEE ALSO: Endangered Species; Endangered Species Act (ESA); Indigenous Peoples; Mercury; Overfishing.

BIBLIOGRAPHY. T. Endo et al., "Mercury Contamination in the Red Meat of Whales and Dolphins Marketed for Human Consumption in Japan," *Environmental Science and Technology* (v.37, 2003); S.R. Kellert, *American Perceptions of Marine Mammals and Their Management* (School of Forestry and Environmental Studies, Yale University, 1999); William S. Perrin, Bernd Wursig, and J.G.M. Thewissen, eds., *Encyclopedia of Marine Mammals* (Academic Press, 2002); N. Scott and E.C.M. Parsons, "A Survey of Public Opinion in Southwest Scotland on Cetacean Conservation Issues," *Aquatic Conservation* (v.15, 2005); J.R. Twiss, Jr., and R.R. Reeves, eds., *Conservation and Management of Marine Mammals* (Smithsonian Institution Press, 1999).

E.C.M. Parsons
George Mason University
A. Romero and S. Kannada
Arkansas State University
Naomi A. Rose
Humane Society International

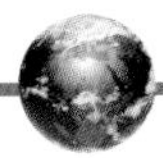

Wheat

WHEAT IS A cereal grass that is one of the most important sources of food in the world. Wheat belongs to the Poaceae family of the genus *Triticum*. Although wheat exists in numerous different species, it characteristically appears with long slender leaves, heads with a large number of small flowers that yield the seeds, and hollow stems. The most important varieties include *Triticum durum*, which is used to make various types of pasta, *Triticum aestivum*, which is used for bread, and *Triticum compactum*, which is used for baking cakes and biscuits.

Archaeological excavation reveals that wheat was first used in agriculture some 10,000 years ago. The first large-scale wheat farming took place in the Middle East and spread from there to Europe, northern Africa, and across Asia to China. It is not entirely clear whether migrants took the concept of farming wheat with them as they moved, or if the idea and necessary technology arose independently. Farmers developed new strains of wheat better adapted to local environmental conditions and tastes. In addition to crossbreeding, technological improvements included the seed drill, the animal-powered plough, and the use of fertilizers. However, it was not until the 20th century that large-scale, systematic attempts to improve wheat agriculture were made, specifically crossbreeding and the testing of environmental variables. The introduction of Japanese strains of wheat into the Americas is of particular significance, since these varieties helped to improve overall yield by a large amount. Also, attempts to reduce the effects of pests such as locusts, aphids, sawfly, and the wheat bug have improved wheat yields.

Moving into the 21st century, this type of research led to the creation of genetically modified (GM) wheat strains, notably by the American corporation Monsanto. The GM wheat products met with strong consumer resistance in Europe and Canada, although few regulations controlled their use in the United States. According to the Food and Agricultural Organization (FAO) of the United Nations, global wheat production in 2006 will reach 617 million tons, which is slightly below the record output of almost 632 million tons reached in 2004.

The introduction of Japanese strains of wheat into the Americas increased overall yield by a large amount.

The variation is attributed to the effect of weather, especially in important grain growing regions in Ukraine, Russia, and the United States. As climate change affects growing conditions and the increasing scarcity of water makes agriculture more difficult, it is expected that the variability in global yields will increase and the cost of crops will also increase. Approximately 70 percent of wheat supply is used for food, and another 18 percent for animal feed. Nearly every country in the world is involved in trading wheat, either by importing or exporting.

The wheat crop is harvested annually and the grains must be kept in suitable conditions to avoid excessive predation. The quality of the crop depends upon variables including the nature and purity of the soil and of the seeds used for sowing. The husks of the grains are threshed and used for different purposes. Many countries maintain specific guidelines with respect to the quality of different grades of wheat. Higher qualities of wheat are valued for nutritional and culinary purposes.

The inherited celiac disease, or gluten intolerance, can cause difficulty in digesting wheat. Gluten is a protein substance that is present in wheat and other cereal grains. People suffering from this disease are required to follow a particular diet to maintain good health. Such problems may increase in the future as the chemical composition of traditional foods is changed and as pollution increases.

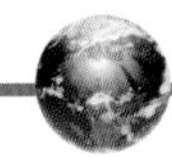

SEE ALSO: Crops; Crossbreeding; Fertilizers; Food; Genetically Modified Organisms (GMOs).

BIBLIOGRAPHY. Derek Berwald, Colin A. Carter, and Guillaume P. Gruere, "Rejecting New Technology: The Case of Genetically Modified Wheat," *American Journal of Agricultural Economics* (v.88/2, 2006); Alain P. Bonjean and William J. Angus, eds., *The World Wheat Book: A History of Wheat Breeding* (Laviosier, 2001); Food and Agriculture Organization, "Global Market Analysis: Wheat," *Food Outlook* (v.1, 2006); Anne Underwood, "The Perils of Pasta," *Newsweek* (v.134/15, 1999).

John Walsh
Shinawatra University

White, Gilbert (1720–93)

NATURALIST GILBERT WHITE was born on July 18, 1720, at a vicarage in Selborne, Hampshire, England, which his family lived in from 1728. Gilbert owned the house from 1763 and lived in it until his death. White graduated with a bachelor of arts from Oriel College, Oxford, in 1743. In 1744 he was elected fellow of Oriel, retaining his fellowship until his death; in the 1750s, he served as junior proctor of the University of Oxford and dean of his college. He was ordained successively deacon and priest in the Church of England and served as curate and vicar of various rural parishes in which his family or college had an interest, but did not seek preferments that would have prevented his living at Selborne.

On January 7, 1751, White began the record of the natural world on which his fame rests. His "Garden kalendar" records the cycle of growth and decay in his garden, with some wider references to natural history such as the migration of birds. From Benjamin Stillingfleet's *Miscellaneous Tracts* (1759) and especially from the "Calendar of flora" in the second edition (1762), White learned about the natural calendar, from which naturalists hoped that the observation of natural phenomena could guide the timing of sowing and reaping, ensuring reliable harvests for the benefit of all. Such a project required accurate identification of natural forms, and White's reading of William Hudson's *Flora Anglica* (1762) persuaded him to adopt Linnaean classification and nomenclature to aid rigorous and easily communicable identification of forms.

Through his brothers Thomas and Benjamin, White made important London scientific acquaintances, including Thomas Pennant. Pennant encouraged Daines Barrington to send White his *Naturalist's Journal* (1767), which was designed for the recording of meteorological phenomena and the behavior of flora and fauna: The correlation of these observations would lead to the discovery of the natural calendar. The idea of using his local knowledge for a general, disinterested purpose appealed to White, who in 1768 put aside the "Kalendar" in favor of the "Naturalist's Journal" of Barrington's design, which he kept until shortly before his death some 25 years later. Barrington's design required comprehensive, systematic, and precise observation and measurement. Barrington invited White to prepare for the Royal Society monographs on the house martin (*Delichon urbica*), swallow (*Hirundo rustica*), swift (*Apus apus*), and bank martin (*Hirundo riparia*); these were published in the *Philosophical Transactions of the Royal Society,* volumes 64 and 65. Barrington further encouraged White to prepare a work based on his journal for publication and suggested Samuel Hieronymus Grimm as illustrator. *The Natural History and Antiquities of Selborne* was published in 1789. White made his last journal entry on June 15, 1793, and died on June 26, 1793.

The Natural History of Selborne consists of three sequences of letters: The two to Pennant and Barrington concern natural history and the rest (unaddressed) concern parish antiquities. While his journals, published posthumously, offer a documentary record, *Selborne* is a literary work whose letters were considerably modified and extended for the press. Initially *Selborne* was valued as a new contribution to natural history, but, as its insights became part of general scientific knowledge, it was read more for the inspiration it gave to aspiring amateur naturalists. With numerous 19th- and 20th-century revised editions, *Selborne* entered the literary canon, a reassuring refuge for those troubled by Darwinism. White's exclusive focus on the local meant that his general discoveries and insights are

few, but it enabled him to fulfill his main aim of encouraging readers to give "a more ready attention to wonders of the Creation." White's emphasis on what is accessible to the enquiring observer in any place or time, combined with his gracious style, give *Selborne* its continuing appeal.

SEE ALSO: Darwin, Charles; Linnaeus, Carl; Nature Writing.

BIBLIOGRAPHY. Paul G.M. Foster, *Gilbert White and His Records: A Scientific Biography* (Christopher Helm, 1988); Paul G.M. Foster, "Gilbert White," in H.C.G. Matthew and Brian Harrison, eds., *Oxford Dictionary of National Biography* (Oxford University Press, 2004); Gilbert White, *The Natural History and Antiquities of Selborne,* Paul G.M. Foster, ed. (Ray Society, 1993).

Elizabeth Baigent
Oxford University

White, Gilbert F. (1911–2006)

GILBERT FOWLER WHITE was born on November 26, 1911, in Chicago and died on October 5, 2006, in Boulder, Colorado. White earned his bachelor's degree in 1932 and his doctorate in 1942 at the University of Chicago. He served in the New Deal administration of Franklin D. Roosevelt as secretary to the Mississippi Valley Committee, National Resources Committee, and the National Resources Planning Board. From 1940–42, he worked in the Bureau of the Budget in the Executive Office of the President. White was a Quaker and a conscientious objector to military service who in 1942 joined the American Friends Service Committee (AFSC), which aided refugees in France. He was interned in Baden-Baden, Germany until 1944 when he was allowed to return to the United States; he continued working with the AFSC until 1946. White married Anne Underwood in 1944, with whom he frequently collaborated on problems of domestic water use in Africa, and they had three children. Underwood died in 1989 and White married Claire Sheridan in 2003.

From 1946–55, White served as president of Haverford College and then returned to the University of Chicago as a professor of geography until 1970 when he left in protest over the university's expulsion of students during anti–Vietnam War protests. White moved to the University of Colorado and from 1970 to 1978 he was a professor of geography, director of the Institute of Behavioral Science, founder and director of the university's Natural Hazards Research and Applications Information Center (from 1976 to 1984 and again from 1992 to 1994), and the Gustavson Distinguished Professor Emeritus of Geography from 1980 until his death in 2006.

Considered the "father of floodplain management," White emerged as a central figure in the field of natural hazards research. He was an established scholar recognized for his contribution to the study of flooding and general advocacy of sound water management; he pioneered the United States's system of identification and classification of adjustment mechanisms for flooding. These adjustments he termed *structural* or *nonstructural*. Structural adjustments were those mechanisms constructed by engineers designed to modify flooding hazards so that people could live comfortably in areas that were subject to periodic flooding. Nonstructural adjustments were those changes made by governments to restrict the use of areas susceptible to floods. White advocated whenever possible the accommodation of, or adaptation to, flood hazards rather than structural solutions, such as dams and levees, that had dominated thinking in the first half of the 20th century.

White promoted an integrated system of floodplain management that had at least seven constituent elements:

> (1) mapping the estimated frequency and magnitude of flooding, (2) planning and regulation of use of vulnerable areas and of areas contributing to flood flows, (3) government support of insurance against flood losses, (4) improvement of flood warning systems and advice and training as to how to respond effectively to warning, (5) research and education as to how to flood proof property against damage, (6) extending the federal program of financial assistance to victims of flood damage to include support for buying out damaged property to support abandonment of severely affected property and movement to

lands beyond the reach of floods, and (7) taking explicit account of the costs and benefits to ecosystems and human recreation of leaving a floodplain completely open to water and silt from natural overflow.

Throughout his career as a citizen-scientist, White's ideas had a tremendous impact on government policy at all levels. He used his mediation skills and position as chair of the University of Chicago's Department of Geography to bring members of the Army Corps of Engineers and the Tennessee Valley Authority together. He assisted the Lyndon B. Johnson administration's 1959 Senate Select Committee on National Water Resources, which led to the creation of the Flood Control Act of 1960 and eventually a Water Resources Council in 1965. In these years he chaired a Ford Foundation mission to advise the United Nations (UN) Mekong River Committee concerned with flood control in that region and later headed a UN task force that studied several major water storage projects in the Zambezi, Senegal, Volta, and Nile River drainages.

White's crowning achievement in these years came from his 1966 appointment as chair of the Task Force on Federal Flood Control Policy. His work resulted in the creation of Congressional House Document 465 and Executive Order 11296 that for the first time mandated that all federal agencies incorporate flood planning into their programs. The task force's broader aim was to create a unified national program for managing not only flood losses and flood control, but also floodplains as ecosystems, and it was slowly being achieved. When Congress established the National Flood Insurance Program (NFIP) in 1968 that created the Federal Insurance Administration—something White's 1966 task force and a parallel task force were largely responsible for—White emerged as both its champion and critic. In the late 1970s, White pushed for the transfer of the NFIP out of the Department of Housing and Urban Development to the new Federal Emergency Management Agency, which was established in 1978 by Congress during the James Carter administration.

Throughout his life, White worked to build communication bridges between various constituencies and served on numerous committees. A past president of the Association of American Geographers, he became involved with the International Geographical Union and Commission on Man and Environment (which he chaired from 1969–76) and the International Council of Scientific Unions's (for which he served as president from 1976–82) Scientific Committee on Problems of the Environment.

White also built institutions that survived. For example, as the chair of a Ford Foundation Resources for the Future institute in the 1970s, he spearheaded a successful multimillion dollar fundraising campaign to help ensure the institute's future in the 21st century. In 1974 he founded the Natural Hazards Research and Application Information Center and for decades worked there with students and colleagues producing a host of socially relevant master's theses and doctoral dissertations in a wide range of natural hazards research. Many of his students have gone on to become leading practitioners in the field. In a career that spanned seven decades, White produced over 400 scholarly papers and earned 50 degrees and honorary awards.

SEE ALSO: Dams; Floods and Flood Control; Hazards; Levees; Locks and Dams; Mekong River.

BIBLIOGRAPHY. Robert. E. Hinshaw, *Living with Nature's Extremes: The Life of Gilbert Fowler White* (Johnson Books, 2006); Robert W. Kates and Ian Burton, eds., *Geography, Resources, and Environment, Volume 1: Selected Writings of Gilbert F. White* and *Volume II: Themes from the Work of Gilbert F. White* (University of Chicago Press, 1986); Task Force on Federal Flood Control Policy, *A Unified National Program for Managing Flood Losses: Report by the Task Force on Federal Flood Control Policy* (House Doc. 465, U.S. Government Printing Office, 1966); Gilbert F. White, *Human Adjustment to Floods: A Geographical Approach to the Flood Problem in the United States*, Ph.D. diss., (University of Chicago, 1942).

Michael Butt, Ph.D.
Independent Scholar

Wilderness

WILDERNESS IS ONE of the most potent and contentious concepts in Western, particularly North

American, framings of society-nature relations. As a symbol of humankind's moral and spiritual condition, it is deeply imbedded in Western philosophy and religion. As an actual place, it lies at the heart of political-economic struggles over the future courses of development and biodiversity protection around the world. When wilderness is debated among scholars and scientists, it is more often about the idea of wilderness than actual wild places.

THE IDEA OF WILDERNESS

The idea of wilderness can be categorized into one of three roughly historical trends: Ancient, classical, and romantic. The origins of the ancient view are traceable to the Neolithic revolution in the Fertile Crescent of the ancient Mediterranean, according to Oelshlaeger. This region experienced the earliest transition of human culture from a reliance on hunting and gathering to the domestication of animals and permanent cultivation. This shift in material existence was accompanied by a shift in the cultural meanings of nature, including a move away from totemic rituals and myth, to animal idolatry and fertility cults, a greater sense of separation of humanity from the natural world, and the rise of the belief that nature could be manipulated to fit the designs and desires of humankind, according to Glacken and Oelshlaeger. The later emergence of monotheism, specifically Judaism, in this region marks a major shift in ancient ideas of nature, with the Hebrew's supreme being, Yahweh, believed to be not of nature, but above nature as its creator.

It is from the Hebrews' Old Testament, particularly as translated into English in the King James Version of the Bible, that some of the most influential and persistent ideas of wilderness are introduced into Western thought. Some historians have argued that the roots of modern environmental problems can be traced to the Old Testament and its negative portrayal of wilderness, as claimed by Nash and White. In particular, the metaphor of the garden as the site of God's grace and the wilderness, into which Adam and Eve are cast, as a spiritual wasteland is seen to have tainted Western attitudes toward nature for centuries. In addition, the Book of Genesis, which portrays people as having been created separate from and having dominion over the rest of God's creation, promotes a sense of nature as having no value or purpose beyond service to humankind. Some would argue, however, that the Hebrew Bible's portrayal of the wilderness is inconsistent and ambiguous and not reducible to either a positive or negative generalization, as claimed by Oelschlaeger. For example, the Old Testament also portrayed wilderness as a spiritual refuge, a place where the Hebrew prophets and their followers could come in more direct contact with their God, Yahweh.

The classical view of wilderness is rooted in the Greco-Roman civilizations, specifically in a blend of Greek rationalism, Roman pastoral aesthetic, and Christianity. Greek philosophers developed several key trends that continue today to influence Western conceptualizations of wilderness, particularly the idea of order, unity, and harmony in nature, a rational approach to observing and categorizing phenomena in nature, a homocentric understanding of the universe, and a strong distinction between the city and the county. According to Glacken, no earlier period in Western thought "revealed such strong, self-consciously expressed contrasts between the urban and the rural as did the Hellenistic." These trends were not formed in isolation, but were combined with elements of Hebrew and early Christian thought to create the "genesis of the idea of wilderness that has ruled Western civilization for these past two millennia," according to Oelschlaeger.

The classical view portrays wilderness as a condition of nature that awaits the transforming hand of civilization to make it productive and useful. This perspective is clear in the pastoral poetry of the 1st century B.C.E. poet Virgil, urging farmers to "mellow your harsh fruits by culture, nor suffer fields to lie idle," as quoted by Glacken. This drive to civilize wilderness takes on overt spiritual dimensions in early Christianity, when proselytizing monks set about converting pagan Europe. Paganism, built upon a reverence of nature and wild places, presented a challenge to early Christian missionaries who viewed pagans' sacred groves as the dwelling places of witches and other agents of the devil. Doing God's work meant simultaneously exercising dominion over nature and converting pagans to Christianity,

both of which entailed cutting down the sacred groves and other places of nature worship and turning them into pasture and field, according to Nash and Oelschlaeger. There remained within Christian thought, however, the ideas of wilderness as a refuge from corrupt civilization, and of exposure to wild nature as a means to come into closer contact with God's creation, perspectives more commonly associated with the romantic view.

The romantic view of wilderness is best characterized as a reaction against modernity among Western artists, poets, writers, and philosophers. These cultural elites were generally situated far from the physical reality of wild nature. They lived in civilized Europe or the coastal cities of the Americas and admired wilderness from a comfortable distance, as noted by Nash.

The philosophical roots of the romantic view are found in the concept of the sublime, a term that until the 18th century was associated with the awesome, fearful, and majestic grandeur of God, explain Cronon and Cosgrove. As the industrial revolution transformed Europe, the sublime gradually came to be associated with remote mountain and rugged coastal landscapes, where the awesome display of wild nature's power could be experienced and contemplated. By the turn of the 19th century, genres of romantic poetry and landscape painting had emerged in which raw, untamed nature was celebrated for its very wildness. Wilderness, rather than being morally degenerate and economically unproductive, as in the classical view, became inspirational, even sacred, and deserving of protection and preservation. This romantic vision of wilderness experiences its fullest expression in the North American conservation movement of the 19th and 20th centuries.

AMERICAN WILDERNESS

The dominant conceptualization of wilderness today is most commonly associated with ideas that developed among North American conservationists. So pervasive is the wilderness idea in North American conservation thought that some have characterized it as an obsession that provides "the dominant ideological underpinning" for a wide range of environmental concerns, as stated by Cronon. To understand how the wilderness became so dominant in U.S. environmental thought, the idea has to be situated within the broader history of European conquest of North America, the role of the frontier in political culture and national identity, the encounter between Native Americans and Europeans, and the development of capitalist social relations in the United States, as pointed out by Cronon, Nash, and Cosgrove.

The Pilgrims and Puritans who spearheaded the European colonization of North America carried with them from Europe a classical perspective on wilderness. For the Puritans and the colonizers that followed in the 17th and 18th centuries, the continent's forests were obstacles to overcome, lands to be tamed, made productive, and civilized. Colonists described the lands beyond the meager coast-bound settlements as a "howling wilderness" that was dark, savage, and filled with dangers to both the physical and spiritual health of the colonizers. In various secular and sacred forms, this perspective remained dominant until the end of the 18th century.

By the latter half of the 19th century, there was a reversal in wilderness thought among cultural and political elites in the United States. Western North America, with its rugged and monumental mountain ranges and expansive deserts and forests, came to be viewed as a landscape of virtue and natural purity as opposed to the decadent and desecrated lands of Europe and North America's eastern seaboard. The writings of Henry David Thoreau, Walt

The history of wilderness is also a human history of conquest, colonization, and forced relocation.

Whitman, and John Muir gave shape to a new romantic vision of American wilderness.

A myth of national identity emerged that positioned Americans as different from European nationalities by virtue of their historical encounter with wilderness. Artists, writers, historians, and politicians began to celebrate North American wilderness as the source of personal characteristics of rugged individualism, self-reliance, and moral virtue, which were said to provide the core of American national identity. Something of a "wilderness cult," according to Nash, emerged that celebrated primitivism and noble savagery, suggesting that periodic wilderness encounters kept America vigorous by cultivating manliness and virility among its citizens. Among American political figures, Theodore Roosevelt is most closely identified with this perspective on wilderness through his many writings and his initiatives as president to establish national parks in the wildest areas of the U.S. West.

Among other historical developments, particularly the closing of the American frontier, the 19th-century cultural shift in the wilderness idea has been linked to new waves of immigrants coming to the United States. A nativist movement arose based on the belief that "since 1880 a new type of person had come to dominate movement into the United States ... unskilled, transient young men, largely from southern and eastern Europe, entering urban industrial employment and keeping a distance from earlier settled Americans," as stated by Cosgrove. According to nativist logic, the new immigrants, having come from a different "racial stock" and never having experienced the transforming influences of the great American wilderness, could not be true Americans.

In combination, the wilderness cult and nativism gave the early conservation movement in the United States pervasive racist and masculine underpinnings. Early 20th-century conservationists were almost universally well-to-do, eastern-based, white males whose concerns about wilderness preservation, masculinity, racial purity, and immigration tended to bleed one into the other, an idea expounded upon by Haraway, Cosgrove, and Cronon.

A new political movement to protect what were widely viewed as the vestiges of a disappearing North American wilderness accompanied the cultural shift in American wilderness thought from a classical to a romantic perspective. Muir and Roosevelt were the primary catalysts for the movement, with Muir being the chief philosopher and promoter of the wilderness movement through his writings about Yosemite, and Roosevelt being the elected official most closely associated with early federal government initiatives to legally protect natural areas. Muir and Roosevelt, however, were on opposite sides of the plan to dam the Tuolumne River in Yosemite National Park and create the Hetch Hetchy Reservoir, an initiative that ultimately set the American wilderness movement on fire when Roosevelt approved the project in 1908. Hetch Hetchy became a rallying cry for conservationists.

A new generation of wilderness advocates arose, led by Aldo Leopold, a visionary forester with the U.S. Forest Service (USFS), and Robert Marshall, a professional forester from a wealthy New York family who also worked for a time with the USFS. Marshall made his principal mark on the movement when he founded the Wilderness Society in 1935, an organization specifically dedicated to promoting the permanent preservation of wilderness for wilderness' sake. In 1964, the movement achieved its goal of establishing a federally designated system of protected wilderness areas. The arguments for the U.S. Wilderness Act of 1964 reflected the romantic wilderness perspective developed in the writings of Thoreau, Muir, and Leopold. Today, there are over 106 million acres officially designated as wilderness under the act, more than half of which is in the state of Alaska.

THE NEW WILDERNESS DEBATE

Since the passage of the 1964 Wilderness Act, the wilderness idea has become more politically charged than ever, sparking what Calicott and Nelson identify as the "great new wilderness debate." The debate is not so much one between those that argue for and those that argue against wilderness preservation. Indeed, many of the critics of the wilderness idea make an effort to state their endorsement for the protection of wild areas, such as Cronon and Calicott. Rather, it is a debate about the idea of wilderness, as developed in the American and, now, the global conservation movement, the values that it reflects, and what it means for the ability to imagine models of sustainable society-nature relationships. Critics such as Calicott have labeled the writings of Thoreau, Muir, Leopold, Marshall, and the U.S. Congressional act that they inspired as the "received wilderness idea." The argument is that these writings provide the foundation for current Western conceptualizations of wilderness, yet they were shaped by ethnocentric, racist, and sexist ideologies and since-discredited scientific models of ecology.

At the heart of scholarly critiques of the wilderness idea is the proposition that wilderness is a social construction or, more specifically, that according to Calicott "the name wilderness socially constructs, as we now say, the landscape, in a way not shared by all social groups." The new wilderness debate is thus one aspect of a larger debate from the 1980s and 1990s surrounding the critiques of the philosophy and methodology of science by a broad range of social constructivists in the humanities and social sciences, subsequently labeled "science wars" by Ross. Among other claims, constructivists argued that nature, as an object of scientific study, is socially constructed. The social construction of nature is a phrase commonly employed to stress the role of representation, discourse, and imagery in defining and framing our knowledge of nature and the natural. As Bird argued, "scientific knowledge should not be regarded as a representation of nature, but rather a socially constructed interpretation of an already socially constructed natural-technical object of inquiry." Wilderness, it is argued, has become synonymous with a distinctly Western conceptualization of nature and so has been a central concept in scholarly debates over socially constructed nature, according to Proctor.

Perhaps the most notable critique of wilderness in the new debate is William Cronon's widely reprinted essay, "The Trouble with Wilderness." Cronon argues that nature in Western conservation thinking is idealized as an empty wilderness, clearly placing human society and nature in separate spheres and leading inevitably to the conclusion that human presence alone is enough to degrade nature. This dualistic vision of society-nature relations is for Cronon "the trouble with wilderness." His constructionist approach suggests two fundamental empirical and theoretical limitations of the wilderness model of nature. First, acceptance of this model would require ignoring the conclusions from the empirical findings of cultural geographers, environmental historians, and archeologists that people have manipulated and shaped nature "for as long as we have a record of their passing." In short, the physical actuality of nature as a vacant wilderness is not supported by geo-historical research. Second, the wilderness-humanity duality leaves no room for considering other, less environmentally destructive theories of human history and society. As Cronon explains the core paradox of wilderness, "if nature dies because we enter it, then the only way to save nature is to kill ourselves."

On the other side of the debate, Foreman has suggested that critics do not understand the science of biodiversity conservation and the importance of wilderness to the maintenance of global biodiversity. Another objection to social constructionist approaches is political. As Hayles asks, "If nature is only a social and discursive construction why fight hard to preserve it?" Soule argues that wilderness critics play into the hands of antienvironmental political initiatives. More generally, wilderness advocates have rejected the philosophical position of constructivism altogether and tried paint it as an extreme fringe perspective. Constructivist arguments have been characterized by Soule and Lease as "certain radical forms of 'postmodern deconstructivism'" that "asserts that all we ever perceive about the world are shadows" and so denies the external existence of nature. Foreman dismisses Cronon and others as "postmodern deconstructionist scholars" a label that, while inaccurate, effectively marginalizes those who may support wild land and biodiversity protection, but question the idea of wilderness. There is "real wilderness" in the world, so the argument goes, that is disappearing fast and is in desperate need of protection, according to Foreman.

THE POLITICAL ECOLOGY OF WILDERNESS

While the idea of wilderness, particularly in North America, continues to be debated, wilderness has become a dominant component of global biodiversity conservation strategies and the promotion of international ecotourism. The Convention on

Biological Diversity (CBD), which resulted from the 1992 Earth Summit, now provides the framework and rationale for international efforts to stem biodiversity loss, focusing on in situ conservation in the wilderness of national parks and protected areas. The international tourism industry uses the wilderness idea to sell ecotourism packages to third world settings, particularly sub-Saharan Africa. In short, the North American wilderness idea has been globalized as the dominant way of thinking about nature.

The transfer of the wilderness idea around the globe has raised questions about the political ecology of wilderness. These include how the relationship between society and nature is defined and conceptualized, how access to land and resources is controlled, and how environmental costs and benefits are distributed. For instance, many of the areas now designated as wilderness were only recently cleared of people who had occupied and transformed the environment over generations, sometimes millennia. Recent studies in North America report similar process of forced relocation, suggesting, "uninhabited wilderness had to be created before it could be preserved," according to Spence. More often than not, both in North America and around the world, the dislocations of resident populations were conducted as part of larger efforts by the state to control or eliminate some of its subjects. Thus the wilderness idea has been labeled "a tool of genocide," as stated by Calicott. Because the history of wilderness is a human history of conquest and colonization, wilderness areas have become enveloped in larger struggles for social justice, historical land claims, and self-determination among indigenous peoples and peasant communities around the world.

SEE ALSO: Conservation; Critical Environmental Theory (or Ecocriticism); Leopold, Aldo; Muir, John; National Parks; Nature, Social Construction of; Nature Writing; Political Ecology; Preservation; Pristine Myth; Religion; Restoration Ecology; Thoreau, Henry David; Wilderness Act (U.S. 1964); Wilderness Society; Wildlife.

BIBLIOGRAPHY. E. Bird, "The Social Construction of Nature: Theoretical Approaches to the History of Environmental Problems," *Environmental Review* (v.11/4, 1987); J.B. Calicott, "Contemporary Criticisms of the Received Wilderness Idea," *USDA Forest Service Proceedings* (v.15/1, 2000); J.B. Calicott and M. Nelson, eds., *The Great New Wilderness Debate: An Expansive Collection of Writings Defining Wilderness from John Muir to Gary Snyder* (University of Georgia Press, 1998); D. Cosgrove, "Habitable Earth: Wilderness, Empire, and Race in America," in D. Rothenberg, ed., *Wild Ideas* (University of Minnesota Press, 1995); D. Cosgrove, *Social Formation and Symbolic Landscape* (University of Wisconsin Press, 1998); W. Cronon, ed., *Uncommon Ground: Toward Reinventing Nature* (W.W. Norton, 1995); D. Foreman, "All Kinds of Wilderness Foes," *Wild Earth* (v.6, 1996); D. Foreman, "The Real Wilderness Idea," *USDA Forest Service Proceedings* (v.15/1, 2000); C. Glacken, *Traces on the Rhodian Shore* (University of California Press, 1967); R. Grove, "Environmental History," in P. Burke, ed., *New Perspectives on Historical Writing* (Penn State University Press, 2001); D. Haraway, "Teddy Bear Patriarchy: Taxidermy in the Garden of Eden, New York City, 1908–1936," *Social Text* (v.4, 1984); N. Hayles, "Searching for Common Ground," in M. Soulé, and G. Lease, eds., *Reinventing Nature? Responses to Postmodern Deconstruction* (Island Press, 1995); R. Nash, *Wilderness and the American Mind* (Yale University Press, 1982); R. Neumann, *Imposing Wilderness: Struggles over Livelihood and Nature Preservation in Africa* (University of California Press, 1998); R. Neumann, "Africa's 'Last Wilderness': Reordering Space for Political and Economic Control in Colonial Africa," *Africa* (v.71/4, 2001); R. Neumann, "Nature-State-Territory: Toward a Critical Theorization of Conservation Eenclosures," in R. Peet and M. Watts, eds., *Liberation Ecologies* (Routlege, 2004); M. Oelschlaeger, *The Idea of Wilderness: From Prehistory to the Age of Ecology* (Yale University Press, 1991); J. Proctor, "The Social Construction of Nature: Relativist Accusations, Pragmatist and Critical Realist Responses," *Annals of the Association of American Geographers* (v.88/3, 1998); A. Ross, ed., *Science Wars* (Duke University Press, 1996); M. Soule and G. Lease, eds., *Reinventing Nature? Responses to Postmodern Deconstruction* (Island Press, 1995); M. Spence, *Dispossessing the Wilderness: Indian Removal and the Making of the National Parks* (Oxford University Press, 1999); L. White, "The Historical Roots of Our Ecological Crisis," *Science* (March 10, 1967).

Roderick P. Neumann
Independent Scholar

Wilderness Act of 1964

THE WILDERNESS ACT of 1964 established the National Wilderness Preservation System in the United States, now comprising more than 680 units with a total of 106 million acres (42.8 million hectares). Federal lands protected under the Wilderness Act are defined as "an area where the earth and its community of life are untrammeled by man, where man himself is a visitor who does not remain." Land designated by the U.S. Congress as wilderness areas are subjected to strict management regimes to provide the utmost protection for natural areas and the biological diversity they support. In recognizing the intrinsic value of ecosystems and their many components, Roderick Nash states that "[W]ilderness is not *for* humans at all, and wilderness preservation testifies to the human capacity for restraint."

The Wilderness Act excludes from those public lands designated as wilderness areas, such items as automobiles, motorcycles, bicycles, hang gliders, motorboats, and activities such as road and building construction. The act also excludes certain types of commercial activity such as logging and mining (the latter after 1984) but permits livestock grazing in certain locations. Lands protected under the Wilderness Act are usually at least 5,000 acres (2,023 hectares) and present recreational activities that offer "outstanding opportunities for solitude or a primitive and unconfined type of recreation."

Howard Zahnister, executive director of the Wilderness Society, and David Brower from the Sierra Club were two of the early advocates during the 1950s calling for a wilderness bill that would provide permanent Congressional protection for federal public lands. Zahnister eventually authored the Wilderness Act of 1964 out of concern that the United States lacked a comprehensive, permanent, and legally binding system to protect wilderness areas, thus leaving large tracts of public lands open to degradation: "Let us be done with a wilderness preservation program made up of a sequence of overlapping emergencies, threats, and defense campaigns."

Democratic Senator Hubert Humphrey from Minnesota introduced Zahnister's wilderness bill into Congress in 1956. Due to proposed restrictions on commercial resource extraction in wilderness areas, the original wilderness bill was heavily opposed by loggers, miners, and ranchers, as well as by U.S. federal agencies including the Forest Service and the National Park Service. The act was passed eight years later, though Zahnister died a few months before its signing. President Lyndon B. Johnson signed the Act on September 3, 1964, with 54 wilderness areas named (9.1 million acres [3.6 million hectares]) in 13 states.

The Wilderness Act is unique in that it represented an effort on the part of the conservation community as well as the U.S. Congress to preserve biological diversity, but also embodied, as Nash describes, the notion of wilderness preservation as a radical act: "It is indeed subversive to the forces that have accelerated modern civilization to power but now threaten its continuation: Materialism, utilitarianism, growth, domination, hierarchy, exploitation."

Wilderness Areas are managed by four federal agencies: the Bureau of Land Management, the Fish and Wildlife Service, the U.S. Forest Service, and the National Park Service. The first unofficial wilderness area was the 558,065-acre (225,841-hectare) Gila Wilderness in the Gila National Forest in New Mexico, created in June 1924 at the urging of conservation pioneer Aldo Leopold. The Gila Wilderness later received permanent protection under the Wilderness Act. There are now a total of 680 Wilderness Areas in the United States, the smallest being Pelican Island in Florida (6 acres [2 hectares]) and the largest being Wrangell–Saint Elias, in Alaska (9,078,675 acres [3,674,009 hectares]). The largest wilderness complex in the contiguous United States is the Frank Church–River of No Return and Gospel-Hump Wildernesses, Idaho (2,572,553 acres [1,041,075 hectares]).

States with the most Wilderness Areas include California (130 units), Arizona (90 units), Nevada (56 units), Alaska (48 units), and Colorado (41 units). States lacking lands with wilderness protection are Connecticut, Delaware, Iowa, Kansas, Maryland, and Rhode Island. Of the entire United States, 4.71 percent—about the size of California—is protected as Wilderness Areas; 54 percent of Wilderness Areas are found in Alaska, and only 2.58 percent of the continental United States is protected in this form.

Citizens have the ability to be influential in affecting which lands are designated for protection under the Wilderness Act by creating their own citizen wilderness proposals and submitting these plans to members of Congress. Citizen-supported nonprofit organizations such as the New Mexico Wilderness Alliance, the Sky Island Alliance, Southern Utah Wilderness Alliance, and the Wilderness Society are just a few of many wilderness advocacy groups that work for increased wilderness protection on public lands.

The National Wilderness Preservation System continues to grow. The Southern Utah Wilderness Alliance presented the most recent successful wilderness proposal when in 2006 legislation was granted providing lasting wilderness protection to 100,000 acres (40,469 hectares) in the Cedar Mountains of Utah.

SEE ALSO: Biodiversity; Conservation; Leopold, Aldo; Preservation; Public Land Management; Restoration Ecology; Wilderness; Wilderness Society.

BIBLIOGRAPHY. George Coggins, Charles Wilkinson, and John Leshy, *Federal Public Land and Resources Law* (Foundation Press, 1993); Dave Foreman and Howie Wolke, *The Big Outside* (Harmony 1992); Roderick Nash, *American Environmentalism: Readings in Conservation History* (McGraw-Hill, 1990); William Rodgers, "The Seven Statutory Wonders of U.S. Environmental Law," in Robert Fischman, Maxine Lipeles, and Mark Squillace, eds., *An Environmental Law Anthology* (Anderson Publishing Company, 1996); Southern Utah Wilderness Alliance, www.suwa.org (cited April 2006); Wilderness.net, www.wilderness.net (cited April 2006).

Andrew J. Schneller
Independent Scholar

Wilderness Society

THE WILDERNESS SOCIETY is a nonprofit environmental organization based in Washington, D.C., with 10 U.S. regional offices. Through science, economic analysis, advocacy, and education, it works to achieve wilderness designation on federal lands. Since its founding in 1935, it has helped add 105 million acres (42.4 million hectares) to the National Wilderness Preservation System.

The Wilderness Act of 1964 was written by former society president Howard Zahniser and signed into law by President Lyndon B. Johnson. The act defined wilderness as "an area where the earth and its community of life are untrammeled by man, where man himself is a visitor who does not remain." The act enables Congress to set aside select units in national forests, parks, wildlife refuges, and other federal lands as areas to be kept permanently unchanged by humans, meaning no roads, mechanized vehicles, resource extraction, or other significant impacts. To date, 106,619,208 acres (43,147,262 hectares) of land have been added to the National Wilderness Preservation System and the Wilderness Society is striving to protect an additional 100 million acres (40.4 million hectares).

The story of the society's founding is that foresters and friends Bob Marshall, Benton Mackaye, Bernard Frank, and Harvey Broome were in a heated debate over how to best save America's wilderness as they drove across the rolling hills of Tennessee. The men got out of the car, scrambled up an embankment, and argued over the philosophy and definition of the new organization that they eventually called the Wilderness Society.

The Wilderness Society's work is guided by a "land ethic," a philosophy of the relationship between people and the land based on the work of Aldo Leopold, one of the society's founding members. Leopold, who believed in preserving the integrity, stability, and beauty of ecosystems, envisioned that the society would help form the cornerstone for the movement needed to save America's vanishing wilderness.

Public support for wilderness has fluctuated over the past century. However, wilderness will always have an unmistakable lure to the human psyche because it provides release for our basic need for creativity, self-sufficiency, and freedom, all that civilization precludes. In Roderick Nash's classic *Wilderness and the American Mind,* he suggests that wilderness itself is a large part of American identity. Wilderness provides an escape from the noise and pollution of urban areas and opportunities for spiritual renewal. Environmentally, tracts

of wilderness land provide a safe haven for wildlife, protect watersheds, and improve air quality. Wilderness Areas also offer amazing vistas and opportunities for outdoor recreation.

The Wilderness Society's current campaigns are to protect the Arctic National Wildlife Refuge and other wildernesses from oil and gas drilling, to stop road building and logging on 58 million acres (23,471,767 million hectares) of forest lands, and to reduce the destruction caused by off-road vehicle use. As urban areas continue to sprawl in an ever-increasing network of roads, traffic, and shopping malls, the value of wilderness to society will increase. The Wilderness Society's long-term mission is to ensure that wild areas will remain preserved and protected for future generations.

SEE ALSO: Arctic National Wildlife Refuge; Land Ethic; Leopold, Aldo; Preservation; Wilderness; Wilderness Act of 1964.

BIBLIOGRAPHY. Aldo Leopold, *A Sand County Almanac* (Oxford University Press, 1968); Roderick Nash, *Wilderness and the American Mind,* 4th ed. (Yale University Press, 2001); The Wilderness Society, www.wilderness.org (cited April 2006); Dyan Zaslowsky and the Wilderness Society, *These American Lands* (Henry Holt, 1986).

Colleen M. O'Brien
University of Georgia

Wild Horses

WILD HORSES ARE horses that roam in wilderness areas of the world. Wild horses, strictly speaking, are horses descended from horses that have never been domesticated. More broadly, horses that have escaped into the wilds or that have been born wild are feral horses. The only remaining wild horse is the Asian wild horse, also known as the dun-colored, black-maned equids of Mongolia, which are a national symbol. The wild Mongolian horse is called *Takhi* in Mongolian, meaning "spirit" or "spiritual." Russian General Nikolai Przhevalsky (1839–88) first identified the Mongolian wild horse as unique, and he went to Mongolia in the 1880s to search for the horse because it was so rare. In 1900 Carl Hagenbeck captured some, which were put into zoos. Some of these reproduced and their stock was re-introduced into the wilds of Mongolia in 1992 after repeated attempts to locate wild stocks from 1960 onward failed. In 2006 the wild population was around 1,500; they were all descended from animals bred in zoos. Przewalski's (from the Polish spelling of Przhevalsky) horses are about the size of large ponies, are muscular in body, and have a heavy head. Their color is usually light brown, with a black tail, mane, and lower legs, but a white muzzle.

All the other horses in the world that are called wild horses are actually feral horses. That is, they are domesticated horses that have escaped into the wilds, or they are descended from horses that were originally domesticated, but which were either abandoned or escaped into wild areas. Christopher Columbus brought horses with him on his second voyage (1493–96) to the New World. By 1600 a number of Spanish horses had escaped into the great open spaces of the sparsely settled America. They soon grew into great herds, which transformed the lifestyle of the Plains Indians into a horse culture. Descendants of some of these horses still roam wilderness areas of the American West. They are often called *mustangs*, which is an English pronunciation of the Spanish word *mesteno* for "stray" or "wild." Many Americans view mustangs as a symbol of America's frontier heritage. Before passage of the Wild Free-Roaming Horse and Burro Act (1971), Western ranchers killed great numbers as nuisances. In 2004, Senator Conrad Burns (Montana) was able to attach a rider to a much larger bill that effectively gutted the 1971 act. Today many environmental, humane, and historical organizations are campaigning for a restoration of full protection for American wild horses.

In Canada few, if any, wild horses are true mustangs. Most are a mixture of feral English breeds including Suffield, Shire, and Clydesdales. Most are located in Alberta. Attempts to reduce or eliminate them have been opposed by numerous animal rights groups or individuals. Australia also has wild or feral horses that are called Brumby horses. By the early 1800s, there were a number of them in the mountains of eastern Australia, and today they also live in the west and other areas of Australia.

Wild horses on the barrier islands of Chesapeake Bay and of North Carolina's Outer Banks were probably originally from Spanish galleons. Others were strays that went feral. In North Carolina the barrier island of Corolla has become home for a few wild horses. The Corolla Wild Horse Fund was formed in 1989 and works to protect them. In Maryland, Assateague Island has a wild horse population that is separated from Virginia's by fences. Because of poor diet in the salt marshes they are about the size of ponies. The same is true for the wild horses of Chincoteague Island (Virginia) National Wildlife Refuge. Both Assateague and Chincoteague Islands have roundups to manage the population.

SEE ALSO: Mongolia; Ranchers; Wild versus Tame.

BIBLIOGRAPHY. Joel Berger, *Wild Horses of the Great Basin: Social Competition and Population Size* (The University of Chicago Press, 1986); Jay Featherly, *Mustangs: Wild Horses of the American West* (Carolrhoda Books, 1986); Ann Weiss, *Save the Mustangs! How a Federal Law Is Passed* (Simon and Schuster, 1974).

Andrew J. Waskey
Dalton State College

Wildlife

ACCORDING TO THE *Oxford English Dictionary* (OED), the usage of *wild*, meaning "of an animal; living in a state of nature; not tame, not domesticated," can be traced back to 725 C.E. By 1440 C.E., the word *wildness*, meaning "the state or character of being wild" or "undomesticated," referred to a particular way of being—a category of behaviors and attributes, but not Kingdom or Phylum-specific ones. The word *wildlife* (or *wild life*), meaning "native flora and fauna of a particular region," dates back only to 1879 C.E., and popular usage of its attributive form (e.g., wildlife conservation) and combinative form (e.g., wildlife park, wildlife sanctuary), began in the mid-1930s and 1960s, respectively (OED). Wildlife, then, originated as a category inclusive of animals *and* plants. As such, wildlife together comprise biodiversity.

Although *wildlife* was meant to refer to the native flora and fauna of a particular region, for nearly half a century, television, film, and a number of prominent organizations have privileged fauna. Wildlife as animals dominates National Geographic documentaries, Marlin Perkins's *Wild Kingdom*, more recent shows on the Discovery Channel and Animal Planet. The U.S. Department of Agriculture describes wildlife as "any living creature, wild by nature, endowed with sensation and power of voluntary motion and including mammals, birds, amphibians and reptiles, which spend a majority of their life cycle on land." The Natural Resources Defense Council describes wildlife as "animals living in the wilderness without human intervention," while the standard forestry glossary describes wildlife as "a broad term that includes nondomesticated vertebrates, especially mammals, birds, and fish."

Humans are not counted among wildlife, although anthropogenic processes certainly impact the life forms that are. Prior to the Neolithic Revolution, all human beings relied on undomesticated plants and animals for survival. Thus, for most of human history, all plants and animals would have been considered "wild" by today's standards. With the advent of agriculture, many species of "wild" plants and animals were domesticated. Over time and through human selection, plants and animals of today's farmlands and pet shops have become quite different from their ancestors. This contrast between domesticated species and their increasingly distant relatives contributed to the creation of the word *wildlife*.

More recently, interest in wildlife and wildlife conservation has increased because many species of wildlife have been driven to extinction or near-extinction due to rapid human growth rates and their concomitant ecological pressures. Non-human species have suffered habitat loss and other threats due to agricultural and urban expansion, deforestation, desertification, pollution, and the introduction of exotic species, also called biopollution.

WILDLIFE CONSERVATION

Wildlife conservation describes various practices to regulate certain species to guarantee their abilities to reproduce and remain plentiful. Conservation

goals may be based on ideals of wildlife's intrinsic value, wildlife's utility in providing goods and services, or some combination thereof. Wildlife has featured prominently in worldviews, or life-ways, since time immemorial. The majority of the world's religions—including major faiths such as Buddhism, Hinduism, Jainism, and Islam, as well as thousands of small-scale, so-called indigenous religions—support spiritual interrelationships among all living beings. An ethic of stewardship obligates many religious practitioners to care for other species, as conveyed through stories, customary laws, rituals, and religious figures. Examples include portrayals of Noah's Ark replete with breeding pairs of all of the world's animals (Book of Genesis, chapters 6–9; also featured in the Torah and the Koran); the Seventh Generation precept of the Haudenosaunee (Six Nations Iroquois Confederacy), which requires that chiefs consider the impacts their decisions will have on the seventh subsequent generation of living beings; the Tsembaga ritual of *kaiko*, described in Roy Rappaport's *Pigs for the Ancestors*, as a homeostatic process, regulating ecological relationships; and the Roman Catholic St. Francis of Assisi, patron saint of animals and environment.

Across the globe, wildlife products have been exchanged within and between communities as parts of tribute and bartering systems. Hunting reserves are a particular form of utilitarian wildlife conservation and date back millennia. As Mulder and Coppolillio describe, historical records indicate that Assyrians had set aside land for hunting reserves by 700 B.C.E. Reserves in India emerged by 500 B.C.E. to provide not only exclusive areas for royal hunts, but also to protect elephants, which served important roles in the war efforts of state expansion. Such reserves connote a utilitarian approach to managing wildlife to ensure the reproduction of certain species desired for elite use and to restrict nonelites' access to the flora and fauna in those reserves. Furthermore, the species within these "protected areas" became what today would be called natural resources—things to be managed and commodified.

The term *wildlife* was preceded by *game*, defined as "the object of the chase; the animal and animals hunted" (traced to 1400 C.E.) or a collective form defined as "wild animals or birds such as are pursued, caught or killed in the chase" (traced to the late 1200s), and later still "the flesh of such animals used for food" (traced to the mid-1800s) (OED). British colonial discourse in Asia and Africa favored *game* well into the 20th century. There were colonial game reserves by the end of the 19th century, game departments shortly thereafter, and game feasts. As Gibson describes with respect to Africa, meat from wild animals and ivory supported early European explorers and colonial troops, as well as comprised a significant portion of the household budget for colonial administrators and early settlers. More recently, fortress conservation (the locking up of land for the preservation of wildlife) and community-based conservation (attempts to implement utilitarian agendas that permit human habitation in and use of biodiverse regions) have been posed as solutions to the ecological "problem" of wildlife management. Particularly with regard to community-based conservation, wildlife conservation programs fit within the nebulous realm of sustainable development.

WILDLIFE CONSERVATION INSTITUTIONS

Conservation efforts began with the creation of protected areas. The formal gazetting of land—and thus wildlife—dates back to hunting reserves and royal forests. The dominance of Western conservation can be traced predominantly to British and U.S. models. The English enclosure movement beginning in the 16th century reached its peak in the 18th and 19th centuries through various Acts of Parliament. Communally-held and open lands were reconfigured by a system of private land management, and the landscape was literally divided by fences, hedges, and walls into units of production and residence, while separate areas existed for "nature." Such divisions of landscape overlapped with 19th-century U.S. westward expansion and its repercussions.

The establishment of Yellowstone National Park in 1872 inspired a sweeping fortress conservation movement based on the national park model to "protect" indigenous flora and fauna. As noted above, the late 19th and early 20th century efforts at wildlife preservation in British colonially-held territories were often oriented toward protecting certain game species for elite

hunters and from indigenous hunters, the latter of whom would be accused of poaching for pursuing those "protected" species.

Wildlife conservation *seems* to stem from the idea that people destroy nature (i.e., wildlife and their habitat in this case) because economic activity has appeared to be incompatible with conservation goals, and yet people are by default the stewards of wildlife and work to save "it" by making it economically productive. The challenge of defining and overseeing wildlife conservation has led to the creation of a variety of institutions, including game and then wildlife departments, government ministries or "parastatals" to manage parks and reserves, and a rapidly growing number of nongovernmental organizations (NGOs). Regulating wildlife internationally has proven particularly difficult.

The Convention for the Preservation of Animals, Birds, and Fish in Africa was the first international conservation treaty. Signed in London in 1900, it served as the foundation for wildlife policies in British colonial Africa, and it was subsequently adopted elsewhere in the world for large-scale conservation efforts. It provided for the gazetting of lands into parks and reserves, and allowed for spin-off legislation regarding trespassing, poaching of protected flora and fauna, and the manners in which natural resources could be exploited—and by whom. By the time the first World Parks Congress convened in 1962, the majority of 10,000 protected areas were in Africa and North America and totaled two million square kilometers of surface area. By the fifth World Parks Congress held in 2003, there were over 100,000 protected areas, totaling over 18 million square kilometers. Those areas include Biosphere Reserves, World Heritage Sites, and other sanctuaries.

The term *wildlife* is often deployed by international conservation organizations and legislation to draw attention to issues of biodiversity and the protection of endangered species. The aforementioned World Parks Congress is a regular gathering sponsored by IUCN, the World Conservation Union, formerly the International Union for the Conservation of Nature and Natural Resources, founded in 1948. IUCN works with representatives from 82 states, 111 government agencies, over 800 nongovernmental organizations, and approximately 10,000 scientists to oversee wildlife management and to "influence, encourage and assist societies throughout the world to conserve the integrity and diversity of nature and to ensure that any use of natural resources is equitable and ecologically sustainable."

Perhaps the best known international wildlife treaty, CITES, or the "Washington" Convention on International Trade in Endangered Species of Wild Fauna and Flora, endeavors to protect certain plants and animals by regulating and monitoring their international trade to prevent such trade from reaching unsustainable levels. The Convention entered into force in 1975, and there are now over 160 parties. The United Nations Environment Program (UNEP) administers CITES. Plants and animals are classified as endangered, vulnerable, or lower risk species, and monitored accordingly. The World Wildlife Fund lists its primary conservation goals as "saving endangered species, protecting endangered habitats and addressing global threats such as toxic pollution, over-fishing and climate change." WWF, in conjunction with IUCN, also enforces CITES through TRAFFIC, the world's largest trade monitoring network.

Another significant international treaty, the Convention on Biological Diversity (CBD) was signed by 150 government leaders at the 1992 Earth Summit. The Convention promotes sustainable development while protecting biodiversity, which it describes as "the fruit of billions of years of evolution, shaped by natural processes and, increasingly, by the influence of humans." The agreement serves as a basis through which to regulate all species, ecosystems, and genetic resources, and it also covers the field of biotechnology. CBD objectives are implemented within the signatory nations through a variety of mechanisms, such as required implementation of national parks and protected areas systems to protect wildlife; green taxes, tax deductions, and/or severe economic penalties for habitat destruction or wildlife poaching; and expansion of the nonprofit sector.

Some indigenous groups have founded explicitly environmental institutions. For example, the First Nations Environmental Network is a national organization of individuals, nonprofit agencies, green technologies and corporations, and Nations working together on environmental issues. FNEN

describes itself as "a circle of First Nations people committed to protecting, defending, and restoring the balance of all life by honoring traditional Indigenous values and the path of our ancestors."

The definition of *wildlife* as "native flora and fauna of a particular region" does not preclude the possibility that the animals that qualify as "wildlife" can also be domesticated for profit (e.g., ranching, sport-hunting) or pleasure (e.g., zoos, private, or pet ownership). Following this logic, if we understand the definition of *wild* as "living in a state of nature; not tame, not domesticated," then the phrase *wild animals* can have an entirely different meaning from *wildlife*, even though members of the same species may comprise each category.

For example, ostriches—which live in an east African national park and have not been intentionally tamed by humans—may be considered as both wild animals and wildlife; but, given the above definitions, ostriches raised and tamed by humans, such as those at various ostrich farms throughout east Africa, qualify as wildlife but not as wild animals. The phrase *wildlife domestication* may sound like an oxymoron, but it is both an implicit characteristic and process of "wildlife conservation" and its conceptual sibling "wildlife resource management"—both of which bear the hallmark of colonial and neo-colonial interests.

SEE ALSO: Animals; Conservation; Convention on International Trade in Species of Wild Fauna and Flora (CITES); Environmental Organizations; Movements, Environmental; National Parks; Poaching; Safaris; Wild versus Tame.

BIBLIOGRAPHY. Raymond Bonner, *At the Hand of Man: Peril and Hope for Africa's Wildlife* (Knopf, 1993); Dan Brockington, *Fortress Conservation: The Preservation of the Mkomazi Game Reserve, Tanzania* (Indiana University Press, 2002); Jennifer Coffman, "The Invention of Wildlife: Managing a Natural Resource in Colonial and Post-Colonial Kenya," in Clark Gibson, ed., *Politicians and Poachers: The Political Economy of Wildlife Policy in Africa* (Cambridge University Press, 1999); John MacKenzie, *The Empire of Nature: Hunting, Conservation, and British Imperialism* (Manchester University Press, 1988); Monique Borgerhoff Mulder and Peter Coppolillo, *Conservation: Linking Ecology, Economics, and Culture* (Princeton University Press, 2005); Roderick Neumann, *Imposing Wilderness: Struggles over Livelihood and Nature Preservation in Africa* (University of California Press, 1988); James O'Connor, *Natural Causes: Essays in Ecological Marxism* (Guilford Press, 1998); David Western and Michael Wright, eds., *Natural Connections: Perspectives in Community-based Conservation* (Island Press, 1994).

Jennifer E. Coffman
James Madison University

Wild versus Tame

THE WORDS *wild* and *tame* go back to ancient Germanic roots, and perhaps earlier still, if they are—respectively—cognate with Latin *ferus* "wild" and *domare* "dominate," as suggested by the *Oxford English Dictionary*. They always had the meanings they have now, and they also were always opposed. The first English reference to *tame*, an Anglo-Saxon gloss of 888 C.E., explicitly opposes them. They are defined in relation to each other. A wolf is wilder than a bad or willful dog, but the latter is wilder than a thoroughly subjugated one; the cur is tame relative to the wolf, wild relative to the good pet. Jasper National Park is wilder than Yosemite, and Yosemite is wilder than Times Square. Naturally occurring species of roses are wilder than hybrid single roses, but the latter, even when they are modern hybrids, seem wilder to gardeners than the huge multi-petaled florists' roses. Formerly cultivated land reverts slowly and gradually to the wild. Tame animals can go wild or feral.

Wild has always had its present double or extended meaning: Natural as opposed to human-managed, and uncontrolled or hyper-reactive as opposed to tranquil and calm. A wild person can be violently emotional or somehow remote mentally from ordinary people. Latin *ferus* has similar extended meanings. *Tame* means controlled by humans; its secondary meaning of dull and ordinary is not attested before 1600. *Wilderness* is a derivative of *wild*, with attributive suffixes. Other languages have equivalent, but not always exactly equivalent, words. Chinese *ye* implies not only "wild" and "wil-

derness," but also "abandoned land." Romance languages usually use words derived not from *ferus*, but from Latin *sylvaticus*, "of the forest," such as: *sauvage* (French), and *selvatico* (Italian). These usually have a negative, even violent connotation, as in the English derivative savage (from the French). Yucatec Maya parallels Latin: *k'aaxil* "of the forest" and *baalche'* "things of the trees" are the nearest equivalent to "wild things." However, in Maya the connotation is good: The Maya love the forest and have strong positive associations with its inhabitants. Tame in Yucatec Maya is *alakbil*, "raised by humans," a close parallel with English.

In short, wild and tame are concepts that are broadly held—every culture feels the need to contrast the home-reared with the natural and uncontrolled, but culture and tradition powerfully influence their connotations. People understand them differently at different times and places. *Wild* and *wilderness* had broadly negative connotations through much of history. Conversely, wildness can be so valued that it is imitated. English landscape architects of the 17th–19th centuries laid out artificial wildernesses, and saw nothing oxymoronic about this. Today, restoration ecologists recreate the wild or the wilderness. In most countries, opinions range from strongly pro-wild (as in the John Muir tradition of conservation), to strongly antiwild. This leads to political debates that often become impassioned. The United States, home of the ideas of conservation, national parks, and national wilderness areas, is also home to a powerful pro-development ethic that defines progress as increasingly radical transformation of natural resources into commodities. Holders of these views come into conflict. The concept of tame inspires less emotion, but it too has positive and negative connotations.

From the ancient Chinese (such as Chuang Tzu and Han Shan) and the Desert Fathers of early Christianity to modern conservationists like John Muir and Edward Abbey, many people have found wild areas to be desirable, or even necessary, for personal renewal and contemplation, and have lived in the wild when possible. Conversely, others have found fulfillment only in destroying the wild to produce a tame world of houses, lawns, and factories. Many appreciate both types of landscapes, and are fulfilled only when they can move from one to the other with some ease. This attitude is now often identified with young educated urbanites, but is by no means confined to them, in the Western world or elsewhere. The Maya, and other Native Americans of Mexico, regard the balance of town and wild as essential to the world—a religious or cosmological necessity. Medieval India had almost the same view, widely expressed in Sanskrit and Tamil poetry. The ancient Celtic peoples idealized the wild more than perhaps any other culture on earth, but they usually (though not always) preferred to live in villages and castles.

It is often said today that there is no real wild or wilderness, because all parts of the planet are affected by humanity. However, this claim ignores the relative nature of the world. The usage of *wild* to mean totally unaffected by human action has never been standard. *Wilderness* has recently been widely used to mean areas thus unaffected, but this is a rather specialized and recent usage. The French philosopher Bruno Latour has argued that we should speak of natures rather than nature in recognition of these considerations, and the same might be argued for wild had not wilds long been used to mean wild places. Tame contrasts in scientific usage with domesticated. Domesticated, formally, refers to organisms that have been significantly changed by human breeding, such that they are genetically different from any population not managed and bred by humans. Wild animals can be tamed, but are not automatically domesticated. The elephants used so widely in ancient and modern times in Africa and Asia, for war and draught, have never been truly domesticated. Many cultivated tree crops are barely, if at all, domesticated, in spite of long histories of orchard use. Usually these are minor, or new crops such as macadamia nuts, but even such ancient crops as commercial olive varieties may have been propagated from naturally occurring trees rather than deliberately bred.

Wild and domestic forms of a given crop routinely interbreed. This introduces valuable new genes, especially for disease and pest resistance, to the domestic stock. For millennia, people have known this, and deliberately let their crops breed with wild relatives. The search for valuable genes in wild populations of wheat, barley, potatoes, and other major crops is a major research industry. Since pests and diseases quickly home in on particular domesticated varieties, plant breeders must constantly find new resistance genes in wild populations, and also in local landraces developed by small traditional communities in isolated areas. Loss of wild and landrace strains would be devastating to world food security. This is a most immediate and urgent practical reason to preserve wild and non-modernized landscapes. Moreover, domesticated forms can escape from tameness and go wild. If they are animals, they are said to be feral; if plants, volunteer. So an organism can be domesticated without being tame. Such organisms, having been bred to flourish among humans, can become pests. The common dandelion owes much of its success as a weed to a long history of being bred as a garden crop.

SEE ALSO: Domestication; Domination of Nature; Landrace; Wilderness; Wilderness Act (U.S. 1964); Wilderness Society; Wildlife.

BIBLIOGRAPHY. E.N. Anderson, *The Political Ecology of a Yucatec Maya Community* (University of Arizona Press, 2005); A.C. Graham, *Chuang Tzu: The Inner Chapters* (George Allen and Unwin, 1981); B. Latour, *Politics of Nature: How to Bring the Sciences into Democracy*, trans. by Catherine Porter (Harvard University Press, 2004); Han Shan, *The Collected Songs of Cold Mountain*, trans. by Red Pine, also known as Bill Porter (Copper Canyon Press, 2000); R. Torrance, ed., *Encompassing Nature* (Counterpoint, 1998).

Eugene Anderson
University of California, Riverside

Wind Energy

THE KINETIC ENERGY in wind can be converted into mechanical or electrical power. For centuries, windmills have converted wind energy into mechanical power to grind wheat into flour. Now, improved turbine technology allows wind generators to affordably produce electricity. Electricity is produced when an electrical conductor moves perpendicularly to a magnetic field. Generators have a conductor (or rotor) that spins inside a magnetic field (or stator). The challenge is to find abundant, low-cost energy to spin the rotor.

Wind generates less than 1 percent of the world's electricity, but is its fastest-growing energy source.

Hydroelectricity is a renewable energy source that uses the potential energy of falling water to turn the rotor. Wind is also a renewable energy source because its supply is not depleted after being used. Generators also use nuclear or fossil fuels to convert water into steam pressure, which turns the rotor. Fossil fuels like coal are essentially nonrenewable because they form over millions of years. Although wind is renewable, it is not always reliable. Wind velocity may fluctuate or fall below the minimum annual average velocity of 13 miles per hour needed to generate electricity.

Wind is created when solar radiation interacts differentially with clouds, vegetation, and water bodies to unevenly heat the earth's surface. Resulting atmospheric pressure differences force air from high to low pressure areas. Wind velocity, density, and temperature determine wind quality. Winds aloft are better than surface winds. Hills, valleys, and other geomorphic features can locally increase or decrease the relative velocity.

Wind energy potential depends on the area swept by the wind, its density, and velocity. Blade length and rotor design define the area component for a turbine. Denser air will also have an impact by generating more momentum. Velocity is the primary

factor in wind power generation because changes in velocity have a cubed effect on power.

PRACTICAL ASPECTS OF WIND POWER

Wind generates less than 1 percent of the world's electricity, but is the world's fastest-growing energy source, with an installed turbine capacity of 58,982 megawatts. Environmental and human factors, however, affect the global and regional patterns of where wind is used to generate electricity. The production of wind energy is still centered in Western Europe but is increasing in developing countries with rapidly-growing economies, such as India.

In the United States, wind energy development is dependent on adequate wind resources, the availability of transmission lines, and federal and state policies and incentives. The potential amount of electricity available from wind is measured by the installed capacity of the turbines. For example, North Dakota is ranked highest in the United States for wind potential, but lacks the transmission lines to move electricity on a large scale to population centers.

Wind energy advocates tout its reduction of greenhouse gas emissions and the income it generates as wind farm operators pay property taxes and rent for the use of local land. Opponents argue that large wind farms are a visual blight and that rotating blades can be noisy and endanger bats and migratory birds. These concerns are like those associated with other large industrial developments.

All energy sources, including wind, affect the environment when they are harnessed for human uses. Consequently, in the current energy policy debate there are no easy answers about which energy source is the best. For the time being, wind energy comprises only a small portion of the American energy portfolio. Its expanded use depends upon continuing technological improvements and public acceptance of its net benefits.

SEE ALSO: Fossil Fuels; Hydropower; Renewable Energy; Solar Energy.

BIBLIOGRAPHY. American Wind Energy Association, *Wind Energy Projects throughout the United States of America*, www.awea.org (cited April 2006); L. Daniels, S. Johnson, and W. Slaymaker, *Harvest the Wind*, www.iira.org (cited April 2006); U.S. Department of Energy, *Wind and Hydropower Technologies Program*, www1 .eere.energy.gov (cited April 2006); World Wind Energy Association, *Worldwide Wind Energy Boom in 2005*, www.wwindea.org (cited April 2006).

Roger Brown and Christopher D. Merrett
Western Illinois University

Wine

WINE IS AN alcoholic beverage made of grape juice through fermentation. A few dozen grape types produced by the vine plant *Vitis vinifera* are of particular interest to wine experts. Popular varieties include cabernet sauvignon, chardonnay, merlot, pinot noir, and zinfandel. Where these grapes grow determines much of a wine's character. Differences in weather and lighting conditions, soil, and temperature are distinguishable in the taste of wine so that the same grapes result in different wines in different regions. Details of the fermentation process and aging further profile the end product—be it red, white, rosé (blush), sparkling, sweet, or dry.

The history of wine is a history of global trade by powerful economic actors. The story begins in Mesopotamia and Caucasia roughly 8,000 years ago. The know-how traveled to Egypt in 3,000 years and reached Greece 1,000 years later. Greek colonizers and merchants introduced wine to the Mediterranean sphere: To present-day Italy, France, Spain, and northern Africa. Imperial Romans then domesticated wine further north, in today's Britain, Germany, and northern France. When the Romans left these areas by the 5th century, the seeds of contemporary vineyards had been planted. These were typically located along rivers, the most important channels of transportation in this era. By the Middle Ages the Christian Church had emerged as the most powerful producer and trader of wine, with monasteries as the most important centers of innovation. Wine dominated the European beverage market until the 18th century, when imports from the colonies (chocolate, coffee, and tea) gained popularity, distilling and preservation techniques of

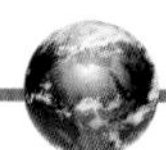

other alcoholic beverages developed, and water in European cities became safer to drink. Competition encouraged innovation, creating the foundations of modern wine. Wine clearly is a culturally specific (Western) product with origins in particular climatic, historical-cultural, and economic conditions.

Over the past century the wine business has industrialized in form and become global in scale. Most of the production still comes from Europe, although the New World has gained power in the market. The leading wine producing countries are France, Italy, Spain, the United States, and Argentina. In each country wines have intensely local and regional roots. The production comes from strictly defined regions that only use certain methods and varieties of grapes. Famous wine regions include Bordeaux in France, La Rioja in Spain, and Napa Valley in California.

The legal definition of these regions, their maximum annual crops, and the alcohol contents of their products exemplify the detailed controls and regulations applied to wine by authorities and specific regulatory bodies. Producers, their organizations, and national governments use regulation to control the quality of wines, to protect their reputation, and to improve their sales. Local and national governments may tax the licensed producers, distributors, retailers, and consumers of wine. This revenue is typically used to cover social costs related to alcohol consumption. International tariffs may apply to the import and export of wine for reasons of market protection and revenue.

Some regulations connect to cultural, social, and moral values, which often dictate where, when, and by whom wine can be purchased and consumed. Many countries have set legal age limits to the consumption of wine and prohibit driving under its influence, but how young is too young, how much is too much, or how strictly the control is enforced varies from one society to another. Social tolerance for visible intoxication also varies significantly both at the level of societies and socioeconomic class, so that one can speak of "drinking cultures." Some religious communities consider the consumption of all alcoholic beverages unacceptable, whereas others shun intoxication but find deep symbolic or metaphorical meaning in wine consumed in sacred rituals. The importance of wine to human social life is reflected in artistic representation—in paintings, literature, and various forms of contemporary popular culture.

Wine thus connects to identity in multiple ways. A globally famous wine is a source of local, regional, or national pride and influences the identity and landscapes of the place associated with the wine. Wine tasters and other specialists have professional group identities supported by expert vocabularies and know-how. Wines profile their consumers: What wine is served, in what situation, to whom, and how may function as a powerful indicator of socioeconomic status, cultural knowledge or background, and lifestyle.

Over the course of tough competition small family-owned wineries have merged into multinational production companies with complex ownership structures. The distance between the producer and the consumer, and between the botanical and the chemical, has grown: Wine is still an agricultural product, but it is increasingly produced under strict technological control and consumed in urban, post-industrial environments far from the original production region. Because of their reputation, consumer identities, and available profit, there is demand for wines to routinely travel over long distances. This movement of trendy consumables may be criticized from a perspective of environmental sustainability. On the other hand, environmentally-aware wine drinkers have grown increasingly interested in the sustainability of production. Interest in local produce and urbanites' recreational needs have reconnected wine consumers and producers through sustainable, moderately scaled forms of wine tourism.

The delicate relationship between wine production and the natural environment shapes the future of the business. Unexpected side effects of global exchange have been well-traveling plant diseases and pests, which have seriously damaged wine production in Europe (especially in the 1870s), the United States (contemporary California), and New Zealand. In some areas, methods of biological or chemical pest control have created further problems by damaging soils and water supplies or by altering the balance of species. At worst, these problems may threaten entire regions and their wine-dependent economies.

SEE ALSO: Cacao; Coffee; Drugs; Green Consumerism; Religion; Soils.

BIBLIOGRAPHY. H. J. de Blij, "Viticulture and Viniculture in the Southeastern United States," *Southeastern Geographer* (v.27, 1987); J. P. Dickenson and J. Salt, "In Vino Veritas: An Introduction to the Geography of Wine," *Progress in Human Geography* (v.6, 1982); Daniel W. Gade, "Tradition, Territory, and Terroir in French Viniculture: Cassis, France, and Appellation Contrôlée," *Annals of the Association of American Geographers* (v.94, 2004); Hugh Johnson, *World Atlas of Wine,* 4th ed. (Chancellor Press, 2002); Tim Unwin, *Wine and the Vine* (Routledge, 1991).

PAULIINA RAENTO
UNIVERSITY OF HELSINKI, FINLAND

Winters Doctrine

WINTERS DOCTRINE IS a legal principle announced by the U.S. Supreme Court in the case of *Winters v. United States* 207 U.S. 564 (1908). The doctrine created a different legal rule for the distribution of water in the arid western states. The rule is that water rights have been reserved to Indian reservations and public lands in order to make sure that there will be enough water to achieve their assigned purposes. The Winters Doctrine contradicts the law of water rights in the western United States that is applied to privately held lands. The rule generally followed in the west is the "state-based appropriative rights" rule. It is the principle that the rights to water derive from the first time that they were put to beneficial use. The federal rule, however, begins its allocation of water based on the date on which the lands in question were reserved for a dedicated purpose. The federal rule means that a rancher may have been using water for years that was not called upon by an Indian reservation or a federal land management project. The rancher or a farm may therefore have been applying the water to beneficial use long before federal claimants began to seek to use the limited water supplies.

The Winters doctrine developed from a case brought by the U.S. government against Henry Winters, John W. Acker, Chris Cruse, Agnes Downs, and others. Winters and his associates were constructing and maintaining dams on the Milk River in the State of Montana. Their upstream activities impeded and reduced the riparian flow downstream to the Fort Belknap Indian Reservation. Winters and his associates lost the case in the federal district court. They were ordered to not interfere in any manner whatsoever with the flow of 5,000 inches of the water of the Milk River. The Circuit Court of Appeals (74 C. C. A. 666, 143 Fed 740) affirmed the decision. In response, Winters, et al., appealed to the U.S. Supreme Court.

The opinion delivered by the Supreme Court noted that the Fort Belknap Indian Reservation had been established on May 1, 1888, when property of the U.S. government was set aside for the Gros Ventre and Assiniboing tribes as an abiding dwelling place. The Court accepted the fact that Winters, et al., had a right to the water under Montana law; it ruled that the federal government had a prior and a different claim that superseded the otherwise lawful state claim. The federal claim to the water was established when Congress created federal lands for special purposes. It did so with the reservation that water is always reserved for the use of the federal facility despite claims of others at a later time.

The Winters decision has meant that all public lands of the federal government are first in line for water uses. This includes national parks, wildlife refuges, national forests, military bases, wilderness areas, or other public purpose areas. Later court cases expanded the rule to include lands set aside by treaty, executive orders, or by federal statutes.

SEE ALSO: Native Americans; Riparian Rights; United States, Mountain West; Water Law.

BIBLIOGRAPHY. Reid Peyton Chambers, *Implementing Winters Doctrine Indian Reserved Water Rights: Producing Indian Water and Economic Development without Injuring Non-Indian Water Users?* (University of Colorado School of Law, 1991); Michael C. Nelson and Bradley L. Booke, *The Winters Doctrine: Seventy Years of Application of Reserved Water Rights to Indian Reservations* (Office of Arid Land Studies, 1977); Jeffrey M. John Shurts, *Indian Reserved Water Rights: The Winters*

Doctrine in its Social and Legal Context, 1880s–1930s (University of Oklahoma Press, 2000).

ANDREW J. WASKEY
DALTON STATE COLLEGE

Wise Use Movement

THE WISE USE movement is a conglomeration of grassroots activists and organizations presenting an alternative philosophy regarding resource extraction and access on U.S. public lands. The movement started in the late 1980s by a handful of influentials such as Ron Arnold, Chuck "Rent-a-Riot" Cushman, and Allan Gottlieb. Arnold has been Executive Vice President of the Center for the Defense of Free Enterprise since 1984, and was honored as the "Father of the Wise Use movement." His blatant and harsh criticism of environmentalists has made him a veritable spokesperson for the movement.

The authors (Maughan & Nilson) identified the seven predominant strategies of the movement:

(1) bills itself as the "true" environmental movement; (2) tries to marginalize environmental groups by highlighting the views and actions of the radical fringe of environmentalism, and in other ways promote the perception that environmentalists are atypical of the public; (3) downplays threats to the environment; (4) tries to form coalitions with interests who perceive they have been harmed or are threatened with harm from environmental policies; (5) forms coalitions with groups that share part of the Old West ideology; (6) stresses the economic costs of environmental policy; and (7) creates the perception that the real goal of environmentalists is attainment of authoritarian power.

Overall, the Wise Use movement is regarded by many in and outside of the movement as "anti-environmental." A quick look at the Wise Use Agenda reveals a philosophy that is in many regards contrary to the policies advocated by environmentalists, conservation organizations, and U.S. federal agencies.

Since there are numerous Wise Use organizations that focus on a myriad of specific issues, the goals stated below are not supported by every Wise Use group. However, at a Multiple Use Strategy Conference in Reno, Nevada in August 1988 sponsored by Ron Arnold's Center for the Defense of Free Enterprise, Wise Use organizations collaborated on their mutual concerns about resource management. As a result of the conference, Alan Gottlieb compiled *The Wise Use Agenda*, a book detailing the goals of the movement. Note that many of the policies presented were subsequently adopted by the George W. Bush administration. The first 10 goals are stated below:

1. Initiation of a Wise Use public education project by the U.S. Forest Service explaining the wise commodity use of the national forests and all federal lands (to reduce the federal deficit).
2. Immediate wise development of the petroleum resources of the Arctic National Wildlife Refuge in Alaska.
3. Advocate the passage of an Inholders Protection Act, giving broader property rights to inholders (persons who own land within the borders or tangent to federal or state lands).
4. Passage of the Global Warming Protection Act that works to remove all decaying matter from national forests to be replaced by young stands of carbon-dioxide absorbing trees.
5. Designate 3 million acres in the Tongass National Forest in Alaska for timber harvest.
6. Open all public lands (including wilderness areas and national parks) to mining and energy production.
7. Assert states' sovereign rights in matters pertaining to water distribution and regulation.
8. Commemorate the one hundredth anniversary of the founding of the Forest Service by calling attention to the commodity use of forests and the homestead settlement of these areas.
9. Increase harvesting of trees in national forests to promote "rural, timber-dependent community stability" through the Rural Community Stability Act. These sales will be exempt from administrative appeal.
10. Create a national timber harvesting system that allows for greater harvesting of timber on public lands.

Many of the individual groups within the Wise Use movement were at one time funded in part by the oil, off-road (recreation), timber, mining, and ranching industries, as well as anti-environmental

politicians. In 2001 the Sierra Club reported that Boise Cascade Company (timber company), DuPont (chemical manufacturers), and Chevron (gas and oil company), at one time funded Wise Use movement conferences. However, many of the corporate interests quickly soured on the wise-use groups' overheated rhetoric (and sometimes aggressive tactics) and pulled their funding.

The Blue Ribbon Coalition, the Mountain States Legal Foundation, and the Center for the Defense of Free Enterprise are among the more prominent Wise Use groups in the United States

The Mountain States Legal Foundation was founded in 1987 by Clark Collins to advocate for motorized access for off-road vehicles on U.S. public lands. Of note, the Coalition joined in a lawsuit against the U.S. Forest Service over the 2001 Roadless Area Conservation Rule. The rule was developed following years of scientific evidence, hundreds of public meetings across the country and 1.6 million public comments. Since 2006 the Forest Service has received more than four million comments on the rule, 95 percent in favor. The rule received the largest turnout of public comments in U.S. Forest Service history.

According to their website, the Center for the Defense of Free Enterprise (CDFE) is highly interested in environmental issues and "was founded by a group of distinguished businessmen, educators, legislators and students who were deeply concerned about the rollback of 200 years of individual rights and the multitude of restrictions being imposed on America's free enterprise system by big government."

SEE ALSO: Bush, George W. Administration; Industry; Timber Industry.

BIBLIOGRAPHY. Ron Arnold, www.cdfe.org (cited April 2006); The Blue Ribbon Coalition, www.sharetrails.org (cited April 2006); Alan Gottlieb, *The Wise Use Agenda: The Citizen's Guide to Environmental Resource Issues* (Free Enterprise Press, 1989); Jennifer Hattam, "Wise-Use movement, R.I.P.?: Friends in the White House, But Few Foot Soldiers," *Sierra* Magazine (May/June 2001); David Helvarg, "Wise Use in the White House," *Sierra* Magazine (September/October 2004); The Heritage Forest Campaign, "Roadless Area Protection Act," www.ourforests.org (cited April 2006); Scott Silver, www.wildwilderness.org (cited April 2006);

Andrew J. Schneller
Independent Scholar

Wittfogel, Karl A. (1896–1988)

KARL A. WITTFOGEL, a German historian and prominent sinologist, is most noted for his theory of the hydraulic civilization. Applied by Wittfogel primarily to ancient Egypt, Mesopotamia, China, the Indus Valley, and regions in pre-Columbian Latin America, he contended that the strong central political control was necessary to control the source and disposition of water. The degree of centralization within these civilizations was extreme to the point of being despotic. All of the civilizations noted in Wittfogel's studies existed in arid regions, where vast irrigation systems supported extensive agricultural operations, with the exception of China. It is on this point that his theory has been sternly criticized. The prominent China scholar, Joseph Needham, argued that early Chinese governments, although exercising central control, were not despotic. Needham, along with other scholars, also correctly pointed out that the most productive agricultural regions in China are not arid, sources of water are widespread, and water control measures are locally administered.

Wittfogel moved to the United States and became a naturalized citizen in 1939, after enduring two years in a Nazi concentration camp for his vocal attacks on fascism in Germany. He served on the faculty at Columbia University before joining the Far Eastern and Russian Institute at the University of Washington in 1947. It was here that Wittfogel completed his most important book, *Oriental Despotism: A Comparative Study of Total Power* (1957), which laid out his schema for the origins of bureaucratic totalitarianism based on the control of a society's water supply.

Wittfogel drew heavily on the works of Karl Marx, Friedrich Engels, and Max Weber in developing his ideas about early non-European societies and their governmental structure. Weber, in particular, was most influential and is credited with introducing Wittfogel to the unique hydraulic-bureaucratic societal structures in South Asia and East Asia.

In his book, *Rivers of Empire: Water, Aridity, and Growth of the American West,* environmentalist Donald Forster invokes Wittfogel's thesis and applies it to the American southwest, a region of aridity and closely managed water sullies. Forster strenuously argues that a true hydraulic society emerged in the southwest as population increased and demands on a limited water supply rapidly increased. In his scenario, a mega-bureaucracy emerged, which included large land-holding agriculturalist and governmental officials who concentrated water rights in the hands of a relatively small number of influential individuals. Forster considers the National Reclamation Act (1902) and the Bureau of Reclamation to be the primary instruments in the creation of the southwest's hydraulic society.

SEE ALSO: Marx, Karl; Socialism; Water; Water Demand.

BIBLIOGRAPHY. Karl A. Wittfogel, *Oriental Despotism: A Comparative Study of Total Power* (Yale University Press, 1957); Orlan Lee, *Bureaucratic Despotism and Reactionary Revolution: The Wittfogel Theory and the Chinese Revolution* (Chinese Materials Center, 1982); G. L. Ulman, *Society and History: Essays in Honor of Karl August Wittfogel* (Mouton, 1978); Donald Worster, *Rivers of Empire: Water, Aridity, and Growth of the American West* (Pantheon Books, 1985).

GERALD R. PITZL, PH.D.
NEW MEXICO PUBLIC EDUCATION DEPARTMENT

Wolves

WOLVES (*CANIS LUPUS*) are mammals of the order Carnivora and belong to the same family (Canidae) of carnivores as coyotes, doges, foxes and jackals. They are digitigrades like other Canidae family members. Wolves are social animals that live in packs with a dominance hierarchy. The members of the pack include wolf pups, several nonbreeding adults, the dominant male (alpha) and his mate (fae). They mate in January and usually have five or six pups about six weeks later. The pups are fed by the pack until they become young adults.

Wolves are keystone predators. What they leave behind feeds other animals such as scavengers. They also keep the populations of ungulates such as bison, caribou, Dall sheep, elk, moose, mountain goats, and musk-oxen at healthy levels. However, many ranchers and sports hunters view the wolf as a menace that should be exterminated.

The gray wolf (timber wolf) is found across the Northern Hemispheres in North America, Europe, and Asia. There were five subspecies, but several have become extinct because of habitat destruction and hunting encouraged by fear of wolves. The Mexican gray wolf (*Canis lupus baileyi)* is a subspecies hunted to near extinction in the last 50 years. The South American maned wolf *(Chrysocyon brachyurus)* resembles a dog with reddish fur.

In the American West, wolves were hunted to extinction partly because of traditional fears of wolves.

Called a *lobo* in Spanish and Portuguese, it is not actually a wolf. It lives in Paraguay, Southern Brazil, and Bolivia east of the Andes.

The Arctic wolf (*Canis lupus arctos*) is a subspecies of the gray wolf. Also called the white wolf or the polar wolf, they range across the Canadian Arctic and Greenland. They are different from the tundra wolf (*Canis lupus albus*) that ranges across the tundra of the northern European and Asian Arctic.

The largest subspecies of the gray wolf is the Russian wolf (*Canis lupis communis*). Its range is north-central Russia where it is hunted legally. In Europe wolves have been nearly driven to extinction. Small numbers exist in the mountains of Italy, Spain, Portugal, Norway, Sweden, and Finland. Larger numbers survive in the Carpathian Mountains.

Africa's only species of wolf is the Ethiopian wolf (*Canis simenis)*. Only a few hundred individuals survive in the alpine ecosystem of the Ethiopian highlands. The Arabian wolf (*Canis lupus Arabia*) is a subspecies of the gray wolf. Numbers have increased in the United Arab Emirates since hunting was banned. The dire wolf was common in North America during the Pleistocene era but is now extinct. It disappeared with other mega-fauna at the end of the last ice age. Specimens have been found in tar pits and fossil beds. The North American red wolf (*Canis rufus*) may have been a descendant of the dire wolf. It was hunted nearly to extinction and was declared biologically extinct in the wild in 1980, but it has since been reintroduced to the wild in the southern Appalachian Mountains.

SEE ALSO: Conservation; Habitat Protection; Native Species; Nature, Social Construction of; Predator/Prey Relations; Public Land Management; Yellowstone National Park.

BIBLIOGRAPHY. John Elder, *The Return of the Wolf: Reflections on the Future of Wolves in the Northeast* (Middlebury College Press, 2000); Thomas McNamee, *The Return of the Wolf to Yellowstone* (Owl Books, 1998); Chris Whit, *Wolves: Life in the Pack* (Main Street Press, 2004); Daniel Wood, *Wolves* (Whitecap Books, 2005).

Andrew J. Waskey
Dalton State College

WEDO: Women's Environment and Development Organization

THE WOMEN'S ENVIRONMENT and Development Organization (WEDO) was founded in 1990 by the late Bella Abzug (former Democratic U.S. congresswoman; 1920–98), and Mim Kelber (1922–2004), a well-known women's rights and peace activist. As noted in their mission statement, the international organization "advocates for women's equality in global policy. It seeks to empower women as decision makers to achieve economic, social and gender justice, a healthy, peaceful planet and human rights for all."

WEDO was inspired by an earlier movement, women in development (WID), which began in the 1970s and was a shift in international aid development policies to emphasize the importance of women's roles in economic development. The WID approach was later institutionalized by aid organizations such as the U.S. Agency for International Development (USAID), and served as impetus for the First United Nations (UN) Conference on Women and Development in Mexico City in 1975. WID was an important foundation for the feminist movement of its time and it helped make women's voices heard in international development projects and policies. It also helped set in motion the concept of ecofeminism, a belief that the connection of women with nature is so strong that it "called upon women to lead an ecological revolution to save the planet."

As women's voices from Northern and Southern countries became heard regarding topics of environment and development (including not just those of well-educated and upper- or middle-class backgrounds—but also women from farming, fishing, and indigenous communities), world leaders and nongovernmental organizations began preparations for the 1992 UN Conference on Environment and Development (UNCED) held at Rio de Janiero, also known as the Earth Summit. Abzug and Kelber, influenced by the infamous Bruntland Report, were concerned that even after two decades of WID policies and discourse, few women occupied positions associated with global policy formulation. As a strategy to attract attention from the global

community and those creating the UNCED agenda, Azbug and Kelber created WEDO to formally organize a conference called the World Congress for a Healthy Planet in Miami in 1991. The 1,500 women there agreed to demand press leaders at UNCED for an equal say in governmental policies created by local and global governments or institutions. They believed that "male-led technologies, wars, and industries are killing people and the planet." Their vision was that more women must be involved in designing policies that linked environment with issues of poverty and social justice and that this would be an important advance toward bringing the "earth's political, economic, social and spiritual systems into healthy balance."

The Healthy Planet conference, as intended, was instrumental in leveraging women's influence on UNCED and ultimately increased women's visibility in decision making regarding policy issues that link environment with issues of poverty, social justice, free trade and international debt. The mechanism that was used to gain influence in UNCED and produced by WEDO, as a result of the Healthy Planet conference, was called "Women's Action Agenda 21," referencing Agenda 21 because this was the title of the document that governments from participating countries at UNCED planned to produce at Rio de Janeiro.

Action Agenda 21 contained a list of specific demands regarding topics of: Democratic rights, diversity, environmental ethics, women and militarism, foreign debt, trade, poverty, land rights, food security, population and women's rights, biotechnology and biodiversity, alternative energy and nuclear power, technology and science, consumer power of women, education, and information. The document also challenged the UN for its lack of gender balance in the organization itself. The WEDO lobbying efforts using Action Agenda 21 were successful in positioning women's issues in Agenda 21 in a majority of the chapters in the document, as well as one chapter specifically dedicated to women's roles in globally sustainable and equitable development.

WEDO continues to stay active in pressing the UN for a "wider gender lens" in all of its activities and policies, particularly the UN Millennium Declaration, a 2000 Declaration agreed upon by 191 governments at the largest gathering ever of world leaders. The organization campaigns for women's rights throughout the world. Its recent activities include: Beijing +10, a 2005 campaign to reaffirm the Original Action Plan of the 1995 Fourth Conference on Women held in Beijing, China; and a global campaign: "Plant a Tree for Peace," commemorating Nobel Laureate Wangari Maathai and connecting environment, human rights, peace and gender equity.

SEE ALSO: Agenda 21; Brundtland Report; Ecofeminism; Gender; Justice; United Nations Conference on Environment and Development (Earth Summit 1992).

BIBLIOGRAPHY. B. Abzug, "The Century of the Woman," *Social Policy* (v.28, 1998); E. Boserup, *Woman's Role in Economic Development* (St. Martin's Press, 1970); R. Langara, "How Beijing Has Been Betrayed: WEDO's New Book Carries Women's Reports on How Governments Have Failed to Act on the BPFA," *Women in Action* (v.1, 2005); C. Merchant, *Radical Ecology: The Search for a Livable World* (Rosenberg, 1992); Women's Environment and Development Organization, www.wedo.org (cited April 2006).

Rebecca Austin
Florida Gulf Coast University

Wood (as energy source)

BIOMASS RECEIVES AND stores energy from the sun. When burned, this energy is released as heat. Wood fuel (commonly referred to as fuelwood) serves a variety of heating purposes although the most common fuelwood-based practices around the world cooking and heating—especially in developing nations.

Calculating how much wood is harvested and burned each year is difficult to determine because fuelwood collection and use occurs predominantly through informal practices. As a result there is a dearth of precise data on wood energy use. This stands in contrast to non-biomass energy sources such as oil and natural gas, which have been subject to more in-depth analysis.

According to the United Nations Food and Agricultural Organization (FAO), biomass accounts for roughly 30 percent of the total energy consumed in developing nations- with fuelwood accounting for approximately half of this amount or 15 percent of the total energy. Other common types of biomass include agricultural matter and animal dung. In some countries, dependence on fuelwood is much higher. For example in Nepal and countries in Sub-Saharan Africa, fuelwood accounts for roughly 80 percent of the total energy requirements.

Developed nations use fuelwood to a much lesser extent, although fuelwood contributes to between 12 to 18 percent of total energy needs in Scandinavian and Central and East European countries primarily because of heating practices during cold winters. Still, dependence on fuelwood is most common in developing nations with the FAO estimating in 1998 that 50 percent of the world's fuelwood was consumed in five countries: Brazil, China, India, Indonesia and Nigeria.

Wood as a source of energy varies in its heating potential. Two major factors determining heat potential are wood density and dryness. Density is determined by tree species type. The potential heat content per kilogram is roughly equal between all types of wood so it is the density of that wood which influences its heat producing capacity. Broadly speaking, wood either comes from softwood or hardwood varieties. Softwood tree species include many conifers while hardwood species are typically broadleaf trees. Softwood trees are typically less dense than slower growing hardwood tree species and are therefore less desirable for fuelwood. There are some hardwoods such as Aspen or Poplar with lower density wood and some softwood trees like Western Larch and Yew with higher density fuelwood. Dryness is another important factor influencing the heating potential of fuelwood. Efficient combustion is greatest in wood that is well dried.

There are many benefits and conveniences associated with fuelwood that make it an optimal fuel choice for rural communities worldwide. In many regions, the most obvious benefit is that wood is free and readily available for individuals to collect. A renewable energy source, fuelwood can be managed in ways that replenish tree stocks and maintain a consistent local supply.

This is a claim that other common domestic fuel types such as kerosene, Liquid Petroleum Gasoline (LPG) and coal cannot easily make. Wood fuel is also a desirable form of household energy because it produces smoke that can serve practical purposes. For example, many households use wood smoke to cure meats while other homes with thatched roofing and siding find smoke a useful mechanism for repelling pests and insects.

Despite the many benefits and conveniences of wood, indoor air pollution from cooking and heating remains a serious global health problem. According to reports by Practical Action, the indoor burning of solid fuels kills 1.6 million people each year. The affected population is comprised predominantly of women and children who usually partake in cooking and heating practices. In India alone, the World Health Organization (WHO) concluded in 2002 that exposure to indoor air pollution contributes to 500,000 deaths and 500 million cases of illness among women and children each year. According to the United Nations Development Program, this places India alongside China as one of the two countries in the world to experience the highest levels of indoor air pollution. Indoor air pollution from wood and other biomass fuels has been described as the "Silent Killer" by a number of international NGOs because for many years, little attention was given to the problem by governments around the world. It is argued by some that the main explanations for such low attention is the politically marginal position of the main victims-third world, rural, poor, women.

With a heavy reliance on fuelwood to satisfy household energy needs, many regions of the world have suffered from a perceived "fuelwood crisis." These concerns were legitimized largely by a series of influential reports during the 1970s and early 1980s by the World Bank, United Nations Development Program, and the FAO. These reports explained how fuelwood shortages were a result of heavy deforestation in developing nations. The fuelwood crisis was predicated on the fuelwood gap theory stating simply that tree removal was outpacing tree regeneration resulting in a fuelwood "deficit."

The basic premise of the theory stated that fuelwood collection was the principal driver of rapid

tree removal and the fuelwood deficit. A number of subsequent studies revealed the fuelwood crisis to be misdiagnosed and largely overstated. In terms of the so-called fuelwood gap, many of the regional studies failed to include trees outside of forests including trees in villages, along roads and in agricultural areas. These studies also overlooked other forms of biomass which constitute fuelwood such as farm-derived woody biomass. As far as blaming heavy deforestation on fuelwood collection, these studies failed to identify a significant underlying factor—large-scale forest clearing for agricultural purposes.

During this time period, community forests were established to ameliorate the fuelwood crisis. The historically inadequate policies of top-down, state-led forest management programs led to the formation of community forestry programs during the 1970s. "Forests for the People" programs were established to cater specifically to the needs of the energy consumers. These programs included the establishment of woodlots in tree deficient regions and cash crop capabilities for local residents. Ultimately many of these projects failed due to inappropriate, persistent top-down 'expert' management, and corruption at the micro politics scale- including community level exclusionary practices.

Despite the alarmist and sweeping inaccuracy of the fuelwood crisis, today there remains ample evidence of localized fuelwood supply irregularities. For example, fuelwood shortages are increasingly common around urban areas. Commonly referred to as treeless halos, fuelwood shortages occur as urban residents and their wood foraging activities are pushed to settlements along the margins of urban areas. In response to local fuelwood shortages, both supply-side and demand-side measures have been implemented worldwide to ensure sustainable fuelwood resources.

From a supply standpoint, participatory forest management, or joint forest management programs have been established in many countries to provide reliable fuelwood resources to local communities. The basic philosophy guiding these programs is to give joint control over forest stewardship and revenue generating potential to the government and communities. Demand side regulation occurs principally through improved cookstove programs. The goal of these programs is to increase the wood burning efficiency of cookstoves by replacing old, inefficient stoves with new models.

SEE ALSO: Energy; Forests; Underdeveloped ("Third") World.

BIBLIOGRAPHY. B. Agrawal, *Cold Hearths and Barren Slopes: The Woodfuel Crisis in the Third World* (Riverdale Company, 1986); G. Leach and R. Mearns, *Beyond the Woodfuel Crisis: People, Land and Trees in Africa* (Earthscan Publications, 1988); D. Pandey. *Fuelwood Studies in India: Myth and Reality* (Center for International Forestry Research, 2002); World Energy Council, *The Challenge of Rural Energy Poverty in Developing Nations* (1999).

Gregory Simon
University of Washington

Workplace Hazards

WORKPLACE HAZARDS ARE the possible causes of physical or mental problems for people working in any particular place. They range from exposure to hazardous chemicals, excessively loud noise, psychological stress, and workplace violence or discrimination. Workplace hazards cause large numbers of deaths and injuries every year. It is estimated that some two million people die from workplace-caused hazards annually, and another 160 million become subject to a disease stemming from their work. Since prevention of accidents and risk generally comes at the initial expense of lowered productivity, employers have a strong disincentive to provide safety systems. Consequently, workers with less power or representation are generally more at risk from workplace hazards. Women, migrants, and temporary workers are some of the groups most likely to be put at risk of workplace hazards.

The nature and severity of workplace hazards varies depending on the type of work, the regulations in force in the geographic area in which the work takes place, the state of new scientific understanding, and the industry concerned. New forms of occupation and working activity, based more on

sedentary work with computers, has brought about the risk of repetitive strain injuries, for example. Previously, hazards would be more likely to be working with dangerous machinery or with chemicals or other materials that lead to negative health outcomes. Some industrial activities, such as mining or deep-sea fishing, are inherently dangerous and even more so when proper regulations of activities are not enforced and policed.

In February 2005, 203 miners were killed in an explosion in China, part of the approximately 6,000 people killed in that industry annually, most of whom worked in poorly-regulated private sector mines. In addition to the possibility of accidents and explosions, miners also face the problem of inhaling dangerous substances leading to disease. Exposure to coal dust, for example, can lead to the lung disease pneumoconiosis, which has claimed thousands of lives and destroyed the health of thousands more. The period between exposure and development of the disease can be around 10 years, which means that it has been very difficult to demonstrate causal links and therefore obtain either better safety equipment or compensation. Figures from the United Kingdom reveal that identification of new cases of the disease continues to run at nearly 1,200 per year, despite the closure of much of the country's coal mining industry.

Other forms of hazard include bloodborne pathogens, HIV/AIDS, exposure to toxic animal droppings, and deliberate attack by terrorists. The extent to which people may be protected from these hazards depends on national-level legislation and the ability and will to enforce those laws, together with the willingness of state governments to enforce international-level safety standards, for example those promoted by the International Labor Organization (ILO). As understanding of mental health develops, so too has the understanding of what workplace conditions can cause mental health problems. Forcing this to happen requires lengthy and often difficult attempts to bring civil or criminal actions against employers who may have much more power to withstand such attempts.

Workers, especially those in vulnerable situations, are often unwilling or scarcely capable of participating in such cases. However, when precedents are set, then it is much more possible for subsequent cases to be prosecuted, when employers may prefer to settle out of court rather than fight a losing battle. Since the majority of the world's population either depends on agriculture or else lives and works in poor urban conditions, it is considered important by many that international rights-based organizations focus on those hazards which are broadly applicable rather than focusing too much on those which are more contentious and expensive to implement.

SEE ALSO: Acquired Immune Deficiency Syndrome (AIDS); Coal; Disease; Fisheries; Hazards; Mining.

BIBLIOGRAPHY. "Chinese Mine Explosion Kills 203," *BBC News* (February 15, 2005); Health and Safety Executive, "Coalworkers' Pneumoconiosis and Silicosis, Chronic Bronchitis and Emphysema," www.hse.gov.uk (cited July 2006); New York Committee for Occupational Safety and Health, www.nycosh.org (cited July 2006).

John Walsh
Shinawatra University

World Bank

THE WORLD BANK is not a single institution but a group of five international finance and regulatory bodies that function to provide financial services and advice to governments around the world. While the bank initially was designed to provide reconstruction loans in the wake of WWII, the primary aims of the group today are economic development, poverty reduction, and the protection of the international investment market.

The motto of the World Bank Group is "a dream of a world without poverty," but historically the goals of the institution were the physical reconstruction of war-devastated countries after World War II. However, by the 1960s these reconstruction efforts were effectively complete and economic development and poverty reduction in poor countries became the primary focus for the Bank's efforts. Today the mission of the Bank is to improve the living standards of people in the developing world through the provision of long-term loans, grants,

and technical assistance designed to help developing countries implement their own poverty reduction strategies. World Bank assistance is evident in everything from broad economic planning to infrastructure development, health and education reforms, and environmental projects.

HISTORY AND OPERATIONAL STRUCTURE

The initial institution in the World Bank Group was the International Bank for Reconstruction and Development (IBRD) which, along with the International Monetary Fund (IMF), was created during the United Nations (UN) Monetary and Financial Conference which took place on July 1–22, 1944, at Bretton Woods in New Hampshire. The Bank formally came into existence on December 27, 1944, commenced operations on June 25, 1946, and made its first loan ($250 million to France) on May 9, 1947.

The primary function of the bank was intended to be the reconstruction of the physical infrastructure of the world after the devastation of World War II. However, the presence of several Latin American countries among the 44 representatives of the UN ensured that the bank's mission statement would include language allowing for future economic development. The articles of membership in the IBRD were ratified by 28 countries on December 27, 1945, and since then the membership of the bank has grown to 184 countries. In subsequent years the IBRD was joined by a series of affiliate agencies, the International Finance Corporation (1956), the International Development Association (1960), the International Center for Settlement of Investment Disputes (1966), and the Multilateral Investment Guarantee Agency (1988).

The headquarters of the World Bank Group (WBG) is in Washington, D.C., and the organization maintains field offices in all the countries in which it operates. Each of the five agencies is an independent entity (although they are all part of the UN system) and is owned by the countries that make up its membership. Each country subscribes to the Bank's basic capital share and while some voting rights are equal for all member countries, others are determined by the financial contribution of the member. Consequently, decisions within the WBG tend to be made by developed countries, while loans and grants are made for projects primarily in developing countries. The primary share holders are the United States (16 percent), Japan (8 percent), Germany, France, and the United Kingdom (4.5 percent each). As any decision requires an 85 percent majority, it is possible for the United States to use veto power to control the decisions of the board.

Each agency in the WBG has different membership, the IBRD is the largest (184 members); and the others have between 140 and 175 members. The group is run by a Board of Directors made up of 24 executive directors representing either a single large country or a group of smaller countries.

By tradition, the president of the World Bank is always a U.S. citizen (while the managing director of the IMF is always a European). Presidents serve for five-year renewable terms. Former Secretary of Defense Robert McNamara was one of the longest-serving directors of the World Bank (1968–81) and initiated a major shift in World Bank priorities toward poverty reduction and the support of developing government efforts to regulate and strengthen local markets. The appointment of William Clausen in 1981 shifted the focus once again, toward free market economics and the removal of government protections for developing economies.

Since the 1980s, both the IMF and the World Bank have maintained this focus on opening developing economies to free market forces and many economists contend that this emphasis has increased the rapid and often inappropriate globalization of third world economies and has benefited the governments and industries of the primary world bank shareholders, rather than of the client states. A final shift in World Bank priorities was signaled during the Clinton administration when James Wolfensohn became president and indicated a new emphasis on battling corruption in client state governments.

CURRENT ACTIVITIES AND PROJECTS

In recent years, the World Bank has moved from general economic development and large scale infrastructure projects toward specific poverty reduction efforts and has increased its efforts to support sustainable development, and small local enterprises that are appropriate to the scale of economic activ-

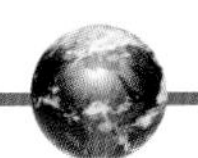

ity in the client countries. Examples of recent project approvals include a water supply and sanitation program for low-income communities in Indonesia; an assessment of labor market conditions in Argentina; and a project to rehabilitate the watershed of the Changjiang and Pearl Rivers in China.

After a series of controversial projects in the 1970s and 1980s, the World Bank has set in place a series of Safeguard Policies which require the Bank to assess the environmental, social, economic, and demographic consequences of each project before it can proceed. In addition, there is an independent institution within the bank, the Independent Evaluation Group, which assesses the impacts and effectiveness of projects once they have been completed. The IEG reports directly to the board and is designed to ensure that the World Bank is accountable to its member governments.

CRITICISMS

One reason for the creation of both the Safeguard Policies and the IEG is long-standing criticism of the World Bank from nongovernmental organizations, member governments, and even from within the institution itself. Early criticisms focused on the lack of environmental and social accountability of bank projects, which, in projects like the Indonesian Transmigration Project of the early 1970s, led to major abuses of both the environment and of local populations. The most frequent recent criticisms however, have emphasized the role of the World Bank in opening up client state economies to the global economy, often to the detriment of local business and governments. These Western-centric practices are often part of "structural adjustment" programs that force free-market liberalization on economies that may not be stable or robust enough to sustain external competition. In addition, such programs often force reductions in public services and increases in external control of the economy. In addition, "structural adjustment" is often undertaken to salvage unpayable loans from developed nations, thereby shifting the burden of risk from the lender to the populations of developing countries. Finally the recent shift to combating corruption in recipient countries has led to a charge that the World Bank has abandoned its traditional non-political stance. However, the bank has countered that reducing corruption is an economic rather than a political goal.

Despite theses criticisms, many of which are valid, the ability of the World Bank to provide below-market rate loans to member countries, many of whom do not have access to traditional global capital markets, can be extremely beneficial when the projects are appropriate and carefully monitored.

SEE ALSO: Debt; Globalization; International Monetary Fund; Underdeveloped (Third) World; World Trade Organization.

BIBLIOGRAPHY. Bahram Ghazi, *The IMF, the World Bank Group, and the Question of Human Rights* (Transnational, 2005); Amy L.S. Staples, *The Birth of Development: How the World Bank, Food And Agriculture Organization, and World Health Organization Have Changed the World 1945–1965* (Kent State University Press, 2006).

Fiona Davidson
University of Arkansas

World Conservation Union (IUCN)

THE INTERNATIONAL UNION for the Conservation of Nature and Natural Resources (IUCN) was founded in 1948 following an international conference in France (Fontainebleau) under the name International Union for Protection of Nature (IUPN). However, the name was changed to IUCN in 1956. The name was once again changed to World Conservation Union in 1990, but it is still called by its old name and acronym. It is still known as the Union or IUCN. The Union's mission is "to influence, encourage and assist societies throughout the world to conserve the integrity and diversity of nature and to ensure that any use of natural resources is equitable and ecologically sustainable."

The World Conservation Union has its headquarter in Gland, Switzerland, and is headed by a director general. The IUCN has offices in 62 different countries with a staff of 1,000. As a result, it brings

together 82 states, 111 governmental agencies, more than 800 nongovernmental organizations (NGOs), and some 10,000 scientists and experts from 181 countries in a unique worldwide partnership, making the IUCN the world's most important conservation network. Aside from a director general in Switzerland, there are three directors who look after global operations, global programs, and global strategies, respectively. There are eight regional directors, each responsible for Meso-America, west/central Asia and north Africa, south America, Asia, eastern Africa, Europe, south Africa, and central Africa. The one in the United States has the designation of executive director. Members within a country or region often organize themselves into national and regional committees to facilitate cooperation and help coordinate the work of the Union. Networks of volunteer scientists and experts are principal sources of guidance on conservation knowledge, policy, and technical advice, and implement parts of the Union's work program. They are divided into the following six commissions:

1. Ecosystem Management: The purpose of the Commission on Ecosystems Management is to ensure the sustainable and efficient management of ecosystems, integrating social, economic, and environmental aims at local, national, and transboundary levels. It consists of almost 500 volunteer ecosystem management experts from around the world.
2. Education and Communication: The Commission on Education and Communication is IUCN's knowledge network concerned with ways to involve people in learning and changes toward more sustainable development through biodiversity and natural resources management. It consists of a network of almost 600 volunteers who are experts in learning, education, communication, capacity building, and change management.
3. Environmental, Economic, and Social Policy: The Commission on Environmental, Economic, and Social Policy consists of professionals who are experts in environmental, economic, social, and cultural factors that affect natural resources and biological diversity. The group of experts provides guidance and support toward effective policies and practices in environmental conservation and sustainable development.
4. Environmental Law: The Commission on Environmental Law consists of volunteers who are experts in environmental law and policy from all over the world. It acts as the principal source of legal technical advice to the Union on all aspects of environmental law.
5. Protected Areas: The Commission on Protected Areas consists of almost 1,200 volunteers involved in promoting the establishment and effective management of a worldwide representative network of terrestrial and marine protected areas. It provides strategic advice to policy makers; helps strengthen capacity and investment in protected areas; and gathers the diverse constituency of protected area stakeholders to address challenging issues.
6. Species Survival: The Commission on Species Survival consists of almost 7,000 volunteers who are experts on plants, birds, mammals, fish, amphibians, reptiles, and invertebrates and are interested in the conservation of biodiversity in plants and animals. The commission provides information on biodiversity conservation, the inherent value of species, their role in ecosystem health and functioning, the provision of ecosystem services, and their support to human livelihoods.

The priorities and work of the commissions are set every four years at the World Conservation Congress, where the members of the Union also elect each 32-member council together with a president, treasurer, and three representatives from each of the eight regions of the Union. The council also includes the chairs of the six commissions. The council operates like a board of directors meeting once or twice a year to direct Union policy, approve finances, and decide on strategy. The council may appoint up to six additional councilors. Accountable to the council, the secretariat is led by a director general and has a decentralized structure with regional, outpost, and country offices around the world.

The IUCN has been criticized in recent years for its slowness in coming to understand the relationship of indigenous rights, environmental justice, and political ecology to successfully implementing conservation efforts. Associated with sometimes-

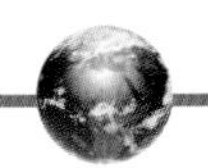

draconian conservation efforts (somewhat unfairly), the IUCN has made recent efforts to broaden its investigation into, and support of, participatory and justice-oriented efforts in conservation. By maintaining databases, assessments, guidelines, and case studies on all environmental issues, and providing scientific understanding of what natural ecosystems provide to humans, the IUCN brings together scientists, policy makers, business leaders and NGOs in ways that increasingly acknowledge the complex social and economic issues that surround conservation problems. The Union is actively engaged in managing and restoring ecosystems, but also is increasingly geared toward improving human lives, economies, and societies where the interests of conservation coincide with the interests of protecting human resources and the rights of local people.

SEE ALSO: Conservation; Environmental Organizations; Extinction of Species; Sustainability.

BIBLIOGRAPHY. P.R. Gil et al., *The Red Book: The Extinction Crisis Face to Face* (Agrupacion Sierra Madre, 2001); Kenton R. Miller and W.V. Reid, *Conserving the World's Biological Diversity* (World Bank, 1990); World Conservation Union, www.iucn.org (cited November 2006).

Vaneeta K. Grover
Independent Scholar

World Health Organization (WHO)

THE WORLD HEALTH Organization (WHO), established in 1948 and with headquarters in Geneva, Switzerland, is the United Nations (UN) specialized agency for health. The organization is perhaps best known for its work to prevent and control epidemics like polio, tuberculosis, malaria, and other diseases. Its constitutional objective is the attainment, for all peoples of the world, of the highest possible level of health. The WHO does not supply health services directly, but rather provides research, advice, training, and funding to assist mainly developing countries to promote health and fight disease.

In terms of structure and processes, the WHO member states appoint delegations to the WHO's supreme decision-making body, the World Health Assembly. This meets once a year, and as well as appointing the director general, supervises the financial policies of the organization and reviews and approves the proposed global budget. The assembly elects 32 members for three-year terms to an executive board. The main functions of the board are to give effect to the decisions and policies of the assembly, to advise it, and generally to facilitate its work. The day-to-day groundwork of WHO is carried out by its secretariat, which is staffed by many thousands of health experts and support staff working in the Geneva headquarters and in the six regional offices.

The WHO member states are grouped into six regions: Africa, the Americas, southeast Asia, Europe and the western Pacific. Each of the six regional offices have a degree of independence. Each is headed by a regional director, while a regional committee for each region sets guidelines for the implementation of all the Health and other policies adopted by the World Health Assembly. The WHO also operates more than 100 country and liaison offices. Each country office includes several health and other experts, as well as various administration staff. The primary functions of WHO country offices include providing leadership and coordination for disaster efforts and being the primary advisor to that country's government in international health issues.

In addition to coordinating international efforts to monitor outbreaks of infectious disease, the WHO organizes specific and focused programs to combat diseases, such as developing and distributing vaccines. Some have been very successful. In 1979, the WHO was able to declare that, due to its activities, smallpox had been eradicated from the world. This was the first disease in history to be completely eliminated by deliberate human design.

Despite the expertise of its public health professionals, however, the WHO has had the unfortunate reputation of being among the UN's worst-run institutions. Its medical and financial policies, notably in relation to HIV/AIDS, have been criticized.

In response, WHO has attempted to make its corporate structure more responsive and flexible. It has also revised its health strategy, adopting an expanded

and more-inclusive approach to health within the context of human development, humanitarian activities, gender equality, and human rights. In the context of the UN Millennium Goals, it has placed a renewed emphasis on the relationships between poverty reduction and health. The WHO failed, however, to meet its "3 by 5" target—a plan to put 3 million AIDS sufferers on antiretroviral treatment by the end of 2005. Progress in meeting its targets within the UN Millennium Goals—notably child mortality and maternal health—have been problematic, mostly in sub-Saharan Africa where public health systems are severely under-resourced or nonexistent.

The WHO publishes statistics and holds comprehensive databases on population health. It also publishes the International Classification of Diseases, which clarifies and universalizes the understanding of disease globally. With regard to research, as well as working with academic collaborating centers based in universities, it both funds and promotes research studies. Meanwhile, it connects with the public through good will ambassadors, the provision of international health information and advice, and events such as World Health Day. As pandemics continue to sweep the developing world and noncommunicable diseases continue to affect the global population, the critical significance of the WHO's original mission continues to be matched by the challenges it faces.

SEE ALSO: Disease; Health; Smallpox; United Nations.

BIBLIOGRAPHY. Gian Luca Burci and Claude-Henri Vignes, *World Health Organization* (Kluwer Law, 2004); M.D. Siddiqi Javed, *World Health and World Politics: The World Health Organization and the UN System* (University of South Carolina Press, 1995).

Gavin J. Andrews
McMaster University
Denis Linehan
University College Cork

World Heritage Sites

AN INTERNATIONAL MOVEMENT for the protection of heritage emerged after World War II, following the decision to build the Aswan High Dam in Egypt. The dam would have flooded the valley containing the ancient Abu Simbel temples. To prevent this, the United Nations Educational, Scientific and Cultural Organization (UNESCO) launched a campaign in 1959 resulting in the successful relocation of the temples. A draft convention on the protection of cultural heritage was subsequently initiated by UNESCO in collaboration with the International Council on Monuments and Sites (ICOMOS). In 1965, a conference at the White House in Washington, D.C., for the first time proposed the linking of cultural and natural heritage, calling for a "World Heritage Trust" that would stimulate international cooperation to protect "the world's superb natural and scenic areas and historic sites for the present and the future of the entire world citizenry." In 1972, a similar proposal from the International Union for Conservation of Nature and Natural Resources (IUCN), also known as The World Conservation Union, was presented to the United Nations (UN) Conference on Human Environment in Stockholm.

Also known as the Convention Concerning the Protection of the World Cultural and Natural Heritage, the World Heritage Convention was adopted by the General Conference of UNESCO in 1972. It came into force on December 17, 1975, as one of the first international conservation conventions and took effect as the World Heritage List. Countries that accept and adhere to the Convention are called State Parties. In all, 181 countries have ratified the Convention or are at various stages of ratification. The Convention is implemented by the World Heritage Committee, which meets once a year and consists of representatives from 21 of the States Parties elected for terms up to six years. The Committee guides the use of the World Heritage Fund and makes the final decision on the inscription of properties in the list. In 2006, the list had 812 properties of cultural and natural heritage, which the World Heritage Committee considers to be of "outstanding universal value." These include 628 cultural, 160 natural, and 24 mixed properties in 137 State Parties. Mixed properties have both cultural and natural attributes, e.g., the Laponian Area in Sweden. The Great Wall in China, Angkor in Cambodia, and the Acropolis in Greece are examples of cultural properties.

The Roman ruins of Volubilis in Morocco are one of many World Heritage Sites on the African continent.

Proposals for the inclusion of properties in the list can only be submitted by signatories to the Convention. In the nomination process, the first step a country takes is to prepare an inventory of important natural and cultural sites, known as the Tentative List. The State Party then selects sites from the Tentative List, collects exhaustive documentation and maps on the sites and prepares a nomination file. The file is evaluated by two Advisory Bodies mandated by the Convention, i.e., ICOMOS and IUCN, the latter providing evaluations of the nominated natural sites. Following nomination and evaluation, the Committee meets once a year and decides which sites are inscribed. Properties listed are considered to be the "common heritage of mankind" and are thus of universal interest and paramount value, the protection of which is the responsibility of all humanity. The Convention calls for such sites to possess "outstanding universal value." A site must also fulfill requirements collectively termed the "conditions of integrity" listed in the Committee's operational guidelines, essentially specifying the long-term conditions a site must meet. To be listed as a natural area, proposed sites must be globally significant and be ecologically viable and protected. Additional criteria that determine a natural site's importance include: Distinctiveness, integrity, naturalness, dependency, and diversity. In the case of cultural sites, significance is determined according to a different set of criteria.

Unfortunately, in times of conflict or war, or due to lack of proper oversight, the basis of the criteria for a site's inscription becomes threatened. Such sites may then be inscribed on the World Heritage in Danger List. As of 2006, 34 properties were in danger from among the 812 properties on the List. These 34 properties include 15 protected areas, e.g., Everglades National Park in the United States; Manas Wildlife Sanctuary in India; and four national parks in the Democratic Republic of the Congo.

In 1994, over two decades after the historic adoption of the Convention, it became apparent that the composition of the list was skewed. There were only 90 natural properties and 304 cultural properties. To rectify this imbalance and to overhaul the framework and methods for defining "World Heritage" and implementing the Convention, the Committee launched the Global Strategy for a Balanced, Representative and Credible World Heritage List. Countries were encouraged to become State Parties in order to ensure geographical representation. Emphasis was placed on nominating and inscribing sites showing coexistence of humans with land, among other attributes. The Committee, at its 28th Session in 2004, reviewed IUCN's assessment that a relatively balanced distribution of regions and wildlife habitats had been achieved. Major gaps remained, however, in the representation of tropical/temperate grasslands, savannas, lakes, tundra and polar systems, and cold winter deserts.

In 2004, IUCN's Review of the World Heritage Network (Review) described the natural and mixed World Heritage Sites as "jewels in the crown" of the world's protected area network. It also laid out the most useful classification and prioritization schemes for revising the Tentative Lists of the State Parties. The schemes are IUCN/Species Survival Commission's habitat analysis, the Udvardy Biogeographic System, WWF Global 200 Ecoregions, and Conservational International's Biodiversity Hotspots. In this Review, the Udvardy biome criteria highlighted cold winter deserts, and tundra and polar systems. The WWF Global 220 Ecoregions approach identified terrestrial ecoregions, e.g., Arctic tundra and

western Ghats, and marine ecoregions, e.g., the Andaman Islands and Tahiti. The IUCN/SSC analysis identified many potential sites around the world, including several grasslands and savanna sites in Africa; subtropical and tropical montane moist forests in India; montane rain forests in New Caledonia and Polynesia (Oceania/Australasia region); the Central Mexican desert areas; desert and coastal areas of Chile and Peru in South America; and the coastal saline wetlands of Europe.

According to the Review, the continent of Africa had the highest number of natural World Heritage Sites (33), followed by Asia (31) and South America (28); Oceania/Australasia, however, had the highest density of World Heritage Sites, approximately one site per 440,000 square kilometers. Of the 126 natural and mixed World Heritage Sites in 2004, 73 had no resident human population, e.g., Kaziranga and Manas in India. Given the widespread presence and dependence of humans on their immediate environment, however, many World Heritage Sites do not preclude human use and are not strict nature reserves, allowing a range of extractive activities. The 2003 World Parks Congress in Durban clearly recognized the interconnectedness of parks and the people living nearby. Among the largest World Heritage Sites with resident human populations are Lake Baikal (88,000 square kilometers) in the Russian Federation, Manu (15,328 square kilometers) in Peru, and the Canadian Rocky Mountain Parks (23,068 square kilometers) in the Provinces of British Columbia and Alberta.

SEE ALSO: Conservation; National Parks; Protected Areas.

BIBLIOGRAPHY. International Union for Conservation of Nature, *Benefits Beyond Boundaries* (IUCN, Gland Switzerland and Cambridge, UK, 2005); Chris Magin and Stuart Chape, *Review of the World Heritage Network. Biogeography, Habitats and Biodiversity* (UNEP World Conservation Monitoring Centre and IUCN, 2004); Jeffrey Sayer, Natarajan Ishwaran, James Thorsell, and Todd Sigaty, "Tropical Forest Biodiversity and the World Heritage Convention," *Ambio: A Journal of the Human Environment* (v.29, 2000); Rahul J. Shrivastava, *Natural Resource Use and Park-People Relations at Kaziranga National Park and World Heritage Site, India* (Florida International University, 2002); Jim Thorsell and Todd Sigaty, *Human Use of World Heritage Natural Sites: A Global Overview* (IUCN: Natural Heritage Program, 1998); United Nations Educational, Scientific and Cultural Organization, "About World Heritage," whc.unesco.org (cited April 2006); Graeme L. Worboys, Michael Lockwood, and Terry D. Lacy, *Protected Area Management: Principles and Practice* (Oxford University Press, 2005).

Rahul J. Shrivastava
Florida International University

World Systems Theory (WST)

WORLD SYSTEMS THEORY (WST) provides a holistic perspective to understand human interaction within a global political economic framework. Though largely heralded by academics in the global South, over the years WST has come under scathing critique by Western social scientists declaring that it is inherently deterministic and overly simplifies human interactions. The theory argues that since the early 19th century, all of humankind has been encompassed within one world system—the capitalist world-economy.

WST espouses that human activities can only be examined within this global system, and that parceled examinations of human political economic and social activity at more focused scales are inherently flawed, because they fail to recognize human processes within their broader context. Thus, WST completely dispels the central role that states, nations, and territorially based political entities (e.g., empires) broadly maintain in the social sciences. The importance of understanding and explaining processes within the capitalist world-system can only be done at the global scale—states, nations, classes, and races are merely the social constructs of systemic, capitalist processes. Different systems come and go, but they are always a product of history. Three different systems have existed—mini-systems, world empires, and the capitalist world system (the latter so called because it is the first system to encompass the entire world's population).

ORIGINS AND DEVELOPMENT

WST originated with, and is still largely associated with, the sociologist Immanuel Wallerstein, who published his seminal work in 1974—*The Modern World System: Capitalist Agriculture and the Origins on the European World-Economy in the Sixteenth Century*. Wallerstein was not the first to argue that state-centric analysis was shortsighted and failed to envelop the true extent of human interaction, but he pioneered synergizing of two disparate, yet intricately congruent, theoretical views of world politics. First, he borrowed heavily from the Annales School of history in France—in particular from the work of Fernand Braudel. Historians of this school were dismayed with 20th-century historians' fetish on the particular details of intra- and interstate political processes and diplomacy. They argued that a holistic approach analyzing political figures within the history of ordinary people was necessary for historical analysis and would be far more useful than case-specific analyses of important events throughout history. The Annales School argued that history operated through *longue durée*—long periods of materialist and economic production that survived regardless of political crises and change.

The second theoretical background interwoven within WST is Marxist theory. In essence, WST is a revision of Marxism itself, attempting to move beyond Marxism's fixation on state economies and its unconvincing portrayal of nationalism in comparison to class. Wallerstein borrows heavily from neo-Marxist critiques of development theories in modern social science. Development theories are state-centric theories that argue states can and do develop linearly. In development theories, all states are provided equal footing, but some states are "behind," "backward," or need to "catch up."

WST arose largely out of Wallerstein's contempt for modernization and development theories that espouse that economic processes operate the same way in all places. WST's counter is that for every state's or region's development, another state or region experiences "underdevelopment." Thus, WST's main critique of such theories is that they are inherently deterministic and fail to take into consideration the context of the greater global economy.

The essence of WST is that since 1800 human interaction—political, economic, and social—has occurred within a global economy—the capitalist system. States are not the containers of human politics, and they are anything but equal. State interactions, as well as the interactions of numerous other political institutions occurring across a variety of geographic scales (i.e., the household, gender, class, race, nation, religion, and transnational institutions such as the United Nations), are all processes occurring within the capitalist world economy. The economy is the structure—everything else is ephemeral.

Among these institutions exist relationships between institutions of the "core" and "periphery." Institutions belonging to the "core" produce high-value goods that can be traded for surplus to "peripheral" institutions producing cheaper, low value goods. This setup creates an inherently unequal system of trade, enlightening what Wallerstein sees as faults in modernization theory. The concept of core and periphery was not originally Wallerstein's, but to these two classes Wallerstein adds the concept of the "semi-periphery." Unlike other social theorists at the time, who largely argued that states and institutions were not mobile between core and periphery, Wallerstein argues that states can move between these positions, albeit rarely. Moreover, he argues that many states belong to the semi-periphery—they are exploited by core states but are often exploiters of peripheral states in the world economy.

Traditionally, most World Systems studies have utilized a quantitative approach. Wallerstein incorporated economic cycles into his theory and often used quantitative analysis of production, trade, and financial statistics to solidify his theory. However, in recent years, WST has increasingly incorporated qualitative analysis—particularly in the disciplines of geography and history.

WST AND THE ENVIRONMENT

WST has increasingly been used to analyze environmental issues by a variety of scientists across numerous disciplines. Arguing that capitalist processes have fended off the major 20th-century resistance to the capitalist system (i.e., global communism), several theorists are now hypothesizing that

the capitalist system may be precipitating its own systemic decline through the ecological ruin resulting from its excesses. For example, World Systems theorists Andrew Jorgenson and Edward L. Kick argue that what is "lacking in the environmental literature ... is a mature long-term historical approach" that only WST can provide.

WST, with its emphasis on core-periphery relationships, has been increasingly used to help explain the roots of environmental degradation. Increasingly, carbon emissions, fossil fuel efficiency, pollution control, and corporate production have been quantitatively analyzed within the capitalist world system context to help understand why peripheral states have continually seen an increase in environmental degradation. World Systems theorists argue that much environmentalist literature fails to see the whole picture—that is, environmental pollution is an outcome of the perpetual quest for more profit in the capitalist world-system, and that pollution affects peripheral states more than the core states due to their poor position within the capitalist economy.

WST rejects the widely accepted notion that "globalization" is characterized by distinct, interacting networks (e.g., political, economic, social, and cultural globalizations) that should be studied individually as additive pieces of a global process. Though the analysis of networks may be useful to distinguish between different types of globalization to better assess different impacts on the environment, WST researchers argue that any such case analysis must be brought back into the broader context of the world system.

As pertains to political and sociological theory, WST has largely passed its peak in academic reverence. Though still used by political geographers, sociologists, and historians, it has been harshly criticized by numerous academics of good repute. In recent decades, Wallerstein has increasingly moved away from analysis of the world-system to focus on philosophies of social science. However, the theory is increasingly being integrated and used as a lens for the analysis of environmental justice, particularly at the global scale. Articles dealing with the environment and capitalism regularly appear in the *Journal of World-Systems Research* and *Review*—the two flagship journals of world-systems research. Edited books using WST to analyze climate and greenhouse gases are a common occurrence as well.

SEE ALSO: Braudel, Fernand; Capitalism; Globalization; Justice; Markets; Marx, Karl.

BIBLIOGRAPHY. Giovanni Arrighi, *The Long Twentieth Century* (Verso, 1994); David Harvey, "The World Systems Theory Trap," *Studies in Comparative International Development* (v.22/1, 1987); Andrew Jorgenson and Edward L. Kick, "Globalization and the Environment," *Journal of World-Systems Research* (v.9/2, 2003); Theda Skocpol, "Wallerstein's World Capitalist System: A Theoretical and Historical Critique," *American Journal of Sociology* (v.82/5, 2003); Peter J. Taylor and Colin Flint, *Political Geography: World-Economy, Nation-State, and Locality*, 4th ed. (Prentice Hall, 2000); Immanuel Wallerstein, *The Modern World System: Capitalist Agriculture and the Origins of the European World-Economy in the Sixteenth Century* (Academic Press, 1974); Immanuel Wallerstein, "The Rise and Future Demise of the Capitalist World System: Concepts for Comparative Analysis," *Comparative Studies in Society and History* (v.16, 1974); Immanuel Wallerstein, *Unthinking Social Science* (Polity Press, 1991).

Ian A. Muehlenhaus
University of Minnesota

World Trade Court

THE WORLD TRADE Court is an appellate court of the World Trade Organization (WTO). However, the term is not accurate, because the World Trade Court is more like an arbitration body than a typical criminal or civil court. Also, the term *World Trade Court* has not yet come into common currency. There has been resistance in some quarters to using the term because it implies a move toward permanent institutionalization and because it cannot issue decisions that are enforceable legal opinions like those in domestic legal systems. The appellate body of the WTO was created to serve as a world trade court in order to aid the WTO in fulfilling its mission—to help producers of goods and services, traders, exporters, and importers conduct their business. This

means that its goal is to ensure just and equitable settlement to all disagreements on trade issues. The work of the appellate body is focused on facilitating the flow of trade between nations. Its major responsibility is to administer WTO trade agreements.

The WTO is headquartered in Geneva, Switzerland. It was created during the Uruguay Round (1986–94) of negotiations on global trade. It began operations on January 1, 1995. As of December 11, 2005, it had a membership of 149 countries. The WTO is the only global organization that applies the rules of trade between nations. It does this on the basis of WTO agreements that have been negotiated and signed as accepted agreements between the major trading nations of the world. The WTO's agreements have all been ratified by its member nations.

The WTO provides a forum for conducting trade negotiations. It handles trade disputes and monitors national trade policies. It also provides technical assistance and training for people from developing countries. Finally, it cooperates with other international organizations. In April of 1994 at Marrakech, Morocco, the Marrakech Agreement (often referred to as the WTO's founding charter) was signed establishing the WTO. The agreement created a mechanism for resolving international trade disputes.

A Dispute Settlement Body (DBS) composed of representatives from all member governments was created. The DBS administers the Understanding on Rules and Procedures Governing Settlement of Disputes (DSU). If a member state believes that another state is engaging in unfair trade practices, then it can bring the dispute to the DBS. A dispute settlement panel composed of three persons who are trade officials will hear a dispute. They meet in secret to avoid political pressures, and their decisions are binding. A state that chooses can appeal a ruling of a panel to the World Trade Court. If a member state appeals to the World Trade Court, then it must do so on the basis of points of law and not on the basis of new evidence or over a dispute about findings of the dispute settlement panel.

The World Trade Court was established by Article 17 of the DSU. It is a permanent body composed of seven persons. They hear appeals from reports issued by the dispute settlement panels. The appellate body can, after reviewing the case, confirm the decision of the original dispute settlement panel, modify its report, or even reverse it. After the appellate body has issued its decision it will be reviewed by the dispute settlement body. The DSB has to accept or reject the appellate body's decision, which will then be binding upon all parties to the case. While binding upon the parties to the case, the WTO does not have an enforcement mechanism. Compliance is voluntary or and requires moral or political persuasion by the parties concerned or by the international community. The appellate body is a standing body also of seven persons and has a permanent seat in Geneva, Switzerland. Members serve four-year terms. The members are chosen for their recognized expertise in law and international trade. They may not be official members in service of any government.

The challenges of free trade for environmental regulation are myriad and many cases heard by the World Trade Court pertain to ecological problems associated with the production of certain traded commodities. In a landmark case, the U.S. federal regulation against the import of shrimp from countries where they are caught with nets that endangered sea turtle populations came under scrutiny. Viewed as an unfair trade restriction by other nations (including India, Malaysia, Pakistan and Thailand), a case was brought before the WTC. In May 1998, the court ruled that the U.S. restrictions represented an embargo that violated free trade agreements. In the aftermath of the decision, U.S. policy was altered so that shrimp imports were handled on shipment-by-shipment basis, rather than a nation-by-nation basis. The compromise, and its implications for national sovereignty in environmental issues, is still being considered by the international business and environmental communities.

Thus, while environmental concerns were important parts of the WTO in its preamble—citing environmental protection, conservation of scarce resources, and sustainable development as WTO goals—the role of the court in adjudicating environmental disputes as trade disputes is somewhat awkward. The DSB has addressed hundreds of cases since 1995. The appellate body has also heard several hundred cases. Among those involving controversial ecological issues was a dispute between the United States and the European Union. The European Union had adopted a number of laws

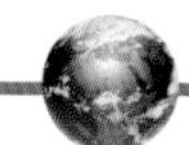

banning genetically modified foods. In 2003, the United States had disputed the laws with a challenge before the WTO. The appellate body's decision in 2006 was complicated. It agreed with the United States that the banning of American genetically modified food was discriminatory. It also dismissed a number of the United States's complaints. Nationalist groups and ecological groups like Green Peace attacked the appellate body's ruling and threatened civil disobedience.

Pesticides used on foods sent into the international market have been inspected with increasing care for food safety. Health-threatening agents such as pesticides, animal and plant diseases, bacterial contaminants, or invasive pests are treated under the SPS Agreement. The SPS agreement puts limits on the food safety policies of member-states in an attempt to prevent them from gaining a competitive edge under the guise of protecting public health. The future of the WTO with regards to environmental policy, therefore, remains important however unclear.

SEE ALSO: Genetically Modified Organisms (GMOs); Trade, Fair; Trade, Free; World Trade Organization.

BIBLIOGRAPHY. Thomas Cottier and Petros C. Mavroidis, eds., *The Role of the Judge in International Trade Regulation: Experience and Lessons for the WTO* (The University of Michigan Press, 2003); Claus-Dieter Ehlermann, "Six Years on the Bench of the 'World Trade Court': Some Personal Experiences as Member of the Appellate Body of the World Trade Organization," *Journal of World Trade* (v.36/4, 2002); Peter Van den Boosche, "From Afterthought to Centerpiece: The WTO Appellate Body and its Rise to Prominence in the World Trading System" (Working Papers, University of Maastricht, Faculty of Law, 2005).

Andrew J. Waskey
Dalton State College

World Trade Organization (WTO)

THE WORLD TRADE Organization (WTO) is an international body responsible for the negotiation, implementation, and enforcement of trade rules. Established in 1995, it presides over a multilateral trading system of which the core international treaty is the General Agreement on Tariffs and Trade (GATT), as legally consolidated in 1994. GATT obligations are informed by a political and ideological commitment to free trade: At the heart of the GATT treaty are the so-called nondiscrimination principles of most-favored-nation status (Article 1) and national treatment (Article 3). Their combined effect is legally to prescribe equality of treatment for imports and exports, such that WTO contracting states subscribe to open, predictable rules of trade.

There are 149 members in the WTO, accounting for over 97 percent of world trade. Meeting at least once every two years, the Ministerial Conference is the governing body of the WTO: conferences have taken place in Singapore (1996), Geneva (1998), Seattle (1999), Doha (2001), Cancun (2003), and Hong Kong (2005). Between ministerial sessions, the WTO General Council—located in Geneva, Switzerland—undertakes key executive functions, including meeting as the organization's Trade Policy Review Body and Dispute Settlement Body. Reporting to the General Council is a council triumvirate addressing goods, services, and intellectual property. There are also numerous specialist committees and working groups dealing individual agreements and thematic areas. One of these is the Committee on Trade and Environment (CTE), which serves as a forum for member states to examine the relationship between trade and environmental protection, with a view to determining whether changes are required in trade rules to enhance their positive interaction. As with all WTO bodies, decision making within the CTE is consensus-based.

The protection and preservation of the environment is embraced as an objective in the Marrakesh Agreement Establishing the World Trade Organization. However, its most important manifestation in international trade rules is still the "general exceptions" clause of GATT first articulated in 1947: under Article 20, trade-restrictive measures may be undertaken for reasons including the protection of human, animal, or plant health and the conservation of exhaustible natural resources. Similar environmental exceptions are also found in specialist WTO agreements, such as those on food safety and

product standards. Not until a 2000 ruling on asbestos products did the WTO Dispute Settlement Body uphold a trade restriction under Article 20 on environment-related grounds. A more significant precedent was set in 2001 when, in its ruling on a U.S. ban of Asian shrimp imports, the Dispute Settlement Body stated that there were circumstances in which WTO members could employ trade restrictive measures under Article 20 in order to prevent serious environmental harm outside their national jurisdiction.

Under the Doha Development Agenda of trade negotiations, launched at the 2001 WTO Ministerial Conference, the CTE was charged with clarifying the relationship between WTO rules and international environmental agreements containing trade measures. There remain legal uncertainties over the rule compatibilities between WTO law and specific trade obligations in international environmental agreements, including those within the 1987 Montreal Protocol on Substances that Deplete the Ozone Layer, the 1989 Basel Convention on the Control of Transboundary Movements of Hazardous Wastes and their Disposal, the 2000 Cartagena Protocol on Biosafety, and the 2001 Stockholm Convention of Persistent Organic Pollutants. While there has not yet been a conflict, the CTE has struggled to improve coordination in this area. It has been suggested by some academic commentators that the robustness of international environmental agreements with specific trade obligations would be strengthened by an amendment to GATT exempting them from WTO rules.

Many environmentalists have campaigned against the WTO, claiming that it is not addressing the harmful ecological effects of its rule making and enforcement. Along with other sections of the global justice movement, they have also leveled the charge that the WTO is undemocratic, with closed negotiations and judicial deliberations dominated by the most powerful states. This is a criticism strongly resisted by the organization, which argues that all its decisions are reached by consensus and ratified in the parliaments of member states. In recent years there have been moves to increase transparency within the WTO, such as improved access to information and meetings with transnational nongovernmental organizations (NGOs). However, member states have resisted proposals to grant transnational NGOs access—as observers—to negotiating forums and dispute settlement hearings. The WTO maintains that the national political processes of member states are the appropriate arenas for NGOs to communicate trade-related environmental concerns, not its own intergovernmental decision making.

There are deep divisions within the WTO over the incorporation of environmental concerns into its negotiations and rulings. Developing countries now constitute two-thirds of the membership of the organization and are generally unsympathetic to the concerns of environmental NGOs, who are perceived as advancing global Northern interests. Indeed, the existence of an environmental protection agenda in the WTO rests on the lobbying of European member states, and developing countries remain fearful of a discriminatory "green protectionism" excluding their exports from European and North American markets. The slow progress of CTE negotiations reflects these divergent positions, as well as the increasing political clout of member states from the global South. Only systemic revisions in trade rule making and interpretation will ensure a more effective consideration of environmental protection issues. Yet, unless environmental protection is linked more directly to the development priorities of developing countries, the scope for advancing these reforms in the WTO will be narrow.

SEE ALSO: Developed (First) World; General Agreement on Tariffs and Trade; Globalization; Justice; Montreal Protocol; Nongovernmental Organizations; Trade, Fair; Trade, Free; Underdeveloped (Third) World.

BIBLIOGRAPHY. Robyn Eckersley, "The Big Chill: The WTO and Multilateral Environmental Agreements," *Global Environmental Politics* (v.4/2, 2004); Michael Mason, *The New Accountability: Environmental Responsibility across Borders* (Earthscan, 2005); Eric Neumayer, "The WTO and the Environment: Its Past Record Is Better Than Critics Believe, but the Future Outlook Is Bleak," *Global Environmental Politics* (v.4/3, 2004); World Trade Organization, www.wto.org (cited April 2006).

Michael Mason
London School of Economics

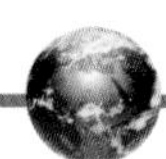

Worldwatch Institute

THE WORLDWATCH INSTITUTE is a nonprofit independent research organization focusing on environmental and social policy. Lester Brown founded the Worldwatch Institute in 1974 with a grant from the Rockefeller Brothers Fund and funding from the United Nations Environment Programme (UNEP). Christopher Flavin has been president since 2001, and Oystein Dahle is chair of the board of directors. Worldwatch Institute has an annual budget of approximately $4 million. Major support comes from foundation grants, the sale of publications, and individual donations. The institute has around 28 full- and part-time staff. An academic advisory panel advises Worldwatch on how it can serve professors and students.

One guiding vision of the institute is to find ways to meet the needs of people without endangering the natural environment or the welfare of future generations. An institute hallmark is to use fact-based and accessible analysis to address key global issues with environmental and development consequences and inform individuals and public officials worldwide about the underlying causes of complex and interconnected problems; the focus is on practical solutions to guide citizens and policy makers.

Four main research areas of the institute are: people, nature, energy, and economy. The institute addresses these areas through a wide range of perspectives (ecology, public health, political science, and economics) to address problems from an interdisciplinary perspective. Unlike related organizations, the Worldwatch Institute has continued to serve as a research institute rather than an interest group.

This agenda follows directly from the long-term leadership of Brown. Brown, a former staffer at the U.S. Department of Agriculture and developer of the Overseas Development Council, has long seen his role as a "synthesizer" of insights drawn from academic disciplines for solving the world's food, population, and economic problems. Early driving forces at the Institute included Eric Eckholm, Jim Fallows, Orville Freeman, and Denis Hayes (coordinator of Earth Day). The institute's founding was part of the third wave of growth in the American public policy research sector.

The Worldwatch Institute disseminates its analysis and viewpoints through three primary and high profile annual publications: *State of the World*, *Vital Signs*, and *World Watch*, a bimonthly magazine, along with other papers and books. *State of the World* was first published in 1984, and has since been translated into over 25 languages worldwide. It is widely read by policy analysts, legislators, world leaders, students, and citizens. Its central purpose is to provide accessible summaries of issues related to the global environment; in recent years each issue has focused on specific topics such as "Redefining Global Security," "The Consumer Society," and "China and India." *Vital Signs* was first published in 1992 and provides analysis of key global trends using graphs and charts to offer visual references; recent topics include energy and climate, resource economics, agricultural resources, and transportation.

World Watch started in 1988. One recent addition to the *World Watch* portfolio of information sources includes Eye on Earth, a new endeavor in partnership with Blue Moon Fund to provide the institute's perspective on current events and global trends. All Worldwatch Institute publications use environmentally sound paper supplies, and the organization pursues other measures to promote sustainability in the office environment.

Central to the current Worldwatch Institute agenda is the analysis of the relationships between environmental degradation, disasters, conflict, and peacemaking. China Watch is a joint initiative of Worldwatch Institute and Beijing-based Global Environmental Initiative (GEI) that reports on agriculture, energy, health, population, water, and the environment. The institute also supports Worldwatch University (for students and teachers), *Environmental Milestones*, a library, online discussions, and a network of global partners.

Critics have charged that though the Worldwatch Institute addresses a vast range of important environmental topics, it remains largely neo-Malthusian in outlook, consistently holding population as the most important driving problem in global ecological degradation, rather than affluence, under-regulation, market failure, capital accumulation, or other global forces. Nevertheless, the institute is powerful in Washington debates because it defines, creates,

and enforces the meaning of "truth" on these core developmental problems. It is an agenda-setter.

SEE ALSO: Nongovernmental Organizations; Policy, Environmental; United Nations Environment Programme.

BIBLIOGRAPHY. C.H., "Briefing," *Science* (v.187/4181, 1975); Timothy W. Luke, "On Environmentality: Geo-Power and Eco-Knowledge in the Discourses of Contemporary Environmentalism," *Cultural Critique* (v.31, 1995); James G. McGann, "Academics to Ideologues: A Brief History of the Public Policy Research Industry," *PS: Political Science and Politics* (v. 25/4, 1992); Public Broadcasting Service, NOVA, "Voices of Concern: Lester Brown," (2003), www.pbs.org (cited January 2007).

Helen Smith and Andrew B. Whitford
University of Georgia

World Wildlife Fund

THE WORLD WILDLIFE Fund is a global conservation network with offices in more than one hundred countries around the world. Scientists, such as Sir Julian Huxley, and political leaders, such as the Duke of Edinburgh, started WWF in Switzerland in 1961. The organization's initial goal was to raise money for conservation. WWF developed slowly for the first two decades. Then, in the 1980s, it grew to become the world's largest private international conservation organization. While many offices were initially started in industrialized states, branches are now located in places such as Bhutan, Malaysia, Pakistan, and Central America. The organization's environment work is not just focused on policy change; WWF runs more than 1,000 field projects annually around the globe.

WWF has identified 200 ecoregions deemed the most critical for conservation, called the Global 200. WWF is an example of international environmental cooperation, but there have been instances of internal conflict within the network. As a result, in 1986, WWF-International changed its name to the World Wide Fund for Nature. The U.S. and Canadian branches continued with the original name. However, all branches continue to use the acronym WWF and the panda symbol, which can contribute to confusion over which branch is behind specific activities. Although many environmental priorities are shared by the larger network there are strategic differences among the offices.

WWF's transnational network involves many important partnerships. WWF and the World Conservation Union cooperate on several campaigns, including efforts to improve co-management with indigenous and traditional peoples in protected areas. Both groups also support an organization called TRAFFIC (Trade Record Analysis of Flora and Fauna in Commerce) that monitors illegal wildlife trade.

Since 1998 the WWF has shared a forest campaign with the World Bank. The Alliance for Forest Conservation and Sustainable Use employs a market-oriented approach that focuses on the promotion of internationally certified sustainable forest extraction. Both groups also seek to remove perverse incentives leading to ecological degradation that frequently exist in policies, institutions, and markets. They highlight poverty reduction as a major environmental concern. They propose solutions based on positive incentives, created through market mechanisms, to support conservation, and finance sustainable local resource extraction. Following this paradigm, WWF has shifted from a nearly exclusive focus on protected areas to eco-friendly production.

WWF's conservation approaches are popular with bilateral and multilateral aid agencies and the private sector. While two decades ago private foundations and individuals provided the majority of funding for WWF, a growing portion of funds currently originate from private firms.

A major focus area for the World Wildlife Fund is sustainable forest management. In the 1980s WWF was a major advocate of Integrated Conservation with Development Programs, which aimed to better involve local populations in protected area initiatives. WWF helped to initiate the Forest Stewardship Council in the early 1990s and has been a key player in the transition from conventional to certified forestry. One of the greatest challenges for forest certification is strengthening consumer demand, so the WWF started the Global Forest Trade Network (GFTN). GFTN is an independent network, made up of more than 500 companies, including some of the biggest lumber suppliers, forest owners, furniture makers, architects, construction companies,

retailers, and investors. Campaigns such as this fit with WWF's focus on sustainable consumption, an strategy that has received criticism from environmental groups that promote less market-oriented approaches to conservation and development.

SEE ALSO: Forests; Sustainable Development; Wilderness; Wildlife.

BIBLIOGRAPHY. Mac Chapin, "A Challenge to Conservationists," *World Watch* (November/December 2004); Fred Pearce, "A Greyer Shade of Green," *New Scientist* (June 21, 2003); Paul Wapner, "World Wildlife Fund and Political Localism," *Environmental Activism and World Civic Politics* (1996, State University of New York Press); WWF Global Network, www.wwf.org (cited March 2006).

Mary M. Brook
University of Richmond

Worster, Donald

DONALD E. WORSTER is a professor of U.S. History and Environmentalism (among other duties and honorary positions) at Kansas University, which is located in the region where he was born. His work has primarily focused on the link between the economic development of the western and southern United States, changes in the physical environment, and the implications arising from this interaction. His general conclusion has been that economic development has resulted in a set of enormous and significant changes in the environment and in the relation between it and the people who determine its uses. The management of water resources in the expansion of American interests into the western portion of the continent led to, among other things, transfer of control over those resources to political and economic elites who retain control into the modern world. As the public sector took over the role of controlling and manipulating the water resources of the West, the seat of American power shifted from the original colony regions to imperially dominated and environmentally transformed virgin lands, from which indigenous peoples were removed. This method of imperial expansion is essentially similar to that previously followed by European powers and has been replicated in the overseas possessions brought into American possession. Ideologically, therefore, the basis of the American state and the wealth that it has brought to its people is based on the presumed right to seize and transform the physical land and the creatures that depend on it.

Ownership and control over the resources of the environment has been customarily accompanied by the professed belief that the land is a resource to be exploited to its maximum economic value with little, if any, consideration for the sustainability of that form of use. This was seen in the Dust Bowl tragedy of the 1930s, when inappropriate use of fertilizers and other chemicals led to the destruction of farmable land and caused widespread hunger and forcible migration. Worster's argument is that insufficient lessons have been learned from the past and that there are real dangers of similar events being repeated in the future. This method of viewing history, which is typically deployed by Professor Worster, helps to highlight lessons for the present and the future.

The models of ecology used by Worster in his work have been faulted by critics, who point to their somewhat over-simplistic qualities. Specifically, the complex ecological dynamics of most ecosystems over time and the complex interactions and responses between human and non-human systems make the straight-forward narratives of books like his *Dust Bowl* less compelling in light of contemporary disequilibrium ecological theory. Nevertheless, Worster has blazed a trail for work in the humanities, uniting history and ecology in a new way.

SEE ALSO: Development; Dust Bowl, U.S.; Environmentalism; Policy, Environmental.

BIBLIOGRAPHY. Donald Worster, *Rivers of Empire: Water, Aridity, and the Growth of the American West* (Oxford University Press, 1992); Donald Worster, *Nature's Economy: A History of Ecological Ideas*, (Cambridge University Press, 1994); Donald Worster, *Dust Bowl: The Southern Plains in the 1930s* (Oxford University Press, 2004).

John Walsh
Shinawatra University

Wright, Frank Lloyd (1867–1959)

FRANK LLOYD WRIGHT (1867–1959) was perhaps the most celebrated U.S. architect and designer, famous for his original concept of houses built in harmony with nature. In the 1930s, Wright defined his concept of "organic architecture" as a respectful interaction and simple reinterpretation of nature, instead of a mere reproduction of it. For instance, natural materials like wood or stone used in houses should look as such, without being transformed or painted. Wright often declared, "Form and function are one," which means, for instance, that a museum should look like a museum and not like a Greek temple.

Among many notable buildings, Wright's Fallingwater in the Laurel Highlands of southwest Pennsylvania illustrates his ecological approach to architecture: In this case, a unique house built in 1939 on a cascade that goes inside the building and crosses the living room. This synthesis of the architecture with the environment is not just decorative; it is the symbol of the integration of nature that feeds the harmony of the life style, the furnishings, and the indoor design. In this case, instead of having a nice view of the outside waterfalls though a window, inhabitants lived in a house that rose over the unchanged cascade. Unlike some of Wright's projects, Fallingwater, also known as the Edgar J. Kaufmann House, still remains with its original furnishings; it has been open to visitors since the Kaufmann family left it in 1964.

Between 1900 and 1919, Wright also introduced the Prairie style, which was later known as the Prairie School, a design approach that was shared with other U.S. architects and followers. Instead of building houses that looked like boxes, the Prairie style favored unity, open plans, and low, horizontal lines that would seem to blend with the flat landscape, with broad open spaces instead of a group of strictly defined rooms. Perhaps influenced by his trip to Japan in 1916, Wright wanted to design

Wright's Prairie style favored open plans, low, horizontal lines, and blending the building with the surrounding landscape. Perhaps the perfect example of the Prairie style is the Robie House in Chicago, built in 1909.

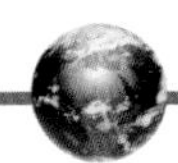

every house as a unified environment in which the interior would be coherent with the natural landscape instead of being similar to the surrounding buildings. Wright recommended the use of natural materials and natural light. Perhaps the perfect example of the Prairie style would be the Robie House in Chicago, built in 1909.

Wright believed that "better homes would create better people." During more than half a century, Wright conceived more than 400 houses, bungalows, factories, theaters, civic centers, and office buildings, mostly in the United States, but also created a few projects in Canada and Japan, like the Imperial Hotel in Tokyo (1922) and the E. H. Pitkin Cottage (1900) on Sapper Island in Desbardts, Ontario. Some of these houses have been remodeled or even demolished. Other houses built by Wright, like the Louis Sullivan Bungalow and the James Charnley Bungalow (both in Ocean Springs, Mississippi), were destroyed by Hurricane Katrina in 2005.

Conceived as a spiral structure that looks like a work of art in itself, the Solomon R. Guggenheim Museum in New York City remains perhaps the most famous building created by Frank Lloyd Wright (between 1943 and 1957). Here, the concept was to allow the visitor to take the elevator to the top of the building and then proceed slowly down the spiral walkway to view the artwork, in order to avoid fatigue.

A prolific writer, and on occasion a teacher and lecturer, Frank Lloyd Wright also published many books and essays: *An Organic Architecture* (1939), *The Natural House* (1954), *A Testament* (1957), and *The Living City* (1958). Already a celebrity before his death at 91, Frank Lloyd Wright has become an American icon.

SEE ALSO: Design (and Ecodesign); Development; Landscape Architecture; Urban Planning.

BIBLIOGRAPHY. Jackie Craven, "Frank Lloyd Wright Buildings Index," architecture.about.com (cited December 2006); Kaufmann Conservation on Bear Run, "Frank Lloyd Wright's Fallingwater, in the Laurel Highlands, Pennsylvania," www.fay-west.com (cited December 2006); Organic Architect, www.organicarchitect.com (cited December 2006); William Allin Storrer, ed., *The Frank Lloyd Wright Companion* (University of Chicago Press, 2006); William Allin Storrer, "The Frank Lloyd Wright Update," www.franklloydwrightinfo.com (cited December 2006); Frank Lloyd Wright, *Frank Lloyd Wright Collected Writings* (Rizzoli, 1994); Frank Lloyd Wright, *Frank Lloyd Wright. An Autobiography* (Pomegranate, 2005); Frank Lloyd Wright and Mike Wallace, *Frank Lloyd Wright. The Mike Wallace Interviews,* audiocassette (HighBridge Company, 1996); Frank Lloyd Wright Foundation, www.franklloydwright.org (cited December 2006); Frank Lloyd Wright Preservation Trust, wrightcatalog.stores.yahoo.net (cited December 2006).

Yves Laberge, Ph.D.
Institut Québécois Des Hautes
Études Internationales
Québec, Canada

Xeriscape

A XERISCAPE (PRONOUNCED "z-ri-scape") is a landscape that has been developed to maximize water conservation. "Xeriscape" is a compound of the Greek word *xeros* ("dry") and *scape* to express the idea of developing a landform that is manageable in a drought-prone area or a dry landscape. Xeriscaping is a creative form of landscaping that uses plants that can derive the maximum benefit possible from the available water.

A xeriscape is not limited to desert plants or to desert or semi-arid regions. Xeriscaping can be used in urban areas; an unusual dry spell will then not kill the plants because they are immune to small droughts, nor will they require watering. The term *xeriscape* was originally coined by Denver Water, a municipal utility that continues to hold the trademark, but it has since become a generically used term for all forms of sustainable, water-conserving landscaping.

In 1977, a severe drought in the western United States made water usage a public and financial issue. Homeowners, landscapers, and others realized that turf grass lawns and other water-intensive plantings were expensive and impractical. However, sand, gravel, and plastic yard coverings were not the answer—xeriscaping was.

Development of a xeriscape takes planning, as does the development of any landscape. But, with a xeriscape the plan seeks to maximize the use of water, including finding ways to retain water and curb runoff. In addition, because water loss due to evaporation is usually greatest in areas with a southern or western exposure, plants used in these areas should be those that need lesser amounts of water and that can withstand higher temperatures. These areas may be appropriate for a drought resistant ground cover, which, if the slope of the area is steep, will retain more moisture and moderate ground temperatures. Another important change in implementing a xeriscape is to replace turf grasses with drought-resistant grasses. Some areas can also be planted with drought-resistant wildflowers.

Success in creating a xeriscape is influenced by good soil preparation. Sandy or heavy clay soils will not retain water as well as a more balanced soil. Modification of the soil so that there are increased pore spaces can be accomplished with mixtures of organic materials, silt, sand, and clays. Since plant roots usually need oxidation, increasing pore spaces will aid plants to grow and withstand drought.

Landscaping with desert plants is one way to create a xeriscape. Cacti are an obvious choice, but to

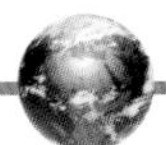

After cacti, there are many other plant choices for xeriscaping—even drought-resistant wildflowers.

cacti can be added a great number of xerophytes, or plants from arid regions. Xerophytes are a class of plants that have adapted to desert environments by some mechanism to prevent water loss or to store water in their leaves. Many Mediterranean plants are xerophytic adaptations that survive summer dryness. Some, like the live oak, have hard, thick, leathery leaves while others have waxy leaf coatings that do not release water easily. Others are succulents, which store water in their leaves.

Some xerophytes can be found in salt marshes, saline soils, or in acid bogs where they have adapted to chemically hostile wet environments. Other xerophytes have evolved at beaches, in sand dunes, and on bare rock surfaces, even in wet regions where local conditions leave some spots dry.

Another way to practice xeriscaping is to reuse water. In California, some cities such as Santa Monica are reusing both treated sewage water and storm runoff. Water from urban runoff is not potable, but it can be used to water plants. A xeriscape can use water that is not potable at a reduced cost compared to using treated water. Besides being a sound horticultural practice, xeriscaping is often cheaper and more environmentally responsible.

SEE ALSO: Gardens; Groundwater; Landscape Architecture; Landscape Ecology; Recycling; Runoff.

BIBLIOGRAPHY. Connie Lockhart Ellefson, *Xeriscape Colorado: The Complete Guide* (Westcliffe Publishers, 2004); Rob Proctor, *Xeriscape Plant Guide: 100 Water-Wise Plants for Gardens and Landscapes* (Fulcrum Publishing, 1999); Gayle Weinstein, *Xeriscape Handbook: A How-to-Guide to Natural, Resource-Wise Gardening* (Fulcrum Publishing, 1999).

Andrew J. Waskey
Dalton State College

Yellow Fever

YELLOW FEVER IS a disease of the flavivirus family affecting humans and monkeys that is transmitted by mosquitoes of the genus *Aedes*. Yellow fever has long been associated with human-created ecological change, both in urban areas and, more recently, in areas of the tropics experiencing rapid deforestation and urbanization. Victims first develop fever, chills, and vomiting followed by internal bleeding and jaundice; the illness may lead to death within two weeks.

Medical historians believe that *Aedes* mosquitoes are native to West Africa and arrived in the Americas via slave ships beginning in the 16th century. The mosquito and the virus spread throughout North and South America, though in temperate regions *Aedes* could not overwinter but was frequently reintroduced through trade. *Aedes* is well adapted to urban environments, breeding in standing water as found in cisterns or tires. The virus exacted high mortality rates and caused social dislocation and economic isolation in the temperate-zone cities where it occurred periodically until 1905.

In the tropical regions where yellow fever was endemic, many experienced moderate cases as children and survived with immunity. Lack of immunity among Europeans and North Americans limited colonial activities in West Africa, foiled French plans to build a canal in Panama, and inhibited U.S. expansion in the Caribbean. These and similar frustrations inspired the development of the field of tropical medicine as an aid to the colonial project. Physicians Walter Reed and Carlos Finley in 1901 discerned that yellow fever was mosquito-borne, and the U.S. Army soon enforced mosquito control measures in Havana and other areas of interest. The same control methods allowed the United States to construct and defend the Panama Canal and brought an end to epidemics in North America.

Yellow fever has been largely eradicated in temperate regions and industrialized countries through quarantine, environmental control, mosquito eradication and, after 1937, vaccination. The Rockefeller Foundation implemented mosquito control measures worldwide in the interwar period, but vaccination delivered the final blow in many areas. Between World War II and the 1960s, public health and military campaigns worldwide used DDT to control *Aedes*. The Pan-American Health Organization attempted to eradicate one *Aedes* species throughout the Americas in the mid-20th century, but lack of U.S. support, and U.S. concern over DDT exposure, doomed the effort.

Yellow fever has reemerged as a threat in tropical regions of Africa and South America and now occurs there as epidemics as well as endemically. Comprehensive vaccination campaigns have protected regions such as The Gambia, but few states have achieved widespread vaccination. Infections have been increasing since the 1980s, with 200,000 cases and 30,000 deaths in 2000. Because several different *Aedes* species can transmit the virus, the disease can move among forest, savanna, village, and city environments. Timber workers have been exposed to the sylvatic form of yellow fever through mosquitoes that feed on forest-dwelling populations of several monkey species. Epidemiologists suspect that rapid, human-caused ecological change has resulted in outbreaks of an intermediate form of the disease at the edge of savannas and forests, and that the growth of cities in Africa and South America will increase epidemics of the urban form. Disease ecology experts in temperate regions fear that global warming will expand the range of *Aedes* mosquitoes and thereby yellow fever.

SEE ALSO: Colonialism; DDT; Disease; Global Warming; Malaria; Mosquitoes.

BIBLIOGRAPHY. Laurie Garrett, *The Coming Plague* (Penguin, 1996); Andrew Learmonth, *Disease Ecology* (Blackwell, 1988); William H. McNeill, *Plagues and Peoples* (Random House, 1976); World Health Organization, "Yellow Fever" (December 2001), www.who.int (cited April 2006).

Dawn Day Biehler
University of Wisconsin, Madison

Yellowstone National Park

ESTABLISHED IN 1872, Yellowstone National Park is now known as the world's first national park. The park's boundaries were set in the northwest corner of Wyoming and narrow portions of southeast Idaho as well as south-central Montana. Famous for spectacular scenery— the Rocky Mountains, explosive active geysers, crystal clear rivers and springs, still-abundant buffalo, elk, eagles, and other wildlife—the mention of Yellowstone invokes an image of pristine nature, unspoiled by humans. However, Native Americans occupied the area as long as 11,000 years ago.

Some of the first European explorers, including members of the Lewis and Clark expeditions in the early 1800s, encountered Native Americans—called the Shoshone—in the area now known as Yellowstone. A member of their expedition, John Colter, remained in the region and documented his winter journey in the mountains. During the first half of the 1800s, aside from the Lewis and Clark expeditions, the territory remained largely unexplored by anyone other than fur traders and Native Americans.

By the 1850s, a few missionaries had begun to explore the area and made detailed records of what they saw; at that time politicians in the east were still in disbelief about the descriptions they had heard of the Yellowstone region. Military excursions into the region were rare, but not unheard of, and no systematic efforts had been made to validate the claims of those who had seen the area. Gold strikes in Idaho in the 1860s brought prospectors and engineers deep into the area.

Finally, a series of formal explorations of the Yellowstone region took place (the Folsom, Washburn, Hayden, and Barlow parties consecutively from 1869 to 1871) supported by the U.S. government, private donors, and the Northern Pacific railway. Explorers, soldiers, and skilled technicians of all kinds—cartographers, zoologists, botanists, artists, and photographers—were sent out to map, record, and photograph the entire region that had been described by early adventurers and missionaries. Even the U.S. Geological Service sent explorers to document the Yellowstone region for its mining potential—as of this time, lawmakers in the east thought of the vast open spaces of the west as resources that could be used for economic advancement of the country.

By the time of the Hadyen and Barlow parties of 1871, photographs and records of the region had stimulated the interest of scientists from the Smithsonian Institution and other prestigious organizations. The country's leaders had become aware of the wonders of the area, then referred to as Great Geyser Basin, and had intended to build a railroad

Nathaniel "National Park" Langford

Nathaniel Pitt Langford, the first superintendent of Yellowstone National Park, became known as "National Park" Langford because of his role in the early establishment of the park.

Langford was born in 1832 at Westmoreland, New York, and moved to Oregon in 1862 after gold was found there. He made his camp on Grasshopper Creek, now in Montana (but then in Dakota). When Montana was organized as a territory, he was appointed collector for internal revenue, a post he held until 1868. In January 1869 President Andrew Johnson nominated him governor of Montana but this was never confirmed by the Senate. In 1870, as a member of the Washburn party, he discovered geysers, or hot springs, in what would become Yellowstone National Park.

Langford became the superintendent of Yellowstone National Park in 1872. There was no money provided to run the park, however, and Langford had no salary. Therefore, he had to spend much of his time working as National Bank Examiner for the Pacific Coast from 1872 until 1884. For this reason he only entered the park twice during his superintendency. On the first occasion, it was during the Second Hayden Expedition in 1872; the second time was two years later, when he had to evict Matthew McGuirk, who claimed to have the rights to the park's hot springs. Eventually Langford was accused of neglecting the welfare of the park; he was removed as superintendent in 1877. He moved to St. Paul, Minnesota, and was director of the Board of Control there. He died in 1911.

into the region. Ironically, it was an employee of Northern Pacific Railroad, Jay Cooke, who promoted the idea of setting aside a reservation or a park for the enjoyment of the public and for its aesthetic and geologic value. In 1872 the U.S. Congress established Yellowstone as the first national park of the country and of the world, setting precedent for human perceptions of nature as wild and uninhabited places.

Unfortunately, the area was not uninhabited, and with the creation of Yellowstone, Native American tribes such as the Shoshone, Blackfoot, and Crow were forcibly evicted from the area or killed. Today, Native Americans who have cultural ties with the park (as with other U.S. national parks) may utilize the park's resources for traditional practices through agreements with Yellowstone Park and the National Park Service Ethnography Program and as protected by the National Historic Preservation Act and the Native American Free Exercise of Religion Act.

Other current issues affecting the park include a variety of scientific and recreational problems, such as fire management, reintroduction of the wolf, and banning of snowmobiles and other motorized vehicles. In 1988, half of Yellowstone National Park was burned by wildfires. Ecologists who criticize the Park Service's forest suppression policies have noted that fires are a natural part of ecosystems and should be allowed to burn naturally because they help regenerate plant life and clear natural fire hazards. The 1988 fires led park managers to reevaluate their fire management policies and some "let burn" policies have been implemented.

Beginning in 1995 wildlife conservationists reintroduced gray wolves into the area known as the Greater Yellowstone Ecosystem (GYE), which extends beyond the boundaries of the national park. A great deal of controversy preceded the release of the wolves because people feared they would damage livestock or even pose a danger to humans. Objectives of the wolf reintroduction were to "restore natural ecological processes" and reduce prey, as well as help control populations of elk and moose that had grown to exceed the capacity of the land to support them. Although access to park areas has traditionally restricted snowmobiles, recent policies—mandated by Congress—have allowed snowmobiles into some areas of the park. Many national and local environmental organizations, such as the Sierra Club and the Greater Yellowstone Coalition, have continued to fight those policies.

SEE ALSO: Fire; National Parks; National Parks Service (U.S.); Native Americans; Preservation; Wolves.

BIBLIOGRAPHY. American Park Network, "Yellowstone: History," www.americanparknetwork.com (cited May 2006); M. Colchester, "Salvaging Nature: Indigenous Peoples and Protected Areas," in K.B. Ghimire and M.P. Pimbert, eds., *Social Change and Conservation* (Earthscan Publications Limited, 1997); A.L. Haines, *Yellowstone National Park: Its Exploration and Establishment* (National Park Service, 1974); G.K. Meffe and C.R. Carroll, *Principles of Conservation Biology* (Sinauer Associates, 1994); D.W. Smith, R.O. Peterson, and D.B. Houston, "Yellowstone After Wolves," *BioScience* (v.53, 2002); L. Smith, "The Contested Landscape of Early Yellowstone," *Journal of Cultural Geography* (v.22, 2004).

REBECCA AUSTIN
FLORIDA GULF COAST UNIVERSITY

Yemen

WHEN THE END of World War I signaled the breakup of the Ottoman Empire, North Yemen became independent. South Yemen, a British protectorate since the 19th century, did not achieve independence until 1967. When South Yemen adopted Marxism in 1970, hundreds of Yemenis fled to the north, setting the stage for dissension that ended only with the unification of the two countries as the Republic of Yemen in 1990. A border dispute with Saudi Arabia was peacefully settled in 2000. However, internal strife in Yemen continued, due in large part to a stagnant economy, ultimately leading to a crisis of debt payments. Following loan rescheduling by the International Monetary Fund, by the end of 2002 Yemen's external debt was 47.9 percent of its GDP, down from 52 percent the previous year.

Yemen is the 14th-poorest nation in the world, with a per capita income of only $800. Over 45 percent of Yemenis live below the national poverty line. More than a third of the labor force is unemployed, and the majority of workers are engaged in subsistence agriculture and herding. Barely a fourth of Yemenis live in urban areas.

A number of social indicators mirror Yemen's status as one of the world's poorest nations, preventing the government from focusing attention on environmental issues. Life expectancy is only 61.5 years for the population of 20,727,063. The combination of an infant mortality rate of 61.5 deaths per 1,000 live births, a fertility rate of 6.67 children per female, an HIV/AIDS rate of 0.1 percent, and exposure to diseases that are common among poor nations produces additional environmental burdens. The low literacy rate (70.5 percent overall and 30 percent for females) makes the dissemination of health and environmental information extremely difficult. The United Nations Development Programme Human Development Reports rank Yemen 151st of 232 nations on overall quality-of-life issues.

Bordering on the Arabian and Red Seas and the Gulf of Aden, Yemen has a coastline of 1,182 miles (1,906 kilometers). The country is located on Bab el Mandeb, the strait that links the Red Sea with the Gulf of Aden and is one of the most active shipping lanes in the world. The Yemeni terrain is composed of narrow coastal plains flanked by mountains and flat-topped hills, with the Arabian Peninsula dissecting the desert plains of the uplands. At least 90 percent of the land area has an arid or hyper-arid climate with high rainfall evaporation rates. Sand and dust storms are frequent in the hot summers. Along the western coast, temperatures tend to be hot and humid. In the mountains of the west, seasonal monsoons occur in direct contrast to the harsh desert conditions of the east. Natural resources include petroleum, fish, rock salt, marble, and small deposits of coal, gold, lead, nickel, and copper. With less than 3 percent arable land, the only fertile soils of Yemen are found in the west.

A study by scientists at Yale University in 2005 ranked Yemen 11th from the bottom in environmental performance, far below the comparable income and geographic groups. Scores were particularly low in the categories of biodiversity and habitat, air quality, and environmental health. Barely a fourth of rural Yemenis have access to safe drinking water, and only 14 percent of this group have access to improved sanitation. In contrast, 68 percent of urban residents have access to safe drinking water, and 76 percent have access to improved sanitation.

Environmental issues arise in Yemen from overexploitation, depletion, and pollution of valuable resources. The lack of freshwater sources, which has created a shortage of potable water, is Yemen's ma-

jor environmental problem. This shortage has been accelerated by the practice of pumping groundwater beyond the sustainable level. Scientists have estimated that without intensive water conservation, existing water basins will disappear by the mid-21st century. There is particular concern over the practice of using scarce water to grow gat, an amphetamine-like narcotic, because it prevents farmers from growing essential food products. The food supply is further threatened by extensive coastal degradation and the loss of fisheries.

Overgrazing and soil erosion are a result of desertification that occurs from the combination of agricultural mismanagement and climatic conditions. Forests are being depleted at a rate of 1.8 percent each year, with each family using an estimated one to two tons of wood a year. Deforestation is accompanied by a loss of biodiversity and habits. Of 66 endemic mammal species, five are endangered, as are 12 of 93 endemic bird species. The government has made some progress in this area by protecting the Socotra archipelago, which is known as the Galapagos of the Indian Ocean.

The Environmental Protection Council is the Yemeni agency charged with promoting environmentalism through the implementation of the National Action Plan, which focuses on strengthening water management, curbing soil degradation, creating sanctuaries, and regulating waste management. Yemen participates in the following international agreements on the environment: Biodiversity, Climate Change, Desertification, Endangered Species, Environmental Modification, Hazardous Wastes, Kyoto Protocol, Law of the Sea, and Ozone Layer Protection.

SEE ALSO: Deforestation; Endangered Species; Fisheries; Land Degradation; Life Expectancy; Overgrazing; Poverty; Soil Erosion; Water Demand.

BIBLIOGRAPHY. Central Intelligence Agency, "Yemen," *The World Factbook,* www.cia.gov (cited April 2006); Timothy Doyle, *Environmental Movements in Minority and Majority Worlds: A Global Perspective* (Rutgers University Press, 2005); Kevin Hillstrom and Laurie Collier Hillstrom, *Africa and the Middle East: A Continental Overview of Environmental Issues* (ABC-CLIO, 2003); Ministry of Planning and International Cooperation, "Environment and Bio-Diversity," www .mpic-yemen.org (cited April 2006); One World, "Yemen: Environment," uk.oneworld.net (cited April 2006); United Nations Development Programme (UNDP), "Human Development Report: Yemen," hdr.undp.org (cited April 2006); UNDP, "Yemen: Natural Resources" www .undp.org (cited April 2006); World Bank, "Yemen," *Little Green Data Book,* lnweb18.worldbank.org (cited April 2006); Yale University, "Pilot 2006 Environmental Performance Index," www.yale.edu/epi (cited April 2006).

Elizabeth Purdy, Ph.D.
Independent Scholar

Yosemite National Park

YOSEMITE NATIONAL PARK is located along the western slopes of California's Sierra Nevada mountain range. Falling under the jurisdiction of the U.S. Department of the Interior, Yosemite spans over 1,158 square miles (3,000 square kilometers) and ranges in elevation from 2,000 to over 13,000 feet (610 to 3,962 meters) above sea level. Though Yosemite is a land of superlatives—containing sheer 3,500-foot (1,067-meter) cliffs, the highest waterfall in North America (Yosemite Falls), and the largest living tree species in the world (the giant sequoia)—it is perhaps most notable for being an icon of the U.S. environmental movement.

In 1864 Abraham Lincoln passed a landmark bill called the Yosemite Grant, which ceded Yosemite Valley and the Mariposa Sequoia Grove to California as a state park. During the late 19th and early 20th centuries, Yosemite National Park became ground zero for a monumental debate between preservationists and conservationists over how to plan for the park's future. During the late 1800s John Muir helped draw the public's attention to Yosemite through his influential writings and environmental activism—including the formation of the Sierra Club.

In response to heavy sheep grazing and the logging of giant sequoia, John Muir and other preservationists advocated for the preservation of Yosemite in its natural state by granting the area federal

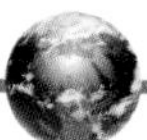

Yosemite's status as a national park has led to 89 percent of the park being designated wilderness area, but also to annual tourist visitation rates that exceed three million.

protection. In 1890 his efforts were rewarded when Yosemite was declared a national park. Meanwhile, Gifford Pinchot, the first chief of the National Forest Service and an important political figure in the conservation movement, had been lobbying to manage Yosemite's natural resources scientifically for productive purposes.

Today, Yosemite can be viewed as a mosaic of compromises between preservationists and conservationists. For example, though a national park, Yosemite contains within its park boundaries a large dam and reservoir in Hetch Hetchy Valley. A conservationist achievement, the O'Shaughnessy Dam at Hetch Hetchy quenched the water needs of the San Francisco Bay Area for many years. The preservationist movement also had notable achievements. With its designation as a national park, approximately 89 percent of the park is designated wilderness area. Yosemite is also home to the headwaters for two designated wild and scenic rivers—the Merced and Tuolumne.

In 1984, the park's federal status expanded when Yosemite was named a world heritage site because of its perceived contribution to California's cultural heritage including the California Gold rush; the broader U.S. National Park movement; and the cultural legacy of over 8,000 years of Native American settlement in the Yosemite region, including by the Miwok and Paiute tribes.

Annual visitation rates in Yosemite exceed three million, with the vast majority of these guests visiting Yosemite Valley. As a result, the Park Service has seen a number of environmental challenges arise. These challenges include: (1) air pollution in the Yosemite Valley from high levels of car and tour bus congestion; (2) the interruption of natural fire regimes and forest regeneration due to heavy fire suppression practices; (3) persistent interaction

between black bears and humans as bears began to rely on human food, causing damage to visitor property and ultimately costing the lives of black bears; (4) the introduction of invasive species bringing significant changes to Yosemite's fragile alpine ecosystems; and (5) habitat fragmentation, especially along the valley floor as roads and buildings are built to accommodate millions of Yosemite sightseers every year.

In 2000, the Yosemite Valley Plan was produced under the guidelines of the National Environmental Policy Act to supplement and modify the goals and proposed strategies found in the 1980 General Management Plan for Yosemite. Aimed at preserving the natural and cultural resources of Yosemite Valley for the use and enjoyment of visitors, the Yosemite Plan is a management response to the park's paradoxical position as a sanctuary for unique wildlife species and fragile habitats and as an overwhelmingly popular tourist destination.

SEE ALSO: Muir, John; National Forest Service (U.S.); National Parks; Pinchot, Gifford; Sierra Club; Tourism; World Heritage Sites.

BIBLIOGRAPHY. John Muir, *My First Summer in the Sierra* (Houghton Mifflin, 1998); Alfred Runte, *Yosemite the Embattled Wilderness* (University of Nebraska Press, 1990); U.S. Department of the Interior, Yosemite National Park, www.nps.gov/yose (cited April 2006).

GREGORY SIMON
UNIVERSITY OF WASHINGTON

Yucca Mountain

YUCCA MOUNTAIN IS the site of a proposed repository for spent nuclear fuel and other radioactive waste material. Yucca Mountain is located in Nye County, Nevada, approximately 100 miles northwest of Las Vegas. The land is under the joint control of the U.S. Department of Energy (DOE), the U.S. Air Force, and the Bureau of Land Management. The proposed repository would occupy 230 square miles of high desert with no permanent settlements within 15 miles. Yucca Mountain was formed by several layers of volcanic rock deposited over 12 million years ago. The rock is identified as "tuff," which is formed ash deposited from volcanic eruptions.

Yucca Mountain has been considered as a potential permanent nuclear waste repository since 1978 when the DOE began studying the geologic character of the site. The DOE received authority to search for a suitable repository under the Nuclear Waste Policy Act. An earlier precedent for the search came from a recommendation by the National Academy of Science in 1957 that suggested that burial of nuclear wastes in deep underground sites would protect the environment and ensure the health and safety of humans. The proposals to use underground burial in general and Yucca Mountain specifically have both met with significant resistance from the outset.

Despite repeated objections to the proposals by politicians and environmentalists alike, repository planning continued into the 21st century and has yet to be finally resolved. In 1983, the DOE studied nine possible sites for the repository and two years later President Ronald Reagan called for further in-depth analysis of three of those sites: Hanford, Washington; Deaf Smith County, Texas; and Yucca Mountain. A congressional amendment to the Nuclear Waste Policy Act in 1987 specified that the DOE would consider only Yucca Mountain. Further progress on the program was made on July 23, 2002, when President George W. Bush authorized the DOE to make formal application to the Nuclear Regulatory Commission for a license to construct the repository. Delivery of the application was specified to be no later than June 30, 2008.

If the Yucca Mountain site receives final approval for repository construction, the total cost is estimated to be as high as $100 billion. In July 2006, the DOE determined that the repository would receive its first shipment of nuclear waste on March 31, 2017. However, the proposal continues to be vociferously opposed. Nevadans in particular are irate over the fact that a site was chosen in their state despite the fact that Nevada has no nuclear power plants within its borders. Concern was raised as well about Environmental Protection Agency (EPA) waste disposal standards, which specified that radiation levels would not exceed established levels

for 10,000 years following the closure of the repository. Court rulings on this provision were found to be inconsistent with earlier recommendations issued by the National Academy of Sciences. Following this ruling the EPA proposed new radiation dosage limits to be effective following the 10,000-year period and extending to one million years following repository closing. No regulatory proposal has ever been made for this long of a period of time.

Nuclear wastes are now stored in containers at 126 sites in the United States, presenting multiple security and safety issues. Nevertheless, alternative storage options, including multiple monitored retrievable burial sites, have largely remained unconsidered. In 2006, a number of politicians suggested that the basic issue of underground storage of nuclear wastes be reexamined. The suggestion has been advanced to find alternatives to the Yucca Mountain underground plan. Political leaders are proposing a moratorium on Yucca Mountain and inviting scientists to come up with other plans for disposing of nuclear wastes. As of 2007, therefore, the viability of Yucca Mountain as a solution to the nuclear waste problem remains unresolved.

SEE ALSO: Bureau of Land Management (BLM); Department of Energy (DOE) (U.S.); Desert; Environmental Protection Agency (EPA); Movements, Environmental; Nuclear Power; Nuclear Regulatory Commission (NRC) (U.S.); United States, Southwest.

BIBLIOGRAPHY. Jeff Johnson, "Yucca Mountain: Building a Final Home for the Nation's Nuclear Waste," *Chemical and Engineering News* (v.80/27); Raymond L. Murray, *Understanding Radioactive Waste* (Battelle Press, 2003); David P. O'Very, Christopher E. Paine, and Dan W. Reicher, eds., *Controlling the Atom in the 21st Century* (Westview Press, 1994); K. S. Shrader-Frechette, *Buying Uncertainty: Risk and the Case against Geological Disposal of Nuclear Waste* (University of California Press, 1993).

Gerald R. Pitzl, Ph.D.
New Mexico Public Education Department

Zambia

FORMERLY A BRITISH colony known as Northern Rhodesia, the Republic of Zambia was created in 1964 after independence from Britain. Copper mining had long been the mainstay of the country's economy, and price reductions in the 1980s and 1990s along with prolonged drought jeopardized the economy of the new country. Contested elections and political corruption threatened political stability until 2002 when an anticorruption campaign was initiated. By 2004 copper prices had recovered, and new mines had begun operation. Other resources with the potential for improving Zambia's economy include: cobalt, zinc, lead, coal, emeralds, gold, silver, uranium, and hydropower.

Economic prosperity in Zambia is hampered by high foreign debt, made more complicated by restructuring through the International Monetary Fund, though multilateral agencies are working with the government to effectuate debt relief. With a per capita income of only $900, Zambia is the 17th-poorest country in the world. Eighty-six percent of the population lives in poverty and nearly half of Zambians are seriously undernourished. The richest 10 percent of the population hold 41 percent of the wealth, while the poorest segments share just over one percent of resources. Around eight percent of land area is arable, but 85 percent of the population are engaged in the agricultural sector, mostly at the subsistence level. Half of the workforce is unemployed. The United Nations Development Programme's Human Development Reports rank Zambia 166 of 232 countries on overall quality of life issues.

Zambia is landlocked but has 11,890 square kilometers of inland water resources. Land borders are shared with Angola, the Democratic Republic of the Congo, Malawi, Mozambique, Namibia, Tanzania and Zimbabwe. The terrain of Zambia is made up of plateaus rising to isolated hills and mountains. Elevations range from 329 meters at the Zambezi River, which forms a riverine boundary with Zimbabwe, to 2,301 meters in the Mafinga Hills. The tropical climate is moderated by the rainy season between October and April. Zambia is subject to periodic droughts, and tropical storms are possible throughout much of the rainy season.

Zambia's population of 11,502,010 is subject to major environmental health hazards. With one of the highest HIV/AIDS adult prevalence rates in the world (16.5 percent), 920,000 Zambians are living with the disease. Approximately 89,000 people have died with HIV/AIDS since 2003. Only 55 percent

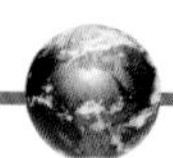

of the population has access to safe drinking water, and only 45 percent has access to improved sanitation. Consequently, Zambians have a very high risk of contracting food and waterborne diseases such as bacterial diarrhea, hepatitis A, typhoid fever, and the water contact disease schistosomiasis. In some areas, Zambians are at high risk for contracting malaria and plague. As a result of high incidences of disease, Zambians have a lower-than-normal life expectancy (40.03 years) and growth rate (2.11 percent), and higher-than-normal infant mortality (86.84 deaths per 1,000 live births) and death rates (19.93 deaths/1,000 population). The high fertility rate of 5.39 children per female presents health hazards for Zambian women.

Centuries of mineral extraction and refining have led to acid rain produced by air pollution. Despite the abundance of water sources, inadequate treatment facilities lead to major health threats to Zambians. Desertification is widespread. Around 42 percent of land area is forested, but deforestation is occurring at a rate of 2.4 percent per year. Soil erosion is extensive as a result of agricultural mismanagement.

The government has protected almost one-third of the land area, including 19 national parks and 31 game management areas. Nevertheless, watersheds contaminated by chemical runoff threaten rhinoceros, elephant, antelope, and large cat populations. The number of endangered species among the 233 mammals that inhabit Zambia is not known, but 11 of the 252 bird species are threatened. A 2006 study by scientists at Yale University ranked Zambia 98 of 132 countries on environmental performance, above the relevant income and geographic groups. The low ranking in environmental health, however, reduced Zambia's overall ranking.

The Zambian Parliament enacted the Environmental Protection and Pollution Control Act of 1990 to provide a framework for environmental policy. In 1992, the Environment Council of Zambia was created and charged with protecting the environment and health of the Zambian population and of animals and plants by controlling pollution and promoting sustainable development. Specific elements of Zambia's environmental policy deal with controlling air pollution, managing water resources, regulating the use of pesticides and toxic substances, and conservation of natural resources. The Copperbelt Environmental Project was established to deal with the impact of copper mining on the environment.

Zambia participates in the following international agreements on the environment: Biodiversity, Climate Change, Desertification, Endangered Species, Hazardous Wastes, Law of the Sea, Ozone Layer Protection, and Wetlands. The government has signed but not ratified the Climate Change–Kyoto Protocol.

SEE ALSO: Acquired Immune Deficiency Syndrome (AIDS); Desertification; National Parks.

BIBLIOGRAPHY. Central Intelligence Agency, "Zambia," *World Factbook,* www.cia.gov (cited April 2006); Timothy Doyle, *Environmental Movements in Minority and Majority Worlds: A Global Perspective* (Rutgers University Press, 2005); Environment Council of Zambia, www.necz.org.zm (cited April 2006); Kevin Hillstrom and Laurie Collier Hillstrom, *Africa and the Middle East: A Continental Overview of Environmental Issues* (ABC-CLIO, 2003); Valentine Udoh James, *Africa's Ecology: Sustaining the Biological and Environmental Diversity of a Continent* (McFarland, 1993); United Nations Development Programme, "Human Development Report: Zambia," hdr.undp.org (cited April 2006); World Bank, "Zambia," lnweb18.worldbank.org (cited April 2006); Yale University, "Pilot 2006 Environmental Performance Index," www.yale.edu/epi (cited April 2006).

Elizabeth Purdy, Ph.D.
Independent Scholar

Zebra Mussels

ZEBRA MUSSELS *(Dreissena polymorpha)* are fingernail-size bivalves indigenous to parts of eastern Europe and the Caspian Sea region of Eurasia. Similar to other mussels, they are planktonic for the first few weeks of life, floating in currents and able to colonize new areas. They then attach to a hard substrate and become sedentary. Females are prolific, producing from 30,000 to over a million eggs during one spawning event.

Zebra mussels spread to western Europe approximately 200 years ago, invading all major rivers via canals. They have since spread to North America, presumably in the ballast water of ocean-going vessels. The mussels were first noticed in 1988 in Lake St. Clair in the Great Lakes region and have since spread to many lakes and river systems in both Canada and the United States. Although they have predators, including several species of freshwater fish and diving ducks, these predators have been unable to stabilize population growth.

In Europe as in North America, the mussels have caused significant ecological and economic harm. Zebra mussels grow together in dense mats attached to a variety of living and nonliving substrates. These dense colonies clog water intake pipes of waterworks, power plants, and other industrial users of water, causing millions of dollars of damage and necessitating the application of chemical treatments or the reconfiguration of the piping of these plants.

Zebra mussels have proven to be a very successful species in their newly colonized territories. They have a formidable ability to filter water, consuming nearly all the available phytoplankton and small zooplankton. They out-compete other species that also feed on microscopic plankton, and in this way, change food web dynamics, impacting larval and juvenile fish as well as other filter feeders. Zebra mussels preferentially remove a variety of nutrients and chemicals from the water column. Phosphorus is removed and sequestered in their shells, changing the phosphorus cycle in aquatic ecosystems where the mussel is found.

North America was home to the greatest biodiversity of freshwater mussels. Many of these populations have declined in numbers or even been extirpated due to combinations of overharvest, pollution, and habitat destruction. Managers are concerned about the threat that zebra mussels pose to many of the remaining native mussel species, such as the endangered winged mapleleaf clam (*Quadrula fragosa*), found in the St. Croix River in the upper Mississippi watershed.

When the zebra mussel was first encountered in North America, there was no regulatory or legal framework in place to stop ballast water introductions of exotic species. In Canada, ballast water introductions are generally addressed through voluntary guidelines under the Canada Shipping Act. In the United States, however, prompted by the negative impacts of the zebra mussel and by concern over the potential for more new invading species, new legislation was passed. In 1990, the U.S. Congress passed the Nonindigenous Aquatic Nuisance Prevention and Control Act, which was reauthorized, amended, and renamed the National Invasive Species Act of 1996.

Despite these measures, new exotic species have continued to appear. Some scientists believe that the Quagga mussel (*D. bugensis*), a close relative of the zebra mussel, may pose a more serious threat to native species in the Great Lakes region of North America than the zebra mussel. The Quagga mussel is larger, does not go dormant in the winter, and has a wider habitat range in which it can live. It appears to compete directly with the zebra mussel, and since 2000, the Quagga has replaced the zebra mussel in many areas of Lake Michigan.

SEE ALSO: Food Webs (or Food Chain); Marine Pollution; Predator/Prey Relations; Species Invasion.

BIBLIOGRAPHY. Renata Claudi and Joseph H. Leach, eds., *Nonindigenous Freshwater Organisms: Vectors, Biology, and Impacts* (Lewis Publishers, 1999); Frank M. D'Itri, ed., *Zebra Mussels and Aquatic Nuisance Species* (Ann Arbor Press, 1997); Thomas F. Nalepa and Donald W. Schloesser, eds., *Zebra Mussels: Biology, Impacts, and Control* (Lewis Publishers, 1993).

Syma Alexi Ebbin
Yale University

Zero Population Growth (ZPG)

ZERO POPULATION GROWTH (ZPG) refers to a state in nature at which the birth rate is equivalent to the death rate, meaning that the population remains exactly at a specific level. There has been speculation about the appropriate level of population that may be sustainable given the restraints of existing economic and environmental resources. In a small state such as Singapore, where severe spatial restraints exist, it has been necessary to limit population growth

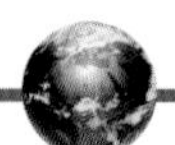

for the sake of maintaining a consistent standard of living. However, the measures necessary to achieve this are believed by many people to infringe civil liberties or religious tenets on birth control. In other countries where space is not so limited, innovations in productivity and technology have so far managed to ensure that some of the more pessimistic predictions of those who fear continued growth in population, notably the 18th-century economist Thomas Malthus and, more recently, Paul Ehrlich, have yet to come to pass.

Despite its often vaunted environmental advantages, demographic decline has led to several unintended social consequences. In some Western countries, the native born populations are in decline, leading to an increasing reliance on a foreign-born immigrant workforce. In China and India, state and family efforts to control growth have led to a disproportionate rate of abortions of female fetuses, resulting in a large overall imbalance in the ratio of boys to girls for an emerging generational cohort.

From 1968 until 2002, Zero Population Growth (ZPG) was also the name of the U.S. organization now called Population Connection. The organization was created to promote the link between environmental problems and growing population levels and how, consequently, a tipping point is likely to arrive after which the impoverishment of the world must follow. This effort brought ZPG into contact with the feminist movement and other progressive causes favoring sexual freedom through contraception. This has led many to consider the ZPG cause to be political and religious in nature. ZPG's message has changed from its original prescription to have no more than two children and to tell other people about that decision.

As an organization, Population Connection addresses a number of different population issues across a range of activities including research, promotion, lobbying, education, and publication. It relies upon the contributions of its subscribing members and also on the sale of merchandise. The organization changed its name in part because Zero Population Growth was seen as a term with negative and draconian connotations. So too, population control and resources conservation are complex and multifaceted and not necessarily achieved by uniformity of behavior. Instead, the organization advocates a sophisticated regime of incentives to encourage people to behave in ways that are economically and environmentally sustainable.

SEE ALSO: Ehrlich, Paul and Anne; Malthus, Thomas; One Child Policy, China; Overpopulation; Population.

BIBLIOGRAPHY. Paul R. Ehrlich, *The Population Bomb* (Buccaneer Books, 1995); Thomas R. Malthus, *An Essay on the Principle of Population*, ed. by Geoffrey Gilbert (Oxford University Press, 1999); Population Connection, www.populationconnection.org (cited July 2006).

John Walsh
Shinawatra University

Zimbabwe

ZIMBABWE COVERS OVER 390,000 square kilometers (about the size of Montana), and is bordered by South Africa on the south, Mozambique on the east, Botswana on the west, and Zambia on the north. Its environment varies from semi-arid regions to moist mountainous areas. It is home to most of the large African mammals, many of which are located in its 11 national parks. Victoria Falls, located along the Zambezi River and the border with Zambia, is one of these parks; during the rainy season, it contains one of the world's largest waterfalls.

Zimbabwe is home to over 12 million people and the population is made up primarily of two ethnic groups—Shona (about 82 percent) and the Ndebele (about 14 percent)—in addition to a small white minority. Zimbabwe was known as Rhodesia until it declared independence from Great Britain in 1980. The British South Africa Company, owned by Cecil Rhodes, colonized the area in the 1880s and the new colony was named in Rhodes's honor. The mission of the company was to exploit the natural resources of Rhodesia for Britain. From the 1880s until independence in 1980, a white-led government ruled Zimbabwe, and the racial segregation in place during that time has been compared to South Africa's apartheid system. Blacks were forbidden to own land outside of reserve areas, and as a result,

Great Zimbabwe

Although Victoria Falls remains the major tourist attraction in the country, the stone buildings of Great Zimbabwe are the most well-known manmade ruins in Africa south of the Sahara Desert. They are the most spectacular of the 150 or so walled remains in the country.

Adam Renders, an American sailor, came across the ruins in 1867, by which time they had been plundered for many years. The first foreigner who was able to write of his experiences was the German Carl Mauch who arrived in 1871 and saw the ruins in a more pristine state than they are in today.

It appears that the stone buildings and walls were the center of a great civilization that existed from the 11th until the 15th century, with a population of as many as 10,000–20,000. Trade was conducted, almost certainly through many intermediaries, with all parts of the world as artifacts from China and India have been found in the remains. Some early historians linked the stone walls to the lands of King Solomon or the Queen of Sheba, but there is no link with any Biblical story.

There were a number of large carved birds at the site; one South African hunter carried one back with him in 1889. It eventually ended up with Cecil Rhodes, who mounted it in the library of his home in Cape Town, subsequently the residence of the South African prime minister. A representation of the bird now appears on the Zimbabwean flag.

With no surviving written records, archaeologists, historians, and ethnologists have long debated the purpose of the stone structures and the walls. Some of the buildings were undoubtedly used to store grain, while others have been identified as the palaces of the king and his family and courtiers. The great mystery is why the civilization ended: Some suggest pestilence, or an environmental catastrophe that exhausted their food supplies.

the limited land in the reserves grew crowded. Until recently, the white minority owned over 70 percent of the arable land, which led to a land redistribution campaign as this system of unequal land access persisted through independence in 1980.

Robert Mugabe was elected the first prime minister in 1980, and in 1987, he declared himself president. When elected, he began a program of land reform based on the "willing seller/willing buyer" system. This continued until 1999 and at this time, he began to use force to remove white farmers from their land to redistribute it to black farmers, a system that was highly criticized by Western countries. Zimbabwe's agricultural base has not recovered since the redistribution and food aid has been needed to feed the country. Mugabe was reelected in 2002 in what most international observers considered a fixed election. Since then, the inflation rate has risen almost 600 percent, with the unemployment rate at 80 percent.

Before the collapse of the agricultural sector, Zimbabwe was known as the breadbasket of southern Africa. Agricultural products included wheat, corn, tobacco, and cotton. The country also contains large deposits of gold and chromite. Before 2000, tobacco—followed by cotton—accounted for the largest export earnings; with the collapse of the agricultural sector, however, gold and cotton now are bigger earners. Complicating Zimbabwe's food shortage and the chaos of the land reform policies is an extremely high rate of HIV/AIDS; it is estimated that about 25 percent of the adult population is HIV positive. The labor shortages due to illness have also contributed to a negative 5 percent gross domestic product growth rate.

SEE ALSO: Acquired Immune Deficiency Syndrome (AIDS); Colonialism; National Parks.

BIBLIOGRAPHY. E. Kalipeni, S. Craddock, J. Oppong, and J. Ghosh, eds., *HIV & AIDS in Africa: Beyond Epidemiology* (Blackwell Publishing, 2004); S. Moyo, "The Political Economy of Land Acquisition and Redistribution in Zimbabwe, 1990–1999," *Journal of Southern African Studies 2000* (v.26, 2000).

Kristina M. Bishop
University of Arizona

Zoology

ZOOLOGY IS AN area of biology that deals with the study of the animal kingdom (or Animalia). One purpose of zoology is to analyze and classify animals. Documents in the Hippocratic Collection state that the earliest attempts to classify animals originated in 400 B.C.E. Hippocrates arranged animals according to their habitat and mode of reproduction; in his *Historia Animalim*, he noted developmental stages of fish and aquatic animals, and studied sexual and asexual reproduction. During Roman times, Pliny the Elder compiled *Historia Naturalis,* a collection of folklore, superstition, and myth that was widely read in the Middle Ages. The Greek physician Galen, who dissected mammals and accurately described their internal features, produced another influential work.

Modern classification of animals began sometime in the 17th and 18th centuries. A system of nomenclature that we still use today—the binomial system of genus and species—was developed by Swedish botanist Carolus Linnaeus, who also established taxonomy as a discipline when he classified animals according to their teeth and toes, and classified birds according to the shape of their beaks. In the 17th century, English scientist Robert Hooke introduced the word *cell,* and after this, the field of embryology evolved. Scientific expeditions dedicated to studying animal life began in the 18th and 19th centuries.

Today, zoology has diversified as a field of science. With new technologies and discoveries, zoology has branched into subjects like biochemistry, genetics, and ecology. These fields deal with different areas of science and apply the acquired knowledge to study the animal kingdom. The main branches of zoology are taxonomy, physiology, and morphology.

Taxonomy deals with animal life based on different divisions. For example, the study of animals with backbones is classified under vertebrate zoology, which is further divided into herpetology, ichthyology, mammalogy, and ornithology. The study of animals that deals with multicellular animals without backbones is called invertebrate zoology, which is further divided into malacology and entomology. Taxonomic groups also subdivide paleontology, a field that deals with fossils. These subdivisions of zoology are aimed at studying the life cycle, distribution, classification, and evolutionary history of a group of animals of a particular animal. Most zoologists specialize in one of these fields and dedicate their research in that specific area.

Physiology deals with the function of body organs. If physiology deals with cellular functions, it is called cellular physiology and is closely connected to molecular biology. A study that deals with the physical connection of animals with their environment is called physiological ecology. This field aims to study animals in different environments like deserts, oceans, and the arctic.

Another study that examines full structures and systems, such as bones, brain, and muscles is called morphology, which generally includes gross morphology. The study that deals with body tissues at the microscopic level is called histology. Cytology examines cells and how cellular components function. With the advent of powerful machines like the electron microscope and the scanning tunneling electron microscope, cytology has made tremendous progress in the field of researching structural detail at levels of high magnification. Methods have been invented for studying neural networks inside the brain, and even studying individual impulses from specific brain areas and even individual brain cells.

The capstone field of zoology is called evolutionary zoology; it is connected to all of the above fields. Evolutionary zoology examines how an animal evolves—through speciation and adaptation—and what happens to the animal in the future as a result.

SEE ALSO: Animals; Ecology; Evolution; Genetics and Genetic Engineering; Linnaeus, Carl; Species.

BIBLIOGRAPHY. Cleveland P. Hickman, Jr., Larry S. Roberts, Allan Larson, and Helen I'Anson, *Integrated Principles of Zoology* (McGraw-Hill Science/Engineering/Math, 2005); Stephen A. Miller and John P. Harley, *Zoology* (McGraw-Hill Science/Engineering/Math, 2004).

Rahul Gladwin
University of Health Sciences, Antigua

Zoos

THE HISTORICAL PREDECESSORS of modern zoos were primarily showcases of empire or poorly-maintained public spectacles, but zoos now promote themselves as institutions dedicated to conservation and education. Animal rights advocates and conservationists today debate the implications of zoos for threatened wildlife species and the ethics of animal captivity. Royal menageries in ancient China, Egypt, and Rome, and their counterparts in early modern Europe, Central America, and South Asia, symbolized the monarch's power to command an extensive and exotic empire. In the 18th and 19th centuries, menageries grew into larger zoological gardens associated with royal scientific societies. These institutions added a layer of scholarly legitimacy to the royal menageries, but they remained a part of the imperial project. With growing popular interest in science, zoos such as the London Zoological Gardens attracted an increasingly broad public, though some were open only to dues-paying members.

Beginning in the late 18th century, animal collections became more accessible to the public. Some impresarios operated traveling zoos, transporting animals from town to town for display in the public square, and later Phineas T. Barnum led the establishment of early circuses in large cities. Public institutions such as New York's Central Park and Philadelphia's Fairmount Park attempted to create more educational exhibits in the 1860s, and though they were extremely popular with urban residents, they lacked the funding to maintain animals safely or to rise above the status of humble entertainment.

Beginning with the National Zoological Park in Washington, D.C., in 1889, and the New York Zoological Park in the Bronx (later known as the Bronx Zoo and now the Wildlife Conservation Park) in 1898, zoos began to take on a conservation mission. Early wildlife advocates such as William Hornaday hoped to use these zoos as arks for disappearing species, most notably bison, from across the American continent, and as stages from which to deliver a conservation message to the public. With public and private funding, they were able to display more species and to construct more spacious, outdoor enclosures to simulate wild habitats and encourage "natural" animal behavior.

Much as in history, animal collections in existence today range from modest petting zoos to the private collections of wealthy and ostentatious individuals. The question remains whether they are sites of spectacle, science, power, conservation, or all of the above. Of the 1,700 animal exhibits in the United States, however, fewer than 200 are accredited by the American Zoo and Aquarium Association (AZA), a group that promotes zoos as a means to advance wildlife conservation and education. These include most major city zoos.

In the 1970s, two new challenges came together to force these zoos to consider their role in the conservation of wild species. One was the passage of the U.S. Endangered Species Act in 1973, which restricted zoos from taking members of listed species, and the other was the growing recognition that zoo collections were becoming inbred. Without taking in new animals, zoos could only expect to see the inbreeding problem worsen. In 1976, striving to become more progressive, AZA adopted a code of ethics governing the treatment and use of zoo animals.

In 1981, the organization began to coordinate Species Survival Plans (SSPs) among its members in order to produce healthier captive populations, rely more on captive breeding rather than capturing new free-ranging animals, and, ideally, benefit the genetic stock of threatened wild populations. American zoos have since integrated their efforts with those of zoo associations around the world, and have created an international database to optimize breeding arrangements. With increasing commitment to the genetic side of conservation, zoos are adding sperm bank to their list of cultural and environmental roles. Zoos now organize breeding exchanges, managing increased reproduction of threatened species with an eye toward introducing zoo-bred individuals back into wild populations. Captive breeding programs have brought a number of species back from the brink of extinction, for example the California condor. SSPs, however, cannot solve the biological difficulties of breeding for many species, and breeding may produce "surplus" offspring that raise another set of ethical issues.

Ethicists concerned about animal rights and conservation also question whether zoos' contributions are adequate for them to truly deserve the label of

conservation institutions, and to justify keeping animals captive. They ask, for example, whether zoos adequately integrate their own activities with conservation needs outside the zoo gates—that is, into what conditions will zoo-bred animals be introduced? If wild habitats are depleted or wild populations unhealthy, zoos' breeding efforts may be wasted. Besides, some argue that very few individuals are ever actually integrated into wild habitats. In the meantime, however, zoo advocates say that they at least maintain an ark of genetic diversity that will last until the time viable wild populations can be supported. They point out that so-called wild populations are actually highly manipulated in a world where reserves must be set aside and managed; therefore, they suggest that to regard zoos as artificial, and zoo animals as wrongly captive, is to overstate the "natural-ness" of the rest of the world.

Zoo advocates also contend that keeping those animals captive is justified because seeing animals in zoos and participating in zoo education programs will ultimately increase public sympathy for conservation causes and thereby benefit animals in the wild. Some say that this is the true contribution of zoos, given the limited effect of SSPs. The Wildlife Conservation Society (WCS), a nonprofit organization based in New York City, is a leading proponent of this view. WCS manages several zoos, including the Bronx's Wildlife Conservation Park, and promotes the use of zoos for education and conservation. WCS facilities deliver carefully designed educational programs to the public, and staff work to support wildlife reserves worldwide.

Meanwhile, ecocritics—scholars of the cultural studies of nature—argue that zoos may reinforce broad cultural conceptions that humans are separate from and superior to nature, rather than encourage concern for nature as zoo advocates claim. Specifically, seeing wild animals contained within enclosures, no matter how naturalistic, gives the impression that nature can be subsumed into an anthropocentric world view.

Frank Buck

Frank Buck (1884–1950) was born in Gainesville, Texas, where he became honorary ringmaster of the Gainesville Community Circus. In 1911, he travelled to South America in search of jaguars and other wild animals. During the 1920s, he operated in Southeast Asia, especially in Singapore, where he captured wild animals for zoos around the world. Buck wrote about his exploits in three books: *Bring 'Em Back Alive* (1930), *Wild Cargo* (1932), and *All in a Lifetime* (1941). He played roles in jungle adventure films such as *Wild Cargo* (based on his book), *Jungle Cavalcade, Jacare,* and *Killer of the Amazon*. Buck was also the subject of a British television series, *Bring 'Em Back Alive*, in the early 1980s. A Frank Buck Zoo remains in existence in Gainesville, Texas.

Frank Buck, his films, and the Gainesville Zoo itself were a part of a larger, deeply fictionalized, romantic, and domineering view of nature typical of their time. Their influence on the form and style of nature documentary cannot be underestimated, however. Later media, including Disney film features and television, from Mutual of Omaha's *Wild Kingdom* to *The Crocodile Hunter*, all drew on the style and popularity of Buck's previous work. Arguably, today's popular nature media, though informed by superior natural history and a less sensationalistic approach, still retain an element of Buck's style. The Discovery Channel and Animal Planet networks are heirs to Frank Buck's natural circus theatricality.

SEE ALSO: Animals; Animal Rights; Anthropocentrism; Conservation.

BIBLIOGRAPHY. Vicki Croke, *The Modern Ark* (Scribner, 1997); R.J. Hoage and William A. Deiss, *New Worlds, New Animals* (Johns Hopkins, 1996); Randy Malamud, *Reading Zoos* (New York University Press, 1998); Bryan G. Norton, Michael Hutchins, Elizabeth F. Stevens, and Terry L. Maple, *Ethics on the Ark* (Smithsonian, 1995); Yi-Fu Tuan, *Dominance and Affection* (Yale, 1984); World Zoo Organization, *The World Zoo Conservation Strategy* (World Conservation Union, 1993).

Dawn Day Biehler
University of Wisconsin, Madison

Resource Guide

BOOKS

Aaron, Sachs. *Eco-Justice: Linking Human Rights and the Environment*. World Watch Paper No. 127. (World Watch Institute, 1995)

Angel, Bradley. *Toxic Threat to Indian Lands: A Greenpeace Report* (Greenpeace, 1991)

Barrow, C.J. *Environmental Management and Development* (Routledge, 2005)

Beacham, Walton, Castronova, Frank V. and Freedman, Bill (eds.). *Beacham's Guide to International Endangered Species* (Beacham Publishing Corporation, 2000)

Beasley, Conger. *Confronting Environmental Racism: Voices from the Grassroots* (South End Press, 1993)

Berry, Brian L., et. al. *The Social Burdens of Environmental Pollution: Comparative Metropolitan Data Source* (Ballinger Publishing Co., 1997)

Bewers, Michael J. *Analysis of Questionnaire Responses* (Global Environment Facility, International Waters Program Study, 2000)

Boserup, Ester. *The Conditions of Agricultural Growth: The Economics of Agrarian Change under Population Pressure* (George. Allen and Unwin, 1965)

Botkin, Daniel *Discordant Harmonies: A New Ecology for the Twenty-first Century* (Oxford University Press, 1992)

Bowles, Ian A. and Pickett, Glenn T. *Reframing the Green Window: An Analysis of the GEF Pilot Phase Approach to Biodiversity and Global Warming and Recommendations for the Operational Phase* (Conservation International and Natural Resource Defense Council, 1994)

Brady, N.C. and Weil, R.R. *The Nature and Properties of soils* (Prentice Hall, 2002)

Brown, Phil and Mikelson, Edwin J. *No Safe Place: Toxic Waste, Leukemia, and Community Action* (University of California Press, 1990)

Bryant, Bunyan (ed.). *Environmental Justice: Issues, Policies, and Solutions* (Island Press, 1995)

Bryant, Bunyan and Mohai, Paul (eds.). *Race and the Incidence of Environmental Hazards: A Time for Discourse* (Westview Press, 1992)

Bryant, Dirk, Nielson, Daniel, and Tangley, Laura. *The Last Frontier Forests: Ecosystems and Economies on the Edge* (World Resources Institute, 1997)

Bullard, Robert D. *Unequal Protection: Environmental Justice and Communities of Color* (Sierra Club Press, 1993)

Byron, William J. *Toward Stewardship: An Interim Ethic of Poverty, Power, and Pollution* (Paulist Press, 1995)

California State Legislature, Senate, Urban Growth Policy Project. *Prosperity, Equity, and Environmental Quality: Meeting the Challenge*

of California's Growth: Final Findings and Recommendations (Senate Publications, 1991)
Castro, Fidel. *Tomorrow Is Too Late: Development And the Environmental Crisis in the Third World* (Ocean Press, 1993)
Chatterjee, Pratap and Finger, Mathias. *The Earth Brokers: Power, Politics and World Development* (Routledge, 1994).
Churchill, Ward. *Struggle for the Land: Indigenous Resistance to Genocide, Ecocide, and Expropriation in Contemporary North America* (Common Courage Press, 1993)
Coggin, Terrance P. and Seidl, John M. *Politics American Style: Race, Environment and Central Cities.* (Prentice-Hall, 1972)
Colorado Journal of International Environmental Law and Politics. *Endangered Peoples: Indigenous Rights and the Environment* (University Press of Colorado, 1994)
Commission for Racial Justice, United Church of Christ. *Toxic Wastes and Race in the United States: A National Report on the Racial and Socioeconomic Characteristics of Communities with Hazardous Wastes Sites* (Public Data Access Inc., 1987)
Conca, Ken and Dabelko, Geoffrey (eds.). *Environmental Peacemaking* (Woodrow Wilson Center Press, 2002)
Crosby, Alfred. *Ecological Imperialism: The Biological Expansion of Europe, 900-1900, New edition* (Cambridge University Press, 2004)
Dana, Alson (ed.). *We Speak for Ourselves: Social Justice, Race and Environment* (Panos Institute, 1990)
Davies, Ben. *Black Market: Inside the Endangered Species Trade in Asia* (Ten Speed Press, 2005)
Diehl, Paul and Petter, Nils Gleditsch (eds.). *Environmental Conflict* (Westview Press, 2001)
Dryzek, John *The Politics of the Earth: Environmental Discourses* (Oxford University Press, 1997)
Durett, Dan. *Environmental Justice: Breaking New Ground* (Committee of the National Institute for the Environment, 1993)
Ecologist Magazine. *Whose Common Future: Reclaiming the Commons* (New Society Publishers, 1993)
Edelstein, Michael R. *Contaminated Communities: The Social and Psychological Impacts of Residential Toxic Exposure* (Westview Press, 1987)
Eichstaedt, Peter H. *If You Poison Us: Uranium and Native Americans* (Red Crane Books Company, 1994)
Fitton, Laura. *A Study of the Correlation between the Siting of Hazardous Waste Facilities and Racial and Socioeconomic Characteristics* (Center for Policy Alternatives, 1992)
Frankel, Otto H. and Soule, Michael. *Conservation and Evolution* (Cambridge University Press, 1981)
Gadgil, Madhav and Guha, Ramachandra. *Ecology and Equity* (Routledge Press, 1995)
Gadgil, Madhav and Guha, Ramachandra. *This Fissured Land: An Ecological History of India* (Oxford University Press, 1992)
Gedicks, Al. *The New Resource Wars: Native and Environmental Struggles Against Multinational Corporations* (South End Press, 1993)
Gerrard, Michael B. *Whose Backyard, Whose Risk: Fear and Fairness in Toxic and Nuclear Waste Siting* (MIT Press, 1994)
Gillroy, J.M., (ed.). *Environmental Risk, Environmental Values, and Political Choices: Beyond Efficiency Trade-offs in Public Policy Analysis* (Westview Press, 1993)
Global Environment Facility. *Achieving the Millennium Development Goals: A GEF Progress Report* (Global Environment Facility, 2005)
Global Land Project (GLP). *Science Plan and Implementation Strategy* (IGBP, 2005)
Goldman, Benjamin A. *Not Just Prosperity: Achieving Sustainability With Environmental Justice* (National Wildlife Federation, Corporate Conservation Council, 1993)
Gould, J. M. *Quality of Life in American Neighborhoods: Levels of Affluence, Toxic Waste, and Cancer Mortality in Residential Zip Code Areas* (Westview Press, 1986)
Griffiths, T. and Robin, L. (eds.). *Ecology and Empire* (University of Washington Press, 1997)
Grinde, Donald A. and Johansen, Bruce E. *Ecocide of Native America: Environmental Destruction of Indian Lands and People* (Clear Light Publishers, 1995)
Guha, Ramachandra and Martinez-Alier, Juan. *Varieties of Environmentalism: Essays North and South* (Earthscan Publications, 1997)
Guha, Ramachandra. *Nature's Spokesman – M. Krishnan and Indian Wildlife* (Oxford University Press, 2002)

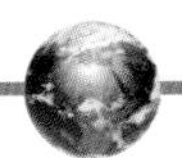

Guruswamy, Lakshman D. and McNeely, Jeffrey A. (eds.). *Protection of Global biodiversity: Converging Strategies* (Duke University Press, 1998)
Hall, Annabel T. *The Egg: An Eco-Justice Quarterly by the National Council of Churches Eco-Justice Working Group* (Cornell University, 1993)
Harrison, D., Jr. *Who Pays for Clean Air: The Cost and Benefit Distribution of Automobile Emission Standards* (Ballinger, 1975)
Held, David, McGrew, Anthony ,Goldblatt, David, and Perraton, Jonathan *Global Transformations: Politics, Economics and Culture* (Stanford University Press, 1999)
Hellawell, John M. *Biological Indicators of Freshwater Pollution and Environmental Management* (Elsevier Applied Science Publishers, 1986)
Hill, David (ed.) *The Quality of Life in America; Pollution, Poverty, Power, and Fear* (Holt, Rinehart and Winston, 1973)
Hofrichter, Richard (ed.). *Toxic Struggles: The Theory and Practice of Environmental Justice* (New Society Publishers, 1993)
Horta, Korinna, Round, Robin and Young, Zoe. *The Global Environment Facility: The First Ten Years – Growing Pains or Inherent Flaws?* (Environmental Defense and Halifax Initiative, 2002)
Houston, Stuart, Ball, Tim and Houston, Mary. *Eighteenth-Century Naturalists of Hudson Bay* (McGill-Queen's University Press, 2003)
Hurley, Andrew. *Environmental Inequalities: Class, Race, and Industrial Pollution in Gary, Indiana, 1945-1980* (University of North Carolina Press, 1995)
Jenny, H., *Factors of Soil Formation: A System of Quantitative Pedology* (McGraw-Hill, 1941)
Jim, Schwab. *Deeper Shades of Green: The Rise of Blue-Collar and Minority Environmentalism in America* (Sierra Club Press, 1994)
Johnston, Barbara R (ed.). *Who Pays the Price? The Sociocultural Context of Environmental Crisis* (Island Press, 1994)
Karliner, Joshua. *Toxic Empire: The Wmx Corporation, Hazardous Waste, and Global Strategies For Environmental Justice* (Political Ecology Group, 1994)
Kasperson, Roger E.(ed.). *Equity in Radioactive Waste Management.* (Oegleschlager, Gunn, and Hain, 1983)
Kroll-Smith, Steven, Brown, Phil and Gunter, Valerie. *Illness and the Environment: A Reader in Contested Medicine* (NYU, 2000)
LaBalme, J. *A Road to Walk, A Struggle for Environmental Justice* (Regulator Press, 1987)
Leopold, Aldo. *A Sand County Almanac and Sketches Here and There* (Oxford University Press, 1949)
Levy, Barry and Sidel, Victor (eds.) *War and Public Health* (American Public Health Association, 2000)
Lotka, Alfred J. *Elements of Mathematical Biology.* (Dover Publications, Inc., 1956).
Mandelker, D.R. *Environment and Equity: A Regulatory Challenge* (McGraw Hill, 1981)
Mann, Eric. *L.A.'s Lethal Air: New Strategies for Policy, Organization and Action.* (Labor/Community Strategy Center, 1991)
Manwaring, Max (ed.). *Environmental Security and Global Stability* (Lexington Books, 2002)
Marsh, George. *Man and Nature* (Harvard University Press,1965)
Meffe, G.K. and Carroll, C.R. *Principles of Conservation Biology* (Sunderland, 1994)
Merchant, Carolyn, (ed.). *Major Problems in American Environmental History, 2nd edition* (Houghton Mifflin Co., 2005)
Merchant, Carolyn. *Radical Ecology: The Search for a Livable World* (Routledge, 1992)
Meyer, Art. *Earth Keepers: Environmental Perspectives on Hunger, Poverty, and Injustice* (Herald Press, 1991)
Millennium Ecosystem Assessment. *Ecosystems and Human Well-being: Synthesis* (Island Press, 2005)
Neuman, Roderick. *Making Political Ecology* (Hodder Arnold, 2003)
Norris, R. *Pills, Pesticides and Profits: International Trade in Toxic Substances* (North River Press, 1982)
Pease, W. and Morello-Frosch, R. *The Distribution of Air Toxics Emissions in the San Franicsco Bay Area: A Preliminary Environmental Justice Analysis* (UC Berkeley, School of Public Health, Environmental Health Program, 1995)
Perrin, Constance. *Everything in its Place: Social Order and Land Use in America.* Princeton University Press, 1992)
Pielou, Evelyn C. *Mathematical Ecology* (John Wiley & Sons, 1977)

Pinchot, G. *Breaking New Ground* (Harcourt, Brace and Company, 1947)
Portney, Kent E. *Siting Hazardous Waste Treatment Facilities: The Nimby Syndrome* (Auburn House, 1991)
Primack, Richard B. *A Primer of Conservation Biology* (Sinauer Associates, Inc., 2004)
Purvis, M. and Grainger, A. (eds.). *Exploring Sustainable Development: Geographical Perspectives* (Earthscan Publications, 2004)
Regan, Richard. *Environmental Racism: An Annotated Bibliography* (Center for Policy Alternatives, 1993)
Ricklefs, Robert E. and Miller, Gary L. *Ecology* (W.H. Freeman & Co., 2000)
Russell, E. *War and Nature* (Cambridge University Press, 2001)
Shapiro, Judith. *Mao's War Against Nature* (Cambridge University Press, 2001)
Smith, J.N., (ed.). *Environmental Quality and Social Justice in Urban America: An Exploration of Conflict and Concord Among Those Who Seek Environmental Quality and Those Who Seek Social Justice* (Conservation Foundation, 1974)
Snow, Donald. *Inside the Environmental Movement: Meeting the Leadership Challenge* (Island Press, 1992)
Sontheimer, Sally (ed.). *Women and the Environment: Crisis and Development in the Third World.* (Monthly Review Press, 1991)
Southwest Organizing Project. *Intel Inside New Mexico: A Case Study of Environmental and Economic Injustice* (Southwest Organizing Project, 1995)
Sparke, Mathew *Introduction to Globalization: The Ties That Bind* (Blackwell, 2006)
Speth, James. *Red Sky at Morning* (Yale University Press, 2004)
Szasz, Andrew. *Ecopopulism: Toxic Waste and the Movement for Environmental Justice* (University of Minnesota Press, 1994)
The World Bank. *Global Environment Facility: Independent Evaluation of the Pilot Phase* (Washington, DC: The World Bank, 1994).
Turner, B.L II, Clark, W., Kates, R.W., Richards, J.F., Mathews, J.T., and Meyer, W.B. (eds.). *The Earth As Transformed By Human Action: Global and Regional Changes in the Biosphere Over the Past 300 Years* (Cambridge University Press, 2003)
United Nations Development Programme, United Nations Environment Programme and United States General Accounting Office. *Siting of Hazardous Waste Landfills and Their Correlation with Racial and Economic Status of Surrounding Communities* (GPO, 1983)
Vallette, J. *The International Trade in Wastes: A Greenpeace Inventory* (Greenpeace, 1989)
VanDeveer, Stacy and Dabelko, Geoffrey D. (eds.) *Protecting Regional Seas: Developing Capacity and Fostering Environmental Cooperating in Europe* (Woodrow Wilson International Center for Scholars, 1999)
Wasserstrom, R.F. and R. Wiles, R. *Field Duty: US Farm Workers and Pesticide Safety* (World Resources Institute, 1985)
Wells, Michael P. *The Global Environment Facility and Prospects for Biodiversity Conservation* (Global Environment Facility, 2005)
Wenz, Peter S. *Environmental Justice* (State University of New York Press, 1988)
Wernette, D. *Minorities, Age Groups, and Substandard Air Quality Areas: An Analysis of Regional and Urban Patterns* (Environmental Assessment and Information Sciences Division Argonne National Laboratory, 1992)
World Commission on Environment and Development (WCED). *Our Common Future* (Oxford University Press, 1987)
Worster, D. *Nature's Economy: A History of Ecological Ideas. 2nd Edition.*
Young, M.D. *Sustainable Investment and Resource Use: Equity, Environmental Integrity, and Economic Efficiency* (Parthenon Publishing Group, 1992)

PERIODICALS

Agroforestry Systems
Association for Temperate Agroforestry
Alternatives
Alternatives Inc.
American Naturalist
Thomson Corporation
Amicus Journal
National Resources Defense Council
Annual Review of Anthropology
Annual Reviews
Annual Review of Ecology and Systematics
Annual Reviews
Audubon

National Audubon Society
Biodiversity and Conservation
Chapman and Hall
Biological Conservation
Elsevier Science
BioScience
American Institute and Biological Sciences
Boston College Environmental Affairs Law Review Boston College
Capitalism, Nature, Socialism
Routledge
Chemical & Engineering News
American Chemical Society
Chemical Week
Chemical Week Associates
Chemistry & Industry
Society of Chemical Industry
Christian Science Monitor
Church of Christ, Scientist
Communications in Soil Science and Plant Analysis Taylor and Francis Group
Comparative Farming Systems
Guilford Press
Conservation Biology
Blackwell Publishing
Critical Reviews in Environmental Science and Technology
Taylor and Francis Group
Crossroads
ASERI Students Association
Demography
Population Association of America
Earth Island Journal
Earth Island Institute
Ecological Economics
International Ecological Economics
Ecology Law Quarterly
University of California, Berkeley
Energy and Environment
Multi-Science Publishing Co. Inc.
Environment
Voyage Publications
Environment and Behavior
SAGE Publications
Environmental Action
American Chemical Society
Environmental Ethics
Center for Environmental Philosophy
Environmental Law
Oxford University Press
Environmental Management
Academic Press
Environmental Politics
Frank Cass
Environmental Science and Technology
Center for Environment and Energy Research
EPA Journal
Environmental Planning Agency
Global Environment Politics
MIT Press
Global Environmental Change
Royal Society of Chemistry
Global Environmental Politics
MIT Press
Harvard International Law Journal
Harvard University
Hazardous Waste
BPI News
Health and Social Policy
Haworth Press
Human and Ecological Risk Assessment
Taylor and Francis Group
Human Ecology
Springer Science and Business Media
International Journal of Health Service
Baywood Publishing
International Journal of Sustainable Development & World Ecology
Taylor and Francis
Journal of Agricultural & Food Information
Extension Journal, Inc.
Journal of Environment and Development
SAGE Publications
Journal of Environmental Economics and Management
Academic Press Inc.
Journal of Environmental Education
Heldref Education
Journal of Environmental Health
Taylor and Francis Group
Journal of Environmental Management
Academic Presss
Journal of Forestry
Oxford University Press
Journal of Geochemical Exploration
Elsevier Science
Journal of Policy Analysis and Management
Wiley InterScience
Journal of Public Health Policy
Palgrave Macmillin
Journal of Risk: Issues in Health and Safety
Franklin Pierce Law Center

Journal of Social Issues
Society for the Psychological Study of Social Issues
Journal of the American Medical Association
American Medical Association
Land Economics
University of Wisconsin Press
Meanjin
Meanjin Company Ltd.
Multinational Monitor
Essential Information Inc.
National Wildlife
National Wildlife Federation
Natural Resources Forum
Blackwell Publishing
Natural Resources Journal
University of New Mexico
Nature
Palgrave Macmillan
New England Journal of Medicine
Massachusetts Medical Society
New Scientist
Reed Business Information Ltd.
Planning
Oxford University Press
Policy Studies Journal
Blackwell Publishing
Polity
Palgrave Macmillan Ltd.
Population and Environment
Center for Environment and Population
Progress in Human Geography
SAGE Publications
Progressive
Progressive Inc.
Quarterly Review of Biology
University of Chicago Press
Sierra
The Sierra Club
Society and Natural Resources
Routlage
Soil Science
Wiley InterScience
The Ecologist
Ecosystems Ltd.
The Professional Geographer
Association of American Geographers
Toxicology and Industrial Health
SAGE Publications
Trends in Ecology and Evolution
Oxford University Press
Utne Reader
Utne Publishing
Waste Age
Prism Business Media Inc.
Water Resources
Springer Science and Business Media
Whole Earth Review
Point Foundation

Chronology

400,000–350,000 B.C.E.: Human beings master the art of lighting and tending fires.

12,000–10,000 B.C.E.: The first Agricultural Revolution occurs in the Fertile Crescent region of the Middle East.

500 B.C.E.–500 C.E.: A significant increase in demand for luxury items accompanies the rise of the Roman Empire.

Ivory, a hard dentine substance notable for its beauty and durability, becomes a valuable commodity. The Romans use it to create such items as musical instruments, statues, furniture, floor coverings, chariots and bird cages.

889–904 C.E.: The first instance of the use of aquaculture, the method of cultivating aquatic organisms in underwater environments, occurs in China, where innovators breed carp fish in flooded rice fields.

1273 C.E.: Shortly after taking power, King Edward I of England engages in one of the first acts of environmental protection in history when he bans the use of coal fires throughout the whole of England.

1347–1350: The Black Death, a disease caused by the transfer of the bacteria *Bacillus pestis* from rats to humans, devastates Europe. Over one-third of its population is left dead.

1690: Colonial Governor William Penn of Pennsylvania forces laborers to engage in one of the first acts of forest conservation when he requires that one acre of forest be preserved for every acre that is cut down.

1789: German chemist Martin Heinrich Klaproth discovers uranium, an element later used in the development of the atomic bomb.

1804: Meriwether Lewis and William Clark embark on their famous expedition to the Pacific Coast and back.

1841: Construction is completed on the 38-mile long Croton River Aqueduct in southeastern New York State. Its opening is attended by U.S. President John Tyler along with three former Presidents.

1849: Following the Mexican-American War, which ended with significant territorial acquisitions by the U.S., the ability to manage land becomes a top national priority, and the Department of Interior is established. The department's duties mainly revolve around the preservation of federal land. The American bison, which at the time was being threatened

with extinction due to excessive hunting, is prominently featured on the logo of the department.

1850: The first use of septic systems, or underground waste treatment centers, occurs in France.

1853: Scientists Alexander Wood and Charles Pravaz invent the hypodermic needle, a device later used to administer preventative vaccinations to such diseases as rabies and polio.

1858: Central Park is constructed in New York City. Designers Frederick Olmstead and Calvert Vaux attempt to conceal the visual din of the surrounding urban ugliness by making the park rich with lush and vibrant greenery.

1862: French scientist Louis Pasteur invents pasteurization, the process of heating food to kill harmful bacteria and viruses.

1862: U.S. President Abraham Lincoln creates the Department of Agriculture, a noncabinet department, whose purpose is to promote increased agriculture production and protect natural resources.

1872: Yellowstone, an area located within the states of Wyoming, Montana, and Idaho, is officially designated by the U.S. federal government as the world's first national park.

1872: Prospectors in the newly settled American West are given free reign to engage in mineral exploration as part of the passage of the U.S. General Mining Law.

1877: In one of the first examples of a government-led effort to reduce pest infestation, Great Britain passes the Destructive Insects Act, providing funds to rid England of the Colorado potato beetle.

1879: The United States Geological Survey is established as a subdepartment of the Department of the Interior. The agency is responsible for "classification of the public lands, and examination of the geological structure, mineral resources, and products of the national domain."

1880: Construction is completed in the Australian city of Warwick on the world's first concrete-arch dam, called the 75-miles Dam.

1883: Concerns that widespread hunting of exotic birds will ultimately lead to their extinction prompts the founding of the American Ornithologist's Union.

1889: U.S. President Grover Cleveland signs into law a bill promoting the head of the Department of Agriculture to a cabinet-level position.

1891: The U.S. Congress passes the Forest Reserve Act, granting the President the authority to designate certain public domain lands as national forest reserves.

1892: Environmental preservationist John Muir founds the Sierra Club, an organization that would increase its membership to 750,000 people by 2006.

1896: The state of Connecticut files suit against Edward M. Greer, who had legally obtained animals in one state but attempted to sell them in another state where it would be considered illegal to do so. The Supreme Court rules that animal rights fall under the jurisdiction of individual states, rather than private interests. The Court's ruling would be later be overturned in the 1979 case of *Hughes v. Oklahoma.*

1897: Congress passes the Forest Organic Act for the purpose of, among other things, to "furnish a continuous supply of timber for the use and necessities of citizens of the United States." It grants the President the authority to select public domain lands as forest preserves. Conservationists argue that forests should be preserved for their natural beauty, and not for timber supply.

1902: President Theodore Roosevelt signs into law the Reclamation Act, providing funds for the irrigation of arid land on the West Coast of the United States.

1902: Various Audubon Societies across the United States combine forces to form the National Audubon Society, which would be instrumental in the coming years in the passage of legislation to protect various bird species from becoming extinct.

1905: The U.S. National Forest Service is created. Over time, the service would be responsible with

overseeing the management of over 150 national forests.

1906: Congress passes the U.S. Antiquities Act, granting the President the authority to designate certain land owned by the federal government as "national monuments" and therefore closed to excavation. Among the notable geological phenomena that were originally designated as "national monuments" is the Grand Canyon.

1910: A massive fire in the northern Rockies of the United States destroys over 12,000 square kilometers of forest.

1914–1918: During World War I, one-fifth of the world's population becomes infected with influenza, resulting in an estimated death toll of 35 million people.

The country most affected by the outbreak of influenza was the United States, which experienced a decrease in its life expectancy rate by 10 years. The term The Great Influenza Pandemic has been coined to describe this period.

1916: The U.S. Congress passes a bill creating the National Park Service, a federal agency that would be responsible for managing all of the U.S. state parks. By 2006, the agency would have control of over 84 million acres of land.

1929–1970: Venezuela exports more oil than any other nation.

1930: General Motors and DuPont introduce Freon, a refrigerant containing the synthetic chemical chlorofluorocarbon which was later found to be harmful to the upper ozone layer.

1931: The Yellow River, the second largest river in China, experiences heavy flooding, resulting in a massive death toll estimated to be between 850,000 and 4,000,000 people.

1931–1936: The Hoover Dam is constructed at a total cost of $49 million on the border between Arizona and Nevada.
The second largest damn ever constructed in the United States, the Hoover Dam would provide over 2,000 megawatts of electricity to millions of people living in the Western United States.

1932: Vincent J. Schaefer and Irving Langmuir are awarded the 1932 Nobel Prize in Chemistry for their pioneering work in manipulating clouds to avert rainfall.

Their method involved supercooling clouds so that they are unable to bond together and produce rain.

1935: As the Dust Bowl reaches its midway point, hundreds of thousands of farmers are given relief when the U.S. Soil Conservation Service is created. Through partnerships with local agencies, the service instructs farmers on the proper way to utilize their farmland without repeating the same mistakes that led to the Dust Bowl's beginnings.

1936: A severe drought in the Chinese province of Sichuan displaces over 30 million farmers and causes the deaths of an estimated 5 million people.

1939: The U.S. Fish and Wildlife Service is created. By 2006, the service's duties would grow to include the management of more than 520 National Wildlife Refuges and over 66 National Fish Hatcheries.

1946: The International Convention for the Regulation of Whaling is held in Washington, D.C. with representatives from 42 nations present. As a result of the conference, the official definition of "whalecatcher" was broadened to include helicopters and ships.

1946: The U.S. Congress passes the Atomic Energy Act, creating the Atomic Energy Commission, the purpose of which is to regulate and monitor the development of nuclear technology. The act states that regulation and monitoring shall be placed under civilian, rather than military, control.

1948: Swiss chemist Paul Herman Muller is awarded the Nobel Prize for his discovery that the chemical DDT (dichlorodiphenyl-trichloroethane) can be used as an insecticide. DDT would help significantly lower the prevalence of malaria in Europe and North America.

1953: Mountaineers Tenzing Norgay and Edmund Hillary became the first people to successfully reach the summit of Mount Everest, which at a height of nearly 30,000 feet, is the highest mountain in the world.

1956: Construction begins on the Glen Canyon Dam in the U.S. state of Arizona. Upon completion, it would become the second highest concrete-arch dam in the country.

1958: The U.S. Congress passes the Food, Drugs, and Cosmetic Act. The act contains a clause called the Delaney Amendment which states that, "the Secretary [of the Food and Drug Administration] shall not approve for use in food any chemical additive found to induce cancer in man, or, after tests, found to induce cancer in animals."

1960: Jane Goodall travels to the Gombe Stream Chimpanzee Reserve in the African country of Tanzinia, beginning her 45-year study of the complex social behavior of the animal from which humans descended.

1962: Cesar Chavez initiates the merger of the National Farm Workers Association and the Agricultural Workers Organization to form the United Farm Workers.

The organization would use such nonviolent methods as fasts, emulating the methods used by Mahatma Gandhi and Martin Luther King, Jr.

1962: Environmental advocate Rachel Carson publishes *Silent Spring*, in which she details the careless handling of hazardous chemicals by major industrial companies. The book facilitated the growth of the modern environmentalist movement.

1962–1971: During the Vietnam War, the U.S. Air Force sprays large quantities of the dioxin chemical "Agent Orange" onto areas of South Vietnam. Later studies concluded that many American veterans of the war were contaminated with the chemical and experienced debilitating effects to their health years later.

To counter claims of negligence, the Dow Chemical company releases a report entitled *Trace Chemistries of Fire*, in which it is alleged that Agent Orange, among other dioxins, is derived naturally from forest fires. The theory is almost immediately debunked by credible experts.

1964: The U.S. Congress passes the Land and Water Conservation Act, providing funds for government-supervised construction of public parks and recreation areas.

1964: U.S. President Lyndon Johnson signs into law the Wilderness Act, creating the National Wilderness Preservation System. The act designates over 9 million acres of land in the U.S. as closed to excavation, as well as officially defining wilderness as "an area where the earth and its community of life are untrammeled by man, where man himself is a visitor who does not remain."

1968: U.S. President Lyndon Johnson signs into law the National Wild & Scenic Rivers Act, enabling the federal government to monitor and correct the destruction of various rivers.

1968: The chemical PCB is found in hundreds of rice paddies in Japan. Over 17,000 cases of PCB infection are reported.

1969: Following significant oil spills in the North Sea, the countries of Belgium, Denmark, France, Germany, Norway, Sweden, and the United Kingdom sign the Bonn Agreement, pledging to offer mutual assistance to fight environmental pollution.

1969: The first oil spill in the United States to gain national attention occurs off the coast of California. Over 200,000 gallons of oil spill from an offshore well.

1970: The League of Conservation Voters, one of the largest environmental lobbying groups in the U.S., is founded.

1970: After touring a region devastated by an oil spill, U.S. Senator Gaylord Nelson of Wisconsin is stricken with a crisis of conscience and proposes on the Senate floor a bill to establish Earth Day as a holiday on April 22 of each year.

1970: The government of West Germany issues the Emergency Program for Environmental Protection.

1970: The National Environmental Policy Act is signed into law by U.S. President Richard Nixon. The preamble of the Act expresses the need "to declare a national policy which will encourage productive and enjoyable harmony between man and his environment."

1970: The Clean Air Act is passed by Congress in an attempt to protect the public from harmful air-

borne contaminants such as sulfur dioxide. The act would be amended in 1977 and again in 1990 to include more stringent regulations.

1970: U.S. President Richard Nixon signs into law a bill creating the Environmental Protection Agency (EPA). The agency provides a federally coordinated effort to enforce environmental regulations.

1970: The most deadly tropical cyclone in history with winds of 120 mph hits East Pakistan. Officials place the death toll at 500,000, attributing the high number of dead to the fact that the cyclone made landfall at a time when most residents in East Pakistan were sleeping.

1970: The National Oceanic and Atmospheric Administration is created for the purpose of obtaining "a better understanding of the intelligent use of the United States' marine resources."

1971: Egyptian and Sudanese laborers complete construction on the Aswan High Dam. Built to prevent overflooding of the Nile River, the dam is 3,600 meters in length and nearly 1000 meters wide.

1972: A week-long blizzard in Iran results in the deaths of 4,000 people, making it the most deadly blizzard of the 20th century.

1972: Congress passes the Ocean Dumping Act, requiring companies to be granted licenses by the Environmental Protection Agency in order to dump industrial, medical, and radioactive waste into U.S. territorial waters. The Act would be amended in 1988 as part of the passage of the Ocean Dumping Act, which outright banned the dumping of the aforementioned substances. An incident where large amounts of medical waste washed up on the shores of New Jersey prompted the Act's revision.

1972: The state legislature of Florida passes a series of laws to protect the Everglades, marshlands located in southeastern Florida that had been threatened due to population growth.

1972: The Federal Insecticide, Fungicide, and Rodenticide Act (FIFRA) is passed by Congress, requiring chemical manufacturers to submit their unlicensed products to the Environmental Protection Agency in order to ascertain its proper usage and any potential safety hazards.

1972: The U.S. Congress bans the usage of the chemical DDT (dichlorodiphenyl-trichloroethane) within the United States after it is revealed that the chemical has a high potential to spread cancer infection.

Despite the banning, the chemical continues to provide revenue for its manufacturers when it begins to be marketed and sold to Third World countries.

1973: The U.S. Congress signs the Endangered Species Act into law. The act equips federal officials with greater tools to combat the potential extinction of certain species of animal.

1973: Growing international concern for the plight of endangered species prompts the creation of the Convention on the International Trade in Endangered Species. The Convention offers funds to protect from extinction some 33,000 species of animals and plants.

1974: Congress passes the Eastern Wilderness Act, enabling the preservation of wilderness areas in the densely populated and industrially polluted Eastern United States.

1974: In her book *Le Feminisme ou la Mort* (*Feminism or Death*), French author Francoise d'Eaubonne posits that the societal instinct to destroy nature for industrial purposes is primarily patriarchal in nature. She encapsulates her philosophy by coining the term ecofeminism.

1974: Through the passage of the Safe Drinking Water Act, the U.S. Environmental Protection Agency is given the authority to enforce safety standards for drinking water at any level of government, whether it be state, local, or federal.

1975: Novelist Edward Abbey publishes *The Monkey Wrench*. The title refers to acts of nonviolent civil disobedience designed to protect the environment, such as tree spiking and billboard graffiti.

1976: The U.S. Congress passes the Toxic Substances Control Act, heavily regulating the usage of the organic compound polychlorinated biphenyl.

1977: Professor Wangari Maathai founds the Green Belt Movement, an organization dedicated to providing environmentalist jobs such as tree planting to poor women in rural Africa.

1977: In response to a sharp increase in the price of oil initiated by the Organization of the Petroleum Exporting Countries, The U.S. Department of Energy is established, bringing together dozens of energy-related organizations and agencies.

1978: The supertanker *Amoco Cadiz* splits in two after accidentally running aground off the coast of France into the Portsall Rocks, spilling 68 million gallons of oil into the surrounding waters. Damages are estimated to be equal to the value of $250,000,000. The French government, however, requests a payment of $2,000,000,000 from the American Oil Company, the owner of the doomed vessel.

1979: A "loss-of-coolant accident" occurs at the Three Mile Island nuclear plant near Harrisburg, Pennsylvania. The accident is later determined to be caused by faulty instrument readings.

1979–1989: The emergence of Japan as an economic superpower following the post-World War II reconstruction period creates a large upper class of Japanese citizens who flaunt their wealth with ivory-made items imported from Africa. As a result of this an other factors, the African elephant population drops nearly in half, from 1,300,000 to 750,000.

1980–2004: The international fish market grows from $15.7 billion to $71 billion.

1980: The U.S. Congress passes the Comprehensive Environmental Resource and Liability Act (CERCLA), giving government the power to identify hazardous waste sites and provide funding to facilitate the cleanup of the sites.

1981: The first confirmed case of the AIDS virus in human beings is reported by the U.S. Center for Disease Control and Prevention. Homosexual men in the city of Los Angeles are found to have unusually large "clusters of pneunomia."
Health officials at first erroneously assume that the disease has something to do with the men's sexual orientation, and name the disease GRID, or Gay-Related Immune Deficiency.

1981–2000: 40 million people are infected with Acquired Immune Deficiency Syndrome (AIDS). Nearly 50 percent of those affected with the virus are left dead.

1983: In an attempt to bridge a compromise between members of the tropical timber industry and conservationists, 54 nations pledge to sign the International Tropical Timber Agreement. Subsequent revisions to the agreement would be made in 1994, 1997, and in 2006, when the total number of countries adhering to the agreement's terms expanded to 180.

1984: The United Nations Educational, Scientific and Cultural Organization designates Yosemite National Park as a World Heritage Site, one of only 10 U.S. national parks to receive such an honor.

1986: 135,000 people are forced to permanently evacuate their homes as a result of a nuclear reactor explosion at the Chernobyl power plant in the Ukraine.

While not causing a significant number of deaths, radioactive material from the explosion continued to drift across the European continent in the coming years.

1986: The Slow Food Movement, an organization dedicated to stopping the spread of culturally damaging fast food restaurants such as McDonald's, is founded in the northern part of Italy. The organization's membership would grow to include 83,000 people by 2006.

1987: The Montreal Protocol becomes open for signature as concerns over the depletion of the ozone layer reach a fevered pitch. The 189 nations that eventually sign the treaty agree to phase out the usage of certain types of substances that are believed to cause ozone depletion.

1988: Members of the World Meteorological Organization and the United Nations Environmental Program combine their respective areas of expertise to form the Intergovernmental Panel on Climate Change (IPCC), for the purpose of studying the causes and potential outcomes of global warming.

1989: The *Exxon-Valdez* oil tanker crashes in Prince William Sound off the coast of Alaska, spilling over 11 million gallons of crude oil. Along with damaging the habitats of countless organisms, the oil spill causes considerable economic damage with estimates ranging up to $43,000,000. The U.S. Congress passes the Oil Pollution Act, requiring oil tankers to meet stringent regulations, in the wake of the incident.

1990: The Human Genome Project, a highly ambitious attempt to identify the purpose of the 20-25,000 genes contained in the human body, is formally launched at a cost of $3,000,000,000 by the U.S. Department of Energy and the U.S. National Institute of Health. Geneticists from Japan, Germany, China, and France make significant contributions to the project.

1990-2006: Following its reunification, Germany invests heavily in wind energy technology through the passage of such legislation as the Renewable Energy Sources Act. It would become the world leader in wind energy power, accounting for nearly half (39 percent) of the world's consumption.

1990: The Intergovernmental Panel on Climate Change (IPCC) concludes after two years of research that global warming poses a serious threat to the environment and that it is likely due to pollution caused by humans.

1991: The U.S. Food and Drug Administration estimates that 36 percent of all food imported to the U.S. is contaminated with harmful amounts of pesticide residue.

1992: The United Nations Conference on Environment and Development is held in the Brazilian city of Rio de Janeiro, with representatives from 172 governments present.

Among the topics discussed at the conference is the feasibility of converting to alternative sources of energy, and the growing problem of worldwide water shortages.

1993: Construction begins on the Three Gorges Dam alongside the Chinese Yangtze River, a river historically prone to dangerous flooding. At 600 feet high and 1.5 miles long, it is the largest hydroelectric dam ever designed.

1994: Mexico, Canada, and the United States enter into the North American Free Trade Agreement, creating the largest free trade area in the world. Environmental advocates complain that the industrial expansion that will result from increased trade will severely harm the environment. In response, the Commission for Environmental Cooperation and the Border Environment Cooperation Commission are created. Both commissions are given operating funds exceeding $1,000,000,000.

1995: A catastrophic heat wave in the U.S. city of Chicago and its surrounding metropolitan areas leaves over 500 people dead.

1995: Members of the Aum Shinrikyo, a Japanese religious cult, release a large quantity of the poisonous gas sarin on 5 separate railway trains in the capital of Tokyo, resulting in a dozen deaths and hundreds of injuries.

1996: In an out-of-court settlement, Pacific Gas & Electric Co. is forced to pay $333 million to the townspeople of Hinkly, California, who had sued the energy giant for contaminating their drinking water with a harmful toxin named hexavalent chromium.

It was the largest settlement ever paid in a direct-action lawsuit in U.S. history. The film *Erin Brockovich,* named after a law clerk who was instrumental in filing the case against PG&E, received 5 Academy Award nominations.

1997: The Kyoto Protocol to the United Nations Framework Convention on Climate Change becomes open for signature. The protocol requires countries to pledge to reduce greenhouse gas emissions by a certain percentage in order to gain membership. The protocol would grow in membership to include 166 countries by 2006.

1999: A startling discovery that 500 tons of animal feed accidentally laced with the carcinogenic chemical PCB were distributed to farms in Belgium prompts farmers to wastefully slaughter a total of over 2 million chickens.

1999: The West Nile Virus, a "flu-like" disease that is contracted through the bites of infected mosquitoes, first appears in the United States in the New York City borough of The Bronx.

1999-2003: The U.S. Center for Disease Control and Prevention reports nearly 15,000 cases of human West Nile Virus infection.

2001: A mere few weeks after the September 11 attacks, letters contaminated with the deadly poison anthrax arrive in the offices of U.S. Senators Tom Daschle and Patrick Leahy, as well as several major news organizations. 17 people are infected, resulting in five deaths.

2004: A 9.0 magnitude earthquake erupts underneath the Indian Ocean, causing a massive tsunami that devastates parts of Indonesia, Thailand, Burma, India, and Sri Lanka. The total death toll is reported to be 200,000 people. A combination of nations including Australia, which pledged to provide funds equal to 25 percent of its gross domestic product, offer a total aid package of $7,000,000,000.

2004: Halliburton Energy Services settles a class-action asbestos lawsuit out of court for a total of $4,200,000,000.

2004: Novelist Michael publishes *State of Fear,* imagining a scenario where ecoterrorists stage a tsunami with the support of media companies in order to bring about climate change legislation.

2005: The Joint United Nations Program on HIV/AIDS estimates that two-thirds of those affected by the AIDS virus live in Sub-Saharan Africa.

2005: The category 3 Hurricane Katrina makes landfall in the southeastern United States. Damages are estimated at $80,000,000,000. 80 percent of the city of New Orleans becomes flooded, leaving a significant portion of the city's population unable to fend for themselves. Television cameras capture images of those waiting to be rescued on the streets of New Orleans, shockingly revealing that most of them are African-American. Many Americans begin to question once again the validity of racial equality.

2005: Hazardous environmental conditions caused by more than a decade of warfare place Afghanistan at the top of the list of the countries with the highest infant mortality rate. According to statistics, approximately 17.5 percent of Afghanis die shortly after being born.

2005: The U.S. Environmental Protection Agency concludes that the herbicide 2,4-D does not pose a risk to human health when it is used in the method described in the product's instruction manual.

2006: Former Vice President Al Gore's film *An Inconvenient Truth*, in which he discusses the effect of global warming, becomes the third highest grossing documentary film of all time.

2006: In its annual report of countries' environmental performance, Yale University ranks New Zealand as the most environmentally friendly nation out of 133 ranked countries. The United States is ranked 28th, nearly the lowest among industrialized nations. Ranking last is the African country of Niger, which also has among the world's lowest GDPs.

2006: Scientists announce the possibility of the avian bird flu mutating into a much more deadly version of itself due to hazardous environmental conditions.

Compiled by Kevin G. Golson
Golson Books, Ltd.

Glossary

Abatement: Reducing the degree or intensity of, or eliminating, pollution.

Absorbed Dose: In exposure assessment, the amount of a substance that penetrates an exposed organism's absorption barriers (e.g. skin, lung tissue, gastrointestinal tract) through physical or biological processes. The term is synonymous with internal dose.

Absorption: The uptake of water, other fluids, or dissolved chemicals by a cell or an organism (as tree roots absorb dissolved nutrients in soil.)

Accident Site: The location of an unexpected occurrence, failure or loss, either at a plant or along a transportation route, resulting in a release of hazardous materials.

Acclimatization: The physiological and behavioral adjustments of an organism to changes in its environment.

Acid Deposition: A complex chemical and atmospheric phenomenon that occurs when emissions of sulfur and nitrogen compounds and other substances are transformed by chemical processes in the atmosphere, often far from the original sources, and then deposited on earth in either wet or dry form. The wet forms, popularly called "acid rain," can fall to earth as rain, snow, or fog. The dry forms are acidic gases or particulates.

Acid Mine Drainage: Drainage of water from areas that have been mined for coal or other mineral ores. The water has a low pH because of its contact with sulfur-bearing material and is harmful to aquatic organisms.

Acid Rain: (See: acid deposition.)

Acidic: The condition of water or soil that contains a sufficient amount of acid substances to lower the pH below 7.0.

Activated Carbon: A highly adsorbent form of carbon used to remove odors and toxic substances from liquid or gaseous emissions. In waste treatment, it is used to remove dissolved organic matter from waste drinking water. It is also used in motor vehicle evaporative control systems.

Active Ingredient: In any pesticide product, the component that kills, or otherwise controls, target pests. Pesticides are regulated primarily on the basis of active ingredients.

Acute Exposure: A single exposure to a toxic substance which may result in severe biological harm

or death. Acute exposures are usually characterized as lasting no longer than a day, as compared to longer, continuing exposure over a period of time.

Acute Toxicity: The ability of a substance to cause severe biological harm or death soon after a single exposure or dose. Also, any poisonous effect resulting from a single short-term exposure to a toxic substance.

Adaptation: Changes in an organism's physiological structure or function or habits that allow it to survive in new surroundings.

Administered Dose: In exposure assessment, the amount of a substance given to a test subject (human or animal) to determine dose-response relationships. Since exposure to chemicals is usually inadvertent, this quantity is often called potential dose.

Adsorption: Removal of a pollutant from air or water by collecting the pollutant on the surface of a solid material; e.g., an advanced method of treating waste in which activated carbon removes organic matter from wastewater.

Adulterants: Chemical impurities or substances that by law do not belong in a food, or pesticide.

Advanced Wastewater Treatment: Any treatment of sewage that goes beyond the secondary or biological water treatment stage and includes the removal of nutrients such as phosphorus and nitrogen and a high percentage of suspended solids.

Advisory: A nonregulatory document that communicates risk information to those who may have to make risk management decisions.

Aeration: A process which promotes biological degradation of organic matter in water. The process may be passive (as when waste is exposed to air), or active (as when a mixing or bubbling device introduces the air).

Aerobic: Life or processes that require, or are not destroyed by, the presence of oxygen.

Aerosol: A finely divided material suspended in air or other gaseous environment.

Afforestation: Conversion of land to forest cover where forests have not historically occurred.

Agent: Any physical, chemical, or biological entity that can be harmful to an organism (synonymous with stressors.)

Agricultural Pollution: Farming wastes, including runoff and leaching of pesticides and fertilizers; erosion and dust from plowing; improper disposal of animal manure and carcasses; crop residues, and debris.

Agricultural Waste: Poultry and livestock manure, and residual materials in liquid or solid form generated from the production and marketing of poultry, livestock or fur-bearing animals; also includes grain, vegetable, and fruit harvest residue.

Agroecosystem: Land used for crops, pasture, and livestock; the adjacent uncultivated land that supports other vegetation and wildlife; and the associated atmosphere, the underlying soils, groundwater, and drainage networks.

Air Pollutant: Any substance in air that could, in high enough concentration, harm man, other animals, vegetation, or material. Pollutants may include almost any natural or artificial composition of airborne matter capable of being airborne. They may be in the form of solid particles, liquid droplets, gases, or in combination thereof.
Generally, they fall into two main groups: (1) those emitted directly from identifiable sources and (2) those produced in the air by interaction between two or more primary pollutants, or by reaction with normal atmospheric constituents, with or without photoactivation. Exclusive of pollen, fog, and dust, which are of natural origin, about 100 contaminants have been identified. Air pollutants are often grouped in categories for ease in classification; some of he categories are: solids, sulfur compounds, volatile organic chemicals, particulate matter, nitrogen compounds, oxygen compounds, halogen compounds, radioactive compound, and odors.

Air Pollution Control Device: Mechanism or equipment that cleans emissions generated by a source (e.g., industrial smokestack, or an automobile exhaust system) by removing pollutants that would otherwise be released to the atmosphere.

Air Pollution: The presence of contaminants or pollutant substances in the air that interfere with human health or welfare, or produce other harmful environmental effects.

Air Quality Standards: The level of pollutants prescribed by regulations that are not be exceeded during a given time in a defined area.

Airborne Particulates: Total suspended particulate matter found in the atmosphere as solid particles or liquid droplets. Chemical composition of particulates varies widely, depending on location and time of year. Sources of airborne particulates include: dust, emissions from industrial processes, combustion products from the burning of wood and coal, combustion products associated with motor vehicle or non-road engine exhausts, and reactions to gases in the atmosphere.

Algae: Simple rootless plants that grow in sunlit waters in proportion to the amount of available nutrients. They can affect water quality adversely by lowering the dissolved oxygen in the water. They are food for fish and small aquatic animals.

Algal Blooms: Sudden spurts of algal growth, which can affect water quality adversely and indicate potentially hazardous changes in water chemistry.

Alkaline: The condition of water or soil which contains a sufficient amount of alkali substance to raise the pH above 7.0.

Alkalinity: The capacity of bases to neutralize acids. An example is lime added to lakes to decrease acidity.

Allergen: A substance that causes an allergic reaction in individuals sensitive to it.

Alternative Fuels: Substitutes for traditional liquid, oil-derived motor vehicle fuels like gasoline and diesel. Includes mixtures of alcohol-based fuels with gasoline, methanol, ethanol, compressed natural gas, and others.

Anaerobic Decomposition: Reduction of the net energy level and change in chemical composition of organic matter caused by microorganisms in an oxygen-free environment.

Anaerobic: A life or process that occurs in, or is not destroyed by, the absence of oxygen.

Animal Dander: Tiny scales of animal skin, a common indoor air pollutant.

Animal Studies: Investigations using animals as surrogates for humans with the expectation that the results are pertinent to humans.

Antarctic "Ozone Hole": Refers to the seasonal depletion of ozone in the upper atmosphere above a large area of Antarctica.

Anti-Degradation Clause: Part of federal air quality and water quality requirements prohibiting deterioration where pollution levels are above the legal limit.

Anti-Microbial: An agent that kills microbes.

Aquifer: An underground geological formation, or group of formations, containing water. Are sources of groundwater for wells and springs.

Aquitard: Geological formation that may contain groundwater but is not capable of transmitting significant quantities of it under normal hydraulic gradients. May function as confining bed.

Architectural Coatings: Coverings such as paint and roof tar that are used on exteriors of buildings.

Area Source: Any source of air pollution that is released over a relatively small area but which cannot be classified as a point source. Such sources may include vehicles and other small engines, small businesses and household activities, or biogenic sources such as a forest that releases hydrocarbons.

Artesian (Aquifer or Well): Water held under pressure in porous rock or soil confined by impermeable geological formations.

Asbestos: A mineral fiber that can pollute air or water and cause cancer or asbestosis when inhaled, and is restricted.

Asbestosis: A disease associated with inhalation of asbestos fibers. The disease makes breathing progressively more difficult and can be fatal.

Attenuation: The process by which a compound is reduced in concentration over time, through absorption, adsorption, degradation, dilution, and/or transformation. an also be the decrease with distance of sight caused by attenuation of light by particulate pollution.

Avoided Cost: The cost a utility would incur to generate the next increment of electric capacity using its own resources; many landfill gas projects' buy back rates are based on avoided costs.

Backflow/Back Siphonage: A reverse flow condition created by a difference in water pressures that causes water to flow back into the distribution pipes of a drinking water supply from any source other than the intended one.

Background Level: 1. The concentration of a substance in an environmental media (air, water, or soil) that occurs naturally or is not the result of human activities. 2. In exposure assessment the concentration of a substance in a defined control area, during a fixed period of time before, during, or after a data-gathering operation.

Backyard Composting: Diversion of organic food waste and yard trimmings from the municipal waste stream by composting hem in one's yard through controlled decomposition of organic matter by bacteria and fungi into a humus-like product. It is considered source reduction, not recycling, because the composted materials never enter the waste stream.

Bacteria: (Singular: bacterium) Microscopic living organisms that can aid in pollution control by metabolizing organic matter in sewage, oil spills or other pollutants. However, bacteria in soil, water or air can also cause human, animal and plant health problems.

Bactericide: A pesticide used to control or destroy bacteria, typically in the home, schools, or hospitals.

Basalt: Consistent year-round energy use of a facility; also refers to the minimum amount of electricity supplied continually to a facility.

Benefit-Cost Analysis: An economic method for assessing the benefits and costs of achieving alternative health-based standards at given levels of health protection.

Best Available Control Measures (BACM): A term used to refer to the most effective measures (according to EPA guidance) for controlling small or dispersed particulates and other emissions from sources such as roadway dust, soot and ash from woodstoves and open burning of rush, timber, grasslands, or trash.

Best Available Control Technology (BACT): The most stringent technology available for controlling emissions; major sources are required to use BACT, unless it can be demonstrated that it is not feasible for energy, environmental, or economic reasons.

Best Demonstrated Available Technology (BDAT): As identified by EPA, the most effective commercially available means of treating specific types of hazardous waste. The BDATs may change with advances in treatment technologies.

Best Management Practice (BMP): Methods that have been determined to be the most effective, practical means of preventing or reducing pollution from nonpoint sources.

Bioaccumulants: Substances that increase in concentration in living organisms as they take in contaminated air, water, or food because the substances are very slowly metabolized or excreted.

Bioconcentration: The accumulation of a chemical in tissues of a fish or other organism to levels greater than in the surrounding medium.

Biodegradable: Capable of decomposing under natural conditions.

Biodiversity: Refers to the variety and variability among living organisms and the ecological complexes in which they occur. Diversity can be defined as the number of different items and their relative frequencies. For biological diversity, these items are organized at many levels, ranging from complete ecosystems to the biochemical structures that are the molecular basis of heredity. Thus, the term encompasses different ecosystems, species, and genes.

Biological Contaminants: Living organisms or derivates (e.g. viruses, bacteria, fungi, and mammal and bird antigens) that can cause harmful health effects

when inhaled, swallowed, or otherwise taken into the body.

Biological Integrity: The ability to support and maintain balanced, integrated, functionality in the natural habitat of a given region. Concept is applied primarily in drinking water management.

Biological Magnification: Refers to the process whereby certain substances such as pesticides or heavy metals move up the food chain, work their way into rivers or lakes, and are eaten by aquatic organisms such as fish, which in turn are eaten by large birds, animals or humans. The substances become concentrated in tissues or internal organs as they move up the chain.

Biological Oxygen Demand (BOD): An indirect measure of the concentration of biologically degradable material present in organic wastes. It usually reflects the amount of oxygen consumed in five days by biological processes breaking down organic waste.

Biologicals: Vaccines, cultures, and other preparations made from living organisms and their products, intended for use in diagnosing, immunizing, or treating humans or animals, or in related research.

Biomass: All of the living material in a given area; often refers to vegetation.

Biome: Entire community of living organisms in a single major ecological area.

Biomonitoring: 1. The use of living organisms to test the suitability of effluents for discharge into receiving waters and to test the quality of such waters downstream from the discharge. 2. Analysis of blood, urine, tissues, etc., to measure chemical exposure in humans.

Bioremediation: Use of living organisms to clean up oil spills or remove other pollutants from soil, water, or wastewater; use of organisms such as non-harmful insects to remove agricultural pests or counteract diseases of trees, plants, and garden soil.

Biosphere: The portion of Earth and its atmosphere that can support life.

Biotechnology: Techniques that use living organisms or parts of organisms to produce a variety of products (from medicines to industrial enzymes) to improve plants or animals or to develop microorganisms to remove toxics from bodies of water, or act as pesticides.

Bloom: A proliferation of algae and/or higher aquatic plants in a body of water; often related to pollution, especially when pollutants accelerate growth.

Bottle Bill: Proposed or enacted legislation which requires a returnable deposit on beer or soda containers and provides for retail store or other redemption. Such legislation is designed to discourage use of throw-away containers.

British Thermal Unit: Unit of heat energy equal to the amount of heat required to raise the temperature of one pound of water by one degree Fahrenheit at sea level.

Brownfields: Abandoned, idled, or under used industrial and commercial facilities/sites where expansion or redevelopment is complicated by real or perceived environmental contamination. They can be in urban, suburban, or rural areas. EPA's Brownfields initiative helps communities mitigate potential health risks and restore the economic viability of such areas or properties.

Bubble: A system under which existing emissions sources can propose alternate means to comply with a set of emissions limitations; under the bubble concept, sources can control more than required at one emission point where control costs are relatively low in return for a comparable relaxation of controls at a second emission point where costs are higher.

Building Related Illness: Diagnosable illness whose cause and symptoms can be directly attributed to a specific pollutant source within a building (e.g., Legionnaire's disease, hypersensitivity, pneumonitis.)

Burial Ground (Graveyard): A disposal site for radioactive waste materials that uses earth or water as a shield.

By-product: Material, other than the principal product, generated as a consequence of an indus-

trial process or as a breakdown product in a living system.

Cadmium (Cd): A heavy metal that accumulates in the environment.

Cap: A layer of clay, or other impermeable material installed over the top of a closed landfill to prevent entry of rainwater and minimize leachate.

Carbon Monoxide (CO): A colorless, odorless, poisonous gas produced by incomplete fossil fuel combustion.

Carcinogen: Any substance that can cause or aggravate cancer.

Carrying Capacity: 1. In recreation management, the amount of use a recreation area can sustain without loss of quality. 2. In wildlife management, the maximum number of animals an area can support during a given period.

Catalyst: A substance that changes the speed or yield of a chemical reaction without being consumed or chemically changed by the chemical reaction.

Catalytic Converter: An air pollution abatement device that removes pollutants from motor vehicle exhaust, either by oxidizing them into carbon dioxide and water or reducing them to nitrogen.

Cells: 1. In solid waste disposal, holes where waste is dumped, compacted, and covered with layers of dirt on a daily basis. 2. The smallest structural part of living matter capable of functioning as an independent unit.

Channelization: Straightening and deepening streams so water will move faster, a marsh-drainage tactic that can interfere with waste assimilation capacity, disturb fish and wildlife habitats, and aggravate flooding.

Chemical Compound: A distinct and pure substance formed by the union or two or more elements in definite proportion by weight.

Chisel Plowing: Preparing croplands by using a special implement that avoids complete inversion of the soil as in conventional plowing. Chisel plowing can leave a protective cover or crops residues on the soil surface to help prevent erosion and improve filtration.

Chlorinated Hydrocarbons: 1. Chemicals containing only chlorine, carbon, and hydrogen. These include a class of persistent, broad-spectrum insecticides that linger in the environment and accumulate in the food chain. Among them are DDT, aldrin, dieldrin, heptachlor, chlordane, lindane, endrin, Mirex, hexachloride, and toxaphene. Other examples include TCE, used as an industrial solvent. 2. Any chlorinated organic compounds including chlorinated solvents such as dichloromethane, trichloromethylene, chloroform.

Chlorination: The application of chlorine to drinking water, sewage, or industrial waste to disinfect or to oxidize undesirable compounds.

Chlorofluorocarbons (CFCs): A family of inert, nontoxic, and easily liquefied chemicals used in refrigeration, air conditioning, packaging, insulation, or as solvents and aerosol propellants. Because CFCs are not destroyed in the lower atmosphere they drift into the upper atmosphere where their chlorine components destroy ozone.

Chronic Exposure: Multiple exposures occurring over an extended period of time or over a significant fraction of an animal's or human's lifetime (usually seven years to a lifetime.)

Circle of Poison: the import of crops from foreign countries that have been contaminated by pesticides produced by the importing country.

Clean Fuels: Blends or substitutes for gasoline fuels, including compressed natural gas, methanol, ethanol, and liquified petroleum gas.

Cleanup: Actions taken to deal with a release or threat of release of a hazardous substance that could affect humans and/or the environment. The term "cleanup" is sometimes used interchangeably with the terms remedial action, removal action, response action, or corrective action.

Clearcut: Harvesting all the trees in one area at one time, a practice that can encourage fast rainfall or snowmelt runoff, erosion, sedimentation of streams and lakes, and flooding.

Climate Change (also referred to as 'global climate change'): The term *climate change* is sometimes used to refer to all forms of climatic inconsistency, but because the Earth's climate is never static, the term is more properly used to imply a significant change from one climatic condition to another. In some cases, *climate change* has been used synonymously with the term, *global warming*; scientists however, tend to use the term in the wider sense to also include natural changes in climate.

Cloning: In biotechnology, obtaining a group of genetically identical cells from a single cell; making identical copies of a gene.

Coastal Zone: Lands and waters adjacent to the coast that exert an influence on the uses of the sea and its ecology, or whose uses and ecology are affected by the sea.

Cogeneration: The consecutive generation of useful thermal and electric energy from the same fuel source.

Columbian Exchange: The interchange of diseases, crop plants, livestock, cultural practices, and people between Eurasia/Africa and North/South America during the period after first contact in 1492.

Combined Sewer Overflows: Discharge of a mixture of storm water and domestic waste when the flow capacity of a sewer system is exceeded during rainstorms.

Commercial Waste: All solid waste emanating from business establishments such as stores, markets, office buildings, restaurants, shopping centers, and theaters.

Commodity Chain: a networked pathway along which a good travels from the site of raw material production, through processing and value-added, ultimately to the consumer as a finished product.

Common Property Resources: those resources owned and managed in common by a group or community, often managed through sophisticated institutions

Communism: a mode of social and economic organization in which communal ownership of productive capital is paramount and profit-seeking is central

Community: In ecology, an assemblage of populations of different species within a specified location in space and time. Sometimes, a particular subgrouping may be specified, such as the fish community in a lake or the soil arthropod community in a forest.

Comparative Risk Assessment: Process that generally uses the judgement of experts to predict effects and set priorities among a wide range of environmental problems.

Compost: The relatively stable humus material that is produced from a composting process in which bacteria in soil mixed with garbage and degradable trash break down the mixture into fertilizer.

Composting: The controlled biological decomposition of organic material in the presence of air to form a humus-like material. Controlled methods of composting include mechanical mixing and aerating, ventilating the materials by dropping them through a vertical series of aerated chambers, or placing the compost in piles out in the open air and mixing it or turning it periodically.

Compressed Natural Gas (CNG): An alternative fuel for motor vehicles; considered one of the cleanest because of low hydrocarbon emissions and its vapors are relatively nonozone producing. However, vehicles fueled with CNG do emit a significant quantity of nitrogen oxides.

Concentration: The relative amount of a substance mixed with another substance. An example is five ppm of carbon monoxide in air or 1 mg/l of iron in water.

Conservation Easement: Easement restricting a landowner to land uses that that are compatible with long-term conservation and environmental values.

Conservation: Preserving and renewing, when possible, human and natural resources. The use, protection, and improvement of natural resources according to principles that will ensure their highest economic or social benefits.

Consumptive Water Use: Water removed from available supplies without return to a water resources system, e.g., water used in manufacturing, agriculture, and food preparation.

Contact Pesticide: A chemical that kills pests when it touches them, instead of by ingestion. Also, soil that contains the minute skeletons of certain algae that scratch and dehydrate waxy-coated insects.

Contaminant: Any physical, chemical, biological, or radiological substance or matter that has an adverse effect on air, water, or soil.

Contamination: Introduction into water, air, and soil of microorganisms, chemicals, toxic substances, wastes, or wastewater in a concentration that makes the medium unfit for its next intended use. Also applies to surfaces of objects, buildings, and various household and agricultural use products.

Contour Plowing: Soil tilling method that follows the shape of the land to discourage erosion.

Conventional Tilling: Tillage operations considered standard for a specific location and crop and that tend to bury the crop residues; usually considered as a base for determining the cost effectiveness of control practices.

Cooling Tower: A structure that helps remove heat from water used as a coolant; e.g., in electric power generating plants.

Cost Recovery: A legal process by which potentially responsible parties who contributed to contamination at a Superfund site can be required to reimburse the Trust Fund for money spent during any cleanup actions by the federal government.

Cost Sharing: A publicly financed program through which society, as a beneficiary of environmental protection, shares part of the cost of pollution control with those who must actually install the controls. In Superfund, for example, the government may pay part of the cost of a cleanup action with those responsible for the pollution paying the major share.

Cost/Benefit Analysis: A quantitative evaluation of the costs which would have incurred by implementing an environmental regulation versus the overall benefits to society of the proposed action.

Cradle-to-Grave or Manifest System: A procedure in which hazardous materials are identified and followed as they are produced, treated, transported, and disposed of by a series of permanent, linkable, descriptive documents (e.g. manifests). Commonly referred to as the cradle-to-grave system.

Cross Contamination: The movement of underground contaminants from one level or area to another due to invasive subsurface activities.

Cryptosporidium: A protozoan microbe associated with the disease cryptosporidiosis in man. The disease can be transmitted through ingestion of drinking water, person-to-person contact, or other pathways, and can cause acute diarrhea, abdominal pain, vomiting, fever, and can be fatal.

Cultural Ecology: A field of research concerned with the interrelationships between humans and their surrounding environment, typically involving detailed small-scale analysis of communities and livelihoods

Cultures and Stocks: Infectious agents and associated biologicals including cultures from medical and pathological laboratories; cultures and stocks of infectious agents from research and industrial laboratories; waste from the production of biologicals; discarded live and attenuated vaccines; and culture dishes and devices used to transfer, inoculate, and mix cultures.

Cumulative Exposure: The sum of exposures of an organism to a pollutant over a period of time.

DDT: The first chlorinated hydrocarbon insecticide chemical name: Dichloro-Diphenyl-Trichloroethane. It has a half-life of 15 years and can collect in fatty tissues of certain animals. EPA banned registration and interstate sale of DDT for virtually all but emergency uses in the United States in 1972 because of its persistence in the environment and accumulation in the food chain.

Decomposition: The breakdown of matter by bacteria and fungi, changing the chemical makeup and physical appearance of materials.

Decontamination: Removal of harmful substances such as noxious chemicals, harmful bacteria or other organisms, or radioactive material from exposed individuals, rooms and furnishings in buildings, or the exterior environment.

Degree-Day: A rough measure used to estimate the amount of heating required in a given area; is defined as the difference between the mean daily temperature and 65 degrees Fahrenheit. Degree-days are also calculated to estimate cooling requirements.

Density: A measure of how heavy a specific volume of a solid, liquid, or gas is in comparison to water. depending on the chemical.

Dermal Toxicity: The ability of a pesticide or toxic chemical to poison people or animals by contact with the skin.

Desertification: The degradation of arid and semiarid lands, resulting in lower levels of productivity and desert-like conditions.

Designer Bugs: Popular term for microbes developed through biotechnology that can degrade specific toxic chemicals at their source in toxic waste dumps or in ground water.

Diffusion: The movement of suspended or dissolved particles (or molecules) from a more concentrated to a less concentrated area. The process tends to distribute the particles or molecules more uniformly.

Dioxin: Any of a family of compounds known chemically as dibenzo-p-dioxins. Concern about them arises from their potential toxicity as contaminants in commercial products. Tests on laboratory animals indicate that it is one of the more toxic anthropogenic (man-made) compounds.

Direct Discharger: A municipal or industrial facility which introduces pollution through a defined conveyance or system such as outlet pipes; a point source.

Disinfectant: A chemical or physical process that kills pathogenic organisms in water, air, or on surfaces. Chlorine is often used to disinfect sewage treatment effluent, water supplies, wells, and swimming pools.

Dispersant: A chemical agent used to break up concentrations of organic material such as spilled oil.

Disposal Facilities: Repositories for solid waste, including landfills and combustors intended for permanent containment or destruction of waste materials. Excludes transfer stations and composting facilities.

Drainage Basin: The area of land that drains water, sediment, and dissolved materials to a common outlet at some point along a stream channel.

Drainage Well: A well drilled to carry excess water off agricultural fields. Because they act as a funnel from the surface to the groundwater below. Drainage wells can contribute to groundwater pollution.

Drainage: Improving the productivity of agricultural land by removing excess water from the soil by such means as ditches or subsurface drainage tiles.

Drawdown: 1. The drop in the water table or level of water in the ground when water is being pumped from a well. 2. The amount of water used from a tank or reservoir. 3. The drop in the water level of a tank or reservoir.

Dredging: Removal of mud from the bottom of water bodies. This can disturb the ecosystem and causes silting that kills aquatic life. Dredging of contaminated muds can expose biota to heavy metals and other toxics. Dredging activities may be subject to regulation under Section 404 of the Clean Water Act.

Ecological Entity: In ecological risk assessment, a general term referring to a species, a group of species, an ecosystem function or characteristic, or a specific habitat or biome.

Ecological Impact: The effect that a man-caused or natural activity has on living organisms and their non-living (abiotic) environment.

Ecological Indicator: A characteristic of an ecosystem that is related to, or derived from, a measure of biotic or abiotic variable, that can provide quantitative information on ecological structure and function. An indicator can contribute to a measure of integrity and sustainability.

Ecological Integrity: A living system exhibits integrity if, when subjected to disturbance, it sustains and organizes self-correcting ability to recover toward a biomass end-state that is normal for that system. End-states other than the pristine or naturally whole may be accepted as normal and good.

Ecological Risk Assessment: The application of a formal framework, analytical process, or model to estimate the effects of human actions(s) on a natural resource and to interpret the significance of those effects in light of the uncertainties identified in each component of the assessment process. Such analysis includes initial hazard identification, exposure and dose-response assessments, and risk characterization.

Ecological/Environmental Sustainability: Maintenance of ecosystem components and functions for future generations.

Ecology: The relationship of living things to one another and their environment, or the study of such relationships.

Ecosystem Structure: Attributes related to the instantaneous physical state of an ecosystem; examples include species population density, species richness or evenness, and standing crop biomass.

Ecosystem: The interacting system of a biological community and its nonliving environmental surroundings.

Ecotone: A habitat created by the juxtaposition of distinctly different habitats; an edge habitat; or an ecological zone or boundary where two or more ecosystems meet.

Effluent: Wastewater—treated or untreated—that flows out of a treatment plant, sewer, or industrial outfall. Generally refers to wastes discharged into surface waters.

Emergency and Hazardous Chemical Inventory: An annual report by facilities having one or more extremely hazardous substances or hazardous chemicals above certain weight limits.

Emission Cap: A limit designed to prevent projected growth in emissions from existing and future stationary sources from eroding any mandated reductions. Generally, such provisions require that any emission growth from facilities under the restrictions be offset by equivalent reductions at other facilities under the same cap.

Emission: Pollution discharged into the atmosphere from smokestacks, other vents, and surface areas of commercial or industrial facilities; from residential chimneys; and from motor vehicle, locomotive, or aircraft exhausts.

Emissions Trading: The creation of surplus emission reductions at certain stacks, vents or similar emissions sources and the use of this surplus to meet or redefine pollution requirements applicable to other emissions sources. This allows one source to increase emissions when another source reduces them, maintaining an overall constant emission level. Facilities that reduce emissions substantially may "bank" their "credits" or sell them to other facilities or industries.

Endangered Species: Animals, birds, fish, plants, or other living organisms threatened with extinction by anthropogenic (man-caused) or other natural changes in their environment. Requirements for declaring a species endangered are contained in the Endangered Species Act.

End-of-the-pipe: Technologies such as scrubbers on smokestacks and catalytic convertors on automobile tailpipes that reduce emissions of pollutants after they have formed.

Engineered Controls: Method of managing environmental and health risks by placing a barrier between the contamination and the rest of the site, thus limiting exposure pathways.

Enlightenment: The historical philosophical movement in the 1700s and after emphasizing scientific methods, empiricism, and rationality.

Environmental Determinism: A largely discredited body of theory that explains human culture, history, and race solely with reference to climatic and topographical conditions

Environmental Equity/Justice: Equal protection from environmental hazards for individuals, groups,

or communities regardless of race, ethnicity, or economic status.
This applies to the development, implementation, and enforcement of environmental laws, regulations, and policies, and implies that no population of people should be forced to shoulder a disproportionate share of negative environmental impacts of pollution or environmental hazard due to a lack of political or economic strength levels.

Environmental Indicator: A measurement, statistic or value that provides a proximate gauge or evidence of the effects of environmental management programs or of the state or condition of the environment.

Environmental Justice: The fair treatment of people of all races, cultures, incomes, and educational levels with respect to the development and enforcement of environmental laws, regulations, and policies.

Environmental Sustainability: Long-term maintenance of ecosystem components and functions for future generations.

Environmental/Ecological Risk: The potential for adverse effects on living organisms associated with pollution of the environment by effluents, emissions, wastes, or accidental chemical releases; energy use; or the depletion of natural resources.

Epidemiology: Study of the distribution of disease, or other health-related states and events in human populations, as related to age, sex, occupation, ethnicity, and economic status in order to identify and alleviate health problems and promote health.

Equilibrium: In relation to radiation, the state at which the radioactivity of consecutive elements within a radioactive series is neither increasing nor decreasing.

Ethanol: An alternative automotive fuel derived from grain and corn; usually blended with gasoline to form gasohol.

Eutrophication: The slow aging process during which a lake, estuary, or bay evolves into a bog or marsh and eventually disappears. During the later stages of eutrophication the water body is choked by abundant plant life due to higher levels of nutritive compounds such as nitrogen and phosphorus. Human activities can accelerate the process.

Evapotranspiration: The loss of water from the soil both by evaporation and by transpiration from the plants growing in the soil.

Exotic Species: A species that is not indigenous to a region.

Exposure: The amount of radiation or pollutant present in a given environment that represents a potential health threat to living organisms.

Externality: The cost or benefit in an action, transaction, or exchange that is borne by parties external to that transaction.

Fecal Coliform Bacteria: Bacteria found in the intestinal tracts of mammals. Their presence in water or sludge is an indicator of pollution and possible contamination by pathogens.

Feng Shui: Geomantic principles for the arrangement of the built environment to assure good fortune and health.

Filtration: A treatment process, under the control of qualified operators, for removing solid (particulate) matter from water by means of porous media such as sand or a man-made filter; often used to remove particles that contain pathogens.

Floodplain: The flat or nearly flat land along a river or stream or in a tidal area that is covered by water during a flood.

Fluoridation: The addition of a chemical to increase the concentration of fluoride ions in drinking water to reduce the incidence of tooth decay.

Food Chain: A sequence of organisms, each of which uses the next, lower member of the sequence as a food source.

Food Web: The feeding relationships by which energy and nutrients are transferred from one species to another.

Fossil Fuel: Fuel derived from ancient organic remains; e.g. peat, coal, crude oil, and natural gas.

Free Trade Associations/Zones: Groups of nations or regions where tariff and quotas barriers are reduced or eliminated to spur increased economic activity.

Fresh Water: Water that generally contains less than 1,000 milligrams-per-liter of dissolved solids.

Fuel Efficiency: The proportion of energy released by fuel combustion that is converted into useful energy.

Fugitive Emissions: Emissions not caught by a capture system.

Fungicide: Pesticides which are used to control, deter, or destroy fungi.

Fungus (Fungi): Molds, mildews, yeasts, mushrooms, and puffballs, a group of organisms lacking in chlorophyll (i.e., are not photosynthetic) and which are usually non-mobile, filamentous, and multicellular. Some grow in soil, others attach themselves to decaying trees and other plants whence they obtain nutrients. Some are pathogens, others stabilize sewage and digest composted waste.

Genetic Engineering: A process of inserting new genetic information into existing cells in order to modify a specific organism for the purpose of changing one of its characteristics.

Geographic Information System (GIS): A computer system designed for storing, manipulating, analyzing, and displaying data in a geographic context.

Giardia Lamblia: Protozoan in the feces of humans and animals that can cause severe gastrointestinal ailments. It is a common contaminant of surface waters.

Global Warming: An increase in the near surface temperature of the Earth. Global warming has occurred in the distant past as the result of natural influences, but the term is most often used to refer to the warming predicted to occur as a result of increased emissions of greenhouse gases. Scientists generally agree that the Earth's surface has warmed by about 1 degree Fahrenheit in the past 140 years. The Intergovernmental Panel on Climate Change (IPCC) recently concluded that increased concentrations of greenhouse gases are causing an increase in the Earth's surface temperature and that increased concentrations of sulfate aerosols have led to relative cooling in some regions, generally over and downwind of heavily industrialized areas.

Gray Water: Domestic wastewater composed of wash water from kitchen, bathroom, and laundry sinks, tubs, and washers.

Greenhouse Effect: The warming of the Earth's atmosphere attributed to a buildup of carbon dioxide or other gases; some scientists think that this buildup allows the sun's rays to heat the Earth, while making the infrared radiation atmosphere opaque to infrared radiation, thereby preventing a counterbalancing loss of heat.

Greenhouse Gas: A gas, such as carbon dioxide or methane, which contributes to potential climate change.

Ground Water: The supply of fresh water found beneath the Earth's surface, usually in aquifers, which supply wells and springs. Because ground water is a major source of drinking water, there is growing concern over contamination from leaching agricultural or industrial pollutants or leaking underground storage tanks.

Ground-Penetrating Radar: A geophysical method that uses high frequency electromagnetic waves to obtain subsurface information.

Ground-Water Discharge: Ground water entering near coastal waters which has been contaminated by landfill leachate, deep well injection of hazardous wastes, septic tanks, etc.

Habitat: The place where a population (e.g., human, animal, plant, microorganism) lives and its surroundings, both living and non-living.

Half-Life: 1. The time required for a pollutant to lose one-half of its original coconcentrationor example, the biochemical half-life of DDT in the environment is 15 years. 2. The time required for half of the atoms of a radioactive element to undergo self-transmutation or decay (half-life of radium is 1620 years). 3. The time required for the elimination of half a total dose from the body.

Hazard Assessment: Evaluating the effects of a stressor or determining a margin of safety for an organism by comparing the concentration which causes toxic effects with an estimate of exposure to the organism.

Heat Island Effect: A "dome" of elevated temperatures over an urban area caused by structural and pavement heat fluxes, and pollutant emissions.

Herbicide: A chemical pesticide designed to control or destroy plants, weeds, or grasses.

High Seas: Portions of the ocean beyond the limits of national jurisdictions as defined by the Third United Nations Convention on the Law of the Sea.

High-Level Nuclear Waste Facility: Plant designed to handle disposal of used nuclear fuel, high-level radioactive waste, and plutonium waste.

High-Level Radioactive Waste (HLRW): Waste generated in core fuel of a nuclear reactor, found at nuclear reactors or by nuclear fuel reprocessing; is a serious threat to anyone who comes near the waste without shielding.

High-Risk Community: A community located within the vicinity of numerous sites of facilities or other potential sources of envienvironmental exposure/ health hazards which may result in high levels of exposure to contaminants or pollutants.

Host: 1. In genetics, the organism, typically a bacterium, into which a gene from another organism is transplanted. 2. In medicine, an animal infected or parasitized by another organism.

Hydrocarbons (HC): Chemical compounds that consist entirely of carbon and hydrogen.

Hydrogeology: The geology of ground water, with particular emphasis on the chemistry and movement of water.

Hypersensitivity Diseases: Diseases characterized by allergic responses to pollutants; diseases most clearly associated with indoor air quality are asthma, rhinitis, and pneumonic hypersensitivity.

Ignitable: Capable of burning or causing a fire.

Incineration: A treatment technology involving destruction of waste by controlled burning at high temperatures; e.g., burning sludge to remove the water and reduce the remaining residues to a safe, non-burnable ash that can be disposed of safely on land, in some waters, or in underground locations.

Incinerator: A furnace for burning waste under controlled conditions.

Indicator: In biology, any biological entity or processes, or community whose characteristics show the presence of specific environmental conditions. 2. In chemistry, a substance that shows a visible change, usually of color, at a desired point in a chemical reaction. 3. A device that indicates the result of a measurement; e.g. a pressure gauge or a moveable scale.

Indirect Discharge: Introduction of pollutants from a non-domestic source into a publicly owned waste-treatment system. Indirect dischargers can be commercial or industrial facilities whose wastes enter local sewers.

Indirect Source: Any facility or building, property, road or parking area that attracts motor vehicle traffic and, indirectly, causes pollution.

Indoor Air Pollution: Chemical, physical, or biological contaminants in indoor air.

Industrial Waste: Unwanted materials from an industrial operation; may be liquid, sludge, solid, or hazardous waste.

Infectious Agent: Any organism, such as a pathogenic virus, parasite, or or bacterium, that is capable of invading body tissues, multiplying, and causing disease.

Infiltration: 1. The penetration of water through the ground surface into sub-surface soil or the penetration of water from the soil into sewer or other pipes through defective joints, connections, or manhole walls. 2. The technique of applying large volumes of waste water to land to penetrate the surface and percolate through the underlying soil.

Inorganic Chemicals: Chemical substances of mineral origin, not of basically carbon structure.

Insecticide: A pesticide compound specifically used to kill or prevent the growth of insects.

Interstate Commerce Clause: A clause of the U.S. Constitution which reserves to the federal government the right to regulate the conduct of business across state lines. Under this clause, for example, the U.S. Supreme Court has ruled that states may not inequitably restrict the disposal of out-of-state wastes in their jurisdictions.

Interstate Waters: Waters that flow across or form part of state or international boundaries; e.g., the Great Lakes, the Mississippi River, or coastal waters.

Inversion: A layer of warm air that prevents the rise of cooling air and traps pollutants beneath it; can cause an air pollution episode.

Irradiated Food: Food subject to brief radioactivity, usually gamma rays, to kill insects, bacteria, and mold, and to permit storage without refrigeration.

Irradiation: Exposure to radiation of wavelengths shorter than those of visible light (gamma, x-ray, or ultra- violet), for medical purposes, to sterilize milk or other foodstuffs, or to induce polymerization of monomers or vulcanization of rubber.

Irrigation: Applying water or wastewater to land areas to supply the water and nutrient needs of plants.

Karst: A geologic formation of irregular limestone deposits with sinks, underground streams, and caverns.

Kinetic Energy: Energy possessed by a moving object or water body.

Laboratory Animal Studies: Investigations using animals as surrogates for humans.

Landfills: 1. Sanitary landfills are disposal sites for non-hazardous solid wastes spread in layers, compacted to the smallest practical volume, and covered by material applied at the end of each operating day. 2. Secure chemical landfills are disposal sites for hazardous waste, selected and designed to minimize the chance of release of hazardous substances into the environment.

Landscape Ecology: The study of the distribution patterns of communities and ecosystems, the ecological processes that affect those patterns, and changes in pattern and process over time.

Landscape: The traits, patterns, and structure of a specific geographic area, including its biological composition, its physical environment, and its anthropogenic or social patterns. An area where interacting ecosystems are grouped and repeated in similar form.

Lead (Pb): A heavy metal that is hazardous to health if breathed or swallowed. Its use in gasoline, paints, and plumbing compounds has been sharply restricted or eliminated by federal laws and regulations.

Lethal Dose 50: Also referred to as LD50, the dose of a toxicant that will kill 50 percent of test organisms within a designated period of time; the lower the LD 50, the more toxic the compound.

Lifetime Exposure: Total amount of exposure to a substance that a human would receive in a lifetime (usually assumed to be 70 years).

Limnology: The study of the physical, chemical, hydrological, and biological aspects of fresh water bodies.

Litter: 1. The highly visible portion of solid waste carelessly discarded outside the regular garbage and trash collection and disposal system. 2. leaves and twigs fallen from forest trees.

Low-Level Radioactive Waste (LLRW): Wastes less hazardous than most of those associated with a nuclear reactor; generated by hospitals, research laboratories, and certain industries. The Department of Energy, Nuclear Regulatory Commission, and EPA share responsibilities for managing them.

Marsh: A type of wetland that does not accumulate appreciable peat deposits and is dominated by herbaceous vegetation. Marshes may be either fresh or saltwater, tidal or nontidal.

Medical Waste: Any solid waste generated in the diagnosis, treatment, or immunization of human beings or animals, in research pertaining thereto, or in the production or testing of biologicals, excluding

hazardous waste identified or listed under 40 CFR Part 261 or any household waste as defined in 40 CFR Sub-section 261.4 (b)(1).

Mercury (Hg): Heavy metal that can accumulate in the environment and is highly toxic if breathed or swallowed.

Methane: A colorless, nonpoisonous, flammable gas created by anaerobic decomposition of organic compounds. A major component of natural gas used in the home.

Methanol: An alcohol that can be used as an alternative fuel or as a gasoline additive. It is less volatile than gasoline; when blended with gasoline it lowers the carbon monoxide emissions but increases hydrocarbon emissions. Used as pure fuel, its emissions are less ozone-forming than those from gasoline. Poisonous to humans and animals if ingested.

Microclimate: 1. Localized climate conditions within an urban area or neighborhood. 2. The climate around a tree or shrub or a stand of trees.

Mitigation: Measures taken to reduce adverse impacts on the environment.

Monoculture: Agriculture in which only one crop is planted at a time, usually over a large area

Montreal Protocol: Treaty, signed in 1987, governs stratospheric ozone protection and research, and the production and use of ozone-depleting substances. It provides for the end of production of ozone-depleting substances such as CFCS. Under the Protocol, various research groups continue to assess the ozone layer.

Morbidity: Rate of disease incidence.

Municipal Sewage: Wastes (mostly liquid) orginating from a community; may be composed of domestic wastewaters and/or industrial discharges.

Municipal Solid Waste: Common garbage or trash generated by industries, businesses, institutions, and homes.

Mutagen/Mutagenicity: An agent that causes a permanent genetic change in a cell other than that which occurs during normal growth. Mutagenicity is the capacity of a chemical or physical agent to cause such permanent changes.

Natural Resources: Objects and entities in the material world considered by people to have utility or value, specifically including only those materials not produced through human industry.

Navigable Waters: Traditionally, waters sufficiently deep and wide for navigation by all, or specified vessels; such waters in the United States come under federal jurisdiction and are protected by certain provisions of the Clean Water Act.

Netting: A concept in which all emissions sources in the same area that owned or controlled by a single company are treated as one large source, thereby allowing flexibility in controlling individual sources in order to meet a single emissions standard.

NIMBY: An acronym for "Not in My Backyard" that identifies the tendency for individuals and communities to oppose the siting of noxious or hazardous materials and activities in their vicinity. It implies a limited or parochial political vision of environmental justice.

Nitrate: A compound containing nitrogen that can exist in the atmosphere or as a dissolved gas in water and which can have harmful effects on humans and animals. Nitrates in water can cause severe illness in infants and domestic animals. A plant nutrient and inorganic fertilizer, nitrate is found in septic systems, animal feed lots, agricultural fertilizers, manure, industrial waste waters, sanitary landfills, and garbage dumps.

Nitrification: The process whereby ammonia in wastewater is oxidized to nitrite and then to nitrate by bacterial or chemical reactions.

No Till: Planting crops without prior seedbed preparation, into an existing cover crop, sod, or crop residues, and eliminating subsequent tillage operations.

Nonpoint Sources: Diffuse pollution sources (i.e. without a single point of origin or not introduced into a receiving stream from a specific outlet). The pollutants are generally carried off the land by storm

water. Common nonpoint sources are agriculture, forestry, urban, mining, construction, dams, channels, land disposal, saltwater intrusion, and city streets.

Nuclear Reactors and Support Facilities: Uranium mills, commercial power reactors, fuel reprocessing plants, and uranium enrichment facilities.

Nuclear Winter: Prediction by some scientists that smoke and debris rising from massive fires of a nuclear war could block sunlight for weeks or months, cooling the earth's surface and producing climate changes that could, for example, negatively affect world agricultural and weather patterns.

Nutrient: Any substance assimilated by living things that promotes growth. The term is generally applied to nitrogen and phosphorus in wastewater, but is also applied to other essential and trace elements.

Oil Spill: An accidental or intentional discharge of oil which reaches bodies of water. Can be controlled by chemical dispersion, combustion, mechanical containment, and/or adsorption. Spills from tanks and pipelines can also occur away from water bodies, contaminating the soil, getting into sewer systems and threatening underground water sources.

Open Dump: An uncovered site used for disposal of waste without environmental controls.

Osmosis: The passage of a liquid from a weak solution to a more concentrated solution across a semipermeable membrane that allows passage of the solvent (water) but not the dissolved solids.

Ozone (O_3): Found in two layers of the atmosphere, the stratosphere and the troposphere. In the stratosphere (the atmospheric layer 7 to 10 miles or more above the earth's surface) ozone is a natural form of oxygen that provides a protective layer shielding the earth from ultraviolet radiation.In the troposphere (the layer extending up 7 to 10 miles from the earth's surface), ozone is a chemical oxidant and major component of photochemical smog. It can seriously impair the respiratory system and is one of the most widespread of all the criteria pollutants for which the Clean Air Act required EPA to set standards. Ozone in the troposphere is produced through complex chemical reactions of nitrogen oxides, which are among the primary pollutants emitted by combustion sources; hydrocarbons, released into the atmosphere through the combustion, handling and processing of petroleum products; and sunlight.

Ozone Depletion: Destruction of the stratospheric ozone layer which shields the earth from ultraviolet radiation harmful to life. This destruction of ozone is caused by the breakdown of certain chlorine and/or bromine containing compounds (chlorofluorocarbons or halons), which break down when they reach the stratosphere and then catalytically destroy ozone molecules.

Ozone Hole: A thinning break in the stratospheric ozone layer. Designation of amount of such depletion as an "ozone hole" is made when the detected amount of depletion exceeds 50 percent. Seasonal ozone holes have been observed over both the Antarctic and Arctic regions, part of Canada, and the extreme northeastern United States.

Ozone Layer: The protective layer in the atmosphere, about 15 miles above the ground, that absorbs some of the sun's ultraviolet rays, thereby reducing the amount of potentially harmful radiation that reaches the earth's surface.

Pandemic: A widespread epidemic throughout an area, nation or the world.

Particulates: 1. Fine liquid or solid particles such as dust, smoke, mist, fumes, or smog, found in air or emissions. 2. Very small solids suspended in water; they can vary in size, shape, density and electrical charge and can be gathered together by coagulation and flocculation.

Parts Per Billion (ppb)/Parts Per Million (ppm): Units commonly used to express contamination ratios, as in establishing the maximum permissible amount of a contaminant in water, land, or air.

Pathogens: Microorganisms (e.g., bacteria, viruses, or parasites) that can cause disease in humans, animals and plants.

Peak Electricity Demand: The maximum electricity used to meet the cooling load of a building or buildings in a given area.

Percolating Water: Water that passes through rocks or soil under the force of gravity.

Percolation: 1. The movement of water downward and radially through subsurface soil layers, usually continuing downward to ground water. Can also involve upward movement of water. 2. Slow seepage of water through a filter.

Permeability: The rate at which liquids pass through soil or other materials in a specified direction.

Permit: An authorization, license, or equivalent control document issued by EPA or an approved state agency to implement the requirements of an environmental regulation; e.g., a permit to operate a wastewater treatment plant or to operate a facility that may generate harmful emissions.

Persistence: Refers to the length of time a compound stays in the environment, once introduced. A compound may persist for less than a second or indefinitely.

Pest: An insect, rodent, nematode, fungus, weed or other form of terrestrial or aquatic plant or animal life that is injurious to health or the environment.

Pesticide: Substances or mixture there of intended for preventing, destroying, repelling, or mitigating any pest. Also, any substance or mixture intended for use as a plant regulator, defoliant, or desiccant.

Petroleum: Crude oil or any fraction thereof that is liquid under normal conditions of temperature and pressure. The term includes petroleum-based substances comprising a complex blend of hydrocarbons derived from crude oil through the process of separation, conversion, upgrading, and finishing, such as motor fuel, jet oil, lubricants, petroleum solvents, and used oil.

pH: An expression of the intensity of the basic or acid condition of a liquid; may range from 0 to 14, where 0 is the most acid and 7 is neutral. Natural waters usually have a pH between 6.5 and 8.5.

Photosynthesis: The manufacture by plants of carbohydrates and oxygen from carbon dioxide mediated by chlorophyll in the presence of sunlight.

Plankton: Tiny plants and animals that live in water.

Plastics: Nonmetallic chemoreactive compounds molded into rigid or pliable construction materials, fabrics, etc.

Point Source: A stationary location or fixed facility from which pollutants are discharged; any single identifiable source of pollution; e.g., a pipe, ditch, ship, ore pit, factory smokestack.

Political Ecology: A field of research concerned with the relationship of systems of social and economic power to environmental conditions, natural resources, and conservation.

Pollen: The fertilizing element of flowering plants; background air pollutant.

Pollutant: Generally, any substance introduced into the environment that adversely affects the usefulness of a resource or the health of humans, animals, or ecosystems.

Pollution: Generally, the presence of a substance in the environment that because of its chemical composition or quantity prevents the functioning of natural processes and produces undesirable environmental and health effects.Under the Clean Water Act, for example, the term has been defined as the man-made or man-induced alteration of the physical, biological, chemical, and radiological integrity of water and other media.

Polychlorinated Biphenyls: A group of toxic, persistent chemicals used in electrical transformers and capacitors for insulating purposes, and in gas pipeline systems as lubricant. The sale and new use of these chemicals, also known as PCBs, were banned by law in 1979.

Population at Risk: A population subgroup that is more likely to be exposed to a chemical, or is more sensitive to the chemical, than is the general population.

Porosity: Degree to which soil, gravel, sediment, or rock is permeated with pores or cavities through which water or air can move.

Postcolonialism: An analytical approach to explaining the persisting conditions of exploitation and domination between historical colonial powers and previously colonized parts of the globe. From this perspective colonial habits, power relations, and ways of thinking remain ingrained in current scientific, political and economic relationships.

Precautionary Principle: When information about potential risks is incomplete, basing decisions about the best ways to manage or reduce risks on a preference for avoiding unnecessary health risks instead of on unnecessary economic expenditures.

Primary Waste Treatment: First steps in wastewater treatment; screens and sedimentation tanks are used to remove most materials that float or will settle. Primary treatment removes about 30 percent of carbonaceous biochemical oxygen demand from domestic sewage.

Prior Appropriation: A doctrine of water law that allocates the rights to use water on a first-come, first-served basis.

Producers: Plants that perform photosynthesis and provide food to consumers.

Proteins: Complex nitrogenous organic compounds of high molecular weight made of amino acids; essential for growth and repair of animal tissue. Many, but not all, proteins are enzymes.

Protozoa: One-celled animals that are larger and more complex than bacteria. May cause disease.

Public Water System: A system that provides piped water for human consumption to at least 15 service connections or regularly serves 25 individuals.

Radioactive Decay: Spontaneous change in an atom by emission of of charged particles and/or gamma rays; also known as radioactive disintegration and radioactivity.

Radioactive Waste: Waste that emits energy as rays, waves, streams or energetic particles. Radioactive materials are often mixed with hazardous waste, from nuclear reactors, research institutions, or hospitals.

Radon: A colorless naturally occurring, radioactive, inert gas formed by radioactive decay of radium atoms in soil or rocks.

Rational Choice Theory: A theory of individual decision-making that views human actions as motivated by seeking the most benefit for the least coast.

Recombinant DNA: The new DNA that is formed by combining pieces of DNA from different organisms or cells.

Recycle/Reuse: Minimizing waste generation by recovering and reprocessing usable products that might otherwise become waste (i.e., recycling of aluminum cans, paper, and bottles, etc.)

Red Tide: A proliferation of a marine plankton toxic and often fatal to fish, perhaps stimulated by the addition of nutrients. A tide can be red, green, or brown, depending on the coloration of the plankton.

Reforestation: Conversion of land to forest cover on deforested land.

Remote Sensing: The collection and interpretation of information about an object without physical contact with the object; e.g., satellite imaging, aerial photography, and open path measurements.

Reservoir: Any natural or artificial holding area used to store, regulate, or control water.

Residential Use: Pesticide application in and around houses, office buildings, apartment buildings, motels, and other living or working areas.

Residential Waste: Waste generated in single and multi-family homes, including newspapers, clothing, disposable tableware, food packaging, cans, bottles, food scraps, and yard trimmings other than those that are diverted to backyard composting.

Reuse: Using a product or component of municipal solid waste in its original form more than once; e.g., refilling a glass bottle that has been returned or using a coffee can to hold nuts and bolts.

Reverse Osmosis: A treatment process used in water systems by adding pressure to force water through a semi-permeable membrane. Reverse osmosis removes most drinking water contaminants. Also used in wastewater treatment. Large-scale reverse osmosis plants are being developed.

Ribonucleic Acid (RNA): A molecule that carries the genetic message from DNA to a cellular protein-producing mechanism.

Riparian Habitat: Areas adjacent to rivers and streams with a differing density, diversity, and productivity of plant and animal species relative to nearby uplands.

Riparian Rights: Entitlement of a land owner to certain uses of water on or bordering the property, including the right to prevent diversion or misuse of upstream waters. Generally a matter of state law.

Risk Assessment: Qualitative and quantitative evaluation of the risk posed to human health and/or the environment by the actual or potential presence and/or use of specific pollutants.

Risk Communication: The exchange of information about health or environmental risks among risk assessors and managers, the general public, news media, interest groups, etc.

Risk Management: The process of evaluating and selecting alternative regulatory and non-regulatory responses to risk. The selection process necessarily requires the consideration of legal, economic, and behavioral factors.

Risk: A measure of the probability that damage to life, health, property, and/or the environment will occur as a result of a given hazard.

River Basin: The land area drained by a river and its tributaries.

Rodenticide: A chemical or agent used to destroy rats or other rodent pests, or to prevent them from damaging food, crops, etc.

Sanitation: Control of physical factors in the human environment that could harm development, health, or survival.

Secondary Treatment: The second step in most publicly owned waste treatment systems in which bacteria consume the organic parts of the waste. It is accomplished by bringing together waste, bacteria, and oxygen in trickling filters or in the activated sludge process. This treatment removes floating and settleable solids and about 90 percent of the oxygen-demanding substances and suspended solids. Disinfection is the final stage of secondary treatment.

Sediments: Soil, sand, and minerals washed from land into water, usually after rain. They pile up in reservoirs, rivers and harbors, destroying fish and wildlife habitat, and clouding the water so that sunlight cannot reach aquatic plants. Careless farming, mining, and building activities will expose sediment materials, allowing them to wash off the land after rainfall.

Senescence: The aging process. Sometimes used to describe lakes or other bodies of water in advanced stages of eutrophication. Also used to describe plants and animals.

Septic System: An on-site system designed to treat and dispose of domestic sewage. A typical septic system consists of tank that receives waste from a residence or business and a system of tile lines or a pit for disposal of the liquid effluent (sludge) that remains after decomposition of the solids by bacteria in the tank and must be pumped out periodically.

Sewage: The waste and wastewater produced by residential and commercial sources and discharged into sewers.

Sewer: A channel or conduit that carries wastewater and storm-water runoff from the source to a treatment plant or receiving stream. "Sanitary" sewers carry household, industrial, and commercial waste. "Storm" sewers carry runoff from rain or snow. "Combined" sewers handle both.

Sick Building Syndrome: Building whose occupants experience acute health and/or comfort effects that appear to be linked to time spent therein, but where no specific illness or cause can be identified. Complaints may be localized in a particular room or zone, or may spread throughout the building.

Silviculture: Management of forest land for timber.

Sludge: A semi-solid residue from any of a number of air or water treatment processes; can be a hazardous waste.

Smog: Air pollution typically associated with oxidants.

Smoke: Particles suspended in air after incomplete combustion.

Soil and Water Conservation Practices: Control measures consisting of managerial, vegetative, and structural practices to reduce the loss of soil and water.

Soil Moisture: The water contained in the pore space of the unsaturated zone.

Solid Waste: Nonliquid, nonsoluble materials ranging from municipal garbage to industrial wastes that contain complex and sometimes hazardous substances. Solid wastes also include sewage sludge, agricultural refuse, demolition wastes, and mining residues. Technically, solid waste also refers to liquids and gases in containers.

Species: 1. A reproductively isolated aggregate of interbreeding organisms having common attributes and usually designated by a common name. 2. An organism belonging to belonging to such a category.

Sprawl: Unplanned development of open land.

Spring: Ground water seeping out of the earth where the water table intersects the ground surface.

Stakeholder: Any organization, governmental entity, or individual that has a stake in or may be impacted by a given approach to environmental regulation, pollution prevention, energy conservation, etc.

Stratification: Separating into layers.

Stratigraphy: Study of the formation, composition, and sequence of sediments, whether consolidated or not.

Stratosphere: The portion of the atmosphere 10-to-25 miles above the earth's surface.

Structural Adjustment: A set of policies, typically imposed by multilateral lending agencies like the World Bank and the International Monetary Fund during a national financial crisis, which imposes restrictions on government trade regulations, subsidies, and labor / environmental standards.

Surface Runoff: Precipitation, snow melt, or irrigation water in excess of what can infiltrate the soil surface and be stored in small surface depressions; a major transporter of nonpoint source pollutants in rivers, streams, and lakes.

Surface Water: All water naturally open to the atmosphere (rivers, lakes, reservoirs, ponds, streams, impoundments, seas, estuaries, etc.).

Suspended Solids: Small particles of solid pollutants that float on the surface of, or are suspended in, sewage or other liquids. They resist removal by conventional means.

Swamp: A type of wetland dominated by woody vegetation but without appreciable peat deposits. Swamps may be fresh or salt water and tidal or nontidal.

Systemic Pesticide: A chemical absorbed by an organism that interacts with the organism and makes the organism toxic to pests.

Tailings: Residue of raw material or waste separated out during the processing of crops or mineral ores.

Tailpipe Standards: Emissions limitations applicable to mobile source engine exhausts.

Technology-Based Standards: Industry-specific effluent limitations applicable to direct and indirect sources which are developed on a category-by-category basis using statutory factors, not including water-quality effects.

Teratogenesis: The introduction of nonhereditary birth defects in a developing fetus by exogenous factors such as physical or chemical agents acting in the womb to interfere with normal embryonic development.

Tertiary Treatment: Advanced cleaning of wastewater that goes beyond the secondary or biological stage, removing nutrients such as phosphorus, nitrogen, and most BOD and suspended solids.

Tidal Marsh: Low, flat marshlands traversed by channels and tidal hollows, subject to tidal inundation; normally, the only vegetation present is salt-tolerant bushes and grasses.

Tillage: Plowing, seedbed preparation, and cultivation practices.

Topography: The physical features of a surface area including relative elevations and the position of natural and man-made (anthropogenic) features.

Total Dissolved Solids (TDS): All material that passes the standard glass river filter; now called total filtrable residue. Term is used to reflect salinity.

Toxic Release Inventory: Database of toxic releases in the United States compiled from SARA Title III Section 313 reports.

Toxic Waste: A waste that can produce injury if inhaled, swallowed, or absorbed through the skin.

Toxicity: The degree to which a substance or mixture of substances can harm humans or animals. *Acute toxicity* involves harmful effects in an organism through a single or short-term exposure. *Chronic toxicity* is the ability of a substance or mixture of substances to cause harmful effects over an extended period, usually upon repeated or continuous exposure sometimes lasting for the entire life of the exposed organism. *Subchronic toxicity* is the ability of the substance to cause effects for more than one year but less than the lifetime of the exposed organism.

Transmissivity: The ability of an aquifer to transmit water.

Transpiration: The process by which water vapor is lost to the atmosphere from living plants. The term can also be applied to the quantity of water thus dissipated.

Treatment Plant: A structure built to treat wastewater before discharging it into the environment.

Trust Fund (CERCLA): A fund set up under the Comprehensive Environmental Response, Compensation and Liability Act (CERCLA) to help pay for cleanup of hazardous waste sites and for legal action to force those responsible for the sites to clean them up.

Tundra: A type of treeless ecosystem dominated by lichens, mosses, grasses, and woody plants. Tundra is found at high latitudes (arctic tundra) and high altitudes (alpine tundra). Arctic tundra is underlain by permafrost and is usually water saturated.

Ultraviolet Rays: Radiation from the sun that can be useful or potentially harmful. UV rays from one part of the spectrum (UV-A) enhance plant life. UV rays from other parts of the spectrum (UV-B) can cause skin cancer or other tissue damage. The ozone layer in the atmosphere partly shields us from ultraviolet rays reaching the earth's surface.

Underground Storage Tank (UST): A tank located at least partially underground and designed to hold gasoline or other petroleum products or chemicals.

Unsaturated Zone: The area above the water table where soil pores are not fully saturated, although some water may be present.

Urban Runoff: Storm water from city streets and adjacent domestic or commercial properties that carries pollutants of various kinds into the sewer systems and receiving waters.

User Fee: Fee collected from only those persons who use a particular service, as compared to one collected from the public in general.

Vadose Zone: The zone between land surface and the water table within which the moisture content is less than saturation (except in the capillary fringe) and pressure is less than atmospheric. Soil pore space also typically contains air or other gases. The capillary fringe is included in the vadose zone.

Value-added: A procedure that increases the worth of a product or raw material through transformation and processing.

Vapor: The gas given off by substances that are solids or liquids at ordinary atmospheric pressure and temperatures.

Vector: 1. An organism, often an insect or rodent, that carries disease. 2. Plasmids, viruses, or bacteria

used to transport genes into a host cell. A gene is placed in the vector; the vector then "infects" the bacterium.

Viscosity: The molecular friction within a fluid that produces flow resistance.

Volatile: Any substance that evaporates readily.

Waste Generation: The weight or volume of materials and products that enter the waste stream before recycling, composting, landfilling, or combustion takes place. Also can represent the amount of waste generated by a given source or category of sources.

Waste Treatment Plant: A facility containing a series of tanks, screens, filters and other processes by which pollutants are removed from water.

Waste: 1. Unwanted materials left over from a manufacturing process. 2. Refuse from places of human or animal habitation.

Waste-to-Energy Facility/Municipal-Waste Combustor: Facility where recovered municipal solid waste is converted into a usable form of energy, usually via combustion.

Wastewater: The spent or used water from a home, community, farm, or industry that contains dissolved or suspended matter.

Water Pollution: The presence in water of enough harmful or objectionable material to damage the water's quality.

Water Supplier: One who owns or operates a public water system.

Water Supply System: The collection, treatment, storage, and distribution of potable water from source to consumer.

Water Table: The level of groundwater.

Water Well: An excavation where the intended use is for location, acquisition, development, or artificial recharge of ground water.

Watershed Approach: A coordinated framework for environmental management that focuses public and private efforts on the highest priority problems within hydrologically-defined geographic areas taking into consideration ground and surface flow.

Watershed: The land area that drains into a stream; the watershed for a major river may encompass a number of smaller watersheds that ultimately combine at a common point.

Weight of Scientific Evidence: Considerations in assessing the interpretation of published information about toxicity—quality of testing methods, size and power of study design, consistency of results across studies, and biological plausibility of exposure-response relationships and statistical associations.

Weir: 1. A wall or plate placed in an open channel to measure the flow of water. 2. A wall or obstruction used to control flow from settling tanks and clarifiers to ensure a uniform flow rate and avoid short-circuiting.

Well: A bored, drilled, or driven shaft, or a dug hole whose depth is greater than the largest surface dimension and whose purpose is to reach underground water supplies or oil, or to store or bury fluids below ground.

Wetlands: An area that is saturated by surface or ground water with vegetation adapted for life under those soil conditions, as swamps, bogs, fens, marshes, and estuaries.

Wildlife Refuge: An area designated for the protection of wild animals, within which hunting and fishing are either prohibited or strictly controlled.

Xenobiota: Any biotum displaced from its normal habitat; a chemical foreign to a biological system.

Yard Waste: The part of solid waste composed of grass clippings, leaves, twigs, branches, and other garden refuse.

Yield: The quantity of water (expressed as a rate of flow or total quantity per year) that can be collected for a given use from surface or groundwater sources.

Zooplankton: Small (often microscopic) free-floating aquatic plants or animals.

Appendix

UNITED NATIONS MAIN ENVIRONMENTAL INDICATORS

Environment statistics covering a range of issues related to Water, Air, Waste and Land, are compiled by the United Nations Statistics Division, Department of Economic and Social Affairs. The data are official data supplied by national statistical offices and/or ministries of environment (or equivalent institutions) in countries in response to a biennial UNSD/UNEP questionnaire, sent out in March 2004. They are supplemented by data taken from UNFCCC (United Nations Framework Convention on Climate Change) for data on greenhouse gas emissions, and FAO (Food and Agriculture Organisation of the United Nations, for data on water resources. Data from OECD countries and from most European countries are taken from OECD and from Eurostat. For land area, agricultural area and forest area, all data are from FAO, awaiting the results of an in-depth analysis of the differences between FAO data and country data sent to UNSD/UNEP.

Results show that environment statistics is still in an early stage of development in many countries, and data are often sparse. Information on the data quality and comparability is given at the end of each table.

Data are provided for the following topics:

Water	**Air Pollution**	**Climate Change**	**Waste**	**Land Use**
Water resources	SO2 emissions	Greenhouse gas emissions	Municipal waste collection	Area of country
Public water supply	NOx emissions	CO2 emissions	Municipal waste treatment	Forest area
Waste water		CH4 and N2O emissions	Hazardous waste	Agricultural land

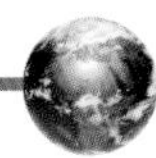

ENVIRONMENTAL INDICATORS

Water

Water resources: long term annual average *last update: April 2007*

	Precipitation	Internal flow	Actual external inflow of surface and ground waters	Total renewable fresh water resources	Total renewable fresh water resources per capita
	mio m3	mio m3	mio m3	mio m3	m3/person
Afghanistan	213 429	55 000	10 000	65 000	2 177
Albania	42 700	26 900	14 800	41 700	13 324
Algeria	211 498	13 900	420	14 320	436
Andorra	406[1]	...	...	...	...
Angola	1 258 793	184 000	0	184 000	11 542
Antigua and Barbuda	500	52	0	52	638
Argentina	1 642 104	276 000	538 000	814 000	21 008
Armenia	17 640	6 317	940	7 257	2 406*
Australia	3 631 000	387 000	0	387 000	19 201*
Austria	92 000	50 000	27 000	77 000	9 402*
Azerbaijan	36 978	10 330	20 573	30 903	3 674*
Bahamas	17 934	20	0	20	62
Bahrain	59	4	112	116	160
Bangladesh	383 832	105 000	1 105 644	1 210 644	8 536
Barbados	600	80	0	80	298
Belarus	136 186	52 938	23 200[2]	76 138	7 805[2*]
Belgium	29 000	12 000	8 000	21 000	2 016*
Belize	39 100	16 000	2 555	18 555	68 789
Benin	117 046	10 300	14 500	24 800	2 939
Bermuda	8	7	0	7	113*
Bhutan	103 400	95 000	0	95 000	43 930
Bolivia	1 258 863	303 531	319 000	622 531	67 799
Bosnia-Herzegovina	52 562	35 500	2 000	37 500	9 598
Botswana	241 825	2 900	11 500	14 400	8 159
Brazil	15 333 391	5 658 600	2 768 672	8 427 271	45 209*
Brunei Darussalam	15 706	8 500	0	8 500	22 738
Bulgaria	67 389	21 000	300	21 300	2 757
Burkina Faso	204 925	12 500	0	12 500	945
Burundi	33 903	3 600	0	3 600	477
Cambodia	344 628	120 570	355 540	476 110	33 836
Cameroon	762 463	273 000	12 500	285 500	17 492
Canada	4 930 000	2 740 000	52 000	2 792 000	86 525
Cape Verde	900	300	0	300	592
Central African Republic	836 662	141 000	3 400	144 400	35 763
Chad	413 191	15 000	28 000	43 000	4 411
Chile	1 151 600	884 000	38 000	922 000	56 581
China	6 172 800	2 840 500	21 400	861 900	2 175*
China, Hong Kong SAR	2 431	1 075	...	...	...
China, Macao SAR	51	...	...	...	...
Colombia	2 974 605	2 112 000	20 000	2 132 000	46754
Comoros	2 000	1 200	0	1 200	1504
Congo	562 932	222 000	610 000	832 000	208057
Costa Rica	149 529	112 400	0	112 400	25975
Cote d'Ivoire	434 676	76 700	4 300	81 000	4462
Croatia	62 912	37 700	67 800	105 500	23180
Cuba	147 965	38 120	0	38 120	3383
Cyprus	4 420	781	0	781	935*
Czech Republic	55 000	15 000	1 000[3]	16 000[3]	1566[3*]
Democratic Republic of the Congo	3 618 119	900 000	383 000	1 283 000	22294
Denmark	38 000	16 000	0	16 000	2946*

Water: *Water resources: long term annual average*

	Precipitation	Internal flow	Actual external inflow of surface and ground waters	Total renewable fresh water resources	Total renewable fresh water resources per capita
	mio m3	mio m3	mio m3	mio m3	m3/person
Djibouti	5 100	300	0	300	378
Dominica	2 577	...	...	...	...
Dominican Republic	68 700	21 000	0	21 000	2 360
Ecuador	582 985	...	0	264 618	20 004*
Egypt	51 400	1 800	85 000	86 800	1 172
El Salvador	56 052	23 212	635	23 847	3 466*
Equatorial Guinea	60 481	26 000	0	26 000	51 637
Eritrea	45 147	2 800	3 500	6 300	1 431
Estonia	30 647	12 044	9 070	21 114	15 879*
Ethiopia	936 005	110 000	0	110 000	1 421
Fiji	47 356	28 550	0	28 550	33 679
Finland	222 000	107 000	3 000	110 000	20 956*
France	488 000	178 000	11 000[4]	189 000[4]	3 124[4]*
French Guiana	260 550	134 000	0	134 000	716 363
Gabon	489 997	164 000	0	164 000	118 511
Gambia	9 099	- 3 423[5]	6 279	2 856	1 883*
Georgia	83 141	46 845	6 931[6]	53 776	12 019*
Germany	307 000	117 000	71 000	188 000	2 274*
Ghana	283 195	30 300	22 900	53 200	2 406
Greece	115 000	60 000	12 000	72 000	6 475*
Greenland	759 000	603 000	0	603 000	10 594 560
Grenada	522	...	...	...	...
Guadeloupe	422	...	...	...	...
Guatemala	217 300	109 200	2 070	111 270	8832
Guinea	405 939	226 000	0	226 000	24037
Guinea-Bissau	56 972	16 000	15 000	31 000	19542
Guyana	513 112	241 000	0	241 000	320812
Haiti	39 966	13 010	1 015	14 025	1645
Honduras	221 434	95 929	0	95 929	13315
Hungary	58 000	6 000	114 000	120 000	11884*
Iceland	200 000	170 000	0	170 000	577130*
India	3 558 800	1 260 540	647 220	1 907 760	1729
Indonesia	5 146 529	2 838 000	0	2 838 000	12739
Iran (Islamic Republic of)	375 790	128 500	9 010	137 510	1978
Iraq	94 677	35 200	61 220	96 420	3347
Ireland	81 000	49 000	1 000	50 000	12054*
Israel	9 200	750	920	1 670	248
Italy	243 000	88 000	8 000	95 000	1635*
Jamaica	22 542	9 404	0	9 404	3548
Japan	649 000	424 000	0	424 000	3310*
Jordan	9 929	680	200	880	154
Kazakhstan	680 408	75 420	34 190	109 610	7394
Kenya	401 906	20 200	10 000	30 200	882
Korea, Dem. People's Rep. of	127 000	67 000	10 135	77 135	3430
Korea, Republic of	127 600	73 100	0	73 100	1529*
Kuwait	2 160	0	20	20	7
Kyrgyzstan	106 500	46 450	0	46 450	8824
Lao People's Dem. Rep.	434 362	190 420	143 130	333 550	56303
Latvia	43 443	18 444	17 748	36 192	15688*
Lebanon	6 900	4 800	37	4 837	1352
Lesotho	23 928	5 230	0	5 230	2914
Liberia	266 286	200 000	32 000	232 000	70661
Libyan Arab Jamahiriya	98 500	600	0	600	103

Water: *Water resources: long term annual average*

	Precipitation	Internal flow	Actual external inflow of surface and ground waters	Total renewable fresh water resources	Total renewable fresh water resources per capita
	mio m3	mio m3	mio m3	mio m3	m3/person
Lithuania	44 010	15 510	8 990	24 500	7141*
Luxembourg	2 300	1 100	700	1 800	3872*
Madagascar	888 192	337 000	0	337 000	18113
Malawi	139 960	16 140	1 140	17 280	1341
Malaysia	948 163	580 000	0	580 000	22882
Maldives	58 592	30	0	30	91
Mali	349 610	60 000	40 000	100 000	7397
Malta	181	67	0	67	167*
Martinique	2 894	...	...	...	...
Mauritania	94 656	400	11 000	11 400	3715
Mauritius	3 700	2 590	0	2 590	2081*
Mexico	1 515 000	424 000	49 000	473 000	4419*
Monaco	2	...	...	...	...
Mongolia	377 370	34 800	0	34 800	13150
Morocco	150 000	29 000	0	29 000	921*
Mozambique	827 200	99 000	117 110	216 110	10919
Myanmar	1 414 594	880 600	165 001	1 045 601	20697
Namibia	235 252	6 160	39 300	45 460	22380
Nepal	220 800	198 200	12 000	210 200	7747
Netherlands	29 770	8 480	81 200[7]	89 680[7]	5522[7]*
New Caledonia	27 831	...	...	...	...
New Zealand	537 000	327 000	0	327 000	81174*
Nicaragua	310 856	189 740	6 950	196 690	35849
Niger	190 810	3 500	30 150	33 650	2411
Nigeria	1 062 336	221 000	65 200	286 200	2176
Norway	471 000	359 000	12 000	371 000	80298*
Oman	26 600	985	0	985	384
Pakistan	393 300	52 400	181 370	233 770	1480
Palestine	120	46	10	56	15
Panama	203 300	147 420	560	147 980	45793
Papua New Guinea	1 454 104	801 000	0	801 000	136059
Paraguay	459 546	94 000	242 000	336 000	54561
Peru	2 233 700	1 616 000	297 000	1 913 000	68399
Philippines	704 340	479 000	0	479 000	5767
Poland	193 000	55 000	8 000	63 000	1635*
Portugal	82 000	39 000	35 000	74 000	7051*
Puerto Rico	18 383	7 100	0	7 100	1795
Qatar	811	51	2	53	65
Republic of Moldova	16 959	1 300	11 500	12 800	3043*
Réunion	7 500	5 000	0	5 000	6368
Romania	154 000	39 415	2 878	42 293	1948*
Russian Federation	7 854 684	4 312 700	194 550	4 507 250	31475
Rwanda	31 932	5 200	0	5 200	575
Saint Helena	237	...	...	...	...
Saint Kitts and Nevis	500	24	0	24	553
Saint Lucia	1 427	...	...	...	...
Samoa	8 496	...	...	...	...
Sao Tome and Principe	3 100	2 180	0	2 180	13928
Saudi Arabia	126 800	2 400	0	2 400	98
Senegal	135 048	26 400	13 000	39 400	3380
Serbia	56 115[8]	12 776[8]	162 600[8]	175 376[8]	...
Seychelles	887	...	...	...	...
Sierra Leone	181 215	160 000	0	160 000	28957

Water: *Water resources: long term annual average*

	Precipitation	Internal flow	Actual external inflow of surface and ground waters	Total renewable fresh water resources	Total renewable fresh water resources per capita
	mio m3	mio m3	mio m3	mio m3	m3/person
Singapore	1 700[9]	830[9]	0[9]	830[9]	192*
Slovakia	37 000	13 000	67 000[10]	80 000[10]	14812[10*]
Slovenia	22 298	7 406	13 496	20 902	10627*
Solomon Islands	87 509	44 700	0	44 700	93565
Somalia	180 075	6 000	7 500	13 500	1641
South Africa	524 600	...	7 273[11]	31 738	669*
Spain	347 000	111 000	0	111 000	2578*
Sri Lanka	112 337	50 000	0	50 000	2410
St. Vincent and the Grenadines	617	...	...	...	...
Sudan	1 043 670	30 000	119 000	149 000	4112
Suriname	380 582	88 000	34 000	122 000	271571
Swaziland	13 678	2 640	1 870	4 510	4368
Sweden	336 000	170 000	11 000	181 000	20019*
Switzerland	60 000	40 000	13 000[12]	53 000	7308*
Syrian Arab Republic	46 700	7 000	39 080	46 080	2420
Tajikistan	98 900	66 300	33 430	99 730	15327
Thailand	832 435	210 000	199 944	409 944	6382
The Former Yugoslav Rep. of Macedonia	15 914	5 400	1 000	6 400	3146
Togo	66 302	11 500	3 200	14 700	2392
Tonga	1 474	...	...	...	...
Trinidad and Tobago	11 300	3 840	0	3 840	2942
Tunisia	36 000	4 170*	...	4 170	413*
Turkey	501 000	227 400	6 900	234 300	3197*
Turkmenistan	78 731	1 360	59 500	60 860	12592
Uganda	284 500	39 000	27 000	66 000	2290
Ukraine	340 970	53 100	86 450	139 550	3002
United Arab Emirates	6 529	150	0	150	33
United Kingdom	268 000	143 000	3 000	146 000	2447*
United Rep. of Tanzania	1 012 191	82 000	9 000	91 000	2374
United States	6 440 000	2 460 000	18 000	2 478 000	8309*
Uruguay	222 865	59 000	80 000	139 000	40136
Uzbekistan	92 299	16 340	55 870	72 210	2715
Venezuela	1 710 094	722 451	510 719	1 233 170	46101
Viet Nam	604 008	366 500	524 710	891 210	10580
Yemen	88 329	4 100	0	4 100	195
Zambia	767 436	80 200	25 000	105 200	9016
Zimbabwe	270 523	14 100	5 900	20 000	1537

Sources:
UNSD/UNEP 2001, 2004 and 2006 questionnaires on Environment statistics, Water section.
OECD/Eurostat 2004 questionnaire on Environment statistics, Water section.
OECD Environmental Data, Compendium 2006, Inland Waters section, marked with "*".
AQUASTAT database of the Food and Agriculture Organization of the United Nations (FAO); http://www.fao.org/ag/agl/aglw/aquastat/main/index.stm. Whenever data were not available for
"Total renewable fresh water resources" variable from UNSD or OECD/Eurostat questionnaire, AQUASTAT data were used.

Footnotes:
1.Data refer to average of 3 meteorological stations.
2.Data only include surface water. Groundwater is excluded.
3.Excludes underground flows.
4.Excludes underground flows. Rhine excluded.
5.The numbers are negative because evapotranspiration covers both waters from precipitations and external inflow of waters. Whereas precipitations covers waters from rains that fall
within the National territory.
6.Data refer to the sum of inflows from Armenia and Turkey.
7.Excludes underground flows, estimated at 2 billion m3.
8.Data refer to the Republic of Serbia without the territory of Kosovo Province.
9.Data refer to years 1990-2005.

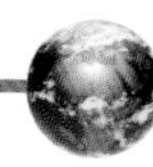

Water: *Water resources: long term annual average*

10.Excludes underground flows (representing 946 million m3).
11.Data refer to inflows from Lesotho and Swaziland.
12.Inflow excludes Liechtenstein (about 1%).

Definitions & Technical notes:
Precipitation refers to the total volume of atmospheric wet deposition (rain, snow, hail, dew, etc) falling on the territory of the country over one year, in millions of cubic metres. Long term annual average is the arithmetic average over at least 20 years. Internal flow is the total volume of river run-off and ground water generated in natural conditions, exclusively by precipitation within the country. The internal flow is equal to precipitation less actual evapotranspiration and can be calculated or measured. Actual external inflow of surface and ground waters refers to the total volume of actual flow of rivers and groundwater, coming from neighboring countries. Total renewable fresh water resources_ = Internal flow + Actual external inflow of surface and groundwaters.

Data Quality:
Countrywide precipitation is usually calculated on the basis of measurements at a selected number of measuring stations within the country. Data is considered to be fairly reliable. Internal flow is the fresh water generated in the country and is usually calculated by subtracting natural evapotranspiration from precipitation. The reliability of the data depends essentially on the estimation method for evapotranspiration. For most countries, actual external inflow of surface and ground water contains only the surface water flow, since ground water flows are often not well known. Surface water flows of inflowing rivers should be measured at the border. Dry countries in particular, tend to have reliable data.

Policy Relevance:
Water is essential to life, for drinking and cooking, for hygiene, agriculture, industrial production and for growing food. In some regions of the world it provides transport for goods and people, giving access to markets and services not available locally. Water scarcity happens when the supply of water is unable to meet the demand. Today, some 460 million people, over eight percent of the world's population, live in areas of water scarcity, mainly the areas situated in Southern Africa, Northern Africa, the Middle East and South Western Asia. Without adequate management of water resources, it is estimated that by 2025 at least 3.5 billion people, almost 50 percent of the world's population, will face water scarcity (International Hydropower Association, Feb 2004: Sustainability Guidelines). Rivers and aquifers are no respecters of political boundaries, and apart from island states, most countries share water resources with their immediate neighbours, and in the case of long rivers, with other countries much further away. This makes water management a particularly sensitive political issue in many parts of the world.
The data shown for total renewable fresh water resources do not represent the volume of water that can be freely exploited by the country for its own needs; countries downstream may rely on a regular inflow in the same rivers, lakes and aquifers to meet their needs. Equally, if the flow of water is seriously reduced, the ecological balance of the water body may be disrupted, affecting the livelihood of fishermen and others, and potentially creating health risks.

ENVIRONMENTAL INDICATORS
Water

Public water supply last update: April 2007

	Latest year available	Total public water supply (PWS) mio m3	Total PWS per capita m3/person	% Population connected to PWS %	Total PWS per capita connected m3/person	PWS delivered to households mio m3	PWS delivered to households mio m3
Algeria	2002	3300	106	...	...	1559[1]	200[2]
Armenia	2004	125	41	88.9	46	91	7
Australia	2004	11337[3]	568[3]	95.0	598	3411[4]	2573[5]
Austria	1999	549	68	90.0[6]	75	383[4]	...
Azerbaijan	2005	498	59	48.0	123	433	...
Belarus	2002	...	...	...	...	794	...
Belgium	2002	340[7]	...	96.0	...	184[4]	95[8]
Belize	2005	6	24	58.9	40	...	...
Bermuda	2005	3	51	10.0[9]	...	2	0
Bolivia	2005	143	16	72.1[10]	...	113	3
Bosnia and Herzegovina	2004	166	43	...	...	114	...
Brazil	2000	16060	93	80.0	117	...	...
British Virgin Islands	2001	...	...	48.4	...	...	...
Brunei Darussalam	2005	169	451	99.0[6]	...	142	27
Bulgaria	2001	424	53	98.7	54	273	54
Canada	1996	5201	175	92.0[11]	...	3272	1928
Chile	2005	933	57	99.8	57	698[6]	221[6]
China, Hong Kong SAR	2005	968	137	99.9	138	...	...
China, Macao SAR	2005	59	129	...	...	25[6]	20[6,12]
Croatia	2004	311	69	...	...	183	...
Cuba	2000	1685	150	95.6[13]	...	...	...
Cyprus	1998	68	88	...	...	...	2[14]
Czech Republic	2004	543	53	92.0	58	472[4]	63[8]
Denmark	2002	380[15]	71[15]	97.0	73	291[4]	48[8]
Ecuador	2000	...	...	...	...	...	2330
Egypt	1996	...	...	...	...	2606[16]	...
Estonia	2001	66	49	71.0	69	...	...
Ethiopia	2001	66[17]	...	...	...	...	...
Finland	2001	408[18]	79[18]	90.0	87	245	50[8]
France	2001	5685	95	99.0	96	3414[4]	...
Gambia	2005	...	...	50.0	...	...	...
Georgia	2004	296	66	80.0	82	374[6]	36[6]
Germany	2004	4729	57	99.0	58	3752	976[8]
Greece	1980	750	...	...	...	670[19]	...
Guadeloupe	1998	36	87	...	...	36	...
Guinea	2005	25	3	...	...	9	2
Hungary	2002	546	55	93.0	59	486[4]	55[8]
Iceland	2003	67[20]	244[20]	95.0	244	30	5[21]
India	2000	...	...	...	...	4200[22]	8000[22]
Ireland	2002	...	...	90.0	...	...	...
Israel	2003	1860	289	...	...	698	117
Italy	1999	5653	98	99.7	99	4882[4]	503[8]
Jamaica	2005	94	35	70.0	51	...	...
Japan	2001	85968[23]	675[23]	97.0[6]	...	16279[4]	12849[8]
Jordan	2004	...	...	...	...	281	38
Kazakhstan	2002	14930	965	...	...	...	...
Korea, Republic of	2003	22275	467	89.0	525	5597[4]	1963[8]
Latvia	2001	302	129	...	...	...	78
Lebanon	2005	...	...	75.6[11]	...	...	...
Luxembourg	2004	33	72	100.0	72	23	13[24]

Water: *Public water supply*

	Latest year available	Total public water supply (PWS)	Total PWS per capita	% Population connected to PWS	Total PWS per capita connected	PWS delivered to households	PWS delivered to households
		mio m3	m3/person	%	m3/person	mio m3	mio m3
Madagascar	2005	77	4	...	...	57	5
Maldives	2005	2[25]	6[25]	...	...	1[25]	0.3[25]
Mali	2002	...	...	66.0	...	...	...
Martinique	1997	40	106	...	...	40	...
Mauritius	2005	214	172	98.7[26]	...	214	...
Mexico	2004	75430[27,28]	714[27,28]	90.0[29]	793	10670[4,27,28]	7298[8,27,28]
Monaco	2005	6[30]	171[30]	100.0	171	2	0.3
Morocco	2002	...	...	...	...	...	79[31]
Nepal	2004	...	...	...	...	94	...
Netherlands	2002	1257	78	100.0	79	990[4,14,32]	218[8,14]
New Zealand	2001	...	...	87.0	...	...	...
Norway	2002	808[33]	179[33]	89.0	201	299	145[8]
Palestine	2005	...	...	90.5	...	152[34]	...
Panama	2005	...	...	83.0	...	...	...
Peru	2001	660[35]	25[35]	...	...	295[26,35]	9[26,35]
Poland	2003	1657	43	85.0	51	1454[4]	203[8]
Portugal	1998	8754	878	92.0[10]	...	680[4]	...
Republic of Moldova	2004	786	186	...	...	118	21
Romania	2001	2462	110	...	...	988	...
Serbia	2004	509	48	72.0	67	368	77
Singapore	2005	506	117	100.0	117	253[4]	253[36]
Slovakia	2004	353	65	85.0	77	166[4]	...
Slovenia	2001	106	54	...	...	88	18
South Africa	2000	17246	392	...	...	3610[37]	...
Spain	2002	4339	106	...	...	3481[4]	459[8]
Sweden	2002	708	80	86.1[19]	...	618[4]	90
Switzerland	2002	1015	142	...	...	720[4]	184[8]
The Former Yugoslav Rep. of Macedonia	2004	363	179	94.0	190	75	31[14]
Trinidad and Tobago	2005	...	...	76.4[6]	...	179	36
Tunisia	2005	...	...	81.0	...	326	...
Turkey	2004	1988	28	74.0	37	1955[4]	27[8]
Ukraine	2004	1955	42	...	...	2300[6]	1586[6]
United Kingdom	2004	6876[38]	116[38]	99.0	117	...	...
United States	2000	...	...	85.0	...	...	...
Yemen	2005	84	4	17.5	23	84	...
Zimbabwe	2005	...	...	...	...	224	...

Sources:
UNSD/UNEP 2001, 2004 and 2006 questionnaires on Environment statistics, Water section
OECD/Eurostat 2004 questionnaire on Environment statistics, Water section.
OECD Environmental Data, Compendium 2006, Inland Waters section.
UN Population Division.

Footnotes:
1. 2004 data.
2. Includes Manufacturing industries, Electricity industry and Other economic activities.
3. Data referring to "Distributed water" in Water account Australia 2004-2005; exclude in-stream and reused water.
4. Data refer to public supply water used by domestic sector.
5. Includes mining, manufacturing, electricity and gas.
6. 2002 data.
7. Water supply: Flanders and Wallonie only.
8. Data refer to public supply water used by all industry activities.
9. This is % provided with the piped water to top up the rainwater tanks. Water is trucked to the remaining households to top up the rainwater tanks.
10. 2003 data.
11. 1999 data.
12. Includes industrial and commercial activities.
13. 2005 data.

Water: *Public water supply*

14. 2001 data.
15. Public supply refers to public and private waterworks.
16. Data refer to the Commercial sector and Households.
17. Data refer to Addis Ababa city only.
18. Public supply: total includes leakages.
19. 1997 data.
20. Total supply: provisional data.
21. Refers to manufacturing industry and electricity production.
22. Data refer to projected water demands.
23. Public water supply includes self-supply and other supply.
24. Average between 1990 and 1995.
25. The figures given here are only pertaining to the water supplied by Male' Water and Sewage Co. Pvt. Ltd.
26. 2000 data.
27. Supply: abstracted volumes of water granted in concessions.
28. Public water supply includes self-supply.
29. Population connected to public water supply: access to safe water for population living in individual housing (of which 65.3% are supplied inside the house).
30. Includes "Water losses during transport".
31. Refers to consumption in urban areas, managed by ONEP.
32. Domestic sector includes agricultural sector (57 million m3 in 1999).
33. Public supply: total includes leakages.
34. Data include quantity of pumped water from wells and springs discharge.
35. The information corresponds to the province of Lima and Constitutional province of the Callao.
36. Data refer to non-domestic consumption, including water delivered to agriculture, manufacturing, electricity industry and other economic activities.
37. Data include Public supply to businesses, institutions, etc.
38. Total public supply includes distribution losses and nonpotable water delivered.

Definitions & Technical notes:
Public water supply refers to water supplied by both public bodies and private companies involved in the collection, purification and distribution of water. Total public water supplied per capita is calculated by dividing the total public water supplied by the total population of the country. Total public water supplied per capita connected is calculated by dividing the total public water supplied by the number of people connected to the public water supply. Percent of Population Connected to Public Water Supply is calculated by dividing the number of people connected to the public water supply by the total population of the country.

Data Quality:
Data on public water supply (PWS) is usually collected from municipalities. Data on population connected to PWS can be obtained through municipalities or through household surveys. Household surveys usually give more accurate results, since they do not rely on sometimes incomplete information about or held by municipalities. Data on amounts of water supplied can significantly vary between countries depending on the extent to which PWS delivers water to industries, businesses, agriculture and power stations in addition to households. Care must be taken when comparing data between countries.

Policy Relevance:
Access to a regular, clean and safe supply of water is essential to maintaining human health, and a key component of sustainable development. Connection to a public water supply not only reduces the risk of water borne diseases, it provides water for drinking, cooking, hygiene and washing, and is associated with improved health, in general. It also relieves women and children of the burden of having to fetch water, giving them time for more productive activities, or for schooling. 'Total public water supply' represents the demand for water from that part of the population that is connected to the public supply system, including any connected industries. The indicator 'total PWS per capita connected' shows the per capita demand for water, when water is readily available. In regions without access to public water supply, use of water tends to be much less. However, the difference between the indicator 'total PWS per capita connected' and the indicator 'total PWS per capita' can be seen as a rough indicator of the unmet demand for good quality water and therefore of how much water would need to be provided, if almost the whole country were provided with access to the public water supply. The extent to which industries are connected to the public water supply will depend on the cost, the required standard of water and the availability of alternatives. Food processing industries will require high quality water, and therefore may prefer to be connected to the public water supply, where the quality is assured. Industries relying on water for cooling are less concerned by the quality of the water, but require large quantities. They may prefer to site their factories close to rivers and lakes so as to extract the water they need, and later return it to the same water body, a few degrees warmer.

ENVIRONMENTAL INDICATORS

Water

Waste water *last update: April 2007*

	Latest year available	Population connected to public waste water collection system	Population connected to public waste water treatment plants
		%	%
Algeria	1998	66.3	3.9
Andorra	2005	100.0	47.7
Argentina	2001	42.5	42.5
Australia	2004	87.0[1]	...
Austria	2002	86.0	86.0
Azerbaijan	2005	30.0	30.0
Belarus	2004	90.8[2]	...
Belgium	2002	82.9[3]	45.9
Belize	2000	15.1	15.1
Bermuda	2005	5.0	5.0
Bolivia	2003	31.4[4]	...
Bosnia-Herzegovina	1990	38.0	...
Brazil	2004	54.1[2]	...
British Virgin Islands	2001	24.5	24.5
Bulgaria	2001	67.9	38.1
Canada	1999	74.3[5]	71.7[5]
Chile	2005	94.9[6]	73.3[7]
China	2004	45.7	32.5
China, Hong Kong SAR	2005	92.9	92.9
China, Macao SAR	1995	99.9	...
Costa Rica	2000	24.8	2.4
Cuba	2005	38.8	...
Cyprus	2000	34.5	34.5
Czech Republic	2004	77.9	71.1
Denmark	2002	87.9	87.9
Dominica	2005	23.0	13.0
Dominican Republic	2000	31.4	12.0[8,9]
Estonia	2000	70.0	69.0
Finland	2002	81.0	81.0
France	2001	81.5	79.4
French Guiana	2001	48.3	27.4
Germany	2004	95.5	93.5
Greece	1997	...	56.2
Guadeloupe	2001	40.9	40.5
Hungary	2002	61.9	57.4
Iceland	2003	89.0	50.0
Ireland	2001	93.0[10]	70.0[10]
Italy	1999	...	68.6
Japan	2003	67.0	67.0
Jordan	2004	97.7	...
Korea, Republic of	2003	78.8[11]	78.8
Kyrgyzstan	2004	27.0	...
Luxembourg	2003	94.8	94.8
Madagascar	2005	...	0.0[12]
Maldives	2005	100.0	...
Malta	2001	100.0	13.0
Martinique	2001	44.4	44.2
Mauritius	2005	23.0	23.0
Mexico	2005	67.6[13]	35.0[13,14]
Monaco	2005	100.0	100.0

Water: *Waste water*

	Latest year available	Population connected to public waste water collection system	Population connected to public waste water treatment plants
		%	%
Netherlands	2004	98.6	98.6
New Zealand	1999	...	80.0
Norway	2004	80.8	75.9
Palestine	2005	44.7	...
Panama	2005	...	37.0
Paraguay	2005	14.1	...
Peru	2004	74.0	...
Poland	2004	59.0[15]	59.0[15]
Portugal	2003	74.0	60.0[16]
Republic of Moldova	2004	60.0	60.0
Réunion	2001	34.8	33.3
Romania	1990	51.4	...
Singapore	2005	100.0	99.9
Slovakia	2003	55.4	52.3
Slovenia	1999	53.0	30.0
Spain	2002	61.8[17]	55.0[17]
Sweden	2002	85.0	85.0
Switzerland	2004	96.7	96.7
The Former Yugoslav Rep. of Macedonia	2000	49.0	5.0
Trinidad and Tobago	2005	20.0	20.0
Tunisia	2004	52.6	46.8
Turkey	2004	65.7[18]	35.0[18]
United Kingdom	2002	97.7[19]	97.5[19]
United States	1996	71.4	...

Sources:
UNSD/UNEP 2001, 2004 and 2006 questionnaires on Environment statistics, Water section
OECD/Eurostat 2004 questionnaire on Environment statistics, Water section.
OECD Environmental Data, Compendium 2006, Inland Waters section

Footnotes:
1. Refers to reticulated sewerage.
2. Excludes Azores and Madeira Islands.
3. OECD secretariat estimates based on regional data.
4. Percentage of Homes with sewage system.
5. OECD secretariat estimates based on MUD Municipal Waste Water Database.
6. Information provided by the sanitary industry that operates in urban sectors to the Supervision of Sanitation Services respect to the percentage of residential buildings connected to the sewage system.
7. Information provided by the sanitary industry that operates in urban sectors to the Supervision of Sanitation Services respect to the percentage of connected residential buildings to the sewage system whose collected water receive treatment.
8. It corresponds to the city of Santo Domingo and represents 350,063 inhabitants.
9. 2005 data.
10. Data refers to agglomerations greater than or equal to 500 population equivalent.
11. Population connected may include population not connected by pipe.
12. No urban wastewater treatment plant.
13. Percentages based on population living in individual housing.
14. Estimates based on treated waste water.
15. Include population not connected by pipe (whose waste water are collected in septic tanks and delivered to urban waste water treatment plants by truck).
16. Public treatment: includes septic tanks (5% in 1998).
17. OECD secretariat estimates.
18. Data based on a sample survey covering 1911 municipalities.
19. Data refer to England and Wales and to the financial year (April to March).

Definitions & Technical notes:
Waste water refers to water that is discharged as being of no further immediate value for the purpose for which it was used. Public waste water collection system means a systems of conduits which collects and conducts urban waste water. Collecting systems are often operated by public authorities or semi-public associations. Population with access to public waste water collection system is the percentage of the population connected to the public sewerage system. Public waste water collection systems may deliver waste water to treatment plants or may discharge it to the environment, without treatment. Public waste water treatment plants refer to municipal treatment plants operated by official authorities or by private companies whose main activity is waste water treatment on behalf of local authorities. The treatment applied can be :
- mechanical, i.e. separates sludge through processes such as sedimentation, flotation, etc.
- biological, i.e. employs aerobic or anaerobic micro-organisms to separate sludges containing microbial mass together with pollutants.

Water: *Waste water*

- advanced, i.e. all treatments that are not considered mechanical or biological, particularly chemical treatments.

Population connected to waste water treatment is the percentage of the resident population whose waste water is treated at public waste water treatment plants.

Data Quality:
Data on population connected to waste water collection and waste water treatment can be obtained from municipalities or through household surveys. Household surveys usually give more accurate results, since they do not rely on sometimes incomplete information about or held by municipalities. In general, data quality can be considered to be fairly good.

Policy Relevance:
Waste water discharged without treatment into rivers or lakes contributes to eutrophication of the water body, affecting the health of the river or lake ecosystem, reducing the viability of the fish, birds and other beneficial organisms, and the livelihoods of populations that rely on these resources. Waste water discharged into the sea without treatment contributes to eutrophication of coastal waters, also affecting ecosystems. Shellfish living near the discharge point will be contaminated, and where these are harvested by the local population, represent a major health risk. Provision of waste water treatment systems is therefore essential for both environmental and public health.

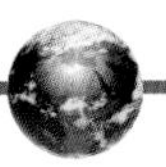

ENVIRONMENTAL INDICATORS

Air Pollution

Emissions of SO_2 from fuel combustion *last update: June 2005*

	Latest year available	SO_2 emissions from fuel combustion	% change since 1990	SO_2 emissions from fuel combustion per capita
		1000 tonnes	%	kg
Algeria*	1995	49.0[1]	...	1.8
Antigua and Barbuda	1990	2.8	..	44.9
Australia	2002	751.0	17.9	38.4
Austria	2002	31.8	-57.3	3.9
Belgium	2002	131.9	-57.3	12.8
Belize	1994	0.5	...	2.6
Bolivia*	2000	10.3	5.0	1.2
Bulgaria	2002	966.9	...	121.4
Cambodia	1994	25.6	...	2.3
Chile	1994	146.2	...	10.4
Colombia	1994	162.5	19.9	4.3
Comoros	1994	0.4	...	0.6
Costa Rica	1996	1.8	...	0.5
Croatia	2002	66.6	-62.4	15.0
Cuba	1996	423.9	1.3	38.5
Czech Republic	2002	230.6	...	22.5
Democratic Republic of the Congo*	1994	0.1	51.7	0.0
Denmark	2002	24.9	-85.9	4.6
Dominican Republic	1994	114.5	51.7	15.1
Estonia	2002	96.3	...	72.0
Ethiopia	1995	13.0	18.2	0.2
Finland	2002	73.1	-54.4	14.1
France	2002	524.2	-57.7	8.8
Georgia*	2002	5.2	-97.9	1.0
Germany	2002	537.6	-89.5	6.5
Greece	2002	491.5	3.4	44.8
Guatemala*	1990	73.7	..	8.4
Haiti	1994	51.3	...	6.9
Honduras	1995	2.1	...	0.4
Hungary	2002	354.9	...	35.8
Iceland	2002	4.4	-12.6	15.3
Ireland	2002	96.0	-47.6	24.6
Israel*	2002	306.4[2]	12.5	48.6
Italy	2002	596.6	-63.9	10.4
Jamaica	1994	98.9	...	40.4
Japan	2002	765.0	-16.4	6.0
Kyrgyzstan	2000	30.2	-73.0	6.1
Latvia	2002	11.7	-87.6	5.0
Lebanon	1994	79.6	...	26.0
Lithuania	2002	45.7	-78.8	13.2
Luxembourg	2002	2.2	...	4.8
Malta	1994	24.3	...	64.9
Mauritius	1995	13.4	...	11.9
Micronesia, Federated States of	1994	0.5	...	5.0
Monaco*	2002	0.0	-27.2	1.3
Morocco	1994	285.0	...	10.8
Netherlands	2002	77.0	-58.2	4.8
New Zealand	2002	51.7	22.2	13.5
Norway	2002	8.8	-59.5	2.0
Peru	1994	105.2	...	4.5

Air Pollution: *Emissions of SO_2 from fuel combustion*

	Latest year available	SO_2 emissions from fuel combustion	% change since 1990	SO_2 emissions from fuel combustion per capita
		1000 tonnes	%	kg
Philippines	1994	433.4	...	6.5
Portugal	2002	269.0	-12.6	26.8
Republic of Moldova	1998	31.7	-88.0	7.4
Romania	2002	632.0	-33.0	28.2
Saint Lucia	1994	0.6	...	4.4
Slovakia	2002	102.3	...	18.9
Slovenia	2002	68.2	-72.4	34.3
Spain	2002	1 909.8	-8.9	46.6
Sri Lanka	1995	41.0	...	2.3
St. Vincent and the Grenadines	1997	0.3	26.8	2.8
Sweden	2002	41.1	-41.3	4.6
Switzerland	2002	13.0	-62.0	1.8
Tajikistan	1998	2.7	-92.1	0.5
Trinidad and Tobago*	1996	8.4[3]	...	6.6
Tunisia	1994	76.4	...	8.7
United Kingdom	2002	959.6	-73.4	16.2
United States	2002	12 469.2	-35.2	42.8
Uruguay	1998	52.3	27.8	15.9
Uzbekistan	1994	247.0	-52.3	11.1
Yemen	1995	2.9	...	0.2

Sources:
UN Framework Convention on Climate Change (UNFCCC) Secretariat (see: http://unfccc.int).
UNSD/UNEP 2004 questionnaire on Environment statistics, Air section.
UN Population Division.
Most data are from UNFCCC, except data for countries with "*" are from UNSD/UNEP 2004 questionnaire.

Footnotes:
1. Emissions from power stations in the north of the country + emissions from main industries + emissions from car traffic in the north of the country.
2. The total refers to emissions from fuel combustion, sectoral approach.
3. Refers to emissions from fuel combustion in energy industries, industry, and transport only.

Definitions & Technical notes:
Data on emissions of SO_2 are usually estimated according to international methodologies on the basis of national statistics on energy, industrial and agricultural production, waste management, etc. The most widely used methodologies are the 1996 Guidelines of the Intergovernmental Panel for Climate Change (IPCC) (see http://www.ipcc-nggip.iges.or.jp/public/gl/invs4.htm) which is the basis for reporting to the UNFCCC. In earlier years the guidelines produced for the UNECE Convention on Long Range Transboundary Air Pollution were widely used in Europe, and are still used in some countries. The main source of SO_2 is burning of fuels, including biomass. Therefore the data shown refer only to emissions from fuel combustion. This covers the combustion of fuels in the energy industries, all other industries and transport (except international aviation and marine transport) as well as small combustion activities such as in commercial, institutional or residential buildings, fuel combustion in agriculture and in all other activities.

Data Quality:
Standardised methods for calculating SO_2 emissions from fuel combustion have been available for many years. The amount of SO_2 emitted is directly related to the sulphur content of the fossil fuels consumed in the country, and the desulphurisation techniques used, if any. Data on emissions from fuel combustion are considered to be reasonable.

Policy Relevance:
SO_2 can be transported over large distances and is partly responsible for acidification of soil and water and for damage to sensitive plants and buildings many kilometres away from the source. The sulphur content of diesel fuels also has an impact on the emissions of particles from diesel engines, and thus impacts on human health. The main anthropogenic source of sulphur dioxide emissions is the combustion of coal, lignite and petroleum products. Some industrial processes also emit sulphur, but these emissions are less well documented, and are therefore not included in this table. While much of the sulphur in petroleum can be removed in the refinery, it is more difficult to remove sulphur from coal and lignite before burning. In this case, other measures can be taken, e.g. scrubbers can be fitted to chimneys at power plants and in large scale industries to remove the SO_2 from the flue gases.

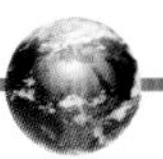

ENVIRONMENTAL INDICATORS

Air Pollution

Emissions of NOx from fuel combustion *last update: June 2005*

	Latest year available	NOx emissions from fuel combustion	% change since 1990 combustion	NOx emissions from fuel per capita
		1000 tonnes	%	kg
Algeria*	1995	177.3[1]	...	6.4
Argentina	1997	703.0	43.2	19.7
Armenia	1990	72.9	..	20.6
Australia	2002	1 611.3	20.5	82.4
Austria	2002	197.9	-1.9	24.4
Barbados	1997	0.1	...	0.2
Belarus	2002	172.0	...	17.3
Belgium	2002	280.7	-19.1	27.3
Belize	1994	2.9	...	14.1
Benin	1995	8.5	...	1.6
Bhutan	1994	0.7	...	0.4
Bolivia*	2000	46.0	46.6	5.5
Bulgaria	2002	114.3	...	14.3
Burkina Faso	1994	4.3	...	0.4
Burundi	1998	10.9	...	1.8
Cambodia	1994	16.7	...	1.5
Cape Verde	1995	0.7	...	1.8
Chile	1994	161.7	...	11.6
Colombia	1994	230.1	-7.9	6.1
Comoros	1994	0.4	...	0.7
Congo	1994	8.0	...	2.8
Costa Rica	1996	24.7	-17.8	6.9
Côte d'Ivoire	1994	114.2	...	8.1
Croatia	2002	92.6	1.9	20.9
Cuba	1996	96.5	-25.0	8.8
Czech Republic	2002	311.3	...	30.4
Democratic Republic of the Congo*	1994	0.1	42.9	0.0
Denmark	2002	197.1	-29.7	36.8
Djibouti	1994	1.4	...	2.6
Dominica	1994	0.4	...	5.8
Dominican Republic	1994	53.4	43.0	7.1
Ecuador	1990	85.9	..	8.4
El Salvador	1994	31.0	...	5.6
Eritrea	1994	0.1	...	0.0
Estonia	2002	46.5	...	34.8
Ethiopia	1995	86.0	13.2	1.5
Finland	2002	210.6	-32.3	40.5
France	2002	1 412.8	-26.4	23.6
Georgia*	2002	16.0	-85.9	3.1
Germany	2002	1 385.9	-48.4	16.8
Greece	2002	314.5	11.0	28.7
Guatemala*	1990	36.9	..	4.2
Guinea	1994	14.2	...	2.0
Guyana	1998	12.0	...	16.0
Haiti	1994	4.6	...	0.6
Honduras	1995	29.9	...	5.3
Hungary	2002	179.0	...	18.0
Iceland	2002	26.0	-2.0	90.6
Indonesia	1994	818.3	...	4.2
Iran (Islamic Republic of)	1994	1 184.4	...	19.3
Ireland	2002	121.2	6.2	31.0

Air Pollution: *Emissions of NOx from fuel combustion*

	Latest year available	NOx emissions from fuel combustion	% change since 1990 combustion	NOx emissions from fuel per capita
		1000 tonnes	%	kg
Israel*	2002	374.1[2]	156.9	59.3
Italy	2002	1 239.3	-35.0	21.6
Jamaica	1994	30.9	...	12.6
Japan	2002	1 854.6	-2.6	14.5
Kazakhstan	1994	165.5	-86.2	9.9
Kenya	1994	46.7	...	1.8
Kyrgyzstan	2000	73.2	-44.8	14.9
Lao People's Dem. Rep.	1990	4.2	0.0	1.0
Latvia	2002	40.1	-51.5	17.2
Lebanon	1994	54.1	...	17.7
Lesotho*	1998	4.9	...	2.8
Lithuania	2002	57.5	-62.9	16.6
Luxembourg	2002	16.2	...	36.2
Mali	1995	6.1	...	0.6
Malta	1994	10.8	...	29.0
Mauritania	1995	6.9	...	3.0
Mauritius	1995	9.8	...	8.7
Mexico	1990	962.8	..	11.6
Micronesia, Federated States of	1994	2.3	...	21.3
Monaco*	2002	0.6	16.1	18.1
Mongolia	1994	2.6	...	1.1
Morocco	1994	152.0	...	5.8
Netherlands	2002	427.7	-27.8	26.6
New Zealand	2002	198.2	47.0	51.5
Nicaragua	1994	17.0	...	3.9
Norway	2002	197.4	-3.0	43.7
Panama*	2002	39.4	...	12.9
Paraguay	1994	2.7	...	0.6
Peru	1994	118.0	...	5.0
Philippines	1994	297.7	...	4.4
Portugal	2002	281.1	12.7	28.0
Republic of Moldova	1998	38.5	-71.8	8.9
Romania	2002	357.2	-28.1	16.0
Saint Lucia	1994	1.3	...	9.7
Senegal	1994	1.2	...	0.1
Seychelles	1995	0.6	...	7.9
Slovakia	2002	101.7	...	18.8
Slovenia	2002	59.6	3.3	30.0
Spain	2002	1 867.4	56.8	45.6
Sri Lanka	1995	59.0	...	3.3
St. Vincent and the Grenadines	1997	0.4	29.5	3.7
Swaziland	1994	7.5	...	8.1
Sweden	2002	228.5	-26.1	25.8
Switzerland	2002	78.4	-48.4	10.9
Tajikistan	1998	9.1	-87.0	1.5
Thailand	1994	271.9	...	4.8
Togo	1995	10.4	...	2.7
Trinidad and Tobago*	1990	35.1[3]	..	28.9
Tunisia	1994	72.0	...	8.2
Turkmenistan	1994	83.6	...	20.4
Tuvalu	1994	0.0	...	0.0
Uganda	1994	23.4	...	1.2
Ukraine	2002	844.6	...	17.3

Air Pollution: *Emissions of NOx from fuel combustion*

	Latest year available	NOx emissions from fuel combustion	% change since 1990 combustion	NOx emissions from fuel per capita
		1000 tonnes	%	kg
United Kingdom	2002	1 579.8	-42.4	26.7
United States	2002	19 043.2	-13.8	65.4
Uruguay	1998	46.2	59.6	14.0
Uzbekistan	1994	242.0	-28.8	10.8
Vanuatu	1994	0.1	...	0.5
Yemen	1995	87.8	...	5.8
Zimbabwe	1994	10.1	9.7	0.9

Sources:
UN Framework Convention on Climate Change (UNFCCC) Secretariat (see: http://unfccc.int).
UNSD/UNEP 2004 questionnaire on Environment statistics, Air section.
UN Population Division.
Most data are from UNFCCC, except data for countries with "*" are from UNSD/UNEP 2004 questionnaire.
Footnotes:
1. Emissions from power stations in the north of the country + emissions from main industries + emissions from car traffic in the north of the country.
2. The total refers to emissions from fuel combustion, sectoral approach.
3. Refers to emissions from fuel combustion in energy industries, industry, and transport only.

Definitions & Technical notes:
Data on emissions of NOx are usually estimated according to international methodologies on the basis of national statistics on energy, industrial and agricultural production, waste management and land use, etc. The most widely used methodologies are the 1996 Guidelines of the Intergovernmental Panel for Climate Change (IPCC) (see http://www.ipcc-nggip.iges.or.jp/public/gl/invs4.htm) which is the basis for reporting to the UNFCCC. In earlier years the guidelines produced for the UNECE Convention on Long Range Transboundary Air Pollution were widely used in Europe, and are still used in some countries. The main source for NOx is burning of fuels, particularly petroleum products. In some countries agriculture and burning of savannas is also an important contributor, but estimating these emissions is more difficult and often data are not available. Therefore the data shown refer only to emissions from fuel combustion. This covers the combustion of fuels in the energy industries, all other industries and transport, except international aviation and marine transport as well as small combustion activities such as in commercial, institutional or residential buildings, fuel combustion in agriculture and in all other activities.

Data Quality:
Although standardised methods for calculating NOx emissions have been available for many years, calculating emissions of NOx is more difficult than for SO_2, as many more parameters need to be taken into account. Therefore the quality of data on NOx emissions is considered to be only fair.

Policy Relevance:
In certain conditions, local NOx emissions in urban areas with high traffic intensity lead to the formation of tropospheric ozone, which affects human health. NOx can be transported over large distances and deposited many kilometres away from the source, contributing to a number of environmental problems, including acidification of soils and eutrophication of soil and water bodies. Emissions arise primarily from the reaction of nitrogen and oxygen during the combustion of fossil fuels, particularly gasoline and diesel fuel, but also from selected production processes. However, road vehicles are the major source. In most developed countries the fitting of catalytic converters to reduce emissions of NOx is obligatory. But with the increased use of road vehicles, particularly for short journeys where the engine has not enough time to reach the temperature needed for the catalyser to function efficiently, NOx emissions are proving more difficult to reduce than SO_2 emissions. This is particularly the case in developing countries where diesel is the most commonly used fuel, the vehicle fleet is old and not well-maintained, and catalytic converters are rare. Together with climatic conditions that favour the production of tropospheric ozone and prevent it from dispersing, this means that urban populations in some developing countries, who also spend more time outdoors and exposed to the smog, suffer more health effects from NOx emissions than the levels of emissions would suggest.

ENVIRONMENTAL INDICATORS

Climate Change

Greenhouse gas emissions *last update: April 2007*

	Latest year available	Total GHG emissions	% change since 1990	GHG emissions per capita
		mio. tonnes of CO_2 equivalent	%	tonnes of CO_2 equivalent/person
Albania	1994	5.53	-22.5	1.72
Algeria	1994	91.61	...	3.35
Antigua and Barbuda	1990	0.39	0.0	6.19
Argentina	1997	279.68	20.6	7.84
Armenia	1990	25.31	0.0	7.14
Australia	2004	529.23	25.1	26.54
Austria	2004	91.30	15.7	11.17
Azerbaijan	1994	42.75	-29.7	5.56
Bahamas	1994	2.20	14.9	7.91
Bahrain	1994	19.47	...	34.34
Bangladesh	1994	45.93	...	0.38
Barbados	1997	4.06	24.8	15.34
Belarus	2004	74.36	-41.6	7.58
Belgium	2004	147.87	1.4	14.22
Belize	1994	6.34	...	30.50
Benin	1995	39.35	...	7.19
Bhutan	1994	1.29	...	0.72
Bolivia	2000	21.46	40.1	2.58
Botswana	1994	9.29	...	6.15
Brazil	1994	658.98	11.1	4.16
Bulgaria	2004	67.51	-41.0	8.68
Burkina Faso	1994	5.97	...	0.60
Burundi	1998	2.00	...	0.33
Cambodia	1994	12.76	...	1.15
Cameroon	1994	165.73	...	12.69
Canada	2004	758.07	26.6	23.72
Cape Verde	1995	0.29	...	0.74
Central African Republic	1994	38.34	...	11.72
Chad	1993	8.02	...	1.26
Chile	1994	54.66	...	3.91
China	1994	4057.31	...	3.36
Colombia	1994	137.49	22.8	3.64
Comoros	1994	0.52	...	0.88
Congo	1994	1.38	...	0.49
Cook Islands	1994	0.08	...	4.22
Costa Rica	1996	10.50	72.2	2.95
Croatia	2004	29.43	-5.4	6.48
Cuba	1996	40.13	-36.9	3.64
Czech Republic	2004	147.11	-25.0	14.38
Dem. Rep. of the Congo	1994	44.53	...	1.03
Denmark	2004	69.62	-1.1	12.86
Djibouti	1994	0.51	...	0.91
Dominica	1994	0.15	-98.8	2.02
Dominican Republic	1994	20.44	...	2.71
Ecuador	1990	30.77	0.0	3.00
Egypt	1990	117.27	0.0	2.10
El Salvador	1994	11.92	...	2.15
Eritrea	2000	0.60	...	0.16
Estonia	2004	21.32	-51.0	15.97
Ethiopia	1995	47.75	11.0	0.83

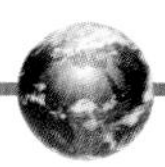

Climate Change: *Greenhouse gas emissions*

	Latest year available	Total GHG emissions	% change since 1990	GHG emissions per capita
		mio. tonnes of CO_2 equivalent	%	tonnes of CO_2 equivalent/person
Fiji	1994	1.39	...	1.83
Finland	2004	81.43	14.5	15.55
France	2004	562.63	-0.8	9.34
Gabon	1994	6.52	...	6.05
Gambia	1993	4.24	...	4.07
Georgia*	2002	19.75	-57.4[1, 2]	3.81
Germany	2004	1015.27	-17.2	12.28
Ghana	1996	13.40	20.1	0.75
Greece	2004	137.63	26.6	12.40
Grenada	1994	1.61	...	19.22
Guatemala	1990	14.74	0.0	1.68
Guinea	1994	5.06	...	0.71
Guyana	1998	3.07	40.8	4.08
Haiti	1994	5.13	...	0.70
Honduras	1995	10.83	...	1.92
Hungary	2004	83.92	-19.4	8.29
Iceland	2004	3.11	-5.1	10.65
India	1994	1214.25	...	1.33
Indonesia	1994	323.26	21.2	1.66
Iran (Islamic Republic of)	1994	385.43	...	6.27
Ireland	2004	68.46	23.1	16.78
Israel*	2000	75.24	...	12.45
Italy	2004	582.52	12.1	10.04
Jamaica	1994	116.23	...	47.44
Japan	2004	1355.17	6.5	10.59
Jordan	1994	21.94	...	5.41
Kazakhstan	1994	219.24	-18.8	13.14
Kenya	1994	21.47	...	0.81
Kiribati	1994	0.03	...	0.39
Korea, Dem. People's Rep.	1990	201.93	0.0	10.12
Korea, Republic of	1990	289.46	0.0	6.75
Kyrgyzstan	2000	15.05	-58.3	3.06
Lao People's Dem. Rep.	1990	6.87	0.0	1.66
Latvia	2004	10.75	-58.5	4.64
Lebanon	1994	15.70	...	5.13
Lesotho	1994	1.82	...	1.10
Liechtenstein	2004	0.27	...	7.89
Lithuania	2004	20.19	...	5.86
Luxembourg	2004	12.72	...	27.71
Madagascar	1994	21.93	...	1.64
Malawi	1994	7.07	...	0.71
Malaysia	1994	136.36	...	6.87
Maldives	1994	0.48	...	1.98
Mali	1995	8.67	...	0.84
Malta	2000	2.85	...	7.32
Mauritania	1995	4.33	...	1.88
Mauritius*	2002	3.36	...	2.78
Mexico	1990	383.08	0.0	4.60
Micronesia, Federated States of	1994	0.25	...	2.36
Monaco	2004	0.10	-7.0	2.87
Mongolia	1998	15.90	...	6.48
Morocco	1994	44.37	...	1.68
Mozambique	1994	8.19	...	0.53

Climate Change: *Greenhouse gas emissions*

	Latest year available	Total GHG emissions	% change since 1990	GHG emissions per capita
		mio. tonnes of CO_2 equivalent	%	tonnes of CO_2 equivalent/person
Namibia	1994	5.60	...	3.51
Nauru	1994	0.04	...	3.81
Nepal	1994	31.19	...	1.52
Netherlands	2004	218.09	...	13.44
New Zealand	2004	75.09	...	18.82
Nicaragua	1994	7.65	...	1.78
Niger	1990	4.86	0.0	0.64
Nigeria	1994	242.63	...	2.51
Niue	1994	4.42	...	2038.75
Norway	2004	54.93	10.3	11.95
Pakistan	1994	160.60	...	1.32
Palau	1994	0.13	...	7.77
Panama	1994	10.69	...	4.09
Papua New Guinea	1994	5.01	...	1.09
Paraguay	1994	140.46	114.6	29.85
Peru	1994	57.58	...	2.46
Philippines	1994	100.87	...	1.51
Poland	2004	388.06	-15.4	10.06
Portugal	2004	84.55[3]	41.0	8.10
Republic of Moldova	1998	10.51	-68.4	2.44
Romania	2004	154.63	-32.8	7.10
Saint Kitts and Nevis	1994	0.16	...	3.71
Saint Lucia	1994	0.89	...	6.42
Samoa	1994	0.56	...	3.41
Senegal	1995	9.57	...	1.15
Seychelles	1995	0.26	...	3.47
Singapore	1994	26.86	...	7.96
Slovakia	2004	51.03	-30.4	9.45
Slovenia	2004	20.06	8.7	10.20
Solomon Islands	1994	0.29	...	0.80
South Africa	1994	379.84	9.4	9.46
Spain	2004	427.90	49.0	10.03
Sri Lanka	1995	29.13	...	1.64
St. Vincent and the Grenadines	1997	0.41	4.6	3.54
Sudan	1995	54.24	...	1.93
Suriname	2003	3.34	...	7.67
Swaziland	1994	2.64	...	2.87
Sweden	2004	69.85	-3.5	7.75
Switzerland	2004	53.02	0.4	7.32
Tajikistan	1998	4.29	-81.9	0.72
Thailand	1994	223.98	...	3.92
The Former Yugoslav Rep. of Macedonia	1998	15.07	-2.4	7.54
Togo	1998	6.28	...	1.47
Tonga	1994	0.23	...	2.32
Trinidad and Tobago	1990	16.39	0.0	13.49
Tunisia	1994	25.14	...	2.85
Turkey	2004	293.81	72.6	4.07
Turkmenistan	1994	52.31	...	12.74
Tuvalu	1994	0.01	...	1.06
Uganda	1994	42.60	...	2.16
Ukraine	2004	413.41	-55.3	8.80
United Kingdom	2004	665.33	-14.3	11.19
United Rep. of Tanzania	1994	39.24	-5.3	1.31

Climate Change: *Greenhouse gas emissions*

	Latest year available	Total GHG emissions	% change since 1990	GHG emissions per capita
		mio. tonnes of CO_2 equivalent	%	tonnes of CO_2 equivalent/person
United States	2004	7067.57	15.8	23.92
Uruguay	1998	33.57	21.4	10.20
Uzbekistan	1994	153.89	-5.7	6.89
Vanuatu	1994	0.30	...	1.79
Venezuela	1999	192.19	...	8.08
Viet Nam	1994	84.45	...	1.18
Yemen	1995	17.87	...	1.18
Zambia	1994	32.77	...	3.59
Zimbabwe	1994	27.59	...	2.40

Sources:
UN Framework Convention on Climate Change (UNFCCC) Secretariat (see: http://unfccc.int).
UNSD/UNEP 2004 questionnaire on Environment statistics, Air section, marked with "*".
UN Population Division.

Footnotes:
1. In 1990, CO_2 and CH_4 data were calculated according to methodology used in the former USSR and the IPCC Guidelines for Greenhouse Gas Inventories, version 1-3 Hadley Centre, UK.
2. In 1990, CO_2 emission refers to emissions from Energy and Industrial processes and CH_4 emission does not include emissions from other fuel combustion.
Definitions & Technical notes:
In this table, greenhouse gases (GHG) refer to carbon dioxide (CO_2), methane (CH_4) and nitrous oxide (N_2O). These three gases account for around 98% of the environmental pressure leading to climate change. Each of these gases has a potential to trap heat in the atmosphere: i.e. methane is 21 times more powerful as a GHG than CO_2, while N_2O is 310 times more powerful. In order to aggregate the three gases to give total emissions of GHG, the data were weighted according to these CO_2 equivalents, also known as Global Warming Potentials. Data on greenhouse gas emissions are usually estimated according to international methodologies on the basis of national statistics on energy, industrial and agricultural production, waste management and land use, etc. The best known and most widely used methodology is the 1996 Guidelines of the Intergovernmental Panel for Climate Change (IPCC) (see http://www.ipcc-nggip.iges.or.jp/public/gl/invs1.htm) which is the basis for reporting to the UNFCCC. The latest revision and update of this guideline is 2006 IPCC Guidelines for National Greenhouse Gas Inventories (see http://www.ipcc-nggip.iges.or.jp/public/2006gl/index.htm).

Data Quality:
Countries should report their greenhouse gas emissions to UNFCCC according to the IPCC Guidelines. The quality of data is regularly checked by UNFCCC for the Annex 1 parties to the Convention that report annually. Non-Annex 1 countries do not report on a regular basis and their data are not subject to the same thorough checking. Data quality depends on the quality of statistics underlying the calculations or estimates and is usually the best for energy related emissions; for other sources, the data should be used with caution when comparing countries.

Policy Relevance:
The Earth's average surface temperature rose by around 0.6°C during the 20th century and most scientific advisors to the world's governments conclude that evidence is growing that most of the warming over the last 50 years is attributable to human activities, such as burning of fossil fuels and deforestation. The resulting increased energy in the weather system is already resulting in increased storms and rainfall in some areas, while others suffer drought. This is expected to increase in future, and while how fast and where this will happen is still controversial, there is consensus in the scientific community that the consequences may be serious. In 1992 the United Nations Conference on Environment and Development, in Rio de Janeiro, adopted the Framework Convention on Climate Change as the basis for global political action. As a result of this convention, commitments to reduce emissions of greenhouse gases were agreed in Kyoto in December 1997. The Kyoto Protocol, which entered into force on 16 February 2005, stipulates that Annex 1 Parties (mainly industrialized countries) shall individually or jointly reduce their aggregate emissions of a "basket" of six greenhouse gases to 5% below 1990 levels by the period 2008-2012. In contrast to this political target the Inter-governmental Panel on Climate Change (IPCC) indicates the need for an immediate 50-70% reduction in global CO_2 emissions in order to stabilise global CO_2 concentrations at the 1990 level by 2100. The World Summit on Sustainable Development (WSSD) held in Johannesburg in 2002 made commitments towards the urgent and substantial increase in the use of renewable (non-carbon) energy sources as well as the setting-up of programmes leading to more sustainable consumption and production patterns, including a reduction in energy use.

ENVIRONMENTAL INDICATORS

Climate Change

CO_2 emissions *last update: April 2007*

	Latest year available	CO_4 emissions	% change since 1990	CO_2 emissions per capita	CO_2 emissions per km²
		mio. tonnes	%	tonne/person	tonne/km²
Afghanistan	2003	0.70	-73.1	0.03	1.08
Albania	2003	3.05	-58.2	0.98	105.92
Algeria	2003	163.95	112.6	5.14	68.83
American Samoa	2003	0.29	2.1	4.72	1467.34
Angola	2003	8.63	85.5	0.57	6.93
Antigua and Barbuda	2003	0.40	32.6	5.01	902.71
Argentina	2003	127.73	16.2	3.36	45.94
Armenia	2003	3.43	...	1.13	115.17
Aruba	2003	2.16	17.2	22.30	11983.33
Australia	2003	371.70	32.3	18.80	48.02
Austria	2003	76.21	24.4	9.40	908.80
Azerbaijan	2003	29.22	...	3.52	337.45
Bahamas	2003	1.87	-4.0	5.96	134.96
Bahrain	2003	21.91	86.8	31.04	31573.49
Bangladesh	2003	34.69	125.5	0.25	240.91
Barbados	2003	1.19	10.7	4.44	2772.09
Belarus	2003	52.59	-48.6	5.30	253.32
Belgium	2003	126.20	6.0	12.20	4133.91
Belize	2003	0.78	149.2	3.01	33.96
Benin	2003	2.05	186.0	0.26	18.16
Bermuda	2003	0.50	-15.6	7.82	9396.23
Bhutan	2003	0.39	202.3	0.19	8.23
Bolivia	2003	7.91	43.6	0.90	7.20
Bosnia and Herzegovina	2003	19.16	...	4.89	374.26
Botswana	2003	4.12	89.7	2.33	7.09
Brazil	2003	298.90	47.3	1.65	35.10
British Virgin Islands	2003	0.08	57.1	3.59	509.93
Brunei Darussalam	2003	4.56	-21.8	12.75	790.63
Bulgaria	2003	53.32	-45.9	6.80	480.74
Burkina Faso	2003	1.04	4.5	0.08	3.80
Burundi	2003	0.24	22.3	0.03	8.48
Cambodia	2003	0.54	18.4	0.04	2.96
Cameroon	2003	3.54	120.3	0.22	7.45
Canada	2003	586.07	27.5	18.50	58.78
Cape Verde	2003	0.14	73.5	0.30	35.71
Cayman Islands	2003	0.30	22.1	7.05	1151.52
Central African Republic	2003	0.25	26.0	0.06	0.40
Chad	2003	0.12	-18.7	0.01	0.09
Chile	2003	58.59	65.6	3.67	77.49
China	2003	4151.41	72.8	3.19	432.58
China, Hong Kong SAR	2003	37.87	44.4	5.50	34454.05
China, Macao SAR	2003	1.87	81.5	4.11	71846.16
Colombia	2003	55.63	-2.2	1.26	48.85
Comoros	2003	0.09	36.9	0.12	39.82
Congo	2003	1.38	17.6	0.37	4.04
Cook Islands	2003	0.03	40.9	1.70	131.36
Costa Rica	2003	6.34	117.0	1.52	124.07
Cote d'Ivoire	2003	5.72	6.1	0.33	17.75
Croatia	2003	23.00	-0.2	5.10	406.81
Cuba	2003	25.30	-21.2	2.25	228.17
Cyprus	2003	7.29	56.6	8.93	788.13

Climate Change: *CO_2 emissions*

	Latest year available	CO_4 emissions	% change since 1990	CO_2 emissions per capita	CO_2 emissions per km²
		mio. tonnes	%	tonne/person	tonne/km²
Czech Republic	2003	127.12	-22.5	12.40	1611.85
Dem. Rep. of the Congo	2003	1.79	-55.0	0.03	0.76
Denmark	2003	60.75	12.0	11.30	1409.71
Djibouti	2003	0.37	3.7	0.48	15.78
Dominica	2003	0.14	137.9	1.76	183.75
Dominican Republic	2003	21.35	122.9	2.47	438.60
Ecuador	2003	23.25	40.1	1.81	81.98
Egypt	2003	139.89	85.1	1.96	139.69
El Salvador	2003	6.55	150.0	0.99	311.44
Equatorial Guinea	2003	0.17	41.9	0.34	5.92
Eritrea	2003	0.70	...	0.17	5.97
Estonia	2003	19.11	-49.9	14.20	423.73
Ethiopia	2003	7.35	147.5	0.10	6.65
Faeroe Islands	2003	0.66	7.0	14.22	474.52
Falkland Islands (Malvinas)	2003	0.05	21.1	15.06	3.78
Fiji	2003	1.12	37.3	1.34	61.29
Finland	2003	73.19	30.0	14.00	216.45
France	2003	408.16	2.8	6.80	740.09
French Guiana	2003	1.00	24.8	5.64	11.17
French Polynesia	2003	0.69	13.6	2.79	173.50
Gabon	2003	1.23	-79.6	0.91	4.58
Gambia	2003	0.28	48.2	0.20	25.06
Georgia	2003	3.73	...	0.82	53.54
Germany	2003	865.37	-14.7	10.50	2423.86
Ghana	2003	7.74	105.4	0.37	32.47
Gibraltar	2003	0.36	495.1	13.04	60500.00
Greece	2003	109.98	30.9	9.90	833.45
Greenland	2003	0.57	2.7	10.03	0.26
Grenada	2003	0.22	84.2	2.17	642.44
Guadeloupe	2003	1.71	33.5	3.88	1004.69
Guam	2003	4.09	80.0	24.95	7444.44
Guatemala	2003	10.71	110.3	0.89	98.37
Guinea	2003	1.34	32.2	0.15	5.45
Guinea-Bissau	2003	0.27	29.2	0.18	7.47
Guyana	2003	1.63	43.9	2.18	7.59
Haiti	2003	1.74	75.0	0.21	62.74
Honduras	2003	6.51	150.9	0.94	58.05
Hungary	2003	60.46	-28.7	6.00	649.88
Iceland	2003	2.18	4.8	7.50	21.17
India	2003	1275.61	87.9	1.19	388.05
Indonesia	2003	295.60	97.7	1.36	155.20
Iran (Islamic Republic of)	2003	382.09	74.8	5.60	231.82
Iraq	2003	73.01	50.2	2.67	166.56
Ireland	2003	44.45	39.8	11.10	632.53
Israel	2003	68.43	106.2	10.57	3089.95
Italy	2003	487.28	13.2	8.40	1617.16
Jamaica	2003	10.74	34.7	4.09	976.89
Japan	2003	1259.43	12.2	9.90	3332.95
Jordan	2003	17.12	67.8	3.16	191.59
Kazakhstan	2003	159.49	...	10.74	58.53
Kenya	2003	8.79	50.7	0.27	15.15
Kiribati	2003	0.03	40.9	0.32	42.70
Korea, Dem. People's Rep.	2003	77.60	-68.3	3.48	643.79

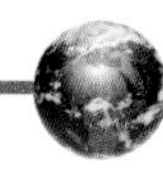

Climate Change: *CO_2 emissions*

	Latest year available	CO_4 emissions	% change since 1990	CO_2 emissions per capita	CO_2 emissions per km²
		mio. tonnes	%	tonne/person	tonne/km²
Korea, Republic of	2003	456.75	89.1	9.62	4588.71
Kuwait	2003	78.60[1]	73.4	31.13[1]	4411.38
Kyrgyzstan	2003	5.33	...	1.04	26.65
Lao People's Dem. Rep.	2003	1.25	445.2	0.22	5.30
Latvia	2003	7.43	-60.2	3.20	115.02
Lebanon	2003	19.00	108.6	5.42	1826.73
Lesotho	1994	0.64	...	0.38	20.95
Liberia	2003	0.46	-0.6	0.14	4.17
Libyan Arab Jamahiriya	2003	50.27	32.9	8.93	28.57
Liechtenstein	2003	0.24	4.3	7.10	1500.00
Lithuania	2003	12.29	-68.4	3.60	188.21
Luxembourg	2003	10.69	-16.2	23.60	4133.80
Madagascar	2003	2.35	148.7	0.13	3.99
Malawi	2003	0.88	47.3	0.07	7.47
Malaysia	2003	156.68	183.0	6.41	475.01
Maldives	2003	0.44	187.0	1.41	1483.22
Mali	2003	0.55	31.0	0.04	0.45
Malta	2003	2.47	10.4	6.20	7806.96
Martinique	2003	1.34	-35.0	3.42	1216.88
Mauritania	2003	2.50	-5.2	0.87	2.44
Mauritius	2003	3.15	115.0	2.58	1544.12
Mexico	2003	416.70	10.9	3.99	212.80
Micronesia, Federated States of	1994	0.24	...	2.20	336.14
Monaco	2003	0.13	44.4	3.80	65000.00
Mongolia	2003	7.99	-20.1	3.09	5.11
Montserrat	2003	0.06	79.4	15.99	598.04
Morocco	2003	37.97	61.4	1.24	85.03
Mozambique	2003	1.57	57.3	0.08	1.96
Myanmar	2003	9.47	121.6	0.19	13.99
Namibia	2003	2.33	38750.0	1.17	2.83
Nauru	2003	0.14	6.8	10.76	6714.29
Nepal	2003	2.95	367.6	0.11	20.08
Netherlands	2003	176.86	11.9	11.00	4258.81
Netherlands Antilles	2003	4.06	236.8	22.70	5073.75
New Caledonia	2003	1.87	16.0	8.20	100.78
New Zealand	2003	34.70	37.1	8.80	128.26
Nicaragua	2003	3.92	47.9	0.74	30.13
Niger	2003	1.21	15.0	0.09	0.95
Nigeria	2003	52.28	15.1	0.42	56.59
Niue	2003	0.00	0.0	2.01	11.54
Norway	2003	43.22	25.6	9.40	112.21
Oman	2003	32.31	214.1	12.87	104.39
Pakistan	2003	114.36	67.8	0.75	143.65
Palau	2003	0.24	3.8	12.30	529.41
Panama	2003	6.03	92.5	1.93	79.92
Papua New Guinea	2003	2.52	3.4	0.44	5.43
Paraguay	2003	4.14	83.0	0.70	10.19
Peru	2003	26.20	24.4	0.96	20.38
Philippines	2003	77.10	75.3	0.96	256.98
Poland	2002	308.28	-35.3	...	985.91
Portugal	2003	64.29	47.4	6.20	698.94
Puerto Rico	2003	2.11	-82.1	0.54	237.18
Qatar	2003	46.26	279.1	63.09	4205.64

Climate Change: *CO_2 emissions*

	Latest year available	CO_4 emissions	% change since 1990	CO_2 emissions per capita	CO_2 emissions per km^2
		mio. tonnes	%	tonne/person	tonne/km^2
Republic of Moldova	2003	7.24	...	1.71	213.88
Réunion	2003	2.48	101.7	3.26	987.65
Romania	2003	111.39	-39.5	5.10	467.26
Russian Federation	1999	1509.00	-36.1	...	88.25
Rwanda	2003	0.60	13.6	0.07	22.86
Saint Helena	2003	0.01	100.0	2.46	38.96
Saint Kitts and Nevis	2003	0.13	93.8	3.02	482.76
Saint Lucia	2003	0.33	100.0	2.06	604.82
Saint Pierre and Miquelon	2003	0.06	-29.3	11.32	268.60
Samoa	2003	0.15	19.8	0.83	53.34
Sao Tome and Principe	2003	0.09	35.3	0.62	95.44
Saudi Arabia	2003	302.88[1]	53.2	12.98[1]	140.90
Senegal	2003	4.85	54.5	0.44	24.63
Serbia and Montenegro	2003	50.02	...	4.76	489.59
Seychelles	2003	0.55	379.8	6.91	1202.20
Sierra Leone	2003	0.65	94.9	0.13	9.10
Singapore	2003	47.88	6.1	11.35	70109.81
Slovakia	2003	43.05	-27.6	8.00	877.98
Slovenia	2003	16.10	0.6	8.20	794.83
Solomon Islands	2003	0.18	9.2	0.39	6.16
Somalia	1997	0.00	...	0.00	0.00
South Africa	2003	364.85	27.6	7.78	298.81
Spain	2003	331.76	45.3	7.90	655.66
Sri Lanka	2003	10.32	174.1	0.51	157.31
St. Vincent and the Grenadines	2003	0.19	142.5	1.65	500.00
Sudan	2003	9.01	67.0	0.26	3.59
Suriname	2003	2.24	23.7	5.05	13.69
Swaziland	2003	0.96	125.2	0.92	55.11
Sweden	2003	56.00	-0.5	6.20	124.45
Switzerland	2003	44.72	0.8	6.20	1083.23
Syrian Arab Republic	2003	49.04	36.6	2.70	264.80
Tajikistan	2003	4.66	...	0.73	32.58
Thailand	2003	246.37	156.9	3.90	480.15
The Former Yugoslav Rep. of Macedonia	2003	10.55	...	5.20	410.10
Timor-Leste	2003	0.16	...	0.20	10.96
Togo	2003	2.20	192.6	0.38	38.74
Tonga	2003	0.11	48.1	1.12	152.61
Trinidad and Tobago	2003	28.70	69.4	22.12	5594.35
Tunisia	2003	20.91	57.4	2.11	127.80
Turkey	2003	220.41	50.5	3.09	281.29
Turkmenistan	2003	43.41	...	9.24	88.94
Turks and Caicos Islands	1990	0.00	...	0.00	0.00
Tuvalu	1994	0.00	...	0.48	178.85
Uganda	2003	1.71	110.2	0.06	7.11
Ukraine	2003	313.14	-57.6	6.60	518.70
United Arab Emirates	2003	135.29	147.0	33.56	1618.24
United Kingdom	2003	557.46	-5.3	9.40	2295.02
United Rep. of Tanzania	2003	3.81	62.9	0.10	4.03
United States	2003	5841.50	16.6	20.00	606.65
United States Virgin Islands	2003	13.55	60.1	121.30	39043.23
Uruguay	2003	4.38	11.9	1.28	25.03
Uzbekistan	2003	123.84	...	4.79	276.80
Vanuatu	2003	0.09	30.9	0.44	7.30

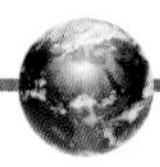

Climate Change: *CO_2 emissions*

	Latest year available	CO_4 emissions	% change since 1990	CO_2 emissions per capita	CO_2 emissions per km²
		mio. tonnes	%	tonne/person	tonne/km²
Venezuela	2003	144.23	22.7	5.59	158.13
Viet Nam	2003	76.24	255.9	0.93	229.86
Western Sahara	2003	0.24	21.8	0.75	0.90
Yemen	2003	17.08	...	0.87	32.35
Zambia	2003	2.20	-10.2	0.19	2.92
Zimbabwe	2003	11.49	-31.1	0.89	29.40

Sources:
UNSD Millennium Development Goals Indicators database (see http://mdgs.un.org/unsd/mdg/Data.aspx)
UN Population Division.
UNSD Demographic Yearbook (see: http://unstats.un.org/unsd/demographic/products/dyb/dyb2004.htm)

Footnotes:
1. Including part of the Neutral Zone.

Definitions & Technical notes:
CO_2 emissions from energy industry, from transport, from fuel combustion in industry, services, households, etc. and industrial processes, such as the production of cement. Changes in how land is used can also result in the emission of CO_2, or in the removal of CO_2 from the atmosphere. However, as there is not yet an agreed method for estimating this, it is not included in the figures for CO_2 emissions. Burning of biomass such as wood and straw also emits CO_2; however, unless there has been a change in land use, it is considered that CO_2 emitted from biomass is removed from the air by new growth, and therefore it should not included in the total for CO_2.

Data Quality:
For Annex 1 countries, data are from UNFCCC. UNFCCC has developed standardised methods for calculating CO_2 emissions, which are widely used. For non-Annex 1 countries, data are from estimates of CO_2 emissions made by the Carbon Dioxide Information Analysis Center (CDIAC) (see: http://cdiac.ornl.gov/). CDIAC acquires or compiles, quality assures, documents, archives, and distributes data and other information concerning carbon dioxide.

Policy Relevance:
See table on greenhouse gas emissions. CO_2 is by far the largest contributor to global warming, and the major source of CO_2 is combustion of fossil fuels. The Inter-governmental Panel on Climate Change (IPCC) indicates the need for an immediate 50-70% reduction in global CO_2 emissions in order to stabilise global CO_2 concentrations at the 1990 level by 2100. Various policy options are available to reduce emissions, including energy efficiency measures and switching to less carbon intensive fuels, e.g. from burning coal and lignite to natural gas. The World Summit on Sustainable Development (WSSD) held in Johannesburg in 2002 made commitments towards the urgent and substantial increase in the use of renewable non-carbon energy sources, such as wind, wave and solar power, but also including biomass. It also urged the setting-up of programmes to promote sustainable consumption and production patterns which should lead to reduced CO_2 emissions.

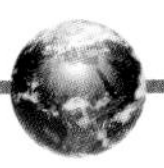

ENVIRONMENTAL INDICATORS

Climate Change

Emissions of other greenhouse gases *last update: April 2007*

	Latest year available	CH_4 emissions	% change since 1990	N_20 emissions	% change since 1990
		mio. tonnes of CO_2 equivalent	%	mio. tonnes of CO_2 equivalent	%
Albania	1994	2.14	...	0.29	...
Algeria	1994	18.77	...	9.13	...
Antigua and Barbuda	1990	0.10	0.0	0.00	0.0
Argentina	1997	87.58	15.2	60.73	-1.6
Armenia	1990	3.21	0.0	0.09	0.0
Australia	2004	117.87	-1.4	24.20	30.0
Austria	2004	7.41	-19.2	5.28	-15.4
Azerbaijan	1994	9.28	-38.9	0.66	-26.6
Bahamas	1994	0.02	0.0	0.31	...
Bahrain	1994	2.94	...	0.04	...
Bangladesh	1994	25.01	...	4.46	...
Barbados	1997	1.81	9.3	0.05	0.0
Belarus	2004	12.65	-16.4	6.72	-34.7
Belgium	2004	7.92	-26.8	11.21	-6.7
Belize	1994	5.57	...	0.17	...
Benin	1995	38.01	...	0.53	...
Bhutan	1994	0.40	...	0.66	...
Bolivia	2000	12.77	36.5	0.92	58.6
Botswana	1994	4.24	...	2.04	...
Brazil	1994	238.73	6.4	166.87	12.4
Bulgaria	2004	9.77	-47.8	4.40	-57.9
Burkina Faso	1994	4.70	...	0.37	...
Burundi	1998	0.94	...	0.91	...
Cambodia	1994	7.77	...	3.67	...
Cameroon	1994	17.71	...	145.25	...
Canada	2004	110.23	34.6	43.98	-2.5
Cape Verde	1995	0.07	...	0.01	...
Central African Republic	1994	11.84	...	26.29	...
Chad	1993	6.94	...	0.77	...
Chile	1994	10.12	...	7.39	...
China	1994	720.03	...	263.81	...
Colombia	1994	48.22	8.2	28.35	15.0
Comoros	1994	0.05	...	0.39	...
Congo	1994	0.58	...	0.12	...
Cook Islands	1994	0.01	...	0.04	...
Costa Rica	1996	3.68	16.5	2.27	1121.5
Cote d'Ivoire	1994	18.68	...	1.70	...
Croatia	2004	3.01	-6.7	3.68	-6.2
Cuba	1996	6.58	-38.4	7.03	-59.7
Czech Republic	2004	10.83	-41.6	8.31	-34.0
Dem. Rep. of the Congo	1994	40.64	...	2.56	...
Denmark	2004	5.79	1.5	7.63	-28.1
Djibouti	1994	0.24	...	0.00	...
Dominica	1994	0.06	...	0.01	...
Dominican Republic	1994	4.66	53.3	0.78	-7.4
Ecuador	1990	10.57	0.0	0.17	0.0
Egypt	1990	22.17	0.0	10.63	0.0
El Salvador	1994	3.11	...	4.09	...
Estonia	2004	1.73	-60.4	0.36	-64.4
Ethiopia	1995	37.46	-0.3	7.44	140.0
Fiji	1994	0.54	...	0.03	...

Climate Change: *Emissions of other greenhouse gases*

	Latest year available	CH_4 emissions	% change since 1990	N_2O emissions	% change since 1990
		mio. tonnes of CO_2 equivalent	%	mio. tonnes of CO_2 equivalent	%
Vanuatu	1994	0.23	...	0.01	...
Venezuela	1999	61.92	...	16.15	...
Viet Nam	1994	48.89	...	10.17	...
Yemen	1995	2.69	...	4.66	...
Zambia	1994	10.70	...	19.47	...
Zimbabwe	1994	7.52	...	2.98	...

Sources:
UN Framework Convention on Climate Change (UNFCCC) Secretariat (see: http://unfccc.int).
UNSD/UNEP 2004 questionnaire on Environment statistics, Air section, marked with "*".

Footnotes:
1. Data calculated according to methodology used in the former USSR and the IPCC Guidelines for Greenhouse Gas Inventories, version 1-3 Hadley Centre, UK.
2. Data does not include emissions from other fuel combustion.

Definitions & Technical notes:
CH_4 emissions: the major sources of CH_4 are leakages during the production and transportation of natural gas and coal mining, livestock rearing, rice cultivation, and decomposition of waste in landfills. N_2O emissions: the major sources of N_2O are agriculture and industrial processes.

Data Quality:
UNFCCC has developed standardised methods for calculating greenhouse gas emissions, which are widely used. However, estimation of emissions from sources other than energy are complex. So while trend figures are fairly good, absolute levels are less reliable.

Policy Relevance:
See table on greenhouse gas emissions. Both CH_4 and N_2O are powerful greenhouse gases, many times more powerful than CO_2. CH_4, otherwise known as methane or natural gas, is a high quality fuel. Because of the different global warming potentials of the two gases, burning methane to produce CO_2 contributes less to global warming than emitting the CH_4 directly to the atmosphere. Measures to reduce leakage of methane from natural gas networks, or to capture fugitive emissions resulting from coal mining and oil and gas extraction will pay for themselves, over time. When waste landfill sites are fitted with methane recuperation systems, the gas recovered can often be used locally for heating or generating electricity, thus reducing the need for other energy sources. Similarly, the use of anaerobic digestors for farmyard manure can provide the farm with a free supply of natural gas, for energy purposes. Emissions of N_2O can be reduced by introducing proper manure handling techniques.

ENVIRONMENTAL INDICATORS

Waste

Municipal waste collection *last update: April 2007*

	Latest year available	Municipal waste collected	Population served by municipal waste collection	Municipal waste collected per capita served
		1000 tonnes	%	kg
Albania	2005	634	...	...
Algeria	2003	8500	80.0	334
Andorra	2005	38	100.0	563
Anguilla	2005	5	100.0	433
Antigua and Barbuda	2005	21[1]	100.0[1]	...
Armenia	2004	376	65.2	191
Australia	2003	8903	...	...
Austria	2004	4588	100.0	562
Azerbaijan	2005	1753[2]	...	...
Belarus	2004	2661	85.0	319
Belgium	2003	4608	100.0	447
Belize	2003	86	51.2	655
Benin	2002	986	23.0	654
Bolivia	2005	751	...	...
Bosnia and Herzegovina	1999	1765	...	...
Brazil	2000	57563	76.0	441
British Virgin Islands	2005	37	...	...
Brunei Darussalam	2002	196	...	...
Bulgaria	2002	3199	81.1	495
Canada	2004	13375[3]	99.0[4]	423[3]
Chile	2005	5459	...	...
China	2003	148565	...	...
China, Hong Kong SAR	2005	6013	100.0	854
China, Macao SAR	2005	163[5]	100.0	354[5]
Colombia	2005	20776	95.0	480
Costa Rica	2002	1280	73.0	428
Croatia	2004	1079	86.0	276
Cuba	2005	4416	75.6	519
Cyprus	2002	500	...	...
Czech Republic	2004	2841[6]	100.0	278
Denmark	2003	3618	100.0	675
Dominica	2005	21	94.0	282
Dominican Republic	2005	1016[7]	...	...
Egypt	2001	14500	...	...
Estonia	2002	524	...	...
Finland	2004	2374[8]	100.0	453
France	2005	33963	100.0	561
French Guiana	2003	110	89.0	695
Georgia	2005	1375	56.0	549
Germany	2004	48434	100.0	586
Greece	2003	4710	100.0	429
Guadeloupe	1999	217	100.0	511
Guatemala	2002	604	30.5	165
Hungary	2003	4387	89.5	496
Iceland	2004	147	100.0	503
India	2001	17569[9]	...	...
Ireland	2005	2847[10]	76.0	903
Israel	2003	5527	...	...
Italy	2005	31677	100.0	545
Jamaica	2004	709	...	...

Waste: *Municipal waste collection*

	Latest year available	Municipal waste collected	Population served by municipal waste collection	Municipal waste collected per capita served
		1000 tonnes	%	kg
Japan	2003	54367[11]	99.8	427
Jordan	2002	2227	...	...
Korea, Republic of	2004	18252	99.3[12]	386
Kuwait	2005	837	...	...
Kyrgyzstan	2004	1602[2]	...	...
Latvia	1999	292	50.0	244
Lebanon	2001	1440	...	...
Lithuania	2002	1000	...	...
Luxembourg	2003	306[13]	100.0	676
Madagascar	2004	341[14]	4.4	...
Maldives	2005	19	...	...
Malta	2003	218	...	...
Martinique	2004	340	100.0	863
Mauritius	2003	351	95.0	303
Mexico	2006	36088	90.0	...
Monaco	2002	40	100.0	1180
Morocco	2003	4710	...	...
Nepal	2002	418	...	...
Netherlands	2004	10161	100.0	626
New Zealand	1999	1541[15]	...	...
Niger	2005	9750	...	...
Norway	2004	1746[3]	99.0	384
Palestine	2001	1350[16]	...	...
Panama	1998	379	...	...
Peru	2001	4740	75.0	240
Poland	2005	9354	...	...
Portugal	2005	5009	100.0	477
Republic of Moldova	2004	1224	...	...
Réunion	2004	461	-999.0	-60
Romania	2002	6865	90.0	341
Russian Federation	2000	207400[2,17]	...	...
Serbia and Montenegro	2005	2890	...	...
Singapore	2005	5088	100.0	1176
Slovakia	2005	1468	100.0	272
Slovenia	2002	862	93.0	467
Spain	2004	22735[18]	...	...
Sri Lanka	2004	1036	...	...
St. Vincent and the Grenadines	2002	38	100.0	317
Sweden	2005	4347	100.0	481
Switzerland	2005	4855	99.0	676
Syrian Arab Republic	2003	7500	...	...
Thailand	2000	13972	...	...
The Former Yugoslav Rep. of Macedonia	2005	2526	...	...
Trinidad and Tobago	2002	425[19]	...	...
Tunisia	2004	1316	65.0	203
Turkey	2004	24237	72.8	461
Ukraine	2004	3235	...	...
United Kingdom	2005	35077	100.0	588
United States	2005	222863	100.0	747
Uruguay	2000	910	...	...
Yemen	2005	1272	...	...
Zambia	2005	389[20]	20.0	167

Waste: *Municipal waste collection*

Sources:
UNSD/UNEP 2001, 2004 and 2006 questionnaires on Environment statistics, Waste section.
OECD/Eurostat 2004 questionnaire on Environment statistics, Waste section
OECD Environmental Data, Compendium 2006/2007, Waste section.
UN Population Division.

Footnotes:
1. Data refer to Antigua only.
2. Unit: thousand cubic meters.
3. Household waste generated only.
4. 1996 data.
5. Data only refer to waste collected from households and sea by a licensed company.
6. Includes amounts undergoing mechanical sorting before treatment/disposal.
7. The information Includes the National District (Capital of the Republic) and the Santo Domingo Province.
8. Data refer to total amounts of municipal waste managed.
9. Total municipal solid waste generated in 299 Class-I cities.
10. Data refer to municipal waste landfilled and recovered (include street cleansing waste).
11. Data refer to waste treated by municipalities and separate collection for recycling by the private sector.
12. 2002 data.
13. Data refer to total amounts of municipal waste managed in the country (exclude exported amounts).
14. For the calculations, only the six important locations of "Faritany" (Antanarivo, Antsiranana, Fianarantsoa, Mahajanga, Toamasia, Toliary) and the cities of Toalagnaro and Nosy-be were taken into account.
15. Data include landfilled household waste and recycled packaging waste.
16. Data refer to solid waste reaching dumping site which was taken from the Dumping Site Survey implemented in 2001.
17. Data refer to municipal waste, specifically rubbish transported by trucks and liquid wastes transported by cesspool trucks.
18. Household and similar waste.
19. Data are from Trinidad and Tobago Solid Waste Management Company Limited (SWMCOL). The landfills managed by SWMCOL collect 85% of solid waste.
20. Data refer to urban population only.

Definitions & Technical notes:
Municipal waste includes household waste and similar waste. The definition also includes bulky waste (e.g. white goods, old furniture, mattresses) and yard waste, leaves, grass clippings, street sweepings, the content of litter containers, and market cleansing waste, if managed as waste. It includes waste originating from: households, commerce and trade, small businesses, office buildings and institutions (schools, hospitals, government buildings). It also includes waste from selected municipal services, e.g. waste from park and garden maintenance, waste from street cleaning services (street sweepings, the content of litter containers, market cleansing waste), if managed as waste. The definition excludes waste from municipal sewage network and treatment, municipal construction and demolition waste. Municipal waste collected refers to waste collected by or on behalf of municipalities, as well as municipal waste collected by the private sector. It includes mixed household waste, and fractions collected separately for recovery operations (through door-to-door collection and/or through voluntary deposits). If data for municipal waste collected are not available, data for municipal waste generated is given, if available. Municipal waste collected per capita served is calculated by dividing the Municipal waste collected by the number of people served by the waste collection system.

Data Quality:
Data on municipal waste collected are usually gathered through surveys of municipalities, which are responsible for waste collection and disposal, or from transport companies that collect waste and transport it to a disposal site. Such surveys deliver fairly reliable data. However, it must be remembered that the figures only cover waste collected by or on behalf of municipalities. Therefore:
- Amounts of waste will vary, depending on how far municipal waste collection covers small industries and the services sector.
- Waste collected by the informal sector, waste generated in areas not covered by the municipal waste collection system or illegally dumped waste are nor included. Caution is therefore advised when comparing countries.

Policy Relevance:
Although on a 'per kilogram' basis, municipal waste is less damaging than hazardous waste, the large number of sources (households, services, small industries), as well as the variety of wastes included and the sheer quantities generated, make the collection and disposal of municipal waste an important issue worldwide. The amount of waste a country generates depends on a number of factors, including GDP, the extent of urbanization, family structures, and lifestyles. Increasing urbanization, economic growth and the move away from traditional family groups have resulted in an increase in the amount of waste generated in recent decades. Waste management, i.e. waste collection and treatment, has become an independent economic sector, as waste becomes an environmental problem of growing concern.

The environmental impacts that are most closely associated with waste are:
- pollution of ground and surface water, through leaching and run-off;
- soil contamination and damage to nature;
- emissions of methane, a powerful greenhouse gas, from landfill sites;
- risks to health due to putrification of food waste,
- emission of dusts, odours and hazardous gases and
- unregulated fires.

The quantity of municipal waste generated will be larger than that collected if large areas of the country are not served by waste collection or if a significant percentage of illegal dumping of waste is suspected. The associated environmental impacts will also be greater, as uncontrolled landfill is generally more environmentally damaging. Some towns and cities rely heavily on the informal sector to collect and recycle household waste, and this may be the sole source of income for whole families, with women and children also actively involved. As this is totally unregulated, the workers are often subject to accidents, to respiratory illnesses, to skin infections and other health problems.

ENVIRONMENTAL INDICATORS

Waste

Municipal waste treatment *last update: April 2007*

	Latest year available	Municipal waste collected	Municipal waste landfilled	Municipal waste incinerated	Municipal waste recycled	Municipal waste recycled
		1000 tonnes	%	%	%	%
Albania	2005	634	95.1	0.0	4.9	0.0
Algeria	2003	8500	99.9	...	0.1	...
Andorra	2005	38	0.0	0.0	0.0	0.0
Anguilla	2005	5	100.0	0.0	0.0	0.0
Antigua and Barbuda	2005	21[1]	100.0[1]	...	...	...
Armenia	2004	376	100.0	...	...	...
Australia	2003	8903	69.7	...	30.3	...
Austria	2004	4588	6.7[2]	21.1	26.5	44.7[3]
Belarus	2004	2661	100.0	...	...	...
Belgium	2003	4608	12.6[5]	35.7	31.3[5]	22.8
Belize	2003	86	100.0	...	...	...
Benin	2002	986	0.0	0.0	...	...
Brazil	2000	57563	62.7	0.3	1.4	4.1
British Virgin Islands	2005	37	0.0	80.3[6]	0.0	0.0
Bulgaria	2002	3199	99.7	...	...	...
Canada	2004	13375[7]	73.3[8]	...	26.8	12.5[9]
Chile	2005	5459	100.0	...	...	...
China	2003	148565	43.1	2.5	...	4.8
China, Hong Kong SAR	2005	6013	56.9	...	43.1	...
China, Macao SAR	2005	163[10]	39.7	...	...	...
Colombia	2005	20776	80.4	...	0.9	...
Croatia	2004	1079	96.1	...	2.5	1.4
Cuba	2005	4416	84.1	0.0	4.8	11.1
Cyprus	2002	500	...	0.0	...	0.0
Czech Republic	2004	2841[11]	79.8	14.0	1.3	3.2
Denmark	2003	3618	5.1	54.0[12]	25.6	15.3
Dominica	2005	21	100.0	...	...	...
Finland	2004	2374[13]	59.9	9.9	30.1[14]	...
France	2005	33963	36.0	33.8	15.8	14.3
French Guiana	2003	110	72.7	...	...	...
Germany	2004	48434	17.7	24.6	33.1	17.1
Greece	2003	4710	91.9	0.0	8.1	0.0
Guadeloupe	1999	217	98.0	2.0	0.0	0.0
Hungary	2003	4387	90.4[15]	5.6	2.7	1.1
Iceland	2004	147	72.1	8.8	15.6	8.8
Ireland	2005	2847[16]	66.1	...	33.9[14]	...
Israel	2003	5527	79.0	0.0	21.0[17]	...
Italy	2005	31677	54.4	12.1[18]	...	33.3[19]
Japan	2003	54367[20]	3.4[21]	74.0	16.8[22]	...
Korea, Republic of	2004	18252	36.4	14.4	49.2	0.0
Kyrgyzstan	2004	1602[4]	100.0[4]	...	...	...
Latvia	1999	292	100.0	...	...	...
Lebanon	2001	1440	41.6	...	3.3	7.6
Lithuania	2002	1000	100.0	0.0	0.0	0.0
Luxembourg	2003	306[23]	18.9	38.9	23.2	19.3
Madagascar	2004	341[24]	100.0[24]	0.0	0.0	0.0
Martinique	2004	340	67.4	31.9	0.0	0.0
Mauritius	2003	351	100.0	...	...	...
Mexico	2006	36088	96.7	0.0	3.3	0.0
Monaco	2002	40	56.5[25]	...	3.7[17]	...
Morocco	2003	4710	90.0	...	10.0	...

Waste: *Municipal waste treatment*

	Latest year available	Municipal waste collected	Municipal waste landfilled	Municipal waste incinerated	Municipal waste recycled	Municipal waste recycled
		1000 tonnes	%	%	%	%
Netherlands	2004	10161	1.7	32.3	25.4	23.5
New Zealand	1999	1541[26]	84.7[27]	...	15.3[28]	...
Niger	2005	9750	64.0	12.0	4.0	...
Norway	2004	1746[7]	25.9	24.7[29]	33.6[30]	15.3
Palestine	2001	1350[31]	100.0[31]	...	...	...
Panama	1998	379	100.0	...	...	...
Peru	2001	4740	65.7	...	14.7	...
Poland	2005	9354	92.2	0.5	3.9	3.4
Portugal	2005	5009	64.1	21.1	8.6	6.3
Réunion	2004	461	89.8	...	...	5.6
Romania	2002	6865	97.5	...	2.5	...
Singapore	2005	5088	15.8	44.8	39.4	0.0
Slovakia	2005	1468	77.9	12.5	1.1	1.4
Slovenia	2002	862	81.1	0.6	10.1	1.3
Spain	2004	22735[32]	51.7	6.7	9.0[33]	32.7
St. Vincent and the Grenadines	2002	38	84.9	0.0	15.1[17]	...
Sweden	2005	4347	4.8	50.2	33.9	10.5
Switzerland	2005	4855	0.5	49.8	33.9[34]	15.9
Syrian Arab Republic	2003	7500	93.9[35]	5.3	1.1[36]	...
Thailand	2000	13972	...	0.8	14.3[17]	...
Tunisia	2004	1316	99.9	...	...	0.1
Turkey	2004	24237	97.8	0.0	0.0	1.4
United Kingdom	2005	35077	64.3	8.4	17.4	9.3
United States	2005	222863	54.3[37]	13.6[38]	23.8	8.4
Uruguay	2000	910	...	...	0.0	0.0
Yemen	2005	1272	100.0	...	...	...

Sources:
UNSD/UNEP 2001, 2004 and 2006 questionnaires on Environment statistics, Waste section.
OECD/Eurostat 2004 questionnaire on Environment statistics, Waste section.
OECD Environmental Data, Compendium 2006/2007, Waste section.

Footnotes:
1. Data refer to Antigua only.
2. Direct delivery without any pretreatment.
3. Includes amounts treated in mechanical-biological facilities.
4. Unit: thousand cubic meters.
5. Includes residues from incineration.
6. Value refers to the main island of Tortola only.
7. Household waste generated only.
8. Data refer to household waste landfilled or incinerated.
9. Composting: from residential and non-residential sources.
10. Data only refer to waste collected from households and sea by a licensed company.
11. Includes amounts undergoing mechanical sorting before treatment/disposal.
12. Data refer to municipal waste incinerated with energy recovery.
13. Data refer to total amounts of municipal waste managed.
14. Includes composting.
15. Excludes residues from other operations (54000 tonnes in 2003).
16. Data refer to municipal waste landfilled and recovered (include street cleansing waste).
17. Data refer to recycling and composting together.
18. Incineration: includes refuse derived fuel.
19. Composting: includes mechanical/biological treatment.
20. Data refer to waste treated by municipalities and separate collection for recycling by the private sector.
21. Direct disposal (excluding residues from other treatments, 6.6 million t.).
22. Data refer to amounts directly recycled (incl. private collection) and recovered from intermediate processing.
23. Data refer to total amounts of municipal waste managed in the country (exclude exported amounts).
24. For the calculations, only the six important locations of "Faritany" (Antanarivo, Antsiranana, Fianarantsoa, Mahajanga, Toamasia, Toliary) and the cities of Toalagnaro and Nosy-be were taken into account.
25. Residues of incineration of waste are landfilled in France.
26. Data include landfilled household waste and recycled packaging waste.

Waste: *Municipal waste treatment*

27. Landfill: household waste excluding construction and demolition waste.
28. Packaging waste only.
29. Excluding residues landfilled.
30. Recycling: waste separately collected (excludes food, park and garden waste which is included in composting).
31. Data refer to solid waste reaching dumping site which was taken from the Dumping Site Survey implemented in 2001.
32. Household and similar waste.
33. Recycling: separate collection.
34. Excludes batteries (2.4 thousand tonnes) and electric and electronic equipment (82.5 thousand tonnes).
35. Data pertains to domestic waste (4,100,000 t/year), municipal rubble and soil (1,000,000 t/year), green waste in coastal towns (40,000 t/year), and building waste (1,900,000 t/year).
36. Data pertains to automobile waste.
37. Landfill: after recovery and incineration.
38. Incineration: after recovery.

Definitions & Technical notes:
Municipal waste includes household waste and similar waste. The definition also includes bulky waste (e.g. white goods, old furniture, mattresses) and yard waste, leaves, grass clippings, street sweepings, the content of litter containers, and market cleansing waste, if managed as waste. It includes waste originating from: households, commerce and trade, small businesses, office buildings and institutions (schools, hospitals, government buildings). It also includes waste from selected municipal services, e.g. waste from park and garden maintenance, waste from street cleaning services (street sweepings, the content of litter containers, market cleansing waste), if managed as waste. The definition excludes waste from municipal sewage network and treatment, municipal construction and demolition waste. Municipal waste collected refers to waste collected by or on behalf of municipalities, as well as municipal waste collected by the private sector. It includes mixed household waste, and fractions collected separately for recovery operations (through door-to-door collection and/or through voluntary deposits). If data for municipal waste collected are not available, data for municipal waste generated is given, if available. Landfill is the final placement of waste into or onto the land in a controlled or uncontrolled way. Municipal waste landfilled includes all amounts going to landfill, either directly, or after sorting and/or treatment, as well as residues from recovery and disposal operations going to landfill. The definition covers both landfill in internal sites (i.e. where a generator of waste is carrying out its own waste disposal at the place of generation) and in external sites. Incineration is the controlled combustion of waste with or without energy recovery. Recycling is defined as any reintroduction of waste material in a production process that diverts it from the waste stream, except reuse as a fuel. Both reprocessing as the same type of product and for different purposes are included. Recycling within industrial plants i.e. at the place where the waste is generated, is excluded. Composting is a biological process that submits biodegradable waste to anaerobic or aerobic decomposition, and that results in a product that is recovered. The sum of the different types of waste disposal may be greater than the total amount of municipal waste collected, as these facilities may be used for other types of waste, or because of double counting due to the landfilling of the residues of incineration, or to the incineration of residues from composting.

Data Quality:
Data on municipal waste collected are usually gathered through surveys of municipalities, which are responsible for waste collection and disposal, or from transport companies that collect waste and transport it to a disposal site. Such surveys deliver fairly reliable data. However, it must be remembered that the figures only cover waste collected by or on behalf of municipalities. Therefore:
- Amounts of waste will vary, depending on how far municipal waste collection covers small industries and the services sector.
- Waste collected by the informal sector, waste generated in areas not covered by the municipal waste collection system or illegally dumped waste are nor included.
Caution is therefore advised when comparing countries.

Policy Relevance:
In many cases, a considerable proportion of municipal waste, particularly glass, paper and metals can be economically recycled. Organic matter can be composted, with or without methane recovery, and used to enrich soil. Another fraction of municipal waste can be burnt as a fuel to generate heat or electricity, preferably in special incinerators that reduce emissions of dioxins and other harmful pollutants. Depending on the type of waste, how the landfill site is constructed and the hydrological conditions, landfilling can lead to environmental problems such as leaching of nutrients, heavy metals and other toxic compounds, emission of greenhouse gases (CH_4 and CO_2) and loss of natural areas. Hence, in the best case, landfill should only be used when other possible waste treatment methods have been exhausted. Some towns and cities rely heavily on the informal sector (scavengers) to recycle waste, and this may be the sole source of income for whole families, with women and children also actively involved. As this is totally unregulated, the workers are often subject to accidents, to respiratory illnesses, to skin infections and other health problems.

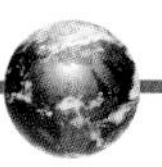

ENVIRONMENTAL INDICATORS

Waste

Hazardous waste generation *last update: April 2007 unit: 1000 tonne*

	1990	1995	2000	2001	2002	2003	2004	2005
Algeria	...	185.0	...	...	...	325.0	...	...
Andorra	...	...	2.6	2.5	2.4	...	0.1[1]	0.1[1]
Armenia	...	...	2.0	1.6	1.2	420.4[2]	544.7[2]	...
Australia	...	...	...	649.1	642.4	...	...	...
Austria	316.8[3]	595.0[3]	1034.8[3]	1025.7[3]	920.2[3]	...	1014.0[3]	...[3]
Azerbaijan	...	27.0	26.6	16.4	9.8	26.9	11.2	...
Belarus	...	90.3	73.0	99.1	116.9	118.5	154.2	...
Belgium	...	1113.5	...	...	...	...	...	...
Belize	...	...	0.8[4]	...	...	...	...	...
Benin	...	...	...	...	1.7[5]	...	...	...
Bosnia and Herzegovina	...	...	34.6	...	...	...	...	...
Brazil	...	...	2858.7[6]	...	...	...	...	...
Bulgaria	...	...	758.0	755.8	...	...	...	...
China	...	...	8300.0[7]	9520.0[7]	10010.0[7]	11700.0[7]	9950.0[7]	11620.0[7]
China, Hong Kong SAR	...	87.6	62.4	62.6	52.6	42.7	34.8	37.6
China, Macao SAR	...	...	0.1[4]	0.2[4]	0.2[4]	...	...	...
Croatia	...	...	26.0	58.3	47.4	48.1	42.3	...
Cyprus	...	50.0	...	...	...	...	...	...
Czech Republic	...	6005.0	2630.0[8]	2817.0[8]	1311.0[9]	1219.0	1447.0	1372.0
Denmark	...	179.0[3]	183.4[3]	200.1[3]	247.5[3]	328.3[3]	342.0[3]	340.5[3]
Estonia	...	7273.0	5965.8	...	...	...	...	...
Finland	...	...	1202.0	827.0	1188.0	...	2349.0	...
France	7000.0	...	9150.0	...	...	...	...	...
Georgia	2000.0[10]	...	...	...	...	...	...	...
Germany	13079.0	...	15542.0	15830.0	19636.0	19515.0	18401.0	...
Greece	450.0	350.0	391.0	326.4	352.7	353.8	...	...
Hungary	4691.0[11]	2274.3[12]	950.9	892.7	543.2	...	...	...
Iceland	...	6.0	7.0	8.0	8.0	8.0	8.0	...
India	...	...	7243.8	...	...	...	...	...
Ireland	66.0	248.0[13]	...	491.7[14]	...	...	673.6[15]	...
Israel	...	...	280.6	324.4	294.3	297.1	340.2	...
Italy	3246.0	2708.0	3911.0	4279.2	5024.5	5439.7	5365.4	...
Jamaica	...	10.0	...	...	...	...	...	...
Japan	2297.0	2883.0	...	...	...	...	...	...
Jordan	...	...	...	...	...	73.6	33.4	...
Kazakhstan	...	72.2[16]	102.5[16]	130.0[16]	137.1[16]	141.9	146.1	100.4
Korea, Republic of	968.3	1622.4	2779.0	2858.0	2914.5	2913.0	...	...
Kyrgyzstan	...	472.7	6204.1	6229.1	6512.8	6421.3	6410.0	...
Latvia	...	48.0	92.7	82.1	...	...	...	...
Lebanon	...	...	...	108.2	...	...	...	...
Lithuania	...	153.0	114.0	111.0	...	...	...	...
Luxembourg	...	200.0	197.1	202.0	227.5	...	...	...
Madagascar	...	...	...	...	1.9[17]	...	...	...
Malta	...	...	5.4	4.4	...	...	...	...
Mauritius	...	...	0.0	...	0.1	0.9[18]	...	...
Mexico	5657.0[19]	...	3706.8[20]	...	...	...	...	...
Monaco	...	0.3	0.3	0.3	0.6	...	...	...
Morocco	...	...	119.0	...	...	...	...	...
Netherlands	1040.0[21]	1004.0[22]	1785.0[22]	...	2159.9[22]	...	...	...
Niger	...	...	...	...	...	...	503.0	554.0
Norway	200.0[23]	650.0	673.0	655.0	...	825.0[24]	940.0	939.0
Palestine	...	...	5.0[25]	16.4[26]	...	12.5[26]	15.1[26]	11.0[26]
Poland	...	3866.0	1601.0[27]	1308.0[27]	1029.0[27]	1339.0[27]	1349.3[27]	1778.9[27]
Portugal	...	668.0	171.6	253.6	204.9	...	...	...

Waste: *Hazardous waste generation*

	1990	1995	2000	2001	2002	2003	2004	2005
Republic of Moldova	178.7	2.7	2.6	1.9	2.2	2.0	0.9	...
Réunion	...	...	9.8[28]	...	...	...	...	...
Romania	...	5710.0	896.7	...	...	...	...	...
Russian Federation	...	83330.3	127545.8	139193.5	...	...	...	...
Serbia and Montenegro	...	...	...	...	...	0.3	0.5	0.9
Singapore	2.3[29]	23.8[29]	29.7[29]	38.4[29]	42.4[29]	43.0[29]	38.2[29]	37.1[29]
Slovakia	...	1352.7	1630.0	1662.8	1441.1	1257.6	1021.2	...
Slovenia	...	170.0	...	67.5	...	...	...	...
Spain	1708.0	3394.0	3063.4	3222.9	3222.9	3222.9	3534.3[30]	...
St. Vincent and the Grenadines	...	...	...	...	0.0	...	...	...
Sweden	154.0	...	1100.0	...	...	...	1353.7	...
Switzerland	...	831.0	1114.5	1134.1	1112.0	...	...	...
Syrian Arab Republic	...	...	...	...	...	0.0	...	...
The Former Yugoslav Rep. of Macedonia	...	...	...	...	...	...	...	4.6
Tunisia	...	...	...	151.0	150.2	...	...	...
Turkey	...		1166.0[31]	...	...	...	1196.0	...
Ukraine	...		81374.9[32]	77513.5[32]	77604.9[32]	79000.9	62910.7	...
United Kingdom	2936.0[33]	2160.0[33]	5419.0[34]	5526.4[34]	5370.0[34]	4991.0[34]	5285.5[34]	...
United States	277339.0	194225.0	...	37033.2[35]	...	27375.8[36]	...	34788.4[36]
Yemen	...	38.2	...	...	...	...	...	...
Zambia	...	...	50.0	...	...	53.8	...	80.0

Sources:
UNSD/UNEP 2001, 2004 and 2006 questionnaire on Environment statistics, Waste section
OECD/Eurostat 2004 questionnaire on Environment statistics, Waste section
OECD Environmental Data, Compendium 2006/2007, Waste section.

Footnotes:
1. Starting with the exports in 2003, the data come from the regular weighing and there have been modifications of management and of the weighing methods.
2. Includes wastes from new mining operations.
3. Data refer to primary waste.
4. Waste from hospitals only.
5. Data refer to biomedical waste.
6. Data refer to hazardous waste collected (sum of collected industrial and septic hazardous waste).
7. Data refer to Industrial Solid Hazardous Waste.
8. Break in time series due to a new Waste Act in 1998. 1998 onwards: data include municipal hazardous waste.
9. Break in time series in 2002 due to a new Waste Act.
10. Data created under the inventory in 1990 reflect the amount of the waste accumulated before 1990. Therefore data do not refer to annually produced waste.
11. 1990: includes red mud.
12. Break in time series in 1995; 1995: excludes red mud.
13. Includes recovery on site.
14. Includes reported and unreported waste, contaminated soil and on-site treatment.
15. Includes reported and unreported waste, contaminated soil (2004: 307 thousand tonnes) and on-site treatment.
16. Data refer to toxic waste.
17. This quantity represents biomedical waste only.
18. Source: Basel Convention
19. Estimates and includes biological infectious waste.
20. Data are based on surveys covering 27280 enterprises; includes biological infectious waste.
21. Includes contaminated soil.
22. Excludes contaminated soil.
23. 1990 data is a rough estimate based on a study carried out in 1988 and exclude on-site treatment.
24. 2003: new type of waste are defined as hazardous in legislation.
25. Data refer to health care private centers waste taken from the Medical Environmental Survey.
26. Data refer to health care centers waste taken from the Environmental Survey for Health Care Centers.
27. 1998 onwards: data refer to a new classification based on the European Waste Catalogue.
28. Special industrial waste: dangerous waste produced by businesses (solvents, hydrocarbon sludges, used oils and batteries, sodium baths, manufacturing scrap, rejects, paints...).
29. Data refer to solid chemical and PVC wastes. Figures prior to 1993 do not include metallic sludge in solid chemical and PVC wastes.
30. Provisional data.
31. Hazardous waste from manufacturing industry.
32. Data refer to industrial hazardous wastes.
33. Production before 1997: defined by the Control of Pollution (Special Wastes) Regulations, 1980.
34. From 1997 onwards: Special wastes as defined by the Hazardous Waste List (94/904/EC) and implemented by the Special Waste Regulations, 1996.
35. From 1997, data exclude waste water.

Waste: *Hazardous waste generation*

36. Reporting requirements have been changed in 2001; includes some waste water.

Definitions & Technical notes:
Hazardous waste is waste that owing to its toxic, infectious, radioactive or flammable properties poses an actual or potential hazard to the health of humans, other living organisms, or the environment. Hazardous waste here refers to categories of waste to be controlled according to the Basel Convention on the Control of Transboundary Movements of Hazardous Wastes and Their Disposal (Article 1 and Annex I). If data are not available according to the Basel Convention, amounts can be given according to national definitions.

Data Quality:
Although countries are asked to report data on hazardous waste according to the categories of the Basel Convention, most countries are not able to do so, and supply data according to national definitions. Some countries have indicated this in footnotes, but it can be assumed that this also applies to other countries. National definitions of hazardous waste may change over time, as national legislation is revised. Therefore the definition of hazardous waste varies greatly from one country to another, and sometime also over time. Moreover, data only refer to wastes declared as hazardous by the generator, or by the company responsible for disposing of the waste. How far this is represents the real amount of hazardous waste generated in the country will depend on how well the sector is regulated and policed. Data quality and comparability are therefore limited and trends should be interpreted with care.

Policy Relevance:
By definition, hazardous waste poses a threat to human and ecological health, often for many years. Correct disposal of hazardous waste is therefore a public and environmental health issue. The amount of hazardous waste generated in a country is closely linked to the country's economy; a highly industrialized country or one with a large mining industry is likely to generate more hazardous waste than a country whose economy is based more on services or on agriculture. In the 1989 Basel Convention on the Control of Transboundary Movements of Hazardous Wastes and their Disposal (http://www.basel.int/), 164 countries agreed to minimize the generation of hazardous waste, to assure sound management of hazardous wastes, to control transboundary movement of hazardous wastes; and to improve institutional and technical capabilities especially for developing countries and countries with economies in transition. At later meetings, Parties agreed to ban on the export of hazardous wastes from OECD to non-OECD countries ('Basel ban'). As a general rule, companies that generate hazardous waste must bear the cost of disposing of it. In many cases this may be easier for the company to do internally, for example through recycling or high temperature incineration. These should be encouraged as they remove the need to transport the waste, and thus the risk of leakage during transport. However, the incinerators must be well regulated and regularly controlled to avoid emissions of toxic by-products from the incineration process.

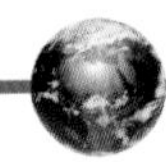

ENVIRONMENTAL INDICATORS

Land Use

Total surface area *last update January 19, 2007*

	Total area km²
Afghanistan	652090
Albania	28748
Algeria	2381741
American Samoa	199
Andorra	468
Angola	1246700[1]
Anguilla	91
Antigua and Barbuda	442
Argentina	2780400
Armenia	29800
Aruba	180
Australia	7741220
Austria	83858
Azerbaijan	86600
Bahamas	13878
Bahrain	694
Bangladesh	143998
Barbados	430
Belarus	207600
Belgium	30528
Belize	22966
Benin	112622
Bermuda	53
Bhutan	47000
Bolivia	1098581
Bosnia and Herzegovina	51197
Botswana	581730
Brazil	8514877[2]
British Virgin Islands	151
Brunei Darussalam	5765
Bulgaria	110912
Burkina Faso	274000
Burundi	27834
Cambodia	181035
Cameroon	475442
Canada	9970610
Cape Verde	4033
Cayman Islands	264
Central African Republic	622984
Chad	1284000
Channel Islands	195
Chile	756096
China	9596961
China: Hong Kong SAR	1099
China: Macao SAR	26
Christmas Islands	135
Cocos (Keeling) Islands	14
Colombia	1138914

Land Use: *Total surface area*

	Total area km²
Comoros	2235
Congo	342000
Cook Islands	236[3]
Costa Rica	51100
Côte d'Ivoire	322463
Croatia	56538
Cuba	110861
Cyprus	9251
Czech Republic	78866
Democratic Republic of the Congo	2344858
Denmark	43094
Djibouti	23200
Dominica	751
Dominican Republic	48671
Ecuador	283561
Egypt	1001449
El Salvador	21041
Equatorial Guinea	28051[4]
Eritrea	117600
Estonia	45100
Ethiopia	1104300
Faeroe Islands	1393
Falkland Islands (Malvinas)	12173[5,6]
Fiji	18274
Finland	338145
France	551500
French Guiana	90000
French Polynesia	4000[7]
Gabon	267668
Gambia	11295
Georgia	69700
Germany	357022
Ghana	238533
Gibraltar	6
Greece	131957
Greenland	2175600
Grenada	344[8]
Guadeloupe	1705[9]
Guam	549
Guatemala	108889
Guinea	245857
Guinea-Bissau	36125
Guyana	214969
Haiti	27750
Holy See	0
Honduras	112088
Hungary	93032
Iceland	103000
India	3287263
Indonesia	1904569
Iran (Islamic Republic of)	1648195

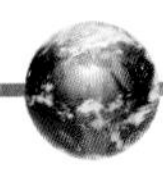

Land Use: *Total surface area*

	Total area km²
Iraq	438317
Ireland	70273
Isle of Man	572
Israel	22145
Italy	301318
Jamaica	10991
Japan	377873
Jordan	89342
Kazakhstan	2724900
Kenya	580367
Kiribati	726[10]
Korea (Dem. People's Republic of)	120538
Korea (Republic of)	99538
Kuwait	17818
Kyrgyzstan	199900
Lao People's Democratic Republic	236800
Latvia	64600
Lebanon	10400
Lesotho	30355
Liberia	111369
Libyan Arab Jamahiriya	1759540
Liechtenstein	160
Lithuania	65300
Luxembourg	2586
Madagascar	587041
Malawi	118484
Malaysia	329847
Maldives	298
Mali	1240192
Malta	316
Marshall Islands	181
Martinique	1102
Mauritania	1025520
Mauritius	2040
Mexico	1958201
Micronesia Federated States of	702
Monaco	2
Mongolia	1564116[2]
Montserrat	102
Morocco	446550
Mozambique	801590
Myanmar	676578
Namibia	824292
Nauru	21
Nepal	147181
Netherlands	41528
Netherlands Antilles	800
New Caledonia	18575[11]
New Zealand	270534[12]
Nicaragua	130000
Niger	1267000

Land Use: *Total surface area*

	Total area km²
Nigeria	923768
Niue	260
Norfolk Island	36
Northern Mariana Islands	464
Norway	385155
Occupied Palestinian Territory	6020
Oman	309500
Pakistan	796095
Palau	459
Panama	75517
Papua New Guinea	462840[13]
Paraguay	406752
Peru	1285216
Philippines	300000
Pitcairn	5
Poland	312685[14]
Portugal	91982[15]
Puerto Rico	8875
Qatar	11000
Republic of Moldova	33851
Réunion	2510
Romania	238391
Russian Federation	17098242
Rwanda	26338
Saint Helena	308
Saint Kitts and Nevis	261
Saint Lucia	539
Saint Pierre and Miquelon	242
Saint Vincent and the Grenadines	388[16]
Samoa	2831
San Marino	61
Sao Tome and Principe	964
Saudi Arabia	2149690
Senegal	196722
Serbia and Montenegro	102173
Seychelles	455
Sierra Leone	71740
Singapore	683
Slovakia	49033
Slovenia	20256
Solomon Islands	28896[17]
Somalia	637657
South Africa	1221037
Spain	505992[18]
Sri Lanka	65610
Sudan	2505813
Suriname	163820
Svalbard and Jan Mayen Islands	62422
Swaziland	17364
Sweden	449964
Switzerland	41284

Land Use: *Total surface area*

	Total area km²
Syrian Arab Republic	185180
Tajikistan	143100
Thailand	513115
The Former Yugoslav Rep. of Macedonia	25713
Timor-Leste	14874
Togo	56785
Tokelau	12
Tonga	747
Trinidad and Tobago	5130
Tunisia	163610
Turkey	783562
Turkmenistan	488100
Turks Caicos Islands	948
Tuvalu	26
Uganda	241038
Ukraine	603700
United Arab Emirates	83600[19]
United Kingdom	242900[20]
United Republic of Tanzania	945087
United States	9629091
United States Virgin Islands	347
Uruguay	175016
Uzbekistan	447400
Vanuatu	12189
Venezuela (Bolivarian Republic of)	912050
Viet Nam	331689
Wallis and Futuna Islands	200
Western Sahara	266000[21]
Yemen	527968
Zambia	752618
Zimbabwe	390757

Source:
United Nations Statistics Division (UNSD)

Footnotes:
1. Including the enclave of Cabinda.
2. Exact reference date unknown.
3. Excluding Niue, shown separately, which is part of Cook Islands, but because of remoteness is administered separately.
4. Comprising Bioko (which includes Pagalu) and Rio Muni (which includes Corisco and Elobeys).
5. Excluding dependencies, of which South Georgia (area 3 755 km2) had an estimated population of 499 in 1964 (494 males, 5 females). The other dependencies namely, the South Sandwich group (surface area 337 km2) and a number of smaller islands, are presumed to be uninhabited.
6. A dispute exists between the governments of Argentina and the United Kingdom of Great Britain and Northern Ireland concerning sovereignty over the Falkland Islands (Malvinas).
7. Comprising Austral, Gambier, Marquesas, Rapa, Society and Tuamotu Islands.
8. Including Carriacou and other dependencies in the Grenadines.
9. Including dependencies: Marie-Galante, la Désirade, les Saintes, Petite-Terre, St. Barthélemy and French part of St. Martin.
10. Including Christmas, Fanning, Ocean and Washington Islands.
11. Including the islands of Huon, Chesterfield, Loyalty, Walpole and Belep Archipelago.
12. Including Campbell and Kermadec Islands (population 20 in 1961, surface area 148 km2) as well as Antipodes, Auckland, Bounty, Snares, Solander and Three Kings island, all of which are uninhabited.
13. Comprising eastern part of New Guinea, the Bismarck Archipelago, Bougainville and Buka of Solomon Islands group and about 600 smaller islands.
14. Includes inland waters as well as part of internal waters.
15. Including the Azores and Madeira Islands.
16. Including Bequia and other islands in the Grenadines.
17. Comprising the Solomon Islands group (except Bougainville and Buka which are included with Papua New Guinea shown separately), Ontong, Java, Rennel and Santa Cruz Islands.

Land Use: *Total surface area*

18. Including the Balearic and Canary Islands, and Alhucemas, Ceuta, Chafarinas, Melilla and Penon de Vélez de la Gomera.
19. Comprising 7 sheikdoms of Abu Dhabi, Dubai, Sharjah, Ajaman, Umm al Qaiwain, Ras al Khaimah and Fujairah, and the area lying within the modified Riyadh line as announced in October 1955.
20. Excluding Channel Islands and Isle of Man, shown separately.
21. Comprising the Northern Region (former Saguia el Hamra) and Southern Region (former Rio de Oro).

Definitions & Technical notes:
Total surface area refers to the total area of the country which comprises land area and inland waters.
0 - magnitude not zero, but less than half of unit employed

ENVIRONMENTAL INDICATORS

Land Use

Forest area *last update: April 2007*

	Forest area in 1990	Forest area in 2005	% change since 1990	% of land area covered by forest in 1990	% of land area covered by forest in 2005
	km²	km²	%	%	%
Afghanistan	13 090	8 670	-33.8	2.0	1.3
Albania	7 890	7 940	0.6	28.8	29.0
Algeria	17 900	22 770	27.2	0.8	1.0
American Samoa	180	180	0.0	91.9	89.4
Andorra	160	160	0.0	35.6	35.6
Angola	609 760	591 040	-3.1	48.9	47.4
Anguilla	60	60	0.0	71.4	71.4
Antigua and Barbuda	90	90	0.0	21.4	21.4
Argentina	352 620	330 210	-6.4	12.9	12.1
Armenia	3 460	2 830	-18.2	12.3	10.0
Aruba	...	...	...	2.2	2.2
Australia	1 679 040	1 636 780	-2.5	21.9	21.3
Austria	37 760	38 620	2.3	45.6	46.7
Azerbaijan	9 360	9 360	0.0	11.3	11.3
Bahamas	5 150	5 150	0.0	51.5	51.5
Bahrain	...	...	...	0.3	0.6
Bangladesh	8 820	8 710	-1.2	6.8	6.7
Barbados	20	20	0.0	4.0	4.0
Belarus	73 760	78 940	7.0	35.6	38.0
Belgium	6 770	6 670	-1.5	22.4	22.0
Belize	16 530	16 530	0.0	72.5	72.5
Benin	33 220	23 510	-29.2	30.0	21.3
Bermuda	10	10	0.0	20.0	20.0
Bhutan	30 350	31 950	5.3	64.6	68.0
Bolivia	627 950	587 400	-6.5	57.9	54.2
Bosnia and Herzegovina	22 100	21 850	-1.1	43.6	43.1
Botswana	137 180	119 430	-12.9	24.2	21.1
Brazil	5 200 270	4 776 980	-8.1	62.2	57.2
British Virgin Islands	40	40	0.0	24.7	24.4
Brunei Darussalam	3 130	2 780	-11.2	59.4	52.8
Bulgaria	33 270	36 250	9.0	30.1	32.8
Burkina Faso	71 540	67 940	-5.0	30.6	29.0
Burundi	2 890	1 520	-47.4	11.3	5.9
Cambodia	129 460	104 470	-19.3	73.3	59.2
Cameroon	245 450	212 450	-13.4	52.7	45.6
Canada	3 101 340	3 101 340	0.0	33.6	33.6
Cape Verde	580	840	44.8	14.3	20.7
Cayman Islands	120	120	0.0	48.4	48.4
Central African Republic	232 030	227 550	-1.9	37.2	36.5
Chad	131 100	119 210	-9.1	10.4	9.5
Channel Islands	10	10	0.0	4.1	4.1
Chile	152 630	161 210	5.6	20.4	21.5
China	1 571 410	1 972 900	25.5	16.8	21.2[1]
Colombia	614 390	607 280	-1.2	59.1	58.5
Comoros	120	50	-58.3	6.5	2.9
Congo	227 260	224 710	-1.1	66.5	65.8
Cook Islands	150	160	6.7	63.9	66.5
Costa Rica	25 640	23 910	-6.7	50.2	46.8
Cote d'Ivoire	102 220	104 050	1.8	32.1	32.7
Croatia	21 160	21 350	0.9	37.8	38.2
Cuba	20 580	27 130	31.8	18.7	24.7

Land Use: *Forest area*

	Forest area in 1990	Forest area in 2005	% change since 1990	% of land area covered by forest in 1990	% of land area covered by forest in 2005
	km²	km²	%	%	%
Cyprus	1 610	1 740	8.1	17.4	18.9
Czech Republic	26 300	26 480	0.7	34.0	34.3
Dem. Rep. of the Congo	1 405 310	1 336 100	-4.9	62.0	58.9
Denmark	4 450	5 000	12.4	10.5	11.8
Djibouti	60	60	0.0	0.2	0.2
Dominica	500	460	-8.0	66.7	61.3
Dominican Republic	13 760	13 760	0.0	28.4	28.4
Ecuador	138 170	108 530	-21.5	49.9	39.2
Egypt	440	670	52.3	0.0	0.1
El Salvador	3 750	2 980	-20.5	18.1	14.4
Equatorial Guinea	18 600	16 320	-12.3	66.3	58.2
Eritrea	16 210	15 540	-4.1	16.0	15.4
Estonia	21 630	22 840	5.6	51.0	53.9
Ethiopia	151 140	130 000	-14.0	13.8	11.9
Faeroe Islands	...	...	...	0.1	0.1
Falkland Islands (Malvinas)	0	0	...	...	...
Fiji	9 790	10 000	2.1	53.6	54.7
Finland	221 940	225 000	1.4	72.9	73.9[2]
France	145 380	155 540	7.0	26.4	28.3
French Guiana	80 910	80 630	-0.3	91.8	91.8[3]
French Polynesia	1 050	1 050	0.0	28.7	28.7
Gabon	219 270	217 750	-0.7	85.1	84.5
Gambia	4 420	4 710	6.6	39.1	41.7
Georgia	27 600	27 600	0.0	39.7	39.7
Germany	107 410	110 760	3.1	30.8	31.7
Ghana	74 480	55 170	-25.9	32.7	24.2
Gibraltar	0	0	...	...	...
Greece	32 990	37 520	13.7	25.6	29.1
Greenland	...	...	...	0.0	0.0
Grenada	40	40	0.0	12.2	12.2
Guadeloupe	840	800	-4.8	49.4	47.2
Guam	260	260	0.0	47.1	47.1
Guatemala	47 480	39 380	-17.1	43.8	36.3
Guinea	74 080	67 240	-9.2	30.1	27.4
Guinea-Bissau	22 160	20 720	-6.5	78.8	73.7
Guyana	151 040	151 040	0.0	76.7	76.7
Haiti	1 160	1 050	-9.5	4.2	3.8
Holy See	0	0	...	...	...
Honduras	73 850	46 480	-37.1	66.0	41.5
Hungary	18 010	19 760	9.7	19.6	21.5
Iceland	250	460	84.0	0.2	0.5
India	639 390	677 010	5.9	21.5	22.8
Indonesia	1 165 670	884 950	-24.1	64.3	48.8
Iran (Islamic Republic of)	110 750	110 750	0.0	6.8	6.8
Iraq	8 040	8 220	2.2	1.8	1.9
Ireland	4 410	6 690	51.7	6.4	9.7
Isle of Man	30	30	0.0	...	...
Israel	1 540	1 710	11.0	7.5	8.3
Italy	83 830	99 790	19.0	28.5	33.9
Jamaica	3 450	3 390	-1.7	31.9	31.3
Japan	249 500	248 680	-0.3	68.4	68.2
Jordan	830	830	0.0	0.9	0.9
Kazakhstan	34 220	33 370	-2.5	1.3	1.2
Kenya	37 080	35 220	-5.0	6.5	6.2

Land Use: *Forest area*

	Forest area in 1990	Forest area in 2005	% change since 1990	% of land area covered by forest in 1990	% of land area covered by forest in 2005
	km²	km²	%	%	%
Kiribati	20	20	0.0	3.0	3.0
Korea, Dem. People's Rep.	82 010	61 870	-24.6	68.1	51.4
Korea, Republic of	63 710	62 650	-1.7	64.5	63.5
Kuwait	30	60	100.0	0.2	0.3
Kyrgyzstan	8 360	8 690	3.9	4.4	4.5
Lao People's Dem. Rep.	173 140	161 420	-6.8	75.0	69.9
Latvia	27 750	29 410	6.0	44.7	47.4
Lebanon	1 210	1 360	12.4	11.7	13.3
Lesotho	50	80	60.0	0.2	0.3
Liberia	40 580	31 540	-22.3	42.1	32.7
Libyan Arab Jamahiriya	2 170	2 170	0.0	0.1	0.1
Liechtenstein	60	70	16.7	40.6	43.1
Lithuania	19 450	20 990	7.9	31.0	33.5
Luxembourg	860	870	1.2	33.2	33.5
Madagascar	136 920	128 380	-6.2	23.5	22.1
Malawi	38 960	34 020	-12.7	41.4	36.2
Malaysia	223 760	208 900	-6.6	68.1	63.6
Maldives	10	10	0.0	3.0	3.0
Mali	140 720	125 720	-10.7	11.5	10.3
Malta	0	...	...	1.1	1.1
Marshall Islands	0	...	...	0.0	...
Martinique	460	460	0.0	43.9	43.9
Mauritania	4 150	2 670	-35.7	0.4	0.3
Mauritius	390	370	-5.1	19.2	18.2
Mexico	690 160	642 380	-6.9	36.2	33.7
Micronesia, Federated States of	630	630	0.0	90.6	90.6
Monaco	0	0	...	...	...
Mongolia	114 920	102 520	-10.8	7.3	6.5
Montserrat	40	40	0.0	35.0	35.0
Morocco	42 890	43 640	1.7	9.6	9.8
Mozambique	200 120	192 620	-3.7	25.5	24.6
Myanmar	392 190	322 220	-17.8	59.6	49.0
Namibia	87 620	76 610	-12.6	10.6	9.3
Nauru	0	0	...	0.0	...
Nepal	48 170	36 360	-24.5	33.7	25.4
Netherlands	3 450	3 650	5.8	10.2	10.8
Netherlands Antilles	10	10	0.0	1.5	1.5
New Caledonia	7 170	7 170	0.0	39.2	39.2
New Zealand	77 200	83 090	7.6	28.8	31.0
Nicaragua	65 380	51 890	-20.6	53.9	42.7
Niger	19 450	12 660	-34.9	1.5	1.0
Nigeria	172 340	110 890	-35.7	18.9	12.2
Niue	170	140	-17.6	66.2	54.2
Northern Mariana Islands	350	330	-5.7	75.3	72.4
Norway	91 300	93 870	2.8	29.8	30.7
Oman	20	20	0.0	0.0	0.0
Pakistan	25 270	19 020	-24.7	3.3	2.5
Palau	380	400	5.3	82.9	87.6
Palestine	90	90	0.0	...	...
Panama	43 760	42 940	-1.9	58.8	57.7
Papua New Guinea	315 230	294 370	-6.6	69.6	65.0
Paraguay	211 570	184 750	-12.7	53.3	46.5
Peru	701 560	687 420	-2.0	54.8	53.7
Philippines	105 740	71 620	-32.3	35.5	24.0

Land Use: *Forest area*

	Forest area in 1990	Forest area in 2005	% change since 1990	% of land area covered by forest in 1990	% of land area covered by forest in 2005
	km²	km²	%	%	%
Pitcairn	40	40	0.0	...	...
Poland	88 810	91 920	3.5	29.2	30.0[4]
Portugal	30 990	37 830	22.1	33.9	41.3
Puerto Rico	4 040	4 080	1.0	45.5	46.0
Qatar	0	...	...	0.0	0.0
Republic of Moldova	3 190	3 290	3.1	9.7	10.0
Réunion	870	840	-3.4	34.9	33.6
Romania	63 710	63 700	0.0	27.8	27.7[4]
Russian Federation	8 089 500	8 087 900	0.0	47.9	47.9
Rwanda	3 180	4 800	50.9	12.9	19.5
Saint Helena	20	20	0.0	...	...
Saint Kitts and Nevis	50	50	0.0	14.7	14.7
Saint Lucia	170	170	0.0	27.9	27.9
Saint Pierre and Miquelon	30	30	0.0	...	...
Samoa	1 300	1 710	31.5	45.9	60.4
San Marino	...	...	...	1.6	1.6
Sao Tome and Principe	270	270	0.0	28.4	28.4
Saudi Arabia	27 280	27 280	0.0	1.3	1.3
Senegal	93 480	86 730	-7.2	48.6	45.0
Serbia	25 590	26 940	5.3	...	...
Serbia and Montenegro	29 010	...	...	25.1	26.4
Seychelles	400	400	0.0	88.9	88.9
Sierra Leone	30 440	27 540	-9.5	42.5	38.5
Singapore	20	20	0.0	3.4	3.4
Slovakia	19 220	19 290	0.4	40.0	40.1
Slovenia	11 880	12 640	6.4	59.0	62.8
Solomon Islands	27 680	21 720	-21.5	98.9	77.6
Somalia	82 820	71 310	-13.9	13.2	11.4
South Africa	92 030	92 030	0.0	7.6	7.6
Spain	134 790	179 150	32.9	27.0	35.9
Sri Lanka	23 500	19 330	-17.7	36.4	29.9
St. Vincent and the Grenadines	90	110	22.2	24.2	27.4
Sudan	763 810	675 460	-11.6	32.1	28.4
Suriname	147 760	147 760	0.0	94.7	94.7
Swaziland	4 720	5 410	14.6	27.4	31.5
Sweden	273 670	275 280	0.6	66.5	66.9
Switzerland	11 550	12 210	5.7	29.2	30.9
Syrian Arab Republic	3 720	4 610	23.9	2.0	2.5
Tajikistan	4 080	4 100	0.5	2.9	2.9
Thailand	159 650	145 200	-9.1	31.2	28.4
The Former Yugoslav Rep. of Macedonia	9 060	9 060	0.0	35.8	35.8
Timor-Leste	9 660	7 980	-17.4	65.0	53.7
Togo	6 850	3 860	-43.6	12.6	7.1
Tokelau	0	0	...	...	...
Tonga	40	40	0.0	5.0	5.0
Trinidad and Tobago	2 350	2 260	-3.8	45.8	44.1
Tunisia	6 430	10 560	64.2	4.1	6.8
Turkey	96 800	101 750	5.1	12.6	13.2
Turkmenistan	41 270	41 270	0.0	8.8	8.8
Turks and Caicos Islands	340	340	0.0	80.0	80.0
Tuvalu	10	10	0.0	33.3	33.3
Uganda	49 240	36 270	-26.3	25.0	18.4
Ukraine	92 740	95 750	3.2	16.0	16.5
United Arab Emirates	2 450	3 120	27.3	2.9	3.7

Land Use: *Forest area*

	Forest area in 1990	Forest area in 2005	% change since 1990	% of land area covered by forest in 1990	% of land area covered by forest in 2005
	km²	km²	%	%	%
United Kingdom	26 110	28 450	9.0	10.8	11.8
United Rep. of Tanzania	414 410	352 570	-14.9	46.9	39.9
United States	2 986 480	3 030 890	1.5	32.6	33.1
United States Virgin Islands	120	100	-16.7	35.0	27.9
Uruguay	9 050	15 060	66.4	5.2	8.6
Uzbekistan	30 450	32 950	8.2	7.4	8.0
Vanuatu	4 400	4 400	0.0	36.1	36.1
Venezuela	520 260	477 130	-8.3	59.0	54.1
Viet Nam	93 630	129 310	38.1	28.8	39.7
Wallis and Futuna Islands	60	50	-16.7	...	...
Western Sahara	10 110	10 110	0.0	3.8	3.8
Yemen	5 490	5 490	0.0	1.0	1.0
Zambia	491 240	424 520	-13.6	66.1	57.1
Zimbabwe	222 340	175 400	-21.1	57.5	45.3

Source:
Food and Agriculture Organization of the United Nations (FAO).
Millennium Indicators Database.

Footnote:
1. Hong Kong SAR of China and Macao SAR of China included.
2. The land area of Finland is still slightly increasing due to the postglacial crustal uplift. On the other hand, the construction of artificial lakes for generating hydro power has decreased the land area during the past 50 years. The land area of Finland is thus not constant. Furthermore, a significant error was discovered in the land area statistics on 1.1.2000, maintained by the National Land Survey of Finland. This erroneous area (30 459, 1000 ha) is also in the records by FAOSTAT. These are the reasons that the official land area by the National Land Survey of Finland on 1.1. 2004 (30 447.4, 1000 ha) is used instead of that by FAOSTAT.
3. The total land area decreased with 28 000 ha (in 1995) due to construction of an artificial lake.
4. Total land area is not constant corrected every year in FAOSTAT.

Definitions & Technical notes:
Forest includes natural forests and forest plantations. It is used to refer to land with a tree canopy cover of more than 10 per cent and area of more than 0.5 ha. Forests are determined both by the presence of trees and the absence of other predominant land uses. The trees should be able to reach a minimum height of 5 m. Young stands that have not yet but are expected to reach a crown density of 10 percent and tree height of 5 m are included under forest, as are temporarily unstocked areas. The term includes forests used for purposes of production, protection, multiple-use or conservation (i.e. forest in national parks, nature reserves and other protected areas), as well as forests stands on agricultural lands (e.g. windbreaks and shelterbelts of trees with a width of more than 20 m), and rubberwood plantations and cork oak stands. The term specifically excludes stands of trees established primarily for agricultural production, for example fruit tree plantations. It also excludes trees planted in agroforestry systems.

Data Quality:
Although there is an agreed and clear definition of forest, not all countries apply this definition. In many northern countries, areas with a crown cover of less than 20% are not considered as real forest land. 'Temporarily unstocked areas' refer to areas that have been designated as forest area, but not yet planted, or more often, areas where storm or fire has removed a large part of the forest cover. Unless aggressively restocked with trees, such areas can take a long time to re-establish forests naturally.

Policy Relevance:
It is difficult to overstate the importance of forests. They provide a habitat for a wide range of biodiversity, which provide food and medicines for local peoples, and provide fuelwood, when other cheap alternatives are not available. When exploited commercially, wood and other forest products can bring much needed foreign currency, as can tourism that exploits the unique wildlife and flora of the forest. But even if not exploited commercially, forests play an important role in maintaining an areas integrity: they naturally regulate water, absorbing rainwater during a storm and, by preventing quick evaporation, can help recharge groundwater. They maintain humidity, thus providing a micro-climate in which crops grown nearby can thrive. Forests are also instrumental in preventing floods and holding soil in place, thus preventing the loss of life and livelihoods landslides can bring. A growing forest acts as a lung removing CO_2 from the atmosphere, thus partly off-setting the emissions of CO_2 released during the combustion of fossil fuels. The percentage of land area covered by forest is one of the MDG indicators.

ENVIRONMENTAL INDICATORS

Land Use

Agricultural land *last update: April 2007*

	Latest year available	Agricultural area	% change since 1990	% of total land area	Arable land	Land under permanent crops	Land under permanent pasture
		km²	%	%	km²	km²	km²
Afghanistan	2003	380 480	0.0	58.3	79 100	1 380	300 000
Albania	2005	11 230	0.2	41.0	5 780	1 220	4 230
Algeria	2003	399 560	3.3	16.8	75 450	6 700	317 410
American Samoa	2003	50	25.0	25.0	20	30	...
Andorra	2005	260	0.0	55.3	10	...	250
Angola	2003	575 900	0.3	46.2	33 000	2 900	540 000
Antigua and Barbuda	2003	140	0.0	31.8	80	20	40
Argentina	2003	1 287 470	1.1	47.0	279 000	10 000	998 470
Armenia	2005	13 900	...	49.3	4 950	600	8 350
Aruba	2003	20	0.0	10.5	20	...	...
Australia	2005	4 451 490	-4.2	57.9	494 020	3 400	3 954 070
Austria	2005	32 630	-6.8	39.6	13 870	660	18 100
Azerbaijan	2005	47 586	...	57.6	18 432	2 215	26 939
Bahamas	2003	140	16.7	1.4	80	40	20
Bahrain	2005	100	25.0	14.1	20	40	40
Bangladesh	2005	90 150	-10.2	69.3	79 550	4 600	6 000
Barbados	2003	190	0.0	44.2	160	10	20
Belarus	2005	88 600	...	42.7	54 550	1 160	32 890
Belgium	2005	13 860	...	45.8	8 440	230	5 190
Belize	2003	1 520	20.6	6.7	700	320	500
Benin	2003	34 670	52.7	31.3	26 500	2 670	5 500
Bermuda	2003	10	0.0	20.0	10	...	...
Bhutan	2005	5 920	37.0	12.6	1 590	180	4 150
Bolivia	2003	370 870	4.6	34.2	30 500	2 060	338 310
Bosnia and Herzegovina	2005	21 470	...	41.9	10 000	970	10 500
Botswana	2003	259 800	-0.2	45.8	3 770	30	256 000
Brazil	2003	2 636 000	9.1	31.2	590 000	76 000	1 970 000
British Virgin Islands	2003	90	0.0	60.0	30	10	50
Brunei Darussalam	2005	250	92.3	4.7	140	50	60
Bulgaria	2005	52 650	-14.5	48.5	31 730	2 010	18 910
Burkina Faso	2003	109 000	13.8	39.8	48 400	600	60 000
Burundi	2003	23 450	10.4	91.3	9 900	3 650	9 900
Cambodia	2005	53 560	-0.1	30.3	37 000	1 560	15 000
Cameroon	2003	91 600	-0.1	19.7	59 600	12 000	20 000
Canada	2003	675 050	-0.4	7.4	456 600	64 550	153 900
Cape Verde	2003	740	8.8	18.4	460	30	250
Cayman Islands	2003	30	0.0	11.5	10	...	20
Central African Republic	2003	51 490	2.9	8.3	19 300	940	31 250
Chad	2003	486 300	0.7	38.6	36 000	300	450 000
Chile	2003	152 420	-4.1	20.4	19 820	3 250	129 350
China	2005	5 554 880	4.7	59.8	1 426 880	128 000	4 000 000
China, Hong Kong SAR	2005	70	-12.5	6.7	50	10	10
Colombia	2005	425 570	-5.6	38.4	20 040	16 090	389 440
Comoros	2003	1 470	14.8	65.9	800	520	150
Congo	2003	105 470	0.2	30.9	4 950	520	100 000
Cook Islands	2003	60	0.0	25.0	40	20	...
Costa Rica	2003	28 650	0.9	56.1	2 250	3 000	23 400
Cote d'Ivoire	2003	199 000	5.1	62.6	33 000	36 000	130 000
Croatia	2005	26 950	...	48.2	11 100	1 160	14 690
Cuba	2003	66 550	-1.3	60.6	30 630	7 250	28 670
Cyprus	2005	1 430	-11.7	15.5	1 000	390	40

Land Use: *Agricultural land*

	Latest year available	Agricultural area	% change since 1990	% of total land area	Arable land	Land under permanent crops	Land under permanent pasture
		km²	%	%	km²	km²	km²
Czech Republic	2005	42 590	...	55.1	30 470	2 380	9 740
Dem. Rep. of the Congo	2003	228 000	-0.3	10.1	67 000	11 000	150 000
Denmark	2005	25 890	-7.1	61.0	22 370	70	3 450
Djibouti	2003	17 010	30.9	73.4	10	...	17 000
Dominica	2003	230	27.8	30.7	50	160	20
Dominican Republic	2003	36 960	3.0	76.4	10 960	5 000	21 000
Ecuador	2005	75 520	-3.7	27.3	13 480	12 140	49 900
Egypt	2005	35 200	32.9	3.5	30 000	5 200	...
El Salvador	2003	17 040	17.5	82.2	6 600	2 500	7 940
Equatorial Guinea	2003	3 340	0.0	11.9	1 300	1 000	1 040
Eritrea	2003	75 320	...	74.6	5 620	30	69 670
Estonia	2005	8 340	...	19.7	5 910	120	2 310
Ethiopia	2003	317 690	...	31.8	110 560	7 130	200 000
Faeroe Islands	2003	30	0.0	2.1	30	...	...
Falkland Islands (Malvinas)	2003	11 300	-5.0	92.9	...	...	11 300
Fiji	2003	4 600	12.2	25.2	2 000	850	1 750
Finland	2005	22 660	-5.5	7.4	22 340	60	260
France	2005	295 690	-3.3	53.8	185 070	11 280	99 340
French Guiana	2003	230	9.5	0.3	120	40	70
French Polynesia	2003	450	4.7	12.3	30	220	200
Gabon	2003	51 600	0.1	20.0	3 250	1 700	46 650
Gambia	2003	7 790	22.3	77.9	3 150	50	4 590
Georgia	2005	30 060	...	43.3	8 020	2 640	19 400
Germany	2005	170 300	-5.6	48.8	119 030	1 980	49 290
Ghana	2003	147 350	16.9	64.8	41 850	22 000	83 500
Greece	2005	83 590	-9.4	64.8	26 270	11 320	46 000
Greenland	2003	2 350	0.0	0.6	...	...	2 350
Grenada	2003	130	0.0	38.2	20	100	10
Guadeloupe	2003	460	-13.2	27.2	200	50	210
Guam	2003	200	0.0	36.4	20	100	80
Guatemala	2003	46 520	8.6	42.9	14 400	6 100	26 020
Guinea	2003	124 500	3.6	50.7	11 000	6 500	107 000
Guinea-Bissau	2003	16 300	8.9	58.0	3 000	2 500	10 800
Guyana	2003	17 400	0.5	8.8	4 800	300	12 300
Haiti	2003	15 900	-0.4	57.7	7 800	3 200	4 900
Honduras	2003	29 360	-11.6	26.2	10 680	3 600	15 080
Hungary	2005	58 640	-9.4	65.4	46 000	2 070	10 570
Iceland	2005	22 810	0.0	22.8	70	...	22 740
India	2005	1 801 800	-0.5	60.6	1 596 500	100 000	105 300
Indonesia	2005	478 000	6.0	26.4	230 000	136 000	112 000
Iran (Islamic Republic of)	2005	616 000	1.8	37.6	161 000	15 000	440 000
Iraq	2003	100 190	4.5	22.9	57 500	2 690	40 000
Ireland	2005	42 270	-25.2	61.4	12 150	20	30 100
Israel	2005	5 170	-10.7	23.9	3 170	750	1 250
Italy	2005	146 940	-12.7	50.0	77 440	25 390	44 110
Jamaica	2003	5 130	7.8	47.4	1 740	1 100	2 290
Japan	2005	46 920	-17.6	12.9	43 600	3 320	...
Jordan	2005	10 120	-2.7	11.5	1 840	860	7 420
Kazakhstan	2005	2 075 980	...	76.9	223 640	1 360	1 850 980
Kenya	2003	265 120	2.0	46.6	46 500	5 620	213 000
Kiribati	2003	370	-5.1	50.7	20	350	...
Korea, Dem. People's Rep.	2005	30 500	21.1	25.3	28 000	2 000	500
Korea, Republic of	2005	18 930	-13.1	19.2	16 350	2 000	580

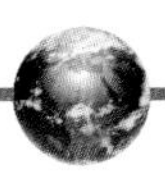

Land Use: *Agricultural land*

	Latest year available	Agricultural area	% change since 1990	% of total land area	Arable land	Land under permanent crops	Land under permanent pasture
		km²	%	%	km²	km²	km²
Kuwait	2005	1 540	9.2	8.6	150	30	1 360
Kyrgyzstan	2005	107 450	...	56.0	12 840	720	93 890
Lao People's Dem. Rep.	2005	19 590	18.0	8.5	10 000	810	8 780
Latvia	2005	17 340	...	27.8	10 920	130	6 290
Lebanon	2003	3 290	3.8	32.2	1 700	1 430	160
Lesotho	2003	23 340	0.6	76.9	3 300	40	20 000
Liberia	2003	26 020	-0.2	27.0	3 820	2 200	20 000
Libyan Arab Jamahiriya	2003	154 500	0.0	8.8	18 150	3 350	133 000
Liechtenstein	2005	90	-10.0	56.3	40	...	50
Lithuania	2005	28 370	...	45.3	19 060	400	8 910
Luxembourg	2005	1 290	...	49.8	600	20	670
Madagascar	2003	275 500	0.8	47.4	29 500	6 000	240 000
Malawi	2003	44 400	17.8	47.2	24 500	1 400	18 500
Malaysia	2003	78 700	8.9	24.0	18 000	57 850	2 850
Maldives	2003	140	55.6	46.7	40	90	10
Mali	2005	394 790	23.0	32.4	48 000	400	346 390
Malta	2005	100	-23.1	31.3	90	10	...
Marshall Islands	2005	140	...	77.8	20	80	40
Martinique	2003	320	-17.9	30.2	100	110	110
Mauritania	2003	397 500	0.2	38.8	4 880	120	392 500
Mauritius	2005	1 130	0.0	55.7	1 000	60	70
Mexico	2003	1 073 000	3.8	56.2	248 000	25 000	800 000
Micronesia, Federated States of	2003	470	...	67.1	40	320	110
Mongolia	2003	1 305 000	3.9	83.3	11 980	20	1 293 000
Montserrat	2003	30	0.0	30.0	20	...	10
Morocco	2003	303 760	0.1	68.1	84 840	8 920	210 000
Mozambique	2003	485 800	1.9	62.0	43 500	2 300	440 000
Myanmar	2003	112 930	8.3	17.2	100 930	8 880	3 120
Namibia	2003	388 200	0.4	47.2	8 150	50	380 000
Nepal	2005	42 220	1.7	29.5	23 570	1 300	17 350
Netherlands	2005	19 210	-4.2	56.7	9 080	330	9 800
Netherlands Antilles	2003	80	0.0	10.0	80	40	...
New Caledonia	2003	2 490	7.3	13.6	60	18 720	2 390
New Zealand	2003	172 350	-0.7	64.3	15 000	...	138 630
Nicaragua	2003	69 760	10.6	57.5	19 250	2 360	48 150
Niger	2003	385 000	16.5	30.4	144 830	170	240 000
Nigeria	2003	726 000	0.7	79.7	305 000	29 000	392 000
Niue	2003	80	14.3	30.8	30	40	10
Norfolk Island	2003	10	0.0	25.0	...	...	10
Northern Mariana Islands	2003	130	...	28.3	60	20	50
Norway	2005	10 260	5.1	3.4	8 590	...	1 670
Oman	2003	10 800	0.0	3.5	370	430	10 000
Pakistan	2005	270 700	4.4	35.1	212 750	7 950	50 000
Palau	2003	90	...	19.6	40	20	30
Palestine	2005	3 720	-1.3	61.8	1 070	1 150	1 500
Panama	2003	22 300	5.0	30.0	5 480	1 470	15 350
Papua New Guinea	2003	10 500	15.8	2.3	2 250	6 500	1 750
Paraguay	2003	248 360	6.6	62.5	30 400	960	217 000
Peru	2003	212 100	-2.9	16.6	37 000	6 100	169 000
Philippines	2003	122 000	9.5	40.9	57 000	50 000	15 000
Poland	2005	159 060	-15.4	51.9	121 410	3 780	33 870
Portugal	2005	38 150	-3.7	41.7	15 340	7 740	15 070

Land Use: *Agricultural land*

	Latest year available	Agricultural area	% change since 1990	% of total land area	Arable land	Land under permanent crops	Land under permanent pasture
		km²	%	%	km²	km²	km²
Puerto Rico	2005	2 230	-48.7	25.1	710	420	1 100
Qatar	2005	710	16.4	6.5	180	30	500
Republic of Moldova	2005	25 180	...	76.6	18 480	2 980	3 720
Réunion	2003	490	-23.4	19.6	350	40	100
Romania	2005	145 130	-1.7	63.1	92 880	5 400	46 850
Russian Federation	2005	2 156 800	...	13.2	1 217 810	18 000	920 990
Rwanda	2003	19 350	3.0	78.4	12 000	2 700	4 650
Saint Helena	2003	120	20.0	38.7	40	...	80
Saint Kitts and Nevis	2003	100	-16.7	27.8	70	10	20
Saint Lucia	2003	200	-4.8	32.8	40	140	20
Saint Pierre and Miquelon	2003	30	0.0	13.0	30	...	...
Samoa	2003	1 310	6.5	46.3	600	690	20
San Marino	2003	10	0.0	16.7	10	...	...
Sao Tome and Principe	2003	560	33.3	58.3	80	470	10
Saudi Arabia	2003	1 737 980	40.7	80.8	36 000	1 980	1 700 000
Senegal	2003	81 570	0.8	42.4	24 600	470	56 500
Serbia and Montenegro	2005	55 900	...	54.8	35 050	3 170	17 680
Seychelles	2003	70	16.7	15.2	10	60	...
Sierra Leone	2003	28 450	3.7	39.7	5 700	750	22 000
Singapore	2005	8	-60.0	1.2	6	2	...
Slovakia	2005	19 410	...	40.4	13 910	260	5 240
Slovenia	2005	5 080	...	25.2	1 760	270	3 050
Solomon Islands	2003	1 170	8.3	4.2	180	590	400
Somalia	2003	440 710	0.1	70.3	10 450	260	430 000
South Africa	2003	996 400	2.9	82.0	147 530	9 590	839 280
Spain	2005	290 300	-4.7	58.2	137 000	49 300	104 000
Sri Lanka	2003	23 560	0.7	36.5	9 160	10 000	4 400
St. Vincent and the Grenadines	2003	160	14.3	41.0	70	70	20
Sudan	2003	1 346 000	9.2	56.6	170 000	4 200	1 171 800
Suriname	2003	890	1.1	0.6	580	100	210
Swaziland	2003	13 920	9.8	80.9	1 780	140	12 000
Sweden	2005	32 190	-5.8	7.8	27 030	30	5 130
Switzerland	2005	15 250	-24.5	38.1	4 100	240	10 910
Syrian Arab Republic	2005	140 080	3.8	76.2	48 730	8 690	82 660
Tajikistan	2003	42 550	...	30.4	9 300	1 270	31 980
Thailand	2003	184 870	-13.5	36.2	141 330	35 540	8 000
The Former Yugoslav Rep. of Macedonia	2005	12 420	...	48.8	5 660	460	6 300
Timor-Leste	2003	3 400	6.9	22.9	1 220	680	1 500
Togo	2003	36 300	13.8	66.7	25 100	1 200	10 000
Tonga	2003	300	-6.3	41.7	150	110	40
Trinidad and Tobago	2003	1 330	1.5	25.9	750	470	110
Tunisia	2003	97 840	13.2	63.0	27 900	21 400	48 540
Turkey	2005	412 230	3.9	53.6	238 300	27 760	146 170
Turkmenistan	2003	329 660	...	70.2	22 000	660	307 000
Turks and Caicos Islands	2003	10	0.0	2.3	10	...	...
Tuvalu	2003	20	0.0	66.7	...	20	...
Uganda	2003	124 620	4.2	63.2	52 000	21 500	51 120
Ukraine	2005	413 040	...	71.3	324 520	9 010	79 510
United Arab Emirates	2003	5 590	96.1	6.7	640	1 900	3 050
United Kingdom	2005	169 560	-6.9	70.1	57 290	470	111 800
United Rep. of Tanzania	2003	481 000	1.5	54.4	40 000	11 000	430 000

Land Use: *Agricultural land*

	Latest year available	Agricultural area	% change since 1990	% of total land area	Arable land	Land under permanent crops	Land under permanent pasture
		km²	%	%	km²	km²	km²
United States	2005	4 147 780	-2.9	45.3	1 744 480	27 300	2 376 000
United States Virgin Islands	2003	60	-45.5	17.1	20	10	30
Uruguay	2003	149 550	0.9	85.4	13 700	420	135 430
Uzbekistan	2003	272 590	...	64.1	47 000	3 400	222 190
Vanuatu	2003	1 470	5.0	12.1	200	850	420
Venezuela	2003	216 400	-1.0	24.5	26 000	8 000	182 400
Viet Nam	2005	95 920	42.6	30.9	66 000	23 500	6 420
Wallis and Futuna Islands	2003	60	0.0	42.9	10	50	...
Western Sahara	2003	50 050	0.0	18.8	50	...	50 000
Yemen	2003	177 340	0.2	33.6	15 370	1 320	160 650
Zambia	2003	352 890	0.1	47.5	52 600	290	300 000
Zimbabwe	2003	205 500	1.9	53.1	32 200	1 300	172 000

Source:
Food and Agriculture Organization of the United Nations (FAO).

Definitions & Technical notes:
Agricultural area refers to the sum of area under arable land, permanent crops, and permanent pastures. Arable land refers to land under temporary crops (double-cropped areas are counted only once), temporary meadows for mowing or pasture, land under market and kitchen gardens and land temporarily fallow (less than five years). The abandoned land resulting from shifting cultivation is not included in this category. Data for "Arable land" are not meant to indicate the amount of land that is potentially cultivable. Land under permanent crops refers to land cultivated with crops that occupy the land for long periods and need not be replanted after each harvest, such as cocoa, coffee and rubber; this category includes land under flowering shrubs, fruit trees, nut trees and vines, but excludes land under trees grown for wood or timber. Land under permanent pastures refers to land used permanently (five years or more) for herbaceous forage crops, either cultivated or growing wild (wild prairie or grazing land). % change since 1990 and Agricultural area as a % of total land area in 2002 are calculated by UNSD based on FAO data.

Data Quality:
FAO promotes national censuses of agricultural land use every 10 years, with varying degrees of success. Standardised definitions exist but can pose problems when land is used for multiple purposes. In many parts of the world, for example, livestock graze in orchards and among other permanent crops. Moreover, land removed from production under set-aside schemes intended to reduce overproduction, is not always reflected adequately in the figures. Agricultural surveys and censuses are generally confined to farmland. However, in many countries common land is used for grazing and may or may not be included in the figures for permanent pastures.

Policy Relevance:
The ability of a given country to produce enough food to feed its own people will depend largely on the climate, on the availability of fertile land and on competing uses for that land. In many parts of the world, forest, wetlands and other natural land is still being cleared for conversion to agriculture, while in others, there are moves to return agricultural land to nature. And everywhere, cities and towns continue their sprawl in river valleys, often the areas with the most fertile soils.

Index

Note: Page numbers in **boldface** refer to volume numbers and major topics. Article titles are in **boldface.**

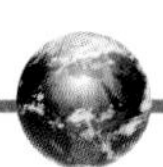

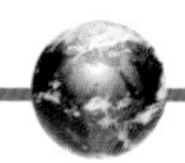

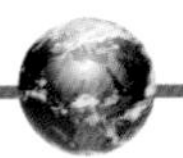

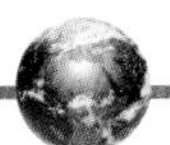

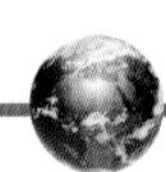

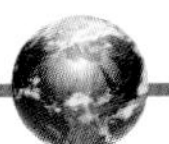

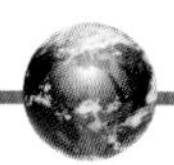

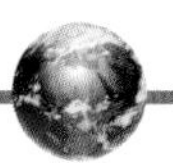

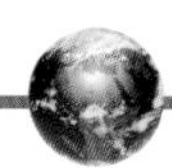

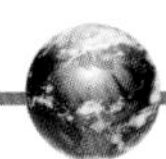

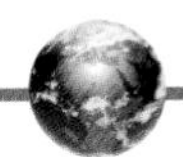

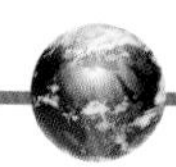

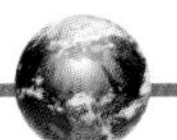

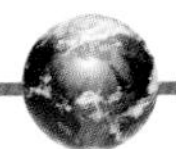

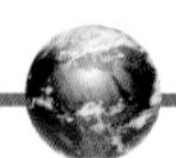

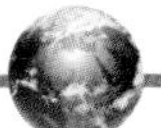

T

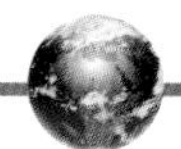

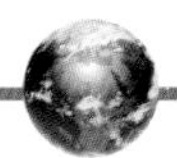

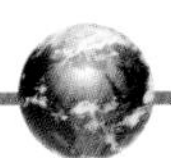

Photo Credits

Kenneth Gabrielsen Photography: 3, 35, 41, 47, 56, 63, 104, 136, 158, 190, 203, 209, 259, 274, 295, 303, 331, 334, 383, 401, 404, 418, 445, 485, 515, 530, 543, 555, 556, 572, 593, 609, 619, 648, 661, 676, 689, 690, 699, 724, 730, 745, 755, 774, 800, 806, 855, 882, 925, 943, 1001, 1006, 1015, 1024, 1031, 1032, 1091, 1100, 1103, 1129, 1155,1195, 1197, 1210, 1218, 1246, 1263, 1266, 1271, 1274, 1297, 1364, 1375, 1445, 1474, 1505, 1520, 1527, 1537, 1555, 1578, 1618, 1654, 1670, 1733, 1747, 1780, 1806, 1823, 1866, 1873, 1946, 1998, 2004; Sherry Collins: 912; Photos.com: 13, 43, 79, 91, 99, 119,127, 148, 181, 197, 247, 280, 318, 411, 424, 448, 459, 507, 627, 655, 670, 719, 762, 791, 842, 887, 893, 901, 907, 963, 967, 975, 986, 1000, 1003,1038, 1049, 1056, 1059, 1064, 1069, 1079, 1118, 1124, 1142, 1149, 1163, 1170, 1180, 1187, 1191, 1206, 1225, 1282, 1289, 1301, 1306, 1312, 1321, 1328, 1339, 1369, 1372, 1394, 1406, 1430, 1451, 1463,1483, 1489, 1509, 1549, 1557, 1575, 1590, 1601, 1613, 1633, 1639, 1645, 1646, 1690, 1707, 1719, 1745, 1770, 1818, 1857, 1868, 1871, 1873, 1887, 1888, 1893, 1909, 1925, 1940, 1951, 1956, 1968, 1974, 1985; MorgueFile: 43 (photo by Sanjay Pindiyatah), 86, 170 (photo by Zach Carter), 187 (photo by David Ellis), 200 (photo by Mary Thorman), 219 (photo by Manuel Jose Alves), 242, 288, 359 (photo by Derek Lilly), 368 & 982 (photos by Balázs Metzger), 374 (photo by Mary R. Vogt), 434 (photo by Kristen Rasmussen), 476, 479, 519 (photo by Mary Thorman), 538 (photo by Kenn Kiser), 601 (photo by Kevin Connors), 635 (photo by Clara Natoli), 705 (photo by Kabir Bakie), 713, 779 (photo by Arturo Delfin), 813; National Parks Service, U.S. Department of the Interior: 500, 585, 625; U.S. Air Force Photo: 498; U.S. Library of Congress: 49, 311, 328, 494, 630, 644, 874, 917, 960; iStockphoto.com: 1011, 1995; Boston University School of Public Health: 768; EER/Earthquake Engineering Research Institute: 853; United States Department of Agriculture Photos: 16, 30, 69, 162, 174, 228, 522, 739, 748, 822, 863, 932; House.gov, Congressman Mario Diaz-Balart Photo: 9; U.S. Fish & Wildlife Service Photo: 23, 348, 440, 562, 792, 829, 994; U.S.G.S Denver Microbeam Laboratory Photo: 77; Federal Emergency Management Agency (FEMA) Photo Library: 684; National Oceanic and Atmospheric Administration (NOAA) Photo : 112, 394, 640; National Institutes of Health (NIH) Photo: 143; National Cancer Institute (NCI) Photo: 214; Center for Disease Control (CDC) Photo: 470. National Aeronautics and Space Administration (NASA) Photos: 235, 266, 786, 838, 845, 953.